On the Origin of Human Masticatory Function

Andrea Papini

On the Origin of Human Masticatory Function

A Darwinian Perspective on Human Chewing Actions

Springer

Andrea Papini
Studio Dentistico
Prato, Italy

ISBN 978-3-031-98663-5 ISBN 978-3-031-98664-2 (eBook)
https://doi.org/10.1007/978-3-031-98664-2

This Springer imprint is published by the registered company Springer Nature Switzerland AG
The registered company address is: Gewerbestrasse 11, 6330 Cham, Switzerland

Drawing by Marco Boggioni

How can man succeed in recognizing himself
how he has been shaped by nature,
amidst all the changes that the succession of ages and things
has had to bring about in his original constitution,
and to separate what circumstances and his progress
have added or modified from his primitive state?

— Jean-Jacques Rousseau.
Discourse on the origin and foundations of inequality among men. 1755

Foreword 1

Within the cultural framework of every dentist, gnathology should unquestionably be an essential element. Despite often being perceived as complex, abstract, and hyper-specialized, the reality is that every dentist faces its principles on a daily basis and in nearly every procedure of their professional practice.

Gnathology doesn't only concern the therapy of ATM dysfunctions, but also establishes the static and dynamic relationships that occur among individual teeth or entire arches. This discipline should encompass a robust dental knowledge, elucidating the functional characteristics of the masticatory system, regardless of the extensive genotypic and phenotypic diversity and the adaptations that the masticatory apparatus perpetually undergoes.

The anatomical and functional model of the human stomatognathic apparatus, while sharing foundations with the mammalian class and primate order, truly characterizes our species. From an evolutionary viewpoint, it's surprisingly recent. Thus, to thoroughly grasp the physiological aspects of its diverse components, a profound comprehension of how this system has adapted to its present morphology and functionality is imperative.

The study of evolutionary development holds a unique opportunity within the stomatognathic domain, unlike other anatomical regions, as teeth are parts of the body that fossilize and preserve most easily. This provides an unparalleled chance for anatomical comparison. Hence, significant interest has arisen from paleoanthropologists in the scientific field, focusing on both tooth form and masticatory function. It is from this shared interest that a fruitful dialogue between the dental and paleoanthropological worlds can emerge.

This work aims to suggest a different perspective that frames dentistry not only in a clinical and operational manner but also broadens its scope to encompass the evolutionary journey of our species. This broader perspective seeks answers that only an interdisciplinary view can provide. Science can offer a unified vision only when it connects closely related disciplines (François Jacob). The extreme specialization of current scientific research levels tends to disregard the numerous connections between seemingly separate areas of knowledge. Otto Veit's words in his 1937 essay "The Age of Technology" come to mind: Contact with adjacent fields of knowledge has been progressively lost, and the increasingly intense specialization demands a progressively lower level of culture.

It is precisely within these "border zones" between diverse fields of knowledge that shared interests intersect, offering numerous and often unforeseen avenues for exploration. The multidisciplinary approach may pose difficulties due to the diverse languages employed and the distinct investigative tools used. Nonetheless, it presents a formidable prospect that can facilitate transcending the compartmentalized culture among different knowledge domains, thus activating an essential synergy of knowledge for scientific advancement.

Centro di Odontoiatria e Stomatologia F. Perrini, Pistoia, Italy Nicola Perrini

Foreword 2

And the shark, it has teeth
And it wears them in its face
The Ballad of Mack the Knife
The Threepenny Opera. 1928
— Bertolt Brecht

The skull is the creation of God,
while the jaw is the work of the Devil
— Yoel Rak

Having healthy teeth is well-known to be essential. In my profession as a paleoanthropologist—and therefore as a student of human evolution—having access to healthy teeth (mostly fossils) for study is absolutely crucial. Among all the organs and tissues constituting the body, teeth are the most likely to be preserved and fossilized. This is because they are composed of the hardest and most resistant materials that an organism, especially a vertebrate, can produce—namely, the enamel that covers the crown, the dentin that forms a significant part of the internal structure, and the cement that strengthens the roots. In all three cases, we are dealing with compounds of hydroxyapatite, the same substance that forms the mineral part of bones but is chemically organized to confer even greater hardness and resistance compared to bones. This results in teeth often being the sole remains of an organism after everything else has long decomposed. Even in the case of colossal creatures like dinosaurs, what predominantly endures in the fossil record are their teeth. In some fossil species, teeth represent the only preserved remains.

Among scholars, there is a saying attributed to the eminent paleontologist George Gaylord Simpson: the evolution of vertebrates, as depicted in fossils, appears as a story of teeth influencing the evolution of other teeth to produce their descendant teeth.

It might seem like a situation to despair about ever making sense of. However, for evolution researchers, all of this translates into a remarkable stroke of luck. Teeth, in fact, are not only the most durable parts of a vertebrate (including humans) but also the most informative, concerning a myriad of aspects related to lifestyle habits and the existence of now-extinct organisms and species.

This is due to the fact that teeth, and the jawbones that encase them, are primarily connected to an essential organic function for survival—feeding and chewing. As a result, they undergo various forms of adaptation, both biological and environmental. Their shape and structure also bear the marks of the long evolutionary history—phylogeny—that has differentiated each of the countless species of vertebrates that have inhabited the Earth. Moreover, their form is influenced by ontogenetic developmental factors that every individual must face to complete their growth journey from birth to adulthood and ultimately to death.

In the specific context of human evolution, we can only attempt to compile an inevitably incomplete list of the diverse information that isolated fossilized teeth or those embedded within jawbones can offer to researchers. For example, teeth and the masticatory system provide crucial data concerning dietary adaptations, enabling the reconstruction of the habitual diet of ancestral hominids, even those who lived millions of years ago. This is possible through the analysis of specific variants of chemical elements, known as isotopes, which constitute enamel and dentin. Such analyses also yield insights into survival strategies, interactions with the surrounding habitat, and the journeys individuals undertook throughout their lives. Teeth hold chemical traces (known as isotopic "signatures") of the diverse places in which our ancestors lived and stayed. The study of dental morphology and anatomical characteristics enables the identification and categorization of varied populations and species, facilitating the reconstruction of evolutionary and phylogenetic connections that bind them together. Counting the visible growth lines in a thin section of a tooth using a microscope allows for determining the age of the tooth's owner at the time of death. It also offers insight into the rate at which the transition from juvenile stages to adulthood occurred within their species. This facilitates comparisons of ontogenetic development rates and bodily growth across different hominid species. Furthermore, it provides the means to discern the age of weaning in extinct hominids—for instance, comparing an *Australopithecus* to a modern human or a chimpanzee.

Access to this information is imperative because the longer our ancestors took to complete their growth, the more time and opportunity their brains had to develop the complexity of structures and behaviors that bring them closer to our own mental faculties. Additionally, the degree of dental wear and observable pathologies offer insights into the ancient populations' lifestyles and health conditions. It's even possible to make inferences about the social structure and level of conflict within fossil species communities based on the proportions of canines in males and females. Furthermore, it provides the means to determine when significant behavioral shifts occurred, such as the systematic adoption of food cooking, by examining the reduction in overall dental and mandibular-facial bone sizes. Moreover, the pulp chambers within teeth represent a valuable source of uncontaminated ancient DNA. This DNA can be analyzed using state-of-the-art sequencing techniques to reconstruct the comprehensive genetic profile of extinct human species, like the Neanderthals. This provides detailed insights, unavailable solely through fossils, about their skin color, hair, eye color, and the potential genetic interactions with our species before their extinction. Lastly (though I'm undoubtedly missing something), teeth can also be used to date other associated fossil remains (including the teeth themselves) by examining the levels of fluoride and the racemization of amino acids they contain.

Therefore, it's easy to understand how the study of the masticatory system and its connections with other morpho-functional systems, not limited to the cranium but encompassing the entirety of the body, stands as a cornerstone in grasping each distinct phase that, over a prolonged evolutionary journey, has culminated in the establishment of our present-day biology, behaviors, and abilities.

A curious paradox arises here, as the increasingly in-depth study of each of the aforementioned aspects, often aided by progressively sophisticated methodologies, might (perhaps inevitably) result in an extreme specialization of knowledge. This specialization runs the risk of diminishing the comprehensive or "systemic" viewpoint, which is the only one capable of accounting for the interactions among the diverse organic components forming the basis of biological processes and, consequently, of evolution itself.

The late Stephen Jay Gould, following Adolf Seilacher's morpho-dynamic vision applied to the study of biological forms (Konstruktions-Morphologie), clearly demonstrated how understanding the living reality that surrounds us, of which we are a part, necessitates an integrated consideration of functional aspects (such as those related to dental occlusion). These facets are intertwined with adaptation, natural selection, and, at the individual level, habits. This must be coupled with structural and morphological aspects originating from ontogenetic development processes. Historical aspects are also crucial, reflecting the phylogenetic heritage of species, which has diversified over millions of years through often contingent and unpredictable evolutionary pathways.

Anticipator of this integrated and morpho-dynamic approach applied to the study of masticatory structures was Melvin Moss, who developed the functional matrix hypothesis in 1962 to explain how different cranio-dental components, during their growth, closely interact with each other, exchanging information on how to achieve functional equilibrium mainly through epigenetic means. Donald H. Enlow also played a significant role in this concept's development, applying it to every phase of bone development and remodeling in the face and skull. In the field of evolution and paleoanthropology, fundamental contributions were made in the early years of the last century. Sergio Sergi, a prominent figure in the Roman anthropological school, played a pivotal role in quantitatively shaping the morpho-dynamic study of how bones, teeth, and other cranial structures interact. Pioneering works by Antoine Delattre and notably André Leroi-Gourhan could also be mentioned.

Leroi-Gourhan, a versatile anthropologist, archaeologist, and ethnologist, further developed this comprehensive morphological perspective to encompass the investigation of artifacts and human culture. More recently, individuals such as Yoel Rak, Paul O'Higgins, and Daniel E. Lieberman have maintained an avid interest in an integrated, morpho-dynamic, and morpho-functional perspective of human evolution. They have channeled their research toward the study of fossil materials. This approach contrasts with the prevailing trend in other fields of anthropological sciences, which often lean toward a reductionist and deterministic viewpoint, even when considering the study of the masticatory apparatus.

The perspective adopted by these latter authors considers every biological structure as part of an interconnected system, in which positive or negative relationships with other structures, both proximal and distant, do influence the possibilities of ontogenetic and evolutionary development.

Andrea Papini's work aligns fully with the tradition of Darwinian studies in a broad and comprehensive sense. Personally, I've had the occasion to draw from Dr. Papini's insights multiple times for my own work, along with those of our friend and colleague, Dr. Marco Boggioni.

These insights have been primarily applied to the study of cranio-dental morphology in Neanderthals and our earliest hominid ancestors. Hence, I consider it a privilege to present this book to the community of scholars, students, and enthusiasts. Dr. Andrea Papini has accomplished this work by capitalizing primarily on the extensive collection of osteological artifacts, of which I am a curator at the Anthropology and Ethnology section of the Museum of Natural History at the University of Florence.

Department of Anthropology and Ethnology
University of Florence's Academic Museum System
Florence, Italy

Fabio Di Vincenzo

Acknowledgments

Considering writing this section of the book has been perhaps the most demanding task of the entire work, because establishing a definitive hierarchy that adequately expresses the gratitude for individual contributions to personal growth is challenging. At times, even the mere gaze of a stranger can ignite an awareness within us that a thousand words from an instructor could never evoke.

We all harbor within us small or large fragments of the many individuals we've encountered in our life: loves, friends, mentors, colleagues, even strangers. Who we are today, with our achievements that carry significance at least to us, and our failures that are even more significant, we owe to the countless encounters through which we have exchanged life's experiences. To single out individuals for gratitude within a book inadvertently means sidelining others, that's why I find it unjust to compile a ranking.

I earnestly hope that those who know me understand how much I owe to them for shaping me into the person I am.

I thank you all.

Introduction

> Nothing ever dies in a life.
> Everything survives.
> We, together, live and survive.
> — Pier Paolo Pasolini

The awareness that many components of our present body have origins in a past so ancient that conceiving it becomes a challenge can evoke a sense of vertigo, akin to contemplating the vast expanses of cosmic space. Yet, it is this perspective that realigns us with the true "Place of Man in Nature." This understanding urges us to embrace our responsibilities for preserving biodiversity and our small yet astonishing planet's environment.

> Knowing our history from a time predating our humanity is crucial, especially in this era where we stand as unrivaled masters of the planet. Sharing this narrative is vital, extending beyond the scientific community which currently comprises only a few thousand people globally (Manzi, 2017).

In humans, the stomatognathic apparatus performs various sophisticated functions, including respiration, sucking, nonverbal communication, and most recently, speech. Amid these functions, not all are strictly essential for survival, except for the most fundamentally significant, chronologically the first, without which we would be destined to die: feeding. This function, feeding, is vital as it combines two distinct processes—mastication and swallowing—supplying nourishment to the digestive system. This enables the assimilation of crucial nutrients for energy, maintenance, growth, and reproduction (Reilly et al., 2001).

The intricate interweaving of the stomatognathic apparatus's diverse functions, its inherent morphological changes throughout life, and the considerable phenotypic variability linked to profound dietary shifts in the last century have led to various approaches in studying its functionality. Consequently, within the field of dentistry, there exists an interpretation that lacks universal agreement and often becomes convoluted.

For a dentist seeking to delve deeper into the realm of gnathology, once they've thoroughly explored the anatomical aspects, they're confronted with a bewildering array of functional analyses and therapeutic approaches. Consequently, they might either distance themselves from this dental branch, deeming it overly complex, or wholeheartedly align themselves with one of the various schools of thought. However, neither of these solutions proves wise. On one hand, lacking a comprehensive functional understanding of occlusion jeopardizes the success of all dental procedures performed on our patients. On the other hand, uncritically adhering to a single school contradicts the scientific approach that should consistently govern our therapeutic decisions.

Even the most basic dental procedure, such as a routine filling, must be executed while taking into account the specific features of occlusion. This ensures the prevention of any disturbances in the occlusal relationship. A dentist who operates according to these principles should rightfully be considered a gnathologist. In fact, practicing dentistry correctly becomes unfeasible without a profound understanding of gnathology.

There are numerous concepts, techniques, and philosophies concerning occlusion (Rangarajan et al., 2015), and the necessity of being able to chew (and swallow) in all potential mouth conditions has contributed to the difficulty of identifying common principles.

An exceptional adaptability is necessary to overcome any obstacles that could hinder the process of feeding overcoming any obstacles that could impede the feeding process necessitates exceptional adaptability.

Our masticatory apparatus took shape with the emergence of mammals around 250 million years ago, representing just one of the evolutionary traits that have developed within this class of vertebrates. Since that time, while sharing a common functional basis, numerous changes in the anatomical and functional framework of the stomatognathic system have unfolded among various mammalian species.

Within this variation on the theme, there exist unique attributes within the primate order. These distinctive characteristics are exemplified by a strange type of hairless naked hominid sitting naked, precariously balanced, on a small peripheral branch of the grand coral of life.

So, ***how*** should our stomatognathic system operate during chewing?

Moreover, ***why*** do specific anatomical and functional traits set apart our species' chewing process, rendering them crucial for proper physiological function?

The *why* and the *how*: in the realm of modern dentistry, the indispensable aid of digital technology sometimes takes on the form of a veritable invasion. There's a growing concern that technological advancement, rather than enhancing the diagnostic process, might veer toward substituting it with pre-packaged and standardized programs (along with predetermined mental patterns). These could guide decisions primarily based on adopting the quickest and least problematic protocols.

This induced intellectual passivity relieves us from confronting the first of the two fundamental questions in therapy: the *why* and the *how*.

Chronologically, the how is generally the first question that confronts us when we encounter an unexpected problem: how can we solve it and navigate through the challenging situation? The *how* is necessary to discover a solution, yet it doesn't shed light on the underlying causes that brought about the issue. Without understanding the *how*, the *why* tends to remain a mere abstract speculation.

However, the *how* primarily leads to developing the technical aspect of resolution—namely, a set of procedures designed to function with minimal wastage. These are guided by an adept, a specialist who adheres to a standardized protocol. Nevertheless, the *how* often overlooks the link between causal events and their origins.

The second question, which only chronologically follows the first, is the *why*: understanding *why* is crucial for comprehending the causes of the problem, selecting the most appropriate path for its resolution, and initiating actions to prevent its recurrence in the future. Without understanding the *why*, the *how* presents itself with the obtuse rigidity of operational protocols that do not encompass anything beyond the limited set of variables under consideration.

To paraphrase Nietzsche, if we possess a *why*, we can navigate any *how*. Without the *why*, the how ceases to be a mere operational means and becomes a self-referential end, justifying itself with itself and absolving the executor of any responsibility, as others have already conceived the "best" solution for them.

The *how* finds expression through therapy, while the *why* enables diagnosis. The *how* without the *why* is a [mis]guided exercise in style; the *how* without a *why* gives rise to monsters.

This work aims to provide (of course, partial) answers about why our stomatognathic system is structured in this way and, based on this understanding, how it should function.

The common thread of this book is consistently reinforced by references to publications in scientific journals. This approach aims to avoid personal speculations or unsupported statements solely relying on the authority of eminent colleagues, whether real or presumed.

As cautioned by Walter Benjamin, "*Quotations are like bandits on the roadside, who leap out armed and seize assent from the idle passerby*." Consequently, readers are encouraged to explore their specific interests by referring to the bibliography associated with the research addressing the various topics covered in this presentation.

Given the vast output of contemporary scientific literature, this book has chosen to emphasize research based on criteria stemming from intellectual foundations, methodological execution, rigorous reasoning, and the reliability of data. This preference stands in contrast to evaluating articles solely based on the number of citations they receive or the quantity of publications by a given author. These latter bibliometric indicators rely on counts alone, which cannot adequately measure the quality of a scientific contribution or indicate its significance or its capacity to influence the research conducted by other scholars.

Assessment and measurement are distinct concepts, and quality, due to its inherently multifaceted nature, cannot be predetermined. Quality is something that is evaluated rather than quantified (Fasanella & Colarusso, 2017). Relying solely on contemporary quantitative data analysis tools to determine the quality of scientific work presents the risk of generating distorted outcomes due to the selection and specificity of the parameters employed.

The conceptual aberration of evaluating the quality of work based solely on metrics like the impact factor of the publishing journal, citation frequency, or an author's publication count becomes evident when considering that, under such a perspective, an obscure young German physicist, employed as a second-class worker at the patent office in Bern, would have been unable to revolutionize the world of physics in 1905. His works on special relativity, Brownian motion, light quanta, and molecular dimensions, including groundbreaking contributions that earned him the Nobel Prize, would have been overshadowed.

In the realm of science, progress primarily emerges from the continuous introduction of novel ideas that require validation, testing, and refinement by the scientific community. This progression should not be assessed solely based on the count of citations. Without the initial spark of intellectual originality, we run the risk of blinding us by directing the powerful beam of research directly into our eyes.

"Science evolves because quite often an aspect still veiled about things suddenly comes to light, not primarily due to the arrival of novel equipment, but rather owing to a fresh manner of observing and considering objects from an altered standpoint. This perspective is necessarily guided by a certain idea of the so-called reality and always implies a specific representation of the unknown, that area beyond what logical, experimental, or computational arguments authorize us to believe," states François Jacob, the Nobel laureate in Medicine in 1965.

To observe and truly comprehend the complete picture, be it in art or medicine, demands the ability to, while remaining conscious of them, liberate ourselves from the details. These details must eventually be reassembled like pieces of a puzzle to craft a global view. As Leonardo da Vinci aptly noted, "...details make perfection..." Yet, this perfection necessitates a broader perspective that reconstitutes individual particulars, imbuing them with a sense they wouldn't possess when fragmented.

In the context of any subject, no single article or book can ever claim comprehensiveness. The objective of a written work should be to empower the reader to uncover the boundaries of their own understanding. This serves as a necessary springboard for them to leap toward new insights. Reading an article or opening a book signifies opening oneself to new horizons. It's a hopeful endeavor that seeks to stimulate the reader's curiosity, enabling them to enrich their knowledge.....Enjoy your reading!

The Collection of the Anthropology Museum at the University of Florence

The collection of human skeletal remains preserved at the Anthropology Museum of the University of Florence is the largest in Italy and undoubtedly one of the most significant collections worldwide.

The collection originated with the establishment of the Museum in 1869 by Paolo Mantegazza, who sought to gather osteological and paleoanthropological collections for the study of human evolution and species variation.

His efforts were so extensive that they resulted in an impressive collection of artifacts, including human remains from the Neolithic, Chalcolithic, Bronze, and Iron Ages, serving as significant evidence of Italian paleoanthropology. This collection also features numerous Etruscan remains. Furthermore, there is a modern osteological collection comprising more than 7000 specimens, acquired from scientific expeditions conducted in the late nineteenth and early twentieth centuries across Italy and various European and non-European countries.

This collection provides a rare insight into the variability of contemporary populations.

The collection, documented by a handwritten chronological catalog, has recently been digitized and is subject to ongoing attention from Italian and foreign scholars for research in auxology, human ecology, biogeography, prehistory, pathology, and paleopathology (Moggi Cecchi & Stanyon, 2014).

References

Fasanella, A., & Colarusso, S. (2017). *Co-authorship e reti di collaborazione: un caso di studio.* Sapienza Università di Roma.

Manzi, G. (2017). *Ultime notizie sull'evoluzione umana.* Il mulino.

Moggi Cecchi, J., & Stanyon, R. (2014). *Il Museo di Storia Naturale dell'Università degli Studi di Firenze. Le collezioni antropologiche ed etnologiche/The Museum of Natural History of the University of Florence. The Anthropological and Ethnological Collections: Le collezioni antropologiche ed etnologiche.* Firenze University Press.

Rangarajan, V., Gajapathi, B., Yogesh, P. B., Ibrahim, M. M., Kumar, R. G., & Karthik, P. (2015). Concepts of occlusion in prosthodontics: A literature review, part I. *The Journal of the Indian Prosthodontic Society, 15*(3), 200.

Reilly, S. M., McBrayer, L. D., & White, T. D. (2001). Prey processing in amniotes: Biomechanical and behavioral patterns of food reduction. *Comparative Biochemistry and Physiology Part A: Molecular & Integrative Physiology, 128*(3), 397–415.

Contents

1 **What Are Teeth for?** 1
References 8

2 **Features of the Stomatognathic System in Mammals** 11
2.1 Tooth Shape and Count 14
2.1.1 Diphyodonty 14
2.1.2 Heterodonty 15
2.1.3 The Tribosphenic Molar 17
2.1.4 Dental Roots 21
2.2 Shape and Number of Muscles 22
2.2.1 Shape and Arrangement Shift in Jaw Adductors 22
2.2.2 The Lips and Cheeks Are Structured 27
2.3 Shape and Count of Jaw Bones 36
2.3.1 Response of Bone Tissue to Mechanical Stresses 36
2.3.2 Structuring the Tmj and the Ossicular Chain of the Middle Ear 46
2.3.3 Formation of the Bony Palate 49
2.3.4 Development of the Zygomatic Arch 54
2.3.5 The Development of the Mandible with its Processes 56
2.4 Formation of the Temporo-Mandibular Joint 86
2.4.1 Mediolateral Translation 89
2.4.2 Antero-Posterior Translation 92
2.5 Formation of the Middle Ear 93
2.6 Formation of the Periodontium 96
2.6.1 Alveolar Bone 99
2.6.2 Periodontal Ligament 99
2.6.3 Cementum 99
2.7 Modifications of Dental Tissues 107
2.7.1 Enamel 107
2.7.2 Dentin 109
2.7.3 Pulp 112

2.8 Modification of Cervical Vertebrae 115
References. 127

3 Tooth Wear. 139
3.1 Abrasive Wear 148
3.1.1 Interproximal Wear 154
3.1.2 Occlusal Wear 163
3.2 Structural Fundamentals of Enamel Wear 167
3.2.1 Millimeter Level 169
3.2.2 Micrometer Level. 171
3.2.3 Nanometric Level. 176
3.3 Masticatory Cycle 180
3.4 Helical Plane and Inversion of Compensation Curves. 194
References. 214

4 The Stomatognathic System in Human Evolution: Bipedalism and Encephalization 225
4.1 Bipedalism 226
4.2 Encephalization 249
4.3 Occlusion Determinants. 275
4.4 Reduction of the Postglenoid Process 277
4.5 Presence of the Civinini Canal. 287
4.6 Tilt of the Glenoid Fossa Plane 304
4.7 The Structuring of the Temporo-Mandibular Ligament. 317
4.7.1 Oblique Portion of the Temporomandibular Ligament 320
4.7.2 Horizontal Portion of the Temporomandibular Ligament. . . . 326
4.8 Modification of Dental Volumes and Disappearance of the C/P3 *Honing Complex*. 332
4.8.1 Modification of Dental Volumes 332
4.8.2 Disappearance of the C/P3 *Honing Complex*. 336
4.9 Reduction in Mandibular Length. 363
4.10 Expansion of the Upper Jaw 364
4.11 Modification of the Occlusal Plane 365
4.12 Formation of Overbite 367
References. 368

5 Masticatory Function 381
5.1 General Criteria for a Homeostatic Occlusion. 385
5.1.1 Static Occlusion. 385
5.1.2 Dynamic Occlusion (Chewing) 392
5.2 Biomechanics of Muscle Attachments. 408
5.3 Mandibular Movements. 413
5.4 Gothic Arch and Posselt's Scheme 419
5.5 Retrusive Laterality 429
References. 446

Conclusion 453

Afterword of the Author 455

The Communication 457

Index 461

Chapter 1
What Are Teeth for?

Abstract The first chapter explores the evolution and function of teeth, establishing a foundation for understanding oral anatomy through an evolutionary lens. It begins with the idea that nothing in biology makes sense except in the light of evolution, highlighting how the form and function of teeth are shaped by selective adaptations. Initially, dental structures served functions other than chewing, such as immobilizing prey. Only with the evolution of mammals did chewing become a central function, accompanied by profound anatomical transformations in the orofacial region.

The text delves into the origin of teeth in vertebrates—from jawless fish to the first enamel structures found in conodonts—and traces the emergence of biomechanical innovations like tribosphenic teeth in mammals. Mastication, alongside endothermy, marked a turning point in mammalian evolution, enabling greater metabolic efficiency and access to new ecological niches. The diversification of tooth forms became essential to meet rising energy demands, contributing to the extraordinary adaptability of mammals. The entire masticatory system evolved to enhance energy assimilation, establishing teeth as fundamental tools not only for nutrition but also for survival and evolutionary success.

> In these times of high specialization, very few people have such a deep understanding of two different fields of knowledge that they do not become ridiculous in at least one of the two—Richard P. Feynman

Teodosius Dobzhansky once wrote, 'In biology, nothing makes sense except in the light of evolution.' With this perspective in mind, dismissing the relevance of evolution to medicine in general and dentistry in particular is shortsighted and mistaken. Our bodies weren't pre-designed; they developed through natural selection, involving random genetic variations with varying degrees of reproductive success (Lieberman, 2014).

Gaining knowledge about our evolutionary history enables us to comprehend the reasons behind our body's current form and functions. This understanding is essential for preventing and treating conditions exacerbated by potential evolutionary mismatches—situations where established traits become detrimental due to rapid environmental changes.

A. Papini, *On the Origin of Human Masticatory Function*, https://doi.org/10.1007/978-3-031-98664-2_1

Evolution is a slow and continuous process of adapting to the environment over vast timescales. Throughout this journey, each living species selects the traits that are best suited for its survival. During this process species may gain or lose characters and abilities based on the type of habitat they inhabit, potentially leading to significant changes between the original and present-day organisms (Mancuso & Viola, 2013).

Our anatomy and functions are deeply rooted in our evolutionary history, and to thoroughly understand how and why we are constructed and function as we do, we must know where we come from. We are seldom inclined to ponder the evolutionary journey that has led us to become who we are today: we are the result of an ongoing series of shape mutations adapting to various environmental pressures. From ancient times to the present day, these forces have shaped the morphology of all living beings. As we begin to perceive anatomy as the outcome of evolutionary processes over time, the structures unveil their deep functional significance. Adaptation refers to one or more traits that have evolved to make an organism highly suited to survive and reproduce in its environment (Nazzi & Vianello, 2018).

With the advancements in contemporary dentistry, we sometimes overlook that our oral cavity has been shaped by a remodeling process that defines our unique characteristics. It's crucial for dentists to recognize that the morphology, structure, and physiology of our teeth have evolved over millions of years, starting from a simple conical form and evolving into a complex pattern of cusps and crests (Bergqvist, 2003). Through their varying volumes, countless shapes, and even their tissue structure and histology, teeth reveal how evolution has operated through adaptations connected to their primary and fundamental purpose: chewing. Evolution is a slow process that operates without preconceived plans, but rather through gradual modifications in response to the selective pressures of the environment (Jacob, 1977).

> Natural selection operates not like an engineer, but like a bricoleur, who doesn't know exactly what he will produce, but who repurposes everything he finds, a bricoleur who uses whatever is at hand to create something useful, and without long-term plans, imbues his materials with unforeseen functions to craft a new object. From an old bicycle wheel, he creates a pulley; from a broken chair, he forms the box for the radio; similarly, evolution constructs a wing from a limb, or an earpiece from a fragment of the jaw. (François Jacob, 1977, 1163)

Mastication has continuously shaped the teeth, muscles, and facial bones of every mammalian species in response to various dietary demands. The stomatognathic system of this class of vertebrates has been and remains one of the bodily districts in which, during the evolution game, its countless tactical schemes have been most expressed.

In our daily lives, teeth serve various roles, yet if you were to ask someone, 'What are teeth for?' the answer you'd inevitably receive is, '…for chewing!' However, the fundamental correctness of this answer doesn't account for the fact that during the course of phylogenesis, when the first structures we can categorize as 'dental' emerged, their function was not primarily mastication.

Most fish, amphibians, and reptiles have simple-shaped teeth, often conical, that do not occlude (as shown in Fig. 1.1) and are seldom arranged on a surface for grinding or crushing. Many vertebrates have relied on teeth primarily for ingestion, and the actual breakdown of food is primarily a chemical process.

So, when did teeth, with their distinctive structure unlike other tissues, make their appearance in the animal kingdom?

The mouth seems to be one of the earliest anatomical features to develop in the ancient ancestor. *Saccorhytus coronarius* is the likely forebear of all vertebrates, including reptiles, amphibians, fishes, and humans. This microorganism had a rounded shape and a huge, versatile mouth, and it existed around 540 million years ago (Han et al., 2017).

While some invertebrates, such as snails, possess grinding elements in their oral regions, these structures are not homologous to the teeth of chordates (Pérez-Pérez et al., 2010). Classically, teeth are believed to have originated in fish from placoid scales of the external dermal skeleton. Initially, these plates were located outside the mouth, forming a protective exoskeletal surface surrounding the entrance of the alimentary canal. Only later in evolutionary history were these structures incorporated into the mouth, marking a transition from armor to weaponry. This transformation occurred through a process known as cooptative development and heterotopic transfer, which involves modifying structures with different functions located in other anatomical positions—a phenomenon known as *exaptation* (Gould & Vrba, 1982).

Exaptation refers to the evolutionary process in which existing structures, originally serving a certain function, are modified, transferred, and repurposed for a different function. This concept is fundamental for understanding the evolutionary adaptations of many anatomical structures and was introduced by distinguished paleontologists Stephen Jay Gould and Elizabeth S. Vrba: *'We suggest that features evolved for other uses (or not for any function at all) and later 'co-opted' for their current function should be called exaptations. They serve their current role effectively, but they didn't evolve specifically for it. Their survival owes to attributes that existed for different reasons, thus becoming useful due to their form.'* (Gould & Vrba, 1982).

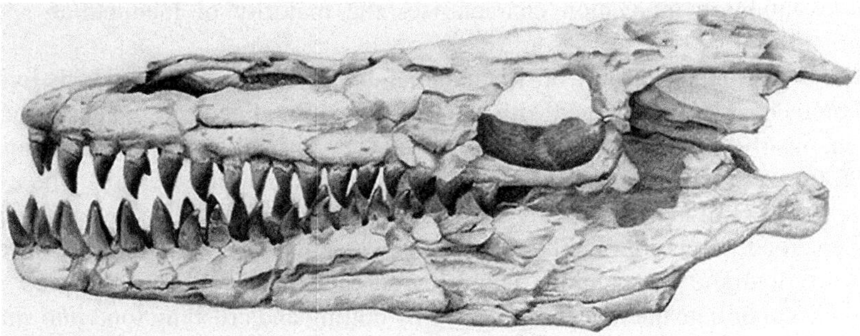

Fig. 1.1 Mosasaurus skull. Drawing by Marco Boggioni

According to current knowledge, the earliest dental structures appeared in ancient jawless vertebrates known as agnathans. These creatures possessed a developed sensory apparatus that enabled them to prey on small, soft-bodied animals (Walker & Liem, 1994). The evolutionary origin of teeth appears to extend far back in time, with the conodonts, recently classified as vertebrates, emerging approximately 500 million years ago. Members of this extinct clade of jawless vertebrates are considered the first vertebrates with a mineralized oropharyngeal feeding apparatus, featuring elements measuring 0.2–2 mm in length, which served a dental function (Jones et al., 2012).

The hyaline enamel of conodonts was composed of flat and elongated crystals of calcium hydroxyapatite, exceptionally large in size and oriented perpendicular to the surface of the element. In contrast to the well-structured prismatic enamel found in mammals, conodont enamel lacked any trace of prisms (Kemp, 2002).

For millions of years, dental structures remained in the form of pointed cones, serving the purpose of intercepting and immobilizing prey, preventing their escape, and directing them into the alimentary canal. In most vertebrates, teeth are indeed simple conical structures, all identical (homodonty), primarily designed to immobilize and coarsely break apart captured prey, facilitating subsequent ingestion. Throughout the evolutionary period of fish, amphibians, and reptiles, teeth were exclusively used for food ingestion. The breakdown of food was purely chemical and occurred solely within the digestive tract (Walker & Liem, 1994). The archaic tooth's basic shape is conical, featuring a single cusp called the protocone on the upper side and the protoconid on the lower side. This pointed shape served as a dagger for piercing prey, rendering it harmless for subsequent ingestion. Therefore, despite our tendency to perceive teeth as chewing instruments, it's been around 250 million years since the emergence of mammals that dental components began to play a role in chewing. The first mammals introduced chewing alongside ingestion. Mastication is the process by which food is mechanically broken down and mixed with saliva to create a bolus, a mixture of small particles, for easy ingestion (Lund & Kolta, 2006). This process allows for the rapid assimilation of nutritional components. While the chemical breakdown of food is a common feature among all animals, the distinctive trait of mammals is the mechanical fragmentation that occurs in the mouth before the chemical breakdown process in the intestines. This significant evolutionary innovation characterizes the majority of mammalian species (Armfield et al., 2013).

The evolutionary journey of the mammalian masticatory system sheds light on how biomechanics intertwined with metabolic function. Each component of the oral system involved in food intake underwent transformations to enhance nutrient absorption, effectively meeting the demands imposed by mastication. The evolutionary success of mammals finds its roots in their high metabolic rate, sustained by efficient food processing facilitated by precise dental occlusion and the development of rhythmic mastication. This efficient masticatory system allowed early mammals to initiate the digestion process by cutting and crushing food into smaller portions, unlike reptiles that swallow larger pieces, relying on uniquely chemical treatment of the bolus (Bhullar et al., 2019).

In nearly all vertebrates, teeth primarily serve the function of channeling food into the digestive tract. However, it's only among mammals that there is a distinct need to break down food into smaller pieces (Lumsden & Osborn, 1977). Although mastication is a hallmark of mammals, it predates their appearance. Interestingly, analogous adaptations for similar specific functions can arise in animals that are phylogenetically distant, but only in certain cases do these solutions become distinctive features. Around 300 million years ago, some ornithopod dinosaurs had developed a system to grind their food (Norman, 1984). These herbivorous reptiles, ranging from medium to large sizes, combined relatively simple adductor muscles with complex dental batteries and mandibular rotations capable of generating alternative modes of transverse grinding for food. This adaptation allowed them to process plant fibers in a manner similar to that employed by present-day mammals (Kaifu et al., 2003). Among present-day reptiles, the Uromastyx lizard displays a sophisticated masticatory cycle resembling that of mammals. This observation suggests that the chewing cycle of mammals evolved by "incorporating" transverse movements into the mandibular dynamics of reptiles (Throckmorton, 1980).

In the animal world, we've seen that similar adaptive solutions for specific functions can arise in phylogenetically distant evolutionary lineages, but they don't always become distinctive characteristics. In mammals, mastication became so firmly established that the entire anatomy of the orofacial region adapted to this new function. Beyond extrastomatognathic features such as the differentiation of spinal muscles and the physiology of endothermic temperature regulation with its exceptionally high metabolic rates (Kemp, 2005), many anatomical and functional parameters used to define and differentiate the class of mammals from that of reptiles pertain to the orofacial region. For example, the structure of the ear canal, the articulation of the mandible, the shape and structure of teeth, the presence of occluding postcanine teeth, and the thorough reorganization of mandibular musculature for their specific function—all these domains underwent modifications intimately tied to the origin of mammals (Rowe, 1988).

Another unique trait, seen even in early mammals, is unilateral and alternating chewing, where only one side of the dentition is used at a time to break down food (Crompton & Hylander, 1986).

Just as the shape of a musical instrument is linked to the sound it can produce, dental forms are connected to an animal's dietary type. Form and function are two sides of the same coin (Walker & Liem, 1994), a principle also applicable to the field of dentistry. In fact, reconstructing one or more dental elements with a sole focus on either aspect would render the effort futile and shallow.

Before the advent of mastication, an unspecialized archaic dental shape corresponded to simple mandibular movements, featuring a rudimentary joint structure that operated with a motion of opening and closing. In this primitive setup, the maximum biting force was simultaneously applied to both sides of the mouth, resulting in the simultaneous crushing of food on both sides (Dean, 1990). In non-mammalian vertebrates, the mandibular-cranial articulation operates as a simple hinge joint, where the quadrate bone on the lower part of the skull projects downward and fits into the concave articulatory bone situated on the back of the mandible. However, the emergence of mammals marked a significant shift. A complex and diversified dental structure accompanied equally intricate musculo-articular movements and an evolved cranio-mandibular articulation. This transformation led to a reversal in the joint dynamics between the mandible and the skull: the concave portion shifted cranially to become the mandibular fossa, while the concave part was formed by the condylar process. Mammals represent a biologically diverse and highly successful class. From the tiny bumblebee bat, weighing less than a small coin, to the colossal blue whale, comparable in weight to a Boeing 747, they inhabit a wide range of environments. Mammals burrow, swim, crawl, leap, run, climb, glide, and fly. Their habitats span from the Arctic to the Antarctic, from the depths of the oceans to the peaks of towering mountains, and from the driest deserts to dense rainforests. Mammals exhibit diverse dietary preferences, with some being herbivores, others consuming fungi, and some being carnivores. Their prey ranges from microscopic plankton to the largest animals on the planet. Some mammals focus on specific foods, while others have voracious appetites, devouring nearly everything they can fit into their mouths.

The extraordinary diversity among mammals can be attributed not only to their teeth but especially to their ability to generate internal body heat, known as endothermy (Ungar, 2017). The emergence and evolution of mammals have been shaped by an ongoing interplay between metabolism and mastication. Nothing is more significant for their existence than their capacity to maintain thermal equilibrium within their bodies despite environmental fluctuations. Mammals thrive in the coldest climates and regions with varying temperatures, remaining active even during cold and dark nights. Endothermy has bestowed mammals with numerous advantages. It creates controlled conditions for chemical reactions within the body, enabling heightened motor activity, faster movement over larger territories, extended migration distances, enhanced endurance for foraging, improved predator avoidance, more effective parental care, longer growth periods, and the development of energy-demanding tissues such as the brain. In essence, without endothermy, mammals could not exist as mammals (Kemp, 2006).

The substantial energy demand associated with maintaining a constant body temperature in mammals necessitated greater catabolic efficiency in the alimentary canal compared to reptiles. This efficiency was achieved through the development of an effective masticatory system (Lumsden & Osborn, 1977). Each structure and function associated with endothermy evolved through incremental changes, forming a gradual interconnection with other evolving traits. It was also thanks to their ability to chew that mammals gained access to entirely new ecological niches. For instance, they could be active during nighttime hours and conquer colder territories that were previously inaccessible (Ungar, 2017).

The increase in catabolic efficiency, made possible through diverse dental forms and masticatory functions, played a pivotal role in meeting the heightened energy demand associated with endothermy (Ungar, 2010). Mastication, which involves the fragmentation of food, enhanced the efficiency of energy assimilation from dietary sources, supporting the high metabolic rate of mammals (Inoue, 2015). Smaller food particles introduced into the alimentary canal can be broken down and absorbed more rapidly. This behavioral plasticity afforded by efficient mastication enabled mammals to exploit new ecological niches, allowing them to adapt their diets during shifts in climate, seasons, or periods of heightened competition, whether within or between species (Fish & Mendel, 1982).

For mammals, endothermy is a critical but energetically costly attribute. A resting mammal typically expends 5 to 10 times more energy compared to ectothermic animals (which rely on external environmental temperatures) of similar size. This energy expenditure can increase by 10 to 15 times during intense activities. Consequently, mammals must extract as many calories as possible from the food they consume. Teeth have become the solution embraced by mammals to grind their food, constituting an integral aspect of their identity. When combined with effective mastication, which reduces food portions from an average size of centimeters to diameters below a millimeter, there is a reduction factor of over 10 times (Fritz et al., 2009). Enzyme reaction rates are directly proportional to the surface area of solid food particles they act upon. Through proper masticatory function, food surfaces can expand by a factor of 100 (Lucas, 2015), allowing for the absorption of more energy from the diet at an accelerated pace.

The mastication of food has brought about a significant transformation in feeding patterns, yielding several key advantages:

- It enhances the efficiency of chemical breakdown for foods that would otherwise be challenging to process solely in the intestine reducing the time required for assimilation.
- It expands the ingested food's surface area available for digestive enzymes. This enlargement facilitates the extraction of energy content from the food.
- It enables the easier reduction of food into swallowable particles, making it more manageable for digestion.
- It allows for the qualitative assessment of ingested material before swallowing.

This last aspect holds great importance, as decisions regarding what to consume are made within the oral cavity prior to it being irrevocably introduced into the alimentary canal. An ingested food that might trigger digestive or absorption complications would subsequently hinder the management of the high-frequency food intake rhythm essential for endothermy.

The development of mastication in dietary habits represented a pivotal functional milestone, bringing about a revolution across the entire maxillofacial framework. As this morpho-functional revolution unfolded, each component of the masticatory system faced a series of intricate biomechanical challenges. The resolution of these challenges led to profound modifications and transformations of pre-existing anatomical structures.

The profound distinction between various feeding modes and the pre-ingestive treatment of food has been the subject of numerous contemporary studies, works, and scientific articles. However, this differentiation was already recognized by Aristotle (Irish & Scott, 2015, 26) over 2000 years ago in his '*De partibus animalium*' In this ancient text, Aristotle observed:

> Of the two separate parts that constitute the head, namely the upper part and the lower jaw, the latter in humans and quadrupeds mammals not only moves upwards and downwards but also from side to side, whereas in fish and birds, the sole movement is up and down. The reason is that the latter motion is what is required for biting and dividing food, while lateral movements serve to reduce substances to a pulp. For such animals, similar to grinding teeth, this lateral motion is in service. However, for those animals that lack grinding, it would be rather useless. Consequently, they invariably lack this motion. (Irish & Scott, 2015, 26)

References

Armfield, B. A., Zheng, Z., Bajpai, S., Vinyard, C. J., & Thewissen, J. (2013). Development and evolution of the unique cetacean dentition. *PeerJ, 1*, e24.

Bergqvist, L. P. (2003). The role of teeth in mammal history. *Brazilian Journal of Oral Sciences, 2*(6), 249–257.

Bhullar, B. A. S., Manafzadeh, A. R., Miyamae, J. A., Hoffman, E. A., Brainerd, E. L., Musinsky, C., & Crompton, A. W. (2019). Rolling of the jaw is essential for mammalian chewing and tribosphenic molar function. *Nature, 566*(7745), 528–532.

Crompton, A. W., & Hylander, W. L. (1986). Changes in mandibular function following the acquisition of a dentary-squamosal jaw articulation. *The Ecology and Biology of Mammal-Like Reptiles, 263*, 282.

Dean, C. (1990). *An introduction to human evolutionary anatomy*. Academic Press.

Fish, D., & Mendel, F. (1982). Mandibular movement patterns relative to food types in common tree shrews (*Tupaia glis*). *American Journal of Physical Anthropology, 58*, 255–269.

Fritz, J., Hummel, J., Kienzle, E., Arnold, C., Nunn, C., & Clauss, M. (2009). Comparative chewing efficiency in mammalian herbivores. *Oikos, 118*(11), 1623–1632.

Gould, S. J., & Vrba, E. S. (1982). Exaptation—A missing term in the science of form. *Paleobiology, 8*(1), 4–15.

Han, J., Morris, S. C., Ou, Q., Shu, D., & Huang, H. (2017). Meiofaunal deuterostomes from the basal Cambrian of Shaanxi (China). *Nature, 542*(7640), 228–231.

Inoue, T. (2015). Neural mechanisms of mastication. *Brain and Nerve = Shinkei kenkyu no shinpo, 67*(2), 141–156.

Irish, J. D., & Scott, G. R. (2015). *A companion to dental anthropology*. Wiley.

Jacob, F. (1977). Evolution and tinkering. *Science, 196*(4295), 1161–1166.

Jones, D., Evans, A. R., Siu, K. K., Rayfield, E. J., & Donoghue, P. C. (2012). The sharpest tools in the box? Quantitative analysis of conodont element functional morphology. *Proceedings of the Royal Society B: Biological Sciences, 279*(1739), 2849–2854.

Kaifu, Y., Kasai, K., Townsend, G. C., & Richards, L. C. (2003). Tooth wear and the "design" of the human dentition: A perspective from evolutionary medicine. *American Journal of Physical Anthropology: The Official Publication of the American Association of Physical Anthropologists, 122*(S37), 47–61.

Kemp, A. (2002). Hyaline tissue of thermally unaltered conodont elements and the enamel of vertebrates. *Alcheringa, 26*(1), 23–36.

Kemp, T. S. (2005). *The origin and evolution of mammals*. Oxford University Press on Demand.

Kemp, T. S. (2006). The origin of mammalian endothermy: A paradigm for the evolution of complex biological structure. *Zoological Journal of the Linnean Society, 147*(4), 473–488.

Lieberman, D. (2014). *The story of the human body: Evolution, health, and disease*. Vintage.

Lucas, P. W. (2015). The masticatory system and its function. In J. D. Irish & G. R. Scott (Eds.), *A companion to dental anthropology*. Wiley.

Lumsden, A. G., & Osborn, J. W. (1977). The evolution of chewing: A dentist's view of palaeontology. *Journal of Dentistry, 5*(4), 269–287.

Lund, J. P., & Kolta, A. (2006). Generation of the central masticatory pattern and its modification by sensory feedback. *Dysphagia, 21*(3), 167–174.

Mancuso, S., & Viola, A. (2013). *Verde brillante: Sensibilità e intelligenza del mondo vegetale*. Giunti.

Nazzi, F., & Vianello, A. (2018). *L'altruismo: competizione e cooperazione dalla biologia all'economia, dalla filosofia alle neuroscienze*. L'altruismo.

Norman, D. B. (1984). On the cranial morphology and evolution of ornithopod dinosaurs. *Symposia of the Zoological Society of London, 52*, 521–547. Cambridge University Press.

Pérez-Pérez, A., Galbany, J., Romero, A., Martínez, L. M., Estebaranz, F., Pinilla, B., & Gamarra, B. (2010). Origen y evolución de los dientes: de los cordados primitivos a los humanos modernos. *Revista española de antropología física, 31*, 167–192.

Rowe, T. (1988). Definition, diagnosis, and origin of Mammalia. *Journal of Vertebrate Paleontology, 8*(3), 241–264.

Throckmorton, G. S. (1980). The chewing cycle in the herbivorous lizard *Uromastix aegyptius* (Agamidae). *Archives of Oral Biology, 25*(4), 225–233.

Ungar, P. S. (2010). *Mammal teeth: Origin, evolution, and diversity*. JHU Press.

Ungar, P. S. (2017). *Evolution's bite: A story of teeth, diet, and human origins*. Princeton University Press.

Walker, W. F., & Liem, K. F. (1994). *Functional anatomy of the vertebrates: An evolutionary perspective*.

Chapter 2
Features of the Stomatognathic System in Mammals

Abstract Chapter 2 examines the evolution of the stomatognathic system in mammals, highlighting how the advent of mastication marked a turning point in craniofacial anatomy. The emergence of chewing triggered a profound structural reorganization involving teeth, muscles, jawbones, joints, and soft tissues. The shape and number of teeth, dental roots, and the development of heterodonty are presented as functional responses to specific biomechanical demands.

The chapter also explores the role of masticatory and facial muscles (lips and cheeks), emphasizing their importance not only in chewing but also in regulating occlusion and enabling social communication. Special attention is given to the functional anatomy of the tongue and its coordination with other elements of the oropharyngeal system.

Additionally, it delves into how bone tissue responds to mechanical stimuli, showing how chewing influences the modeling and remodeling of the skull. The skull is described as a dynamic structure composed of interconnected bony units joined by sutures capable of dissipating forces, with a focus on the relationship between muscle development and skeletal adaptations.

Overall, the chapter illustrates how the masticatory function has been a key factor in shaping the evolution and organization of the mammalian stomatognathic system, deeply interwoven with posture, encephalization, and cranio-cervico-mandibular dynamics.

> Prima digestio fit in ore
> Salerno Medical Institution
> (regarding mammals, author's note)

Anatomy, a biological science that studies the form and structure of living beings (Treccani), is often approached with an almost creationist perspective, as if we have always been as we are, without considering the succession of the countless forms that preceded us, and without bearing in mind how the anatomical changes that have gradually stratified since remote times have been the response to specific functional needs necessary in that evolutionary phase.

In the field of dentistry, function is frequently evaluated with a shortsighted perspective toward the past, as if distorted by an anthropocentric viewpoint, detached

A. Papini, *On the Origin of Human Masticatory Function*,
https://doi.org/10.1007/978-3-031-98664-2_2

from the profound connections that link us to the life forms that have preceded us in evolutionary terms.

The anatomical reasons behind a form are shaped through successive generations by two factors: the function performed by the individual structure and the action of natural selection on the group to which the organism belongs; adaptations are the evolutionary result of the numerous bioengineering constraints that an organ, structure, or even behavior undergoes, in relation to how a certain function beneficial for the organism's survival in its usual environment is executed (Di Vincenzo, 2011).

Compared to other medical disciplines, dentistry holds a privileged position in being able to identify the phylogenetic trajectory of the anatomical structures it deals with: the tissues composing teeth and bones, in fact, often have the unique feature to endure over time, enabling a peculiar anatomical-evolutionary comparison within the human body.

While we will primarily concentrate on the splanchnocranium when considering the development of the stomatognathic system, it's important to note that its evolution cannot be separated from that of the neurocranium.

Throughout the evolutionary history of our stomatognathic apparatus, it's as if certain pivotal "events" have occurred, where the transformations that laid the foundation for our current form and functions seem to coalesce. When we seek to identify the primary factors shaping the evolution of our cranial shape, three predominant influences come to the forefront, three morphogenetic drivers that have affected and molded the structure of both the neurocranium and the splanchnocranium: mastication, bipedalism, and encephalization (Du Brul, 1977). Throughout the entire evolutionary history leading to humans, these three elements have interacted and influenced one another, seemingly linked by associative properties. Chronologically, we can ideally identify these important shifts as three pivotal transitions:

(1) Around 250 million years ago, with the emergence of mammals, anatomical elements dedicated to mastication became structured.
(2) Around 5 to 6 million years ago, as bipedalism developed—initially facultative and later obligatory—adaptations within the stomatognathic apparatus evolved to accommodate the new and challenging musculo-articular balance demanded by a upright bipedal posture.

    This began to define the necessary anatomical-functional relationships to maintain the fragile bipedal equilibrium.
(3) Around 200,000 years ago, as the extreme globular encephalization characteristic of *Homo sapiens* developed, a series of adaptations occurred that reshaped the splanchnocranium in relation to its new interactions with the neurocranium.

In this chapter, we will discuss how a new approach to processing food before ingestion has altered the anatomy of the mammalian stomatognathic apparatus; during the evolutionary transition from reptiles to mammals, the structural changes related to chewing disrupted all the different components of the masticatory system, either modifying them or creating new ones.

As mastication emerged, a profound structural revolution took place within the stomatognathic apparatus: alteration of the Hertwig's sheath, creation of the Tomes' process, formation of the hammer and anvil, heterodontia, development of the bony palate, shaping of the cheeks and lips; all these elements appear so distinct from one another that they seem scarcely related.

However, in reality, these structures formed in mammals in response to the biomechanical demands of a new way of processing food before ingestion.

The masticatory apparatus of mammals differentiated from that of other amniotes (a group of vertebrates comprising the classes of mammals, reptiles, and birds, all terrestrial or secondarily adapted to aquatic life) through the intraoral food processing system.

Within this group, the amnion, a sac-like appendage containing the embryo initially and the fetus subsequently, is suspended in a liquid known as amniotic fluid, offering protection against external impacts, pressures and thermal shocks. Among mammals, a series of morphological and functional features developed, including: transverse movements of the jaw during the masticatory power stroke; precise occlusion; the suspension of teeth in the alveolus by a periodontal ligament; a diphyodont dentition (reduction to only two sets of dental generations); a hard palate; the presence of a single bone (the dentary) in the lower jaw, which articulates with the skull through the temporomandibular joint.

The evolution of these traits facilitated an improvement in the efficiency of food processing within the oral cavity (Ross et al., 2007). These modifications occurred as adaptations of individual components in response to a new set of biomechanical challenges introduced by mastication, and each new adaptation fit within a context that facilitated the overall functionality, influencing and being influenced by it. The alterations developed by the oral cavity to adapt to this new function encompassed:

1. **Shape and number of teeth;**
2. **Shape and number of muscles;**
3. **Shape and number of maxillary bones;**
4. **Formation of the temporomandibular joint;**
5. **Formation of the middle ear;**
6. **Formation of the periodontium;**
7. **Modification of dental tissues.**
8. **Modification of cervical vertebrae.**

We will move on to go through each paragraph mentioned above, seeking to emphasize how chewing has structured the adaptations of each anatomical component listed earlier.

2.1 Tooth Shape and Count

As mentioned earlier, with mastication, teeth ceased to be mere weapons used for seizing and killing prey, transforming into sophisticated tools to execute and regulate the necessary chewing movements for food preparation before entering the digestive tract.

The key macroscopic traits that dental components have developed in mammals are:

- Diphyodonty;
- Heterodonty;
- Tribosphenic molars;
- Dental roots.

2.1.1 Diphyodonty

The majority of non-mammalian vertebrates are polyphyodonts (Bemis et al., 2005), meaning they can replace their dentition multiple times, whereas diphyodonty is the characteristic whereby only two sets of teeth are produced throughout an individual's lifetime: the first being deciduous and the second permanent. Diphyodonty is a distinctive feature of mammals and is used to define this class of vertebrates. As we progress along the evolutionary timeline, the number of dental replacements tends to decrease, moving from fish to amphibians, reptiles, and eventually mammals. For instance, sharks can replace their teeth 200 times (Fig. 2.1), whereas crocodiles do so 45–50 times (Reif, 1984).

Among mammals, which represent the more recent phylogenetic vertebrates, there is a reduction in dental generations: in fact, in the majority of mammals, the dentition is diphyodont, featuring merely two sets of teeth (Rasch et al., 2016).

The need for precise occlusion forms the foundation for excellent masticatory efficiency, which is why the reduction in the number of dental replacements is crucial (Berkovitz & Shellis, 2018):

Fig. 2.1 Shark dentition

> the rationale behind limiting dental replacements to a maximum of two resides in the necessity for dependable occlusion. This guarantees that during mastication, the dental component takes on the role of a true precision instrument. In order to maintain their utmost efficiency, teeth must be capable of enduring significant stresses and constantly adapting to changing biomechanical demands. This fresh functional arrangement made dental stability imperative, a requirement that hadn't been as crucial before; the precision of occlusal contact and the micrometric reciprocal sliding between teeth during masticatory cycles became a vital and indispensable factor for effectively breaking down food. The precision of occlusal contacts, both static and dynamic, necessitated fewer dental replacements. Therefore, driven by the essential need for maintaining reliable occlusion, the mammalian class relies on just one. (Crompton, 1963)

2.1.2 Heterodonty

Heterodonty refers to the presence of elements with varying shapes within the dental arches, and it has played a critical role in the evolution of an efficient masticatory system, broadening dietary adaptability, and enhancing dental structural integrity.

The tooth and its function have evolved in response to environmental changes (Bergqvist, 2003), and the relationship between tooth shape and chewing function is a fundamental key for the acquisition and transformation of food before ingestion.

Our current dental morphology is the product of an evolution that initiated millions of years ago: from the shape of a basic cone to a progressive and more varied complex of cusps, grooves, and crests. However, heterodonty, although one of the characteristics that distinguish the mammalian class, is not exclusive to them: while shape and size variations are relatively uncommon among non-mammalian vertebrates, these changes began to develop earlier in other vertebrate classes as well. The most remarkable aspect of mammalian teeth is the recurring presence of certain forms even in unrelated species: nature consistently presents the same solutions to the biomechanical core challenges of food acquisition and processing (Ungar, 2017).

Therapsid reptiles, which lived around 280 million years ago, already featured a tooth distinct from others, larger in size and similar to canines. Among cynodonts (a variety of synapsids that emerged approximately 250 million years ago), heterodonty became widespread: these reptiles possessed teeth with rows of cusps, ridges, and facets that suggested their function as shearing tools for cutting tough foods (Gow, 1980).

Heterodonty marked the beginning of teeth's functional specialization (Bergqvist, 2003), and the earliest primitive mammals already showcased heterodont dentition with elements of distinct shapes based on diet. Among living mammals, only a few species, such as dolphins, exhibit homodont dentition. A key trait of mammalian evolution is a progressive complexity in the morphology of postcanine teeth: transitioning from a primarily conical form, identical across all teeth in the oral cavity, the prevalence of morphologically differentiated elements became the norm with mammals.

By examining their number, position, and height, multicuspid postcanine teeth are well-configured to withstand high biting forces, provided that multiple cusps come into simultaneous contact to distribute the load (occlusal contact on a single cusp concentrates the forces and reduces resistance). For this reason, one of the criteria that characterizes a physiological occlusion in humans is precisely the even distribution of contacts on all supporting cusps of the posterior teeth in maximum intercuspation.

The position of load points (contact) and the height of cusps, along with their number or area, constitute the primary controlling factors. The structural integrity of complex teeth is upheld when contact points on individual cusps remain adequately distant from their lateral walls, and when cusps do not become excessively tall relative to the width of the teeth (Constantino et al., 2016)

The basic dental formula among placental mammals consists of the following in each quadrant:

- 3 incisors for grasping, tearing, scraping, and gnawing;
- 1 canine for immobilizing or killing preys;
- 4 premolars for piercing and crushing;
- 3 molars for grinding, triturating, and fragmenting: making a total of 11 teeth per quadrant (Fig. 2.2).

This was already the formula of the primordial primate, considered the oldest among our ancestors, Purgatorius, roughly the size of a small rat, which lived around 65 million years ago (Manzi, 2013).

In heterodont dentition, incisors and canines are located at the front of the oral cavity to be used for food grasping, while the posterior teeth, dedicated to masticating ingested food, possess crown shapes designed to concentrate stress during chewing. This aims to create the highest number of fractures in the food and effectively break it down as much as possible (Popowics & Herring, 2006).

Starting from this basic formula, many groups of mammals, such as rodents, have reduced it, while others have expanded it, like the armadillo with 50 teeth, or odontocetes, which can have 60 teeth per arch (Berkovitz & Shellis, 2018). Among primates, who generally have 36 teeth, only catarrhines, unlike other monkeys, have 32 as adults: their dentition includes 2 premolars, and molars erupt by adding one at

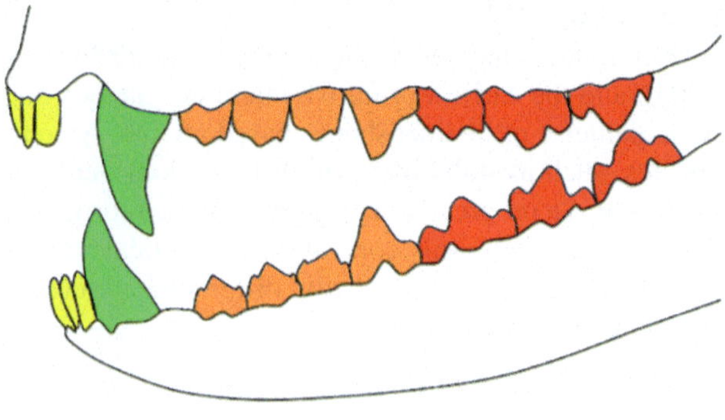

Fig. 2.2 Typical basic dental formula of mammals

a time without replacing any existing elements. These mark the individual growth patterns characteristic of each species and along with them, the major stages of childhood, adolescence, and youth.

In *Homo sapiens*, the eruption timeline changes: around 6 years of age, the first molar defines the transition to the second phase of childhood; approximately by the age of twelve, the second molar emerges, marking the commencement of puberty, and around 18 years old, at the culmination of this period, the third molar comes into view, finalizing skeletal growth.

In anthropoid apes, the sequence remains the same, yet it's within our species that the molar eruption, resembling a metronome, decelerates its speed: in *Homo sapiens*, childhood and youth remarkably extend in contrast to other primates and mammals in general (Manzi, 2013).

2.1.3 The Tribosphenic Molar

The term 'tribosphenic' was coined in 1930 by George Gaylord Simpson, one of the most significant American paleontologists, by blending the Greek words tríbein (to rub) and sphen (wedge), attempting to encompass within a single word the notion of these teeth's dual function, cutting and crushing (Ungar, 2014).

The tribosphenic molar is an exceptionally effective tool with cusps that puncture and serve as pestles, crests that cut like blades, fossae that function as mortars, and grooves that enable the expulsion of processed portions of the bolus: one of the primary functions of this tooth-tool in breaking down tough food is its capacity to penetrate through the food itself.

To quote Alfred Tennyson's words, in a "Nature, red in tooth and claw," teeth are complex instruments designed for fractures (Lucas, 2004), concurrently guiding proper chewing movements (Ungar, 2015). This forced propagation of fractures through food accomplished by the dental tool during chewing occurs via point-like surfaces, the cusps, or through blade-like surfaces, the crests (Lucas, 1979); given the limited morphological options available in terms of shape and geometry, cusps and crests are the biological equivalents and share the same topography as the ideal forms of points and blades.

The resemblance between the actual shapes of teeth and the ideal functional shapes reveals that teeth possess the optimal shape for their function: their morphology as well as their structure best suit geometric and functional requirements. By combining these simple shapes with the vertical and horizontal movements of function and the variety of food textures to which mammals have adapted over the course of evolution, the extensive development of dental forms in mammals has been determined. The stress required to initiate a fracture will vary with the contact area between the cusp and the food: increasing the sharpness of the cusp reduces the force and energy needed for the tooth to penetrate hard food. The smaller the curvature radius of the cuspal tip, the smaller the contact area, resulting in higher stress generated within the food and thus requiring less force and energy to penetrate the substances (Lucas, 1982; Evans & Sanson, 2003). However, this process of cuspal thinning, enhancing their penetration efficiency, cannot exceed a certain limit, beyond which excessive structural fragility would be induced.

The tribosphenic molar is the key to understanding the origin and diversity of the teeth in present-day mammals: it is a very specific type of tooth that marks the entry of mammals into a new adaptive zone; from the archaic tribosphenic molar, the countless molar forms of this class of vertebrates have evolved, and without it, the evolution and diversity of mammals would have taken a significantly different path (Butler, 1972). The emergence of the tribosphenic molar marked a crucial event that shaped the emergence of modern mammals: a new multifunctional complex integrated cutting and grinding into a single chewing action, laying the structure for the subsequent evolution of the diverse dental morphologies they produced (Davis, 2011).

Edward Cope, a nineteenth-century American paleontologist and anatomist, discovered the key to understanding the transition of dental forms from simple cones to complex structures, thus deciphering the evolution and diversity of mammalian molars. Although paleontological evidence, studies on tooth ontogeny, and even an imperfect correspondence in cusp nomenclature (Bels, 2019) have revealed limitations and inconsistencies in Cope's described model, its didactic effectiveness has not waned.

Nevertheless, when attempting to imagine the evolutionary progression of various dental shapes, it's important to keep in mind that the changing dental patterns stemming from the simple conical form should not be visualized as a linear and linked succession of progressively intricate morphologies.

Instead, it should be seen as an eruption of forms with separate trajectories, sometimes parallel and other times divergent, always with the potential to generate new ones—in essence, the bushy form that was so dear to Master Charles Darwin.

From the most archaic model to the form of the tribosphenic molar in mammals, one could envision that on either side of the primitive conical tooth, the *protocone*, two other cusps evolved, mesially and distally—the *paracone* and the *metacone*. This terminology for classifying cusps, predominantly anthropological, assigns the same name to the homologous lower cusps as the upper ones, while appending the suffix "id" (e.g., protocone: upper; protoconid: lower). The tritubercular molar was a dazzling feat of engineering to which evolution subsequently shifted the new cusps away from the original cone, outward for the upper ones and inward for the lower ones, forming two opposing triangular structures: the *trigone*, upper, and the *trigonid*, lower (Ungar, 2017). Subsequently, molars also acquired a low basin called the *talon/-id* (an additional element composed exclusively of hard tissues), which joined the trigone/-id, the original dental structure housing the pulp chamber.

Below, the paraconide was lost, while in the talonid, three cusps were introduced: the *entoconid* on the lingual side, the hypoconid on the vestibular side, and the *hypoconulid* between the last two (Fig. 2.3).

Understanding the trigone/-id and talonid in a molar and visualizing them is not just a mere stylistic exercise, but it is also practical, for instance, in the prosthetic preparation of a molar crown. This prevents excessive exposure of the pulp chamber: the thickness of the hard tissues surrounding the pulp chamber is minimal in the

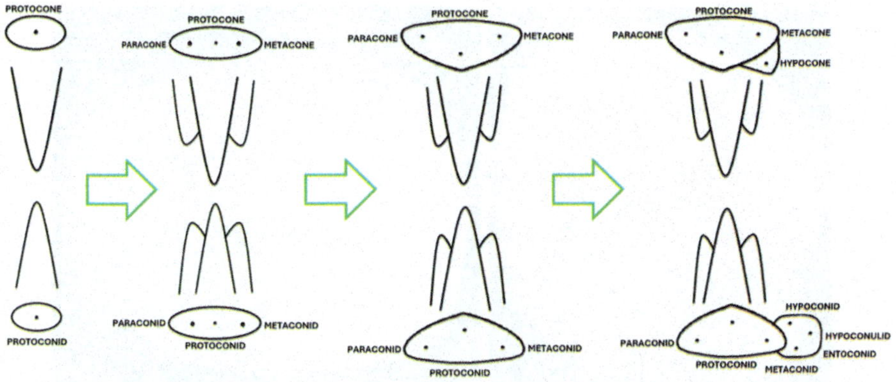

Fig. 2.3 Cope—Osborn Model. Evolution from the monocuspid archaic tooth to the tribosphenic molar

mesio-vestibular corner of the crown, and a preparation that overlooks this aspect could lead to irreversible damage to the dental pulp. Locating the trigone is also valuable in endodontic procedures involving the pulp chamber, as it's situated in the phylogenetically oldest part of the molar, containing the delicate pulp tissue (Fig. 2.4).

Each cusp connected to the others with a sharp ridge to form a V, which pointed palatally in the upper molars and buccally in the lower ones. The crests between the cusps created continuous and jagged blades that functioned as cutting tools. The opposing triangles adapted to each other in a manner where the opposite blades, making contact, transformed the molars into formidable cutting mechanisms (Fig. 2.5).

The process of using a pestle and mortar, involving the alignment of the upper molar's mesiopalatal cusp (protocone) with the central pit of the lower molar as a supporting/working cusp, necessitated a shifting in the sagittal plane of the opposing dental components: across nearly all mammals, the lower molar cusps close in front of the cusps of the opposing molar, exhibiting an anterior translational relationship that Edward Angle termed "first class occlusion" in humans (Angle, 1981). This distinct relationship between opposing masticatory teeth, evident in the earliest mammals of the Upper Triassic, has persisted for over 200 million years, showcasing the efficiency of its function (Osborn, 1982).

Through an evolutionary journey that developed over time, the transition from reptiles' conical teeth led to the ancestral molars of mammals. From this foundational shape, the diverse range of dental forms among present-day mammals emerged: aligning crests and eliminating mortar led to the blade-like molars of carnivores; adjusting dentin and enamel layers gave rise to herbivores' molars; by aligning a fourth cusp with the triangles, raising the basin, and rounding cusps, our own molars came into existence (Ungar, 2017), a feature shared by many omnivores.

The dental form, mirror of its own function, adapts as that function changes: the occlusal surface acquires characteristics such that each morphological component

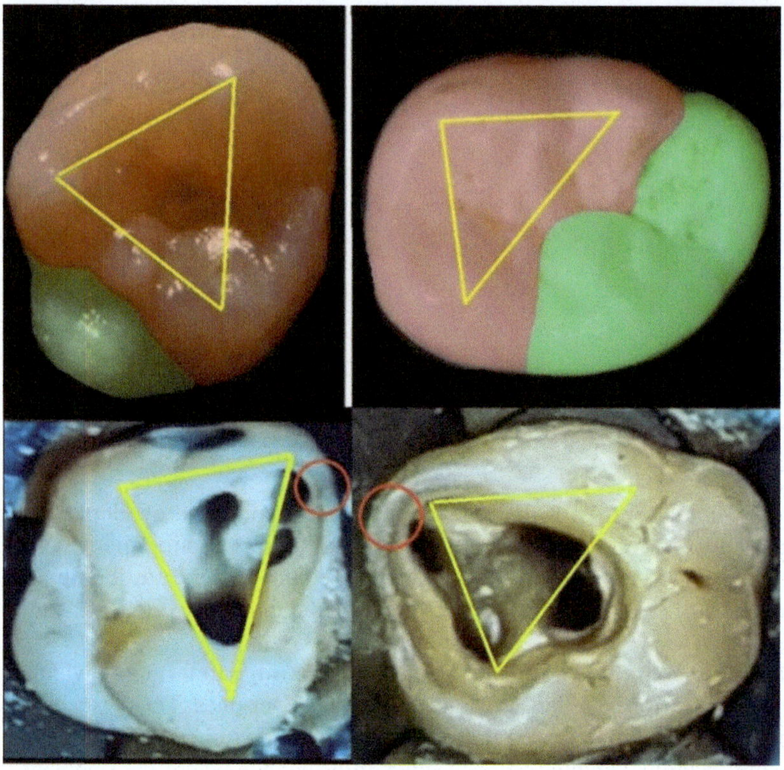

Fig. 2.4 On the left: upper molars; on the right: lower molars. Highlighting the trigon/-id (red) and the talonid (green). Note how the opening of the mesio-vestibular canal is particularly close to the dental surface. Photos below courtesy of Nicola Perrini

plays a role in optimizing the mechanical handling of food; the specific morphology of the tribosphenic molar has broadened the spectrum of dietary options, precisely due to its richness in morphological diversity.

Too many dentists neglect to provide adequate attention to the interactions between teeth and food, and too often, the terms ‘occlusion’ and ‘dental function’ are incorrectly used interchangeably. The importance of dental arches efficiently breaking down food without undergoing harm during the chewing process appears to have been misconceived or considered unimportant (Irish & Scott, 2015).

Dentists deal with dental form on a daily basis, altering, repairing, and preserving these structures. Each dental procedure performed, except perhaps for endodontics, has the potential to impact how the mandible aligns with the cranium, thus influencing—as we will discuss—the relationship between the skull and the vertebral column.

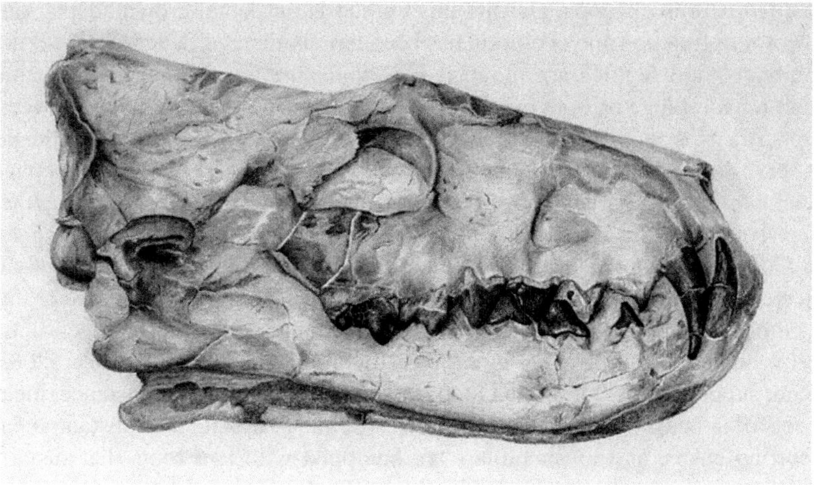

Fig. 2.5 Creodont (Hyenodon horridus). The skull shows the division and sharpening of carnassial teeth between molars and premolars. Drawing by Marco Boggioni

2.1.4 Dental Roots

To withstand and distribute the forces produced during mastication in mammals, dental roots consolidate as distinctive elements (though not solely). The thecodont dentition, consisting of dental roots situated in cavities of the maxillary bones, becomes the standard within this class of vertebrates, even though a similar attachment between root and alveolar bone is also seen in crocodiles and certain fish (e.g., barracuda). The teeth of many fish, amphibians, and reptiles are joined to the alveolar bone by a rigid connection similar to ankylosis, whereas mammals feature "dental gonfosis," an attachment where the dental root fits into a specialized cavity in the alveolar bone, connected through the fibers of the periodontal ligament (McIntosh et al., 2002). The role of the roots is to transmit the forces of chewing from dental cementum to the alveolar bone, and the development of their surface is linked to bite strength, such as the well-developed root of carnivorous mammals and some primates (Berkovitz & Shellis, 2018).

Most periodontal tissues, including alveolar bone, develop in conjunction with the formation of dental roots and originate from the dental follicle (Cho & Garant, 2000); embryologically, all tissues involved in distributing masticatory forces—radicular dentin, cementum, periodontal ligament, and alveolar bone—develop simultaneously and in a coordinated manner as a single anatomo-functional unit: mammalian "dental gonfosis" develop alongside alveolar bone even during the bell stage, and the bone tissue undergoes continuous remodeling in response to dental growth, dental eruption, and maxillary growth (Berkovitz & Shellis, 2018).

Selective pressures on the masticatory apparatus are mainly mechanical in nature, and the morphology of the dental crown varies in accordance with the differing textures of foods. Despite this emphasis on dental crown adaptation, the study of root function and evolution has been largely disregarded. Most teeth in fish, amphibians, and reptiles are monocuspid, while the posterior teeth of mammals exhibit a wide range of both crown and root forms, with a correlation between the number of cusps and the number of roots (Ota et al., 2009). The surface of the dental root serves as an indicator of stress resistance and is linked to load discrepancies: species consuming harder and more resistant foods generally display teeth with larger roots than similar species consuming mechanically less demanding foods. The width of the roots is also tied to load variations on individual elements: teeth subjected to stronger or more frequent forces showcase roots with a greater diameter compared to teeth enduring lower or less frequent loads. Natural selection favors larger roots in teeth exposed to more frequent and intense loads (Spencer, 2003).

Root support is directly linked to the cuspal configuration: for instance, the palatal roots of upper molars exhibit the highest volume to uphold the substantial palatal supporting cusps, and lower molars are equipped with two roots that sustain the trigonid and talonid (Berkovitz & Shellis, 2018). Conversely, supernumerary cusps in a tooth often correspond to supernumerary roots (Butler, 1952), implying the interdependence of the root pattern with cusp structure (Seo et al., 2017; Spencer, 2003).

2.2 Shape and Number of Muscles

Mammals distinguish themselves from other vertebrates by their ability to chew and carry out horizontal jaw movements, crucial for this function. There's something distinct about how mammals chew, and comprehending the shape and function of our teeth becomes difficult without first recognizing this particular characteristic: the ability to perform lateral and anteroposterior horizontal movements of the jaw, known as lateral and protrusive movements.

Most likely, within an early population of mammals, an enhancement of the potential for diverse patterns of mandibular dynamics took place—meaning the ability to perform horizontal jaw movements—resulting in the creation of novel movement types (Fish & Mendel, 1982). This new function required adaptation, achieved by altering the shape and arrangement of the jaw adductor muscles; furthermore, the lips and cheeks evolved to facilitate monolateral and alternated mastication.

2.2.1 Shape and Arrangement Shift in Jaw Adductors

The change in the configuration of jaw adductor muscles represents a pivotal milestone in vertebrate history, playing a fundamental role in the functional morphology and biomechanical stability of the mandible (Irish & Scott, 2015).

Even in terrestrial vertebrates, the muscular mass of the jaws underwent significant growth, resulting in a commensurate increase in the volumes of their supporting structures (Iordansky, 2010). The evolution of mammals was subsequently accompanied by further remodeling and changes in the jaw adductors, allowing for the development of a more efficient masticatory system (Lautenschlager et al., 2016).

Primitive amniotes (a category of vertebrates encompassing the classes of mammals, reptiles, and birds) possessed a single adductor muscle on each side, whose simultaneous contraction in conjunction with a simple hinge rotation lifted the mandible (Liem, 2001); this structural configuration of the musculature, though effective for closing and opening the mouth, did not, however, allow for the lateral and protrusive movements that provide a more precise control of the essential mandibular dynamics for mastication (Ungar, 2017). The transition from reptiles to mammals was crucial for altering the shape, insertions, and contraction vectors of the jaw adductor muscles: these modifications allowed for the movement of the mandible in both anteroposterior and mediolateral directions (Lautenschlager et al., 2016).

In the earliest reptiles, a part of the adductor mass was probably attached to the ventral surface of the mandible beneath its junction with the cranium, likely corresponding to the pterygoid musculature of more advanced forms (Crompton, 1963); subsequently, this original structure evolved into a type of sling comprised of two muscles, one internal and one external (Fig. 2.6), both inserting onto the mandible (Rubidge & Sidor, 2001).

In order to shift from a basic hinged jaw opening to an opening that involved a significant horizontal element, the jaw's closing muscles needed to undergo reorganization.

The individual adductor muscles thus adopted a more complex geometry, becoming capable of expressing different vectors of movement (for example, the external jaw adductor muscle split into two distinct muscles, the temporal and the masseter)

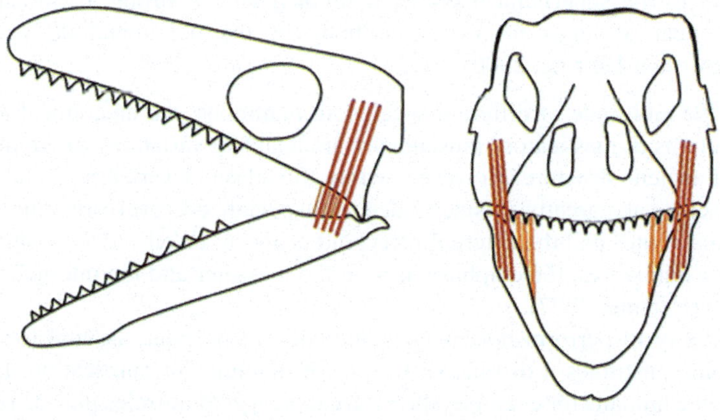

Fig. 2.6 Diagram of adductor mandibular muscles in archaic reptiles: a muscular sling formed by external and internal adductor muscle bundles; a basic model that will also remain in the musculature of mammals in general, and also in humans

(Ungar, 2017). The transformation of the mandibular adductor muscles from their ancestral reptilian configuration to the mammalian type occurred before the full development of the temporomandibular joint (TMJ) and the middle ear in mammals. This suggests that the modification of the mandibular adductor system played a fundamental role in the functional morphology and biomechanical stability of the TMJ (Lautenschlager et al., 2016). This distinct vectoriality expressed by the mandibular adductor muscles resulted not only in increased bite force but also in enhanced efficiency for precise control of mandibular movements, allowing the teeth to be optimally positioned for food grinding (Liem, 2001; Rubidge & Sidor, 2001).

The progressive transformation of musculature, transitioning from the reptilian ancestry to the mammalian form, occurred prior to the emergence of mammals. This process developed during the evolutionary shift and was finalized in archaic mammals with the organization of mandibular adductor muscles (masseters, temporals, and internal pterygoids). The musculature became more complex, with diverse components and orientations, capable of blending a wide range of movements; this rich vectoriality of mandibular movements, distributed three-dimensionally across multiple planes, enabled mammals to adapt mastication to the new energy requirements imposed by endothermy.

These muscles, which are separate and divided into independently controlled parts, allow the mandible greater freedom of movement (Ungar, 2017). The three powerful mandibular adductors are:

1. *Temporal Muscle*: Located above the mandible, it lifts the mandible and, with its posterior fibers, pulls it backwards.
2. *Masseter Muscle*: Positioned on the outside of the mandible, it has two distinct parts. The superficial part lifts and pulls the mandible forwards, while the deep part (sometimes called the zygomaticomandibular) lifts and pulls it backwards.
3. *Internal Pterygoid Muscle*: Located internally to the mandible, with an action similar to the superficial masseter combined with a strong additional medial component, it tends to rotate the body of the hemimandible vestibularly (Popowics & Herring, 2006).

Despite the multitude of different species, in mammals, the anatomical and functional patterns of masticatory musculature are adaptive variations stemming from a common ancient structure. However, within this class of vertebrates, the masticatory muscles vary greatly in size: indeed, in specialized carnivores, the temporal muscle dominates and the external pterygoid is absent (Weijs, 1994), while in herbivores or omnivores, the emphasis lies on the masseter and the internal pterygoid (Fig. 2.7) (Herring, 2007).

The ability to perform horizontal mandibular movements is vital for chewing cycles and constitutes a distinctive feature of mammalian mastication. In certain species, like humans for example, the external pterygoid muscles play a crucial role in these movements through their contractions. Notably, the external pterygoid muscle (PEs), although not an adductor muscle, is indispensable for masticatory function in humans. Through its action, it essentially shifts the mandible on the horizontal

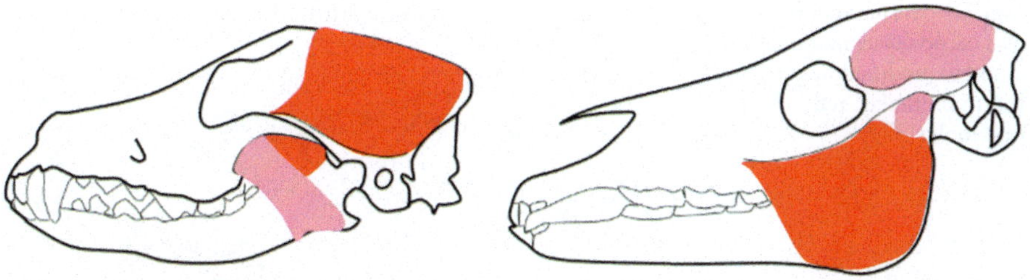

Fig. 2.7 Although the basic structure of mandibular adductor muscles remains almost constant in mammals, the importance of individual muscles varies according to species and their peculiar feeding habits: in carnivores (on the left), the temporal muscle is more developed and powerful compared to the masseter, while in herbivores (on the right), this functional predominance (indicated in the figure by the color red) is reversed

Fig. 2.8 Vector of the temporal muscle oriented posteriorly

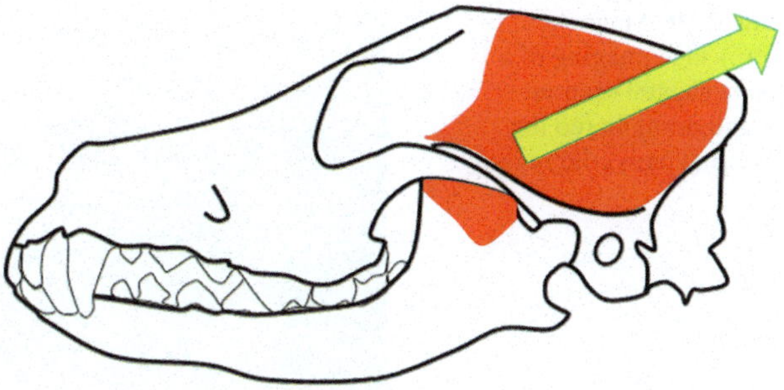

plane: contracting unilaterally draws the mandible to the opposite side, while bilateral contraction moves the mandible forward. Among the varied masticatory musculature, the evolutionary study of the external pterygoids is subject to the highest degree of interpretation, reflecting uncertainties regarding individual attachments and the scarcity of osteological correlates. It is a reduced and less significant muscle in carnivores; for instance, it is extremely small in canids (Ström et al., 1988) and felids: even in the largest tiger, it consists of a few muscle bundles weighing less than a gram. Its small size suggests its limited importance in the essentially vertical chewing pattern of carnivores (Hartstone-Rose et al., 2012). In opossums, considered a living example of archaic mammals, the external pterygoid muscle is much smaller than the internal one (Lautenschlager et al., 2016).

Nevertheless, despite the external pterygoid muscle being small and of lesser importance in carnivores, mediolateral translation remains essential for the opposing action of carnassial teeth. This movement is triggered by the extensively developed and posteriorly situated fibers of the temporal muscle in these mammals (Fig. 2.8).

Concerning carnivorous mammals, it's essential to clarify that this order is not characterized by the existence of well-developed canines, but rather by their carnassial teeth, consisting of the modified fourth upper premolar and the first lower molar (Fig. 2.9).

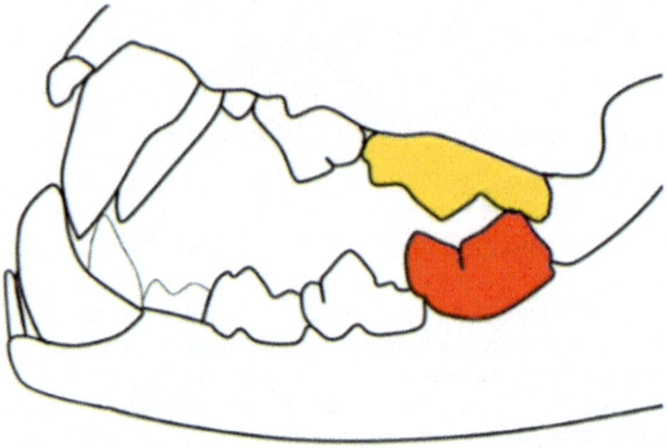

Fig. 2.9 Carnassial teeth of a carnivorous mammal. Orange: modified upper fourth premolar; red: lower first molar

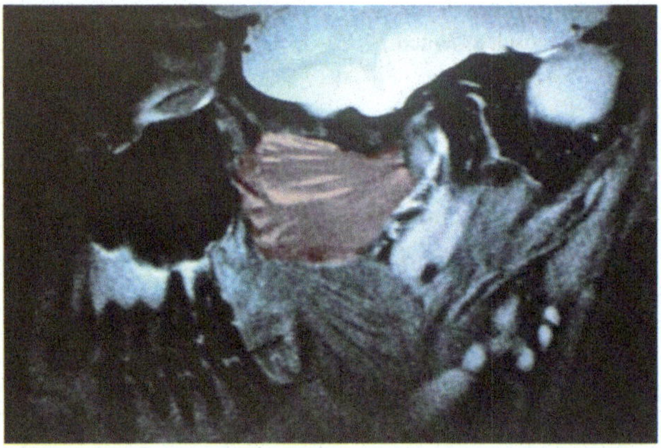

Fig. 2.10 Magnetic resonance imaging of a man, sagittal section: highlighting in red the external pterygoid muscle

In primates, the external pterygoids (PEs) are approximately twice the size of those in primitive mammals, and their development is linked to an enhanced capacity for lateral and anterior translational movements of the mandible and condyles (Weijs, 1994; Hartstone-Rose et al., 2012). In humans, on the other hand, the external pterygoid muscle is notably large (Herring, 2007) (Fig. 2.10), playing a predominant role in determining the horizontal extent of the masticatory cycle.

In mammals, each of the three main adductor muscles follows a unique line of action, characterized by different vector components: anterior, posterior, lateral, and medial. This configuration ensures the complete encasement of the mandible within a muscular sling capable of producing precisely controlled movements, essential for occlusion and mastication (Crompton, 1989).

2.2.2 The Lips and Cheeks Are Structured

The oro-pharyngeal region holds a central role in all oral behaviors of mammals, and to grasp the extent of changes in this region within this vertebrate class, it's crucial to understand the innovations resulting from masticatory function: the evolution of lips, cheeks, and the pharynx in mammals has been marked by the development of a new neuromuscular system, comprising a broad set of striated muscles that contribute to the formation of lips, cheeks, and pharynx (Diogo & Powell, 2019) (Fig. 2.11).

Facial muscles and pharyngeal muscles are characteristic of mammals, and in this class of vertebrates, such musculature shares a phylogenetic correspondence with the primitive jaw (IOIDEA) muscles of other tetrapods, such as the constrictor muscle of the neck in reptiles (Smith, 1972; Diogo et al., 2009). The facial muscles of mammals, innervated by the seventh cranial nerve, are categorized as MIMIC muscles, as they are involved in generating facial expressions during social interactions, in addition to roles in feeding, whisker mobility, hearing, vocalization, and human language (Burrows, 2008). However, even though their action is pivotal in emotional mobility and facial expressiveness, there are considerations that should prompt reflection on the functional origin of facial muscles and their role in masticatory musculature: the first observation is that these muscles appear in the oro-pharyngeal region simultaneously with the onset of mastication; the second observation is that facial muscles are present in ancient mammals with a

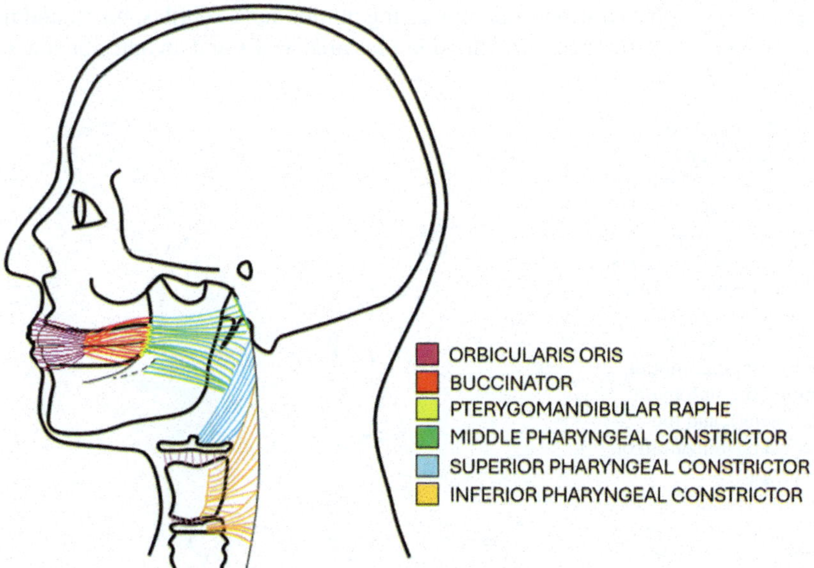

Fig. 2.11 Diagram, in humans, of the connections between the labial, genial, and pharyngeal musculature

fundamental structure that remains largely unchanged across this entire class of vertebrates, including humans: in fact, the facial musculature of contemporary mammals does not exhibit significant differences regarding attachments, configuration, and general architectural appearance.

It's worth noting that the topographic distribution of the trigeminal nerve and the facial nerve significantly coincide (as though trigeminal and facial innervation, largely spanning the same areas, could also collaborate in common functions—chewing, speaking, sucking, facial expressions), and there exist peripheral trigemino-facial communication branches, meaning trigeminal fibers that accompany the branches of the facial nerve (Baumel, 1974) (Fig. 2.12).

The human face is the most adaptable for emotional communication; even though modern humans (alongside gorillas) possess the highest number of 'mimic' muscles, consistent with the significant role played by facial expression in anthropoids in general and modern humans in particular, there appear to be no evident specializations distinguishing this unique capacity of our facial musculature (Burrows, 2008).

Mastication can be seen as a 'synapomorphy' unique to mammals, representing the adoption of an evolutionarily novel feature stemming from a common ancestor, subsequently embraced by practically all species within the class. Similarly, we could also classify facial musculature as a synapomorphy, attributing it a primarily masticatory significance, as it evolved for this purpose and is essential for the processing of the food bolus. It was only subsequently that facial muscles also took on the new function of transmitting emotional expressiveness to facilitate social interactions. These muscles are, in fact, crucial for feeding as well. The presence of cheeks and lips represents a marked divergence from other tetrapods (the superclass of vertebrates with four limbs, encompassing amphibians, reptiles, birds, and mammals), as these structures provide flexible muscular walls and an anterior seal to the

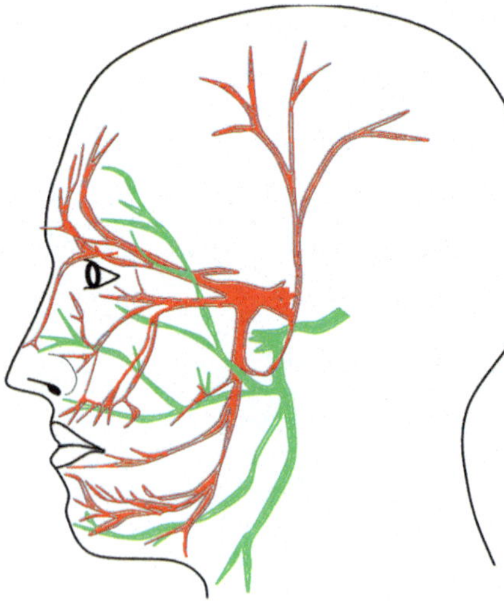

Fig. 2.12 Distribution areas of the trigeminal nerve (light blue) and facial nerve (yellow). The two territories overlap, and the facial nerve innervates, in addition to the 'mimic' musculature, also the posterior belly of the digastric muscle (mouth opening), the stylohyoid (swallowing), and the stapedius (stabilization of the stirrup)

oral cavity. This stands in stark contrast to amphibians and reptiles, whose mouths lack containment when dental arches are separated (Schwenk, 2000).

Most non-mammalian tetrapods lacked or still lack lips, cheeks, and other soft tissues that are helpful in sealing the oral cavity, even though it's important to note that this characteristic, common among mammals, can sporadically be found in other classes too. For instance, ornithischian dinosaurs had muscular fascicles situated beside their arches. While not directly homologous to the buccinator muscles of mammals, they fulfilled a similar function (Galton, 1973). Lips and cheeks, which are indeed distinguishing features of all mammals, are used by newborns to suckle milk from the mother's nipples and subsequently to assist the tongue's muscular control over the food bolus (Liem, 2001).

The musculature of lips and cheeks, in combination with teeth, palate, tongue, pharynx, and masticatory muscles, can be included among the adaptations of mammals for ingestion. While most vertebrates rely on teeth and/or the tongue to seize food, mammals differ in their approach due to their highly mobile lips that encircle the mouth (Wall & Smith, 2001).

In amphibians, reptiles, and birds, the musculature derived from the hyoid arch widens and partially surrounds the neck like a thin sheet, tending to adhere to the skin dermis as well. In mammals, however, this musculature expands superiorly and differentiates into facial muscles (Fig. 2.13), particularly around the eyes, lips, and ears, where they serve to accentuate facial expression (Kardong, 1995).

Present-day reptiles lack facial expression muscles and possess solely (compared to mammals) a musculature innervated by the facial nerve. This comprises:

(1) Neck constrictor muscle: a slim muscular layer exterior to all other hyoid muscles, wrapping around the hyoid and pharyngeal regions. It acts as the central swallowing muscle in reptiles and birds;
(2) Posterior intermandibular muscle, essentially a frontward extension of the preceding one;
(3) Mandible depressor muscle, responsible for mouth opening.

In contrast to the pharyngeal muscles, the mimic muscles have no homologous in reptiles: for instance, even though the posterior belly of the digastric muscle in mammals is innervated by the seventh cranial nerve and serves as a muscle for

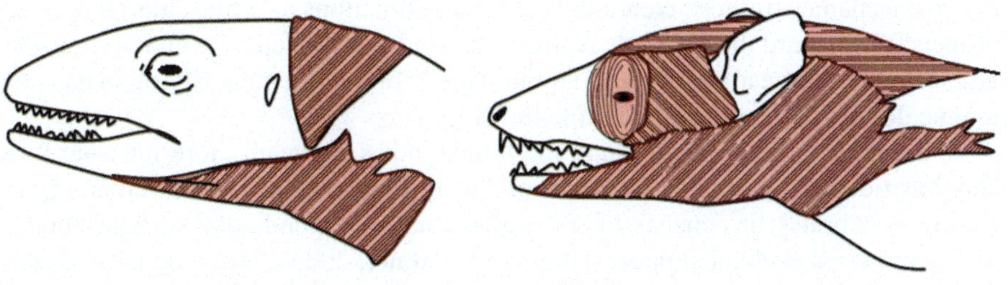

Fig. 2.13 Muscles innervated by the seventh cranial nerve. On the left: monitor lizard (Varanus merrem). On the right: opossum (Didelphis marsupialis)

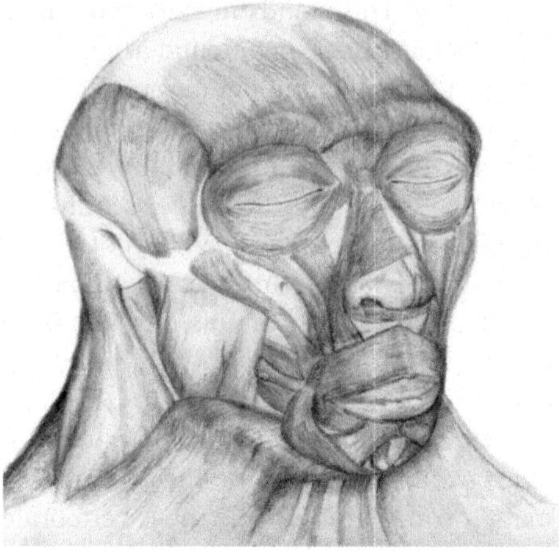

Fig. 2.14 Superficial musculature of the head and neck. Drawing by Marco Boggioni

mouth opening, there is likely no evolutionary correspondence with the reptile's opening muscle (Smith, 1972). Facial muscles (Fig. 2.14) are a distinctive feature of mammals and are vital for mastication: their synchronized movements, in addition to guiding the bolus between the dental arches, aid the tongue in moving food from one side to the other (side switching) (Hiiemae, 1984).

In the feeding process of tetrapods, the tongue holds a predominant function and stands as a significant factor that has shaped the adaptive success of vertebrates in their habitats: most vertebrates rely solely on their mouths for feeding—except for primates and certain raccoon or squirrel species, which also use their hands (Iwasaki, 2002).

The phylogenetic origins of the tongue date back to prehistoric fish, where it emerges for the first time as a rudimentary organ; in terms of its shape, it can be seen as an extension of the hyoid part of the branchial skeleton and can be seen as a modification of the anatomy of the mouth floor. The simplest form of vertebrate tongue, observed in certain birds and snakes, consists of an extension of the hyoid process; this cartilaginous tongue, configured in this manner, can be moved by merely displacing the hyoid bone (Gilbert et al., 2007). In fish, unlike terrestrial vertebrates, the growth of the tongue involves only a slight mucosal rise on the mouth's lower surface, without the presence of voluntary muscles; thus, in the feeding biomechanics of these creatures, the tongue functions as a muscular piston that propels food toward the stomach. A true tongue, characterized by voluntary muscles and mobility, is already present in amphibians, while in the majority of snakes, the tongue doesn't have a significant role in feeding.

In mammals, on the other hand, the tongue takes on much more intricate form and functionality, evolving into one of the most sophisticated muscular organs among vertebrates: it consists of a set of mutually perpendicular muscle bundles without internal skeletal support (Hiiemae & Palmer, 2003). This extensive muscular organ occupies a significant portion of the oral cavity in most mammals, except for certain odontocetes, rodents, and the platypus.

In humans, throughout ontogenetic development, the size of the tongue reflects its physiological function; at birth and during early maturation, the tongue is disproportionately large compared to the surrounding structures: in a newborn, the tongue fills the entire oral cavity, presumably in correlation with its predominant role in breastfeeding. Subsequently, the proportions change: as the child develops, the size of the oral cavity increases more rapidly than the dimensions of the tongue.

The tongues of mammals belong to a group of organs called muscular hydrostats, where muscular contractions generate support for movement. The muscular structure of the tongue serves as the structural basis for hydrostatic deformation, a process that maintains the organ's volume constant throughout (Gilbert et al., 2007).

The tongue performs a crucial role in the process of chewing, initially moving the food backward from the incisors to the post-canine area to position it on the occlusal surface. As the bolus processing continues, the tongue persistently keeps the bolus on the occlusal plane by continuously repositioning it and interacting with the cheeks through buccinator muscle contractions, ultimately achieving a satisfactory degree of food grinding (Hiiemae & Palmer, 2003).

Taking into account the muscle structure of the lips and cheeks in mammals, this is primarily composed of two muscles: the orbicularis, encircling the oral opening, and the buccinator, which lies deeper relative to other facial muscles. The posterior portion of the buccinator muscle situates itself medially to the masseter muscle. Despite being labeled as mimic, both the orbicularis and buccinator muscles should be placed within the masticatory musculature due to their indispensable function in chewing. These two muscles together form a continuous band that externally encircles the dental arches, outlining their outer perimeter. If the tongue, with its centrifugal force, is regarded as a shaping muscular organ of the dental arches, the actions of the orbicularis and buccinator muscles, opposing the pressure exerted by the tongue, aid in shaping the arch perimeter (Fig. 2.15) and in guiding the orientation of teeth within the arch.

The buccinator is a mimetic muscle with bone insertions (Diogo et al., 2009), and even though it attaches superiorly to the skull and inferiorly to the mandible, its action does not elevate the mandible itself, as its fibers have a predominantly horizontal direction. The muscle's anatomy might suggest a retraction action on the corner of the mouth, but in reality, many other facial muscles, like the zygomatic major, are more suitable for this role. Instead, the buccinator muscle works by thickening the entire cheek and acts like a hydrostatic muscular press, pushing the bolus onto the occlusal surfaces. This action is further aided by rhythmic tongue movements that manage to keep the bolus between the dental arches, allowing it to be processed during subsequent chewing cycles.

The buccinator muscle is a flat and thin muscle composed of three bundles that continue anteriorly to form the structure of the orbicularis oris muscle: the upper bundle originates broadly from the bony base beginning at the alveolar process above the molars, extending distally over the alveolar process to the suture between the maxillary and palatine bones. From there, it extends downward towards the inferior surface of the pyramidal process of the palatine bone and continues along a short ligament to the apex of the pterygoid hamulus. The fibers of the middle bundle

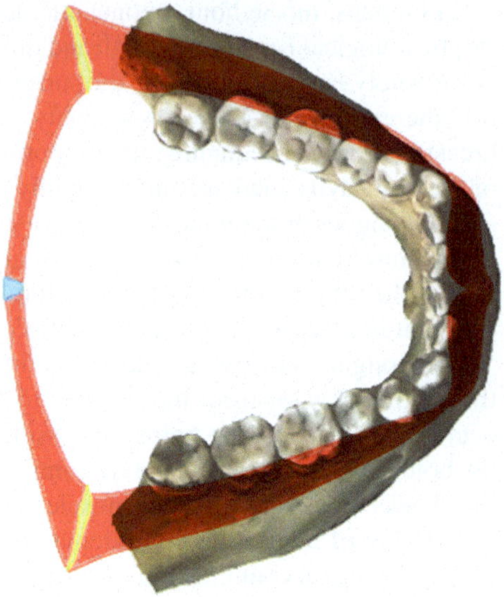

Fig. 2.15 In red: horizontal muscular ring—formed by orbicularis, buccinator, and superior constrictor of the pharynx—delimiting the dental arches, and connecting posteriorly to the first six cervical vertebrae; in yellow: pterygo-mandibular raphe; in blue: pharyngeal raphe

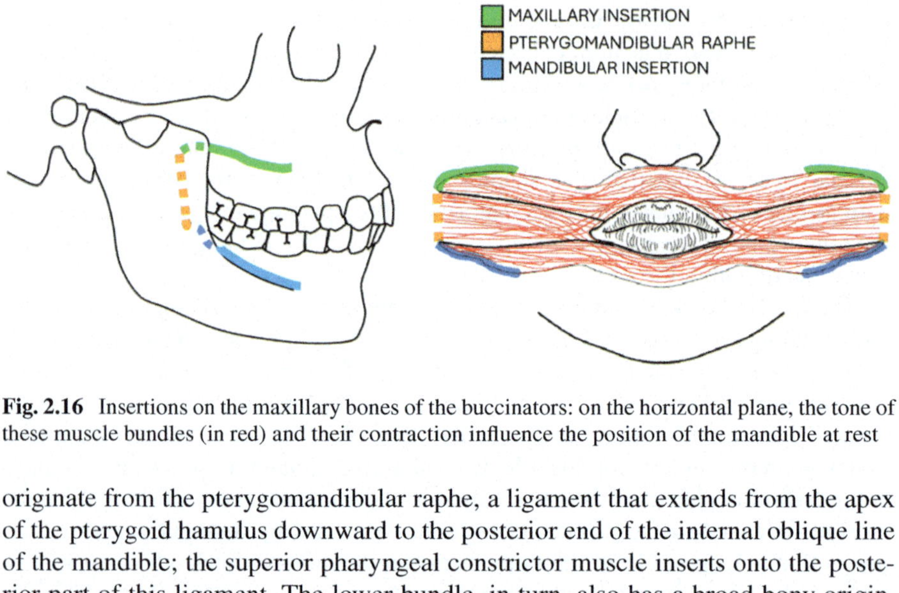

Fig. 2.16 Insertions on the maxillary bones of the buccinators: on the horizontal plane, the tone of these muscle bundles (in red) and their contraction influence the position of the mandible at rest

originate from the pterygomandibular raphe, a ligament that extends from the apex of the pterygoid hamulus downward to the posterior end of the internal oblique line of the mandible; the superior pharyngeal constrictor muscle inserts onto the posterior part of this ligament. The lower bundle, in turn, also has a broad bony origin, starting at the skeletal base of the alveolar process of the molars and extending backward and upward along the internal oblique line. Its bony origin halts at the end of the internal oblique line, where the middle bundle begins.

The upper and lower bundles run continuously from one side to the other, without any crossing points (decussations) (Fig. 2.16).

Anteriorly, the buccinator muscle sends many of its fibers to the orbicularis muscle, contributing to the muscular structure of the lips; these fibers intersect at the corner of the mouth. In primates, the orbicularis muscle plays a role in generating facial expressions and vocalizations, in addition to its involvement in food intake; furthermore, in some non-human primates, it's employed as a grasping and manipulative tool.

During mastication, the buccinator muscle displays significant contractile motion that unfolds right at the outset of the closing motion, subtly preceding the robust contraction of the masseter muscle. Even though the activation is alike on both sides, the buccinator's contraction varies between the working and counterbalancing sides, with more pronounced range and duration noted on the working side in comparison to the counterbalancing side. In the process of swallowing, the buccinator's activity is divided into two stages: an oral stage and a pharyngeal stage; within humans, the buccinator's shortening and thickening exert compressive forces on the alveolar bone and teeth, which increase during chewing and even more during swallowing. Via the pterygomandibular raphe, the buccinator establishes close anatomical and functional ties with the superior pharyngeal constrictor, a muscular-membranous channel primarily constituted by the constrictor muscles. Despite differing origins and innervation compared to the buccinator, these muscles are all characteristic of mammals. Positioned posteriorly to the nasal, oral, and laryngeal cavities, the pharyngeal constrictor muscles form a vertical semicircular canal. During the process of swallowing, these muscles contract, creating a peristaltic wave to propel the bolus into the esophagus.

At the posterior side the pharyngeal tube attaches to the cervical spine through the pharyngeal raphe, a substantial midline tendinous band where the constrictor muscles converge. The pharyngeal raphe that supports the pharynx runs longitudinally and attaches at the top to the pharyngeal tubercle of the occipital bone, positioned anteriorly to the foramen magnum. It then joins the front part of the spine and continues downward to the sixth cervical vertebra. Variations in the tension of the raphe can transmit traction to the cervical vertebrae. During swallowing, the orbicularis, buccinator, and superior and middle pharyngeal constrictor muscles undergo changes in tone and length: the head moves forward, creating tension in the anterior longitudinal ligament of the vertebrae and altering the spinal curves.

In humans, there exists a horizontal muscular sequence (see Fig. 2.16), which includes perioral musculature at the front, encircling the dental arches of the upper jaw and the mandible (and inserting onto these bones). This muscular strip continues posteriorly through the pterygomandibular raphe, with the superior pharyngeal constrictor muscle, subsequently connecting through the pharyngeal raphe to the cervical vertebrae and the suboccipital and neck musculature. The superior pharyngeal constrictor muscle also forms attachments with the tongue and the mandibular bone. The muscular and ligamentous connections between the cervical region and the oral structures constitute a single functional complex, known as the cranio-cervico-mandibular system (Cuccia, 2011), whose components are interconnected in terms of phylogeny, embryology, anatomy, and function.

From a dysfunctional perspective as well, there exists a comorbidity between the stomatognathic system and cervical spine pain syndrome: a significant association exists between the treatment of temporomandibular disorders (TMD) and the

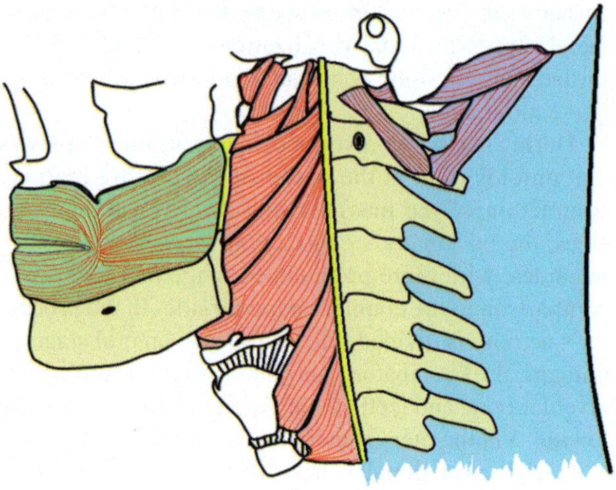

Fig. 2.17 Horizontal muscular chain anatomically and functionally connecting the masticatory apparatus to the cervical spine in general and to C1 and C2 in particular. Green: buccinator and orbicularis; red: pharyngeal constrictor; purple: sub-occipital

reduction of cervical spine pain, along with the improvement of its mobility (Walczyńska-Dragon et al., 2014; Ghodrati et al., 2020). It would be desirable to consider this musculoskeletal system not only for its anatomical connections but also for the functional and dysfunctional relationships among its components, which are so diverse yet closely intertwined (Fig. 2.17).

A dysfunctional condition, such as a fixed one-sided chewing pattern, leads to an asymmetric engagement of musculo-articular structures, resulting in an unfavorable adaptation not only of the masticatory system but also of the posterior musculoskeletal components like those of the pharynx, spine, and suboccipital and cervical region. Every adaptation of biomechanical structures entails an involvement of the organs they encompass, support, and enclose (organs, vessels, nerves), leading to detrimental effects on functionality and well-being. Any postural (and occlusal) disorder is never solely a biomechanical issue.

Refocusing on the impacts of the perioral musculature on the jawbones and dental arches, the repetitive contractions of the orbicularis and buccinator muscles affect the lateral-lateral and antero-posterior alignment of the dental arches, defining a corridor (known as the neutral zone) where the centrifugal forces of the tongue and the centripetal forces of the perioral muscles find equilibrium. After tooth eruption, the muscular strength of the cheeks, lips, and tongue plays a significant role in their labio-lingual and bucco-lingual positioning. Eric Wilfred Fish introduced the concept of the neutral zone in 1931, asserting that natural teeth occupy a 'zone of balance,' and each tooth assumes a position that results from various acting forces.

As each tooth erupts within the arch, it is guided into a specific zone of neutrality that determines its horizontal position; the teeth of mammals are capable of adjusting their position, within certain limits, to adapt to changes in the biomechanical environment of the mouth: if the centrifugal forces exerted by the tongue surpass the centripetal forces applied by the muscular bands of the buccinators and orbicularis, the teeth will move horizontally until the opposing forces reach equilibrium (Fig. 2.18).

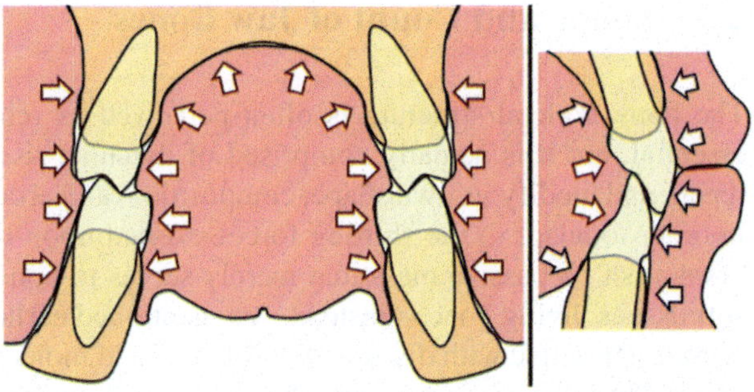

Fig. 2.18 Opposing forces exerted by the tongue musculature and the perioral musculature

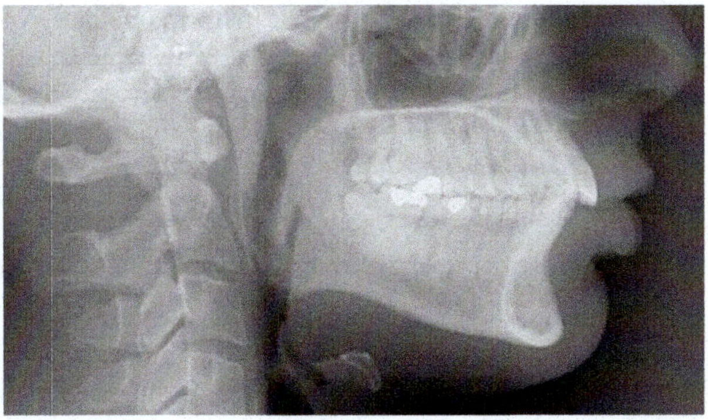

Fig. 2.19 Lateral-lateral projection teleradiography. Note the profile of the labial soft tissues, typical of powerful perioral musculature

Muscle tone and repetitive contractions of the muscular band formed by the buccinator and orbicularis establish the dimensions of the dental arches: the hyperactivity of perioral musculature exerts significant centripetal pressure on the alveolar bones, impacting their perimeter and leading to alterations in occlusal relationships (Rathee & Jain, 2020). The perioral musculature, in opposition to the force exerted by tongue posture and movements, dictates a precise horizontal dimensional relationship between the arches, shaping their perimeter (Oxilia et al., 2018). These forces exert pressure on the palate and the maxillary bones, influencing their morphology as well as tooth inclination (Fig. 2.19).

If the muscular (and occlusal) forces applied are not in balance, regardless of any occlusal pattern, the teeth cannot remain stable in their positions within the arches. Constrictive perioral musculature can hinder the dental arches from expanding to their natural alignment with the skeletal base; skeletal mandibular growth tends to push the chin forward, while the dental arches are restricted by the muscular bands that prevent them from growing proportionately with their skeletal foundation.

2.3 Shape and Count of Jaw Bones

The bony skeletal structure is often perceived as relatively unchanging and less essential, but it is actually comprised of dynamic tissue with the ability to grow, repair, and modify its own shape, composition, and arrangement throughout growth, in order to adapt to the shifting forces exerted upon it (Liem, 2001). The skeletal system isn't a passive mass that merely serves to uphold, safeguard, and transport soft tissues: living bone is instead—as mentioned earlier—a highly dynamic tissue, ceaselessly active with diverse types of cells and molecules that nourish and restructure it. The bone's structure, simultaneously light and strong, determines its mechanical characteristics and allows it to maximize its resistance to stresses by organizing itself with minimal weight. Bone tissue is capable of reacting to the mechanical stimuli it is subjected to, in a balance between deposition and resorption processes that structurally organize it: mechanical stresses tend to induce its renewal and remodeling, in order to best fulfill its mechanical functions.

2.3.1 Response of Bone Tissue to Mechanical Stresses

In mammals, mastication led to an adaptation of cranial bones, which had to bear the stresses imposed by this new function. They structured themselves in a way to best support its biomechanical requirements. Mechanical stresses induce a deformation in the bone that triggers a response, dependent on both the type of bone and the type of stress, as well as the individual's age. Moreover, deformation is proportional to the applied force and the surface area over which it is distributed. Consequently, with the same force, a larger area of impact results in reduced stress and thus deformation. Bones are heavily influenced by genetic factors that govern their growth and development (Parfitt, 1997). However, a combination of intricate and necessary interactions between bone cells and their mechanical surroundings can impact bone morphology, especially during skeletal growth.

During growth, osteocytes reside within the bone matrix, and each of these cells extends dozens of long processes that radiate in all directions from the cell body. These processes establish connections with other osteocytes, forming an intricate structure referred to as the connected cellular network (CCN). This network, akin to a nervous system, runs through the bone, linking the interior cells with each other as well as with the cells of the periosteum on the surface (Cowin, 2007). The CCN detects bone deformations through various mechanisms, and one hypothesis is that, when deformation occurs, the cellular fluid inside the canaliculi undergoes pressure changes that result in its flow (Biewener, 1991). Despite being immersed within and spaced out in the mineralized component, bone cells make contact and mutually anchor. These connections enable compressive, tensile, and hydrostatic forces to activate intracellular mechanotransduction systems that convert mechanical stimuli into biochemical signals. This process governs various aspects of these cells' behavior, such as growth, differentiation, and intracellular modifications. The overall structure of a tissue depends, to some extent, on its biomechanical environment, and the differentiation of stem cells into constituents of a particular tissue is achievable only by adhering to the specific biomechanical attributes that define that tissue. In particular, the generation of new bone tissue can be induced only within a biomechanically suitable environment (Dupont et al., 2011).

Bone remodeling can be influenced by stress only when it is of a dynamic nature (Rubin & Lanyon, 1987), and bone perceives stresses based on various parameters: the magnitude and frequency of stress, as well as its type (tension or compression). Intermittent repetition of moderately intense and short-duration stress leads to a stronger growth response compared to a single release of a more intense force. For the stimulation of new bone formation, the load needs to be cyclic (Duncan & Turner, 1995). The peak deformation on the periosteal surface occurs during vigorous physical activities (like running, jumping, or chewing solid foods): in vertebrates, bone growth is prompted when mechanical stresses cause deformation levels ranging from 0.2% to 0.3% of the bone's dimensions.

As a consequence of the aforementioned, physical exercise affects bone mineralization status: for instance, professional tennis players' arms display an asymmetry in bone cortical thickness between the arm that holds the racket and the other (both right and left arms are genetically identical, but the one that supports the racket is subjected to much greater stress). Among athletes who initiate tennis practice between ages 8 and 14, the dominant arm displays approximately 50% more cortical thickness than the other arm. Conversely, those who begin later, between ages 15 and 30, experience a less pronounced increase, about half of the increment seen in the former group (Krahl et al., 1994; Trinkaus et al., 1994, Ruff et al., 1994, Jones et al., 1977; Currey, 2006). The mineral density found in the bones of the dominant arms among the athletic population is higher compared to the corresponding bones of the non-athletic population (Luddleston et al., 1980).

If the stimulation linked to muscle activity strengthens the bone structure, decreasing the usual levels of stimulation can trigger the opposite process characterized by bone resorption. Among mammals that have lost muscle and tendon function for an extended period (or have lost the nerves innervating these muscles), it's possible to witness up to a 50% loss in the affected bone (Uhthoff & Jaworski, 1978). In the space environment, devoid of gravitational mechanical stimuli, astronauts experience bone mass loss, similar to those who are immobilized for extended periods. In the cranial region, the loss of dental function reduces the weight of the mandible by 50%.

The effects of mechanical stresses change with age in relation to osteoblastic senescence: in adults, bone apposition in response to stimuli tends to occur much less or is even absent compared to growing individuals. The causes of this age-related phenomenon are likely attributed to the fact that mesenchymal cells diminish their capacity to differentiate into osteoblasts, and osteoblasts and osteocytes become less sensitive to mechanical stimuli (Turner et al., 1997).

As a result, throughout a person's life, as with other mammals, there's a tendency to reach peaks of increased bone mass near the end of the growth period. This outcome arises from the interplay of various environmental and genetic factors. Once growth is complete, bone shape and size cannot change significantly, in other words, they cannot radically alter the architecture of the tissue as a whole.

The structure and form of bone also impact the nature of deformations arising from applied forces: structure refers to the material attribute of the bone itself, primarily determined by histology and mineralization level; shape represents the second property that three-dimensionally defines the amount and type of deformation that occurs in a stressed bone.

Since—as mentioned above—deformation is directly proportional to the applied force and inversely proportional to the area over which it's distributed, it follows that the larger the stressed area, the lower the stress, and thus the deformation: a bone, subjected to tension or compression along its axis, increases its cross-sectional area perpendicular to the force, thereby expanding the area over which forces are applied and reducing the generated deformations.

The remodeling of a bone involves morphological changes due to the differentiated growth of its different parts, which adapt to the changing functional requirements from even the soft tissues. Bone growth is regulated by the mechanical stimuli generated by their insertions on the bone (Fig. 2.20).

Influenced by soft tissues, the maxillary bones optimize their mutual relationship, external shape, and internal structure in order to combine strength, lightness, and elasticity, all of which are beneficial for dissipating chewing forces. For instance, the fact that the perimeter of the lower alveolar bones is smaller than that of the upper arch allows for their optimal adaptation to the dynamics of mastication (Fig. 2.21).

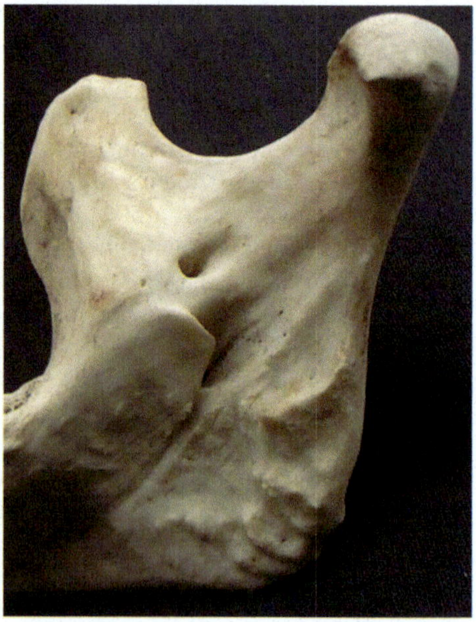

Fig. 2.20 Relief of the bone cortex in the area of insertion of the internal pterygoid muscle: note the double entrance foramen of the inferior alveolar nerve. Skull 3781 from the Museum of Anthropology of the University of Florence

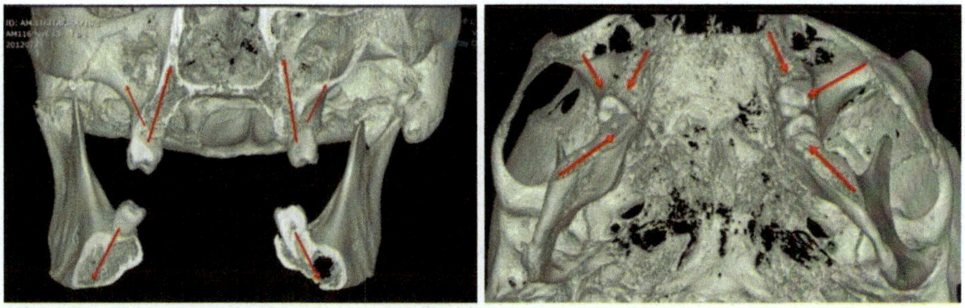

Fig. 2.21 The red arrows indicate the alveolar processes, which receive the dental roots and the bone structures connected to them, oriented to effectively dissipate the powerful forces of bite and chewing. Cone beam CT (cone beam computed tomography) of specimens from the Museum of Anthropology of the University of Florence

In mammals, from a morphological perspective, mandibular shape is closely linked to feeding behavior, unlike the upper jaw, whose functional form isn't solely dictated by its role in feeding. The muscles of mastication originate on the mandible, but not on the upper jaw. They insert on the zygomatic arch, the pterygoid region, and the temporal fossa (Popowics & Herring, 2006).

Also, concerning the cranial structure of mammals, the new muscular dynamics led to the alteration and development of specific muscle insertion areas and distinct shapes for various bone components. The action of masticatory muscles and the structural aspect of cranial bones, aimed at dissipating the developed forces, are intricately intertwined. Phylogenetically, the masticatory muscles are co-adapted to dental morphology with specific motor patterns for individual species, and ontogenetically, they influence cranial growth (Herring, 2007). The cranial structure develops, on a macroscopic level, as a series of interconnected elements that mutually and effectively dissipate the stresses of masticatory forces; the junction points of individual cells are where stress concentrates the most. Each bone cell behaves in the three planes of space like a bow that, by deforming, dampens tensions, discharging them into the ganglia and distributing them throughout the bone volume (Montemurro et al., 2019) (Fig. 2.22).

At a microscopic level as well, each individual trabecular cell bounded by the bony walls behaves like a hollow elastic sphere, capable of dissipating the forces transmitted on its outer surface through its deformation (Fig. 2.23).

As bone dynamically responds to the stresses it undergoes, the structure of an osseous specimen can help us understand the type of activity performed by the individual when they were still alive. Bone reacts to variations in mechanical load to which it is subjected, adding or removing material where necessary to make its architecture as efficient, resistant, and lightweight as possible (Marchi, 2016).

Since the 1700s, the awareness arose that bone is a tissue capable of dynamically adapting to the forces it experiences: Charles Bell (1774–1842), who described the course of the facial nerve and the facial paralysis known as "Bell's Palsy," stated: "The fine network that constitutes the internal structure of bone must be related to

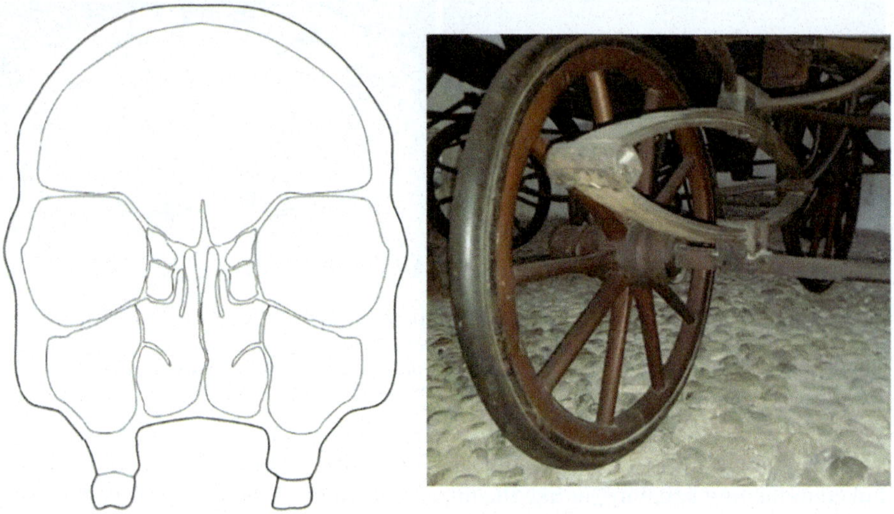

Fig. 2.22 The pneumatization of cranial bones allows them to respond elastically to the intense stresses of masticatory musculature, so that the walls of each individual bone cavity respond by deforming elastically: the bony points connecting the different cells act as the hinges of a crossbow (as in the figure on the right) on which the forces of bite and chewing are discharged and dissipated

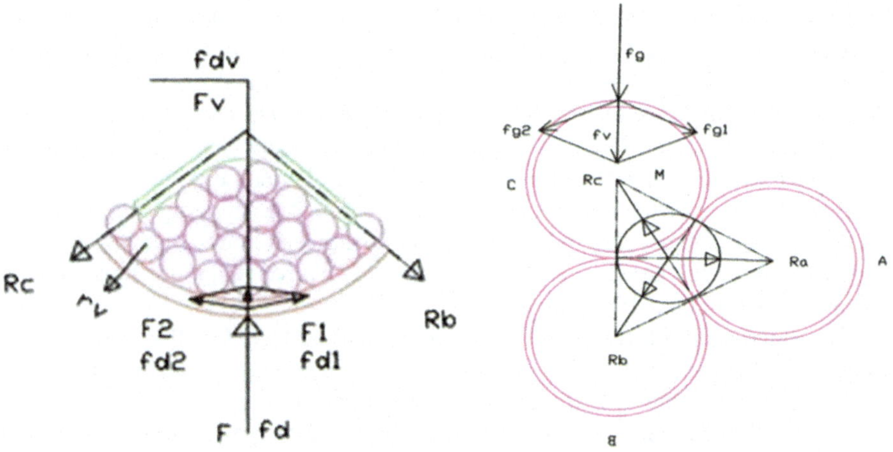

Fig. 2.23 The stress of chewing causes bending stresses on the various bony chambers, elastically deforming them with dilations and compressions that are discharged in the transition zones. Drawing by engineer Vittoriano Mangini

the forces acting upon it." Jeffries Wyman (1814–1874) wrote that the trabeculae of the bones supporting the body are arranged either in the direction of weight or in a way that supports the trabeculae oriented in this direction, and also that the orientation of these trabeculae in certain bones is typically related to the upright position naturally assumed by humans.

Paleontologist Hermann von Meyer (1801–1869) in 1850 studied the human femur in its connection to the pelvis: the femoral head, which supports the body's weight and transfers it outside the center of gravity, possesses numerous curved trabeculae that intersect in perfectly aligned layers to withstand tensile and compressive forces. Karl Cullman (1821–1881), a Swiss engineer, after studying von Meyer's work, created mathematical models of the femur's structure: he noticed that the trabecular structure was identical to the lines of tension and compression produced by the load, indicating that trabeculae form exactly where support is required. Cullman's work was later used by engineer Gustave Eiffel in the design of the eponymous tower, employing a framework of studs and beams resembling the trabeculae of the femoral head. The tower's four support pillars eccentrically bear the entire structure.

Julius Wolff (1836–1902: director of the orthopedic surgery department at the University Polyclinic in Berlin, co-founder of the Berlin Surgical Society, physician in the Prussian army, and recipient of the Iron Cross in 1864, 1866, 1870, 1871) developed the hypothesis of structural adaptation of bone to static and dynamic stimuli through his doctoral thesis on internal bone structure and its modifications in response to loads.

His laws, postulated at the end of the nineteenth century, are as follows:

- Wolff's First Law: Every functional change corresponds to a structural change in the tissue.
- Wolff's Second Law: The distribution and orientation of bone trabeculae in cancellous bone change dynamically in response to varying external loads.
- Wolff's Third Law: Increased forces lead to the formation of new bone tissue, while decreased or absent stimuli lead to the formation of osteoid tissue.

These laws were integrated by the principles of Wilhelm Roux:

1. Principle of functional adaptation: every increase in pressure force stimulates the formation of new bone, while a decrease causes a reduction in bone production.
2. Principle of minimum-maximum design: structures achieve maximum strength with the use of minimal material.

These statements, formulated at the end of the nineteenth century, still maintain their fundamental accuracy today. Numerous experiments on trabecular bone show how bone structure aligns itself according to the compressive orientations described by Wolff (Woo et al., 1981; Cowin, 1986), even though not all bone tissue responds in the same way to the biomechanical stimuli exerted by muscle insertions.

First and foremost, it must be noted that rates of bone growth and remodeling vary significantly across different skeletal sites. Moreover, cortical bone is particularly sensitive to deformation, especially before sexual maturity, both in terms of new bone growth and turnover percentage (Pearson & Lieberman, 2004). In places where form or protection are of primary importance, the architecture of bone tissue will be achieved through growth under direct genetic control. Conversely, in areas where resistance to repetitive loading is crucial (such as the jawbones), only the general form will be achieved as a result of genetically programmed growth: the remaining features will result from functional adaptation. This mechanism ensures that in specific zones, bone architecture is molded and remodeled until the prevailing, if accurate, stimuli correspond to those genetically established for that structure (Lanyon, 1987).

Every region within each bone element is 'genetically programmed' to consider a specific level of cyclic stress as normal, and any departure from this optimal stress environment triggers changes in its homeostatic balance, leading to an increase or decrease in its mass (Rubin & Lanyon, 1987). The frequency of stimulation matters because bone cells subjected to loads quickly become saturated with catabolites, and the full osteogenic potential is attained only after a short exposure to this stimulus; moderate and intermittent loads elicit a more significant growth response—as mentioned before—compared to a single strong force stimulation. As observed in vertebrates, the peak deformation on the periosteal surface during vigorous activities such as running, flying, or consuming solid foods ranges from 0.2% to 0.3%: this threshold suggests that bone growth occurs around these deformation levels (Rubin & Lanyon, 1987; Biewener, 1991).

Bone tissue constantly maintains a balance between deposition and resorption processes, influenced by the intensity and orientation of mechanical strains it endures. In this context, we have previously cited (Uhthoff & Jaworski, 1978, Uhthoff et al., 1985) authors who affirm that under particular conditions (mammals that have lost the use of muscles and tendons for an extended period or those that have lost nerves connected to these muscles), a loss of up to 50% of the affected bone can occur.

The skull stands out from other parts of the skeleton due to its composite structure, a collection of morphologically distinct bone elements that, resembling a complex puzzle, are joined by specific connection points—sutures—which dissipate stress transmission within it. Sutures experience notably higher stress levels compared to adjacent bones, and during mastication, local asymmetrical muscle contractions result in torsion among the various cranial box bones. Throughout maxillary closure, all sutures exhibit tension, leading skulls to be best understood as assemblies of independent components rather than solid structures (Herring & Teng, 2000): each cranial bone functions within a distinct deformation environment, with sutures facilitating load redistribution (Herring et al., 1996) (Fig. 2.24).

This comprehensive overview of morphofunctional adaptability naturally applies to the bones of the maxillofacial area as well. The maxillofacial and tympanic bones continue to grow after birth, primarily as a response to biomechanical stimuli. In response to the new functional demands posed by chewing and under its pressure, the bony architecture of the skull adapted and changed in order to support the muscles of the new chewing function and to effectively dissipate the forces produced by it: new surfaces were formed in order to create new areas of muscle attachment.

Although primitive jawed fishes already possessed well-ossified skulls, the dominant evolutionary trend has led to an increase in flexibility and independent mobility among the various cranial components. Among living vertebrates, a certain level

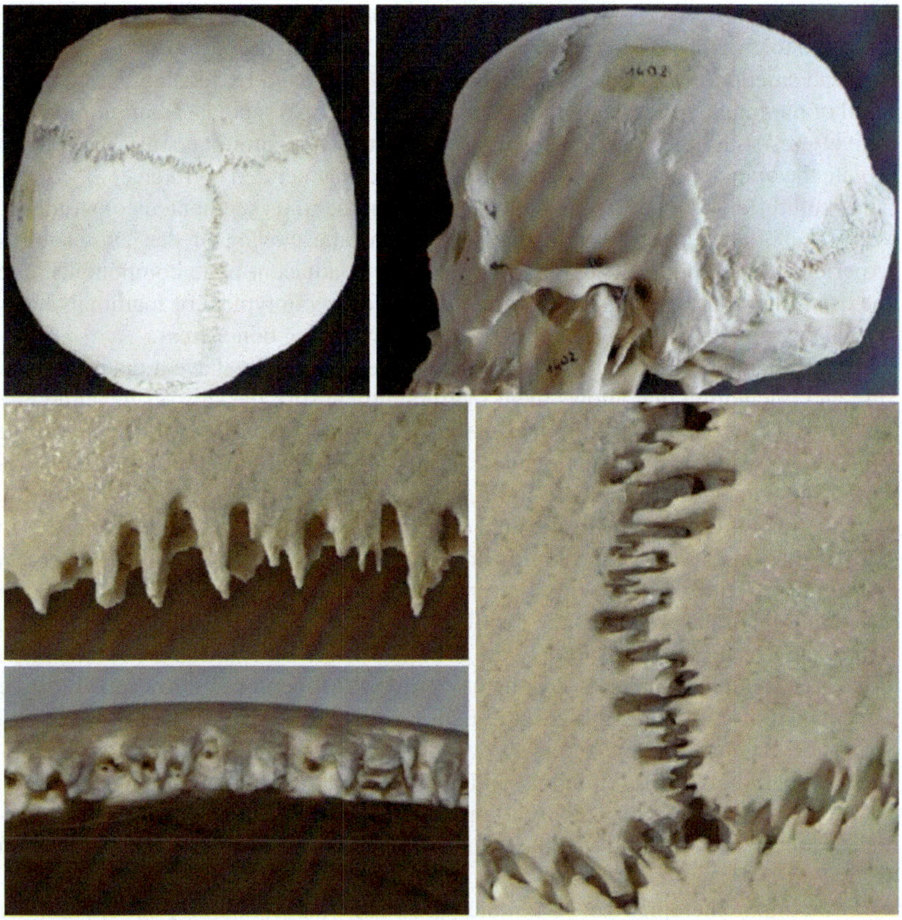

Fig. 2.24 Cranial bones form a set of interconnected individual units through junctions capable of dissipating stresses on the skull. Specimens from the Museum of Anthropology of the University of Florence

of mobility between different cranial parts is common, and rigid skulls are only found in specialized niches that subject the head to significant strains. Cranial intramobility was present in early therapsids and diminished in cynodonts, ancestors of archaic mammals; while all non-archaic mammals have sturdier and tightly sutured skulls that are nonetheless elastic, segmented, and adapted to withstand the intense stresses of chewing. A 'general model' of cranial deformation under strain does not exist, as the distinct cranial regions are deformed in distinct ways by varying muscular loads. Skulls should be examined—as we noted—as combinations of autonomous components rather than solely as solid structures (Herring & Teng, 2000), and the contrast between the 'rigid' skulls of mammals and the 'flexible' ones of other vertebrates seems to be more a matter of quantity than quality (Herring & Liu,

2001). The skull of vertebrates is thus a complex and composite bony structure, designed to protect the central nervous system (CNS) and the organs of vision, hearing, and smell from external and internal forces generated by feeding activity. The skull of mammals suggests that masticatory muscles also serve as postural muscles, actively resisting internal movements of the skull while simultaneously helping to isolate the cranial box and its joints from the forces generated (Wilken et al., 2019). The skull differs from other parts of the skeleton due to its segmented construction, with sutures altering stress transmission. During the closing of the jaws, sutures experience much higher stress levels compared to adjacent bone components, and the asymmetric muscular action of monolateral chewing typical of mammals leads to twisting effects on the cranial box, further stimulating bone stress.

Maxillofacial bones as a whole, and maxillae specifically, have a distinct structure and orientation that allows them to efficiently absorb the stresses generated by chewing. Throughout an individual's life, these bones exhibit significant adaptive plasticity—the capacity of an organism to respond to environmental conditions during its ontogenesis. (Gotthard & Nylin, 1995).

The 'eustress' (stress with positive effects) induced by chewing has short-term and long-term impacts, contributing to the postnatal development of oral tissues based on variations in chewing patterns (Ravosa et al., 2008). Nutrition plays an essential role in the growth of the maxillae from an early age: breastfeeding at birth acts as a powerful functional propellant to stimulate growth, particularly of the mandible. As deciduous incisors erupt, the 'grip/bite' function begins, and with the appearance of deciduous molars, the relationship of intercuspation is established alongside masticatory development. Even in the deciduous dentition, chewing provides essential growth stimulation, and if stimuli are insufficient, proper maxillary development won't occur, leading to reduced space for the correct eruption of permanent teeth and inevitable adjustments to their positions within the dental arch (Fig. 2.25).

Orthodontic therapies can be initiated at an early stage, and they should aim not only at morphological improvement but also, and most importantly, at restoring functional aspects. By the end of the treatment, the goal is to re-establish a

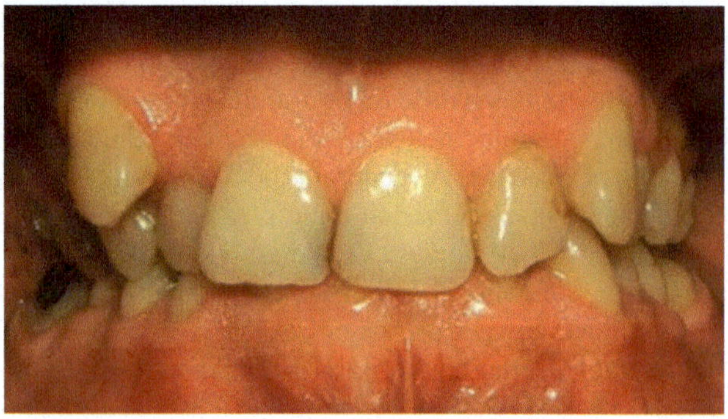

Fig. 2.25 Dental crowding, covered bite, left mandibular lateral deviation, right crossbite: occlusal alterations resulting from developmental alterations of the maxillary bone bases

physiological alternated chewing pattern, which can be effectively achieved with a limited consumption of soft and less solid foods (Limme, 2002).

The developmental phenotypic plasticity, which refers to the ability to respond to external influences, is a characteristic that allows bones to adapt to their mechanical environment (Moss, 1962). The maxillary bones are no exception, as their growth is a result of adapting to functional stimuli, especially those arising from chewing (Carlson, 1985, 1999; Aung et al., 2021). The biomechanical stimulations to which the maxillary bones are subjected mainly result from the forces generated by the action of the masticatory muscles, tongue, and so-called facial muscles. Changes in occlusal patterns have an impact on the stimulations the bone receives, both through the periodontium and muscle attachments (Choi, 2012) (Fig. 2.26).

Not only in humans but throughout the entire class of mammals, feeding is the fundamental activity that has shaped dentition, facial bone structure, and its musculature, with a direct link between chewing activity and diet (Kiliaridis et al., 1988; He, 2004; Beecher & Corruccini, 1981; He, 2004; Corruccini, 1984). Any alteration in occlusal relationships leads to macroscopic and microscopic structural changes in the alveolar bone of the maxillae (Lorenzon, 2010), regulating various aspects of bone growth, including structure, dimensions, and morphology (Bresin et al., 1994; Kiliaridis et al., 1996).

The muscular forces generated during chewing are controlled by a network of muscle-joint and periodontal receptors. The latter have different sensitivities depending on their types: posterior teeth are less sensitive to low-intensity loads than anterior teeth. Nevertheless, every tooth acts as a sensor, continuously offering feedback on the mechanical properties of food throughout each chewing cycle, thereby adjusting muscular forces (Johnsen & Trulsson, 2005).

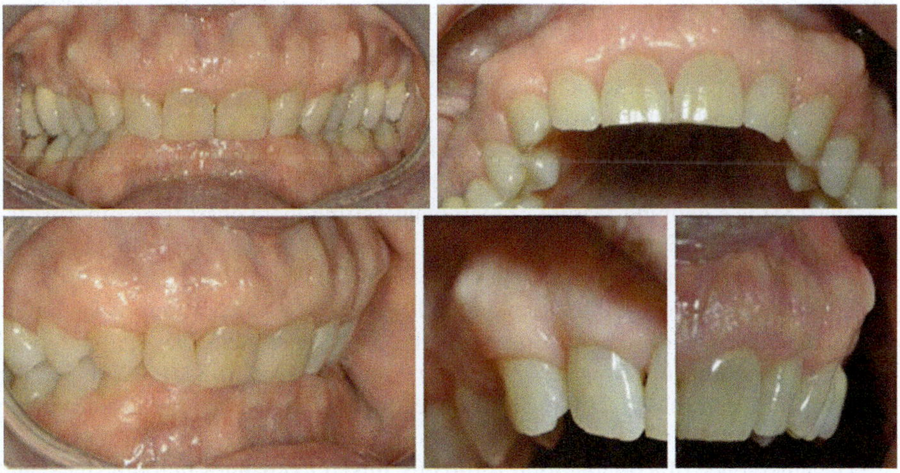

Fig. 2.26 Adaptive modifications of the alveolar bone supporting the incisor group and upper canines in relation to the intense stresses determined by a marked deep bite: note the bony protuberances of the vestibular cortex supporting the dental roots

Only certain bone components, when exposed to deformation levels that exceed an optimal 'environmental distortion' range, can respond plastically by adapting and subsequently increasing or decreasing bone mass (Rubin & Lanyon, 1987). In contrast, in other parts of the skeleton, it's as if the bone was 'genetically programmed' to consider intermittent stresses as 'normal,' remaining essentially stable.

The skeletal structure consists of a series of elements characterized by different functions: in areas where resistance to repetitive loading is necessary, shape primarily results from genetically controlled bone growth. In bone segments where final morphological characteristics are more crucial, shape is also the outcome of adaptation to individual functional needs. Taking the mandibular bone as an example, it comprises various components that undergo remodeling due to function, such as the alveolar, condylar, coronoid, and angular processes. For instance, in a mouse embryo, the mandibular condyle cannot form in the absence of the external pterygoid muscle (Herring & Lakars, 1982). On the other hand, another part of the mandibular bone is not subject to resorption, even in cases of severe atrophy: in cases of edentulism, resorption occurs in the alveolar process, while the basal portion remains relatively intact (Klemetti, 1996; Cawood & Howell, 1988), serving to 'protect' the course of the inferior alveolar nerve and maintain the areas of attachment for the lingual and suprahyoid musculature.

2.3.2 Structuring the Tmj and the Ossicular Chain of the Middle Ear

The temporomandibular joint (TMJ) and the middle ear are two anatomical regions that not only adjoin each other but also share close evolutionary, embryological, anatomical, and functional relationships. The evolution of the middle ear and the TMJ represents pivotal steps in mammalian evolution. The study of their relationship serves as a remarkable example of how classical comparative anatomy, paleontology, and developmental biology have converged to reconstruct how this extraordinary transformation of the temporomandibular joint into the tiny ear ossicles became possible (Anthwal et al., 2013).

The TMJ is a defining feature of mammals, setting them apart from other vertebrates (Rashed, 2015). It possesses a unique structure and functions as a hinge joint responsible for jaw movement during chewing (Porto et al., 2010).

In mammals, the development of a new TMJ involves two portions of the reptilian mandibular bone—the quadrate and the articular bones—being repurposed through a phenomenon known as *exaptation* (Gould & Vrba, 1982) to serve another function. These bone segments are transferred to the adjacent region to contribute to the formation of the middle ear (Maier & Ruf, 2016).

The emergence of a double joint is directly linked to changes in tooth shape and chewing patterns: the need to reduce food into very fine particles for digestion in the alimentary canal necessitated a profound modification of the joint between the skull

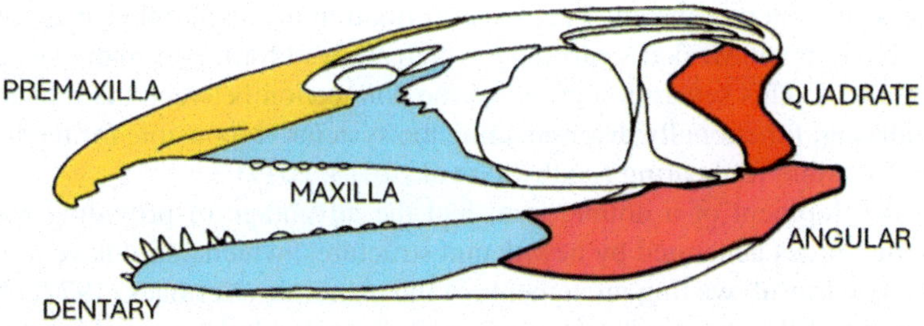

Fig. 2.27 Reptilian articulation between the skull and the lower jaw: the representation is simplified by including the articular bone in the schematization of the angular bone

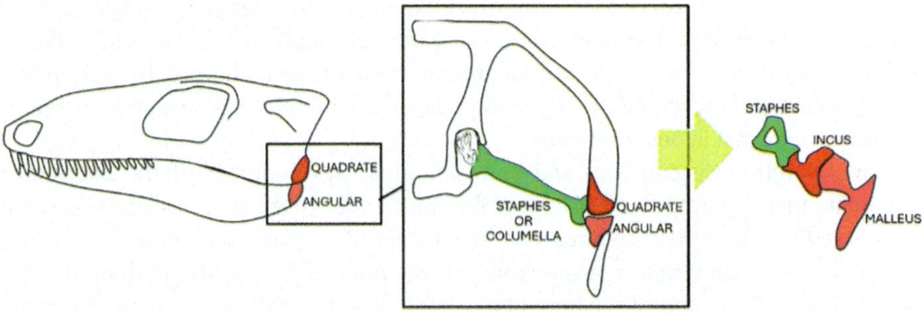

Fig. 2.28 Reptilian articulation between the skull, lower jaw, and stapes, and its evolutionary transformation into the ossicular chain of the mammalian middle ear

and mandible. In non-mammalian vertebrates, the cranio-mandibular joint between the quadrate and articular bones was replaced by a new joint, with these two bones shifting upward and backward to form the two new bones in the middle ear of mammals, the malleus and incus (Benton, 2000; Anthwal et al., 2013).

In reptiles, the joint between the mandible and the skull is formed by the relationship between the mandibular articular bone and the quadrate bone of the skull (Fig. 2.27).

Both of these bones have a role in reptilian hearing, interacting with the stapes. All reptiles and birds have a single ossicle in the middle ear known as the stapes or columella.

The malleus and incus, the two new bones in the mammalian ear, are homologous to the quadrate and articular bones. The incorporation of the reptilian mandibular joint into the middle ear of mammals became possible only with the evolution of a new method of jaw articulation. This reshaping also led to the development of two distinct joint compartments (Maier & Ruf, 2016). The stapes, incus, and malleus of the middle ear are primary characteristics of mammals and are intimately connected with the evolution of a new jaw joint. These two structures evolved together, defining the mammalian skull (Fig. 2.28).

An essential step in the embryological formation of the middle ear in mammals was the breaking of Meckel's cartilage, which allowed for the formation of the malleus and incus. This separation removed the connection between the mandible and the middle ear, functionally distinguishing the systems responsible for feeding from those responsible for hearing (Anthwal et al., 2013, 2017).

The development of a double joint had the advantage of providing resistance against the forces generated by new dental structures, which could have introduced potentially harmful twisting movements of the mandible (Kermack, 1972; Hylander & Crompton, 1986; Kemp, 2005). Separating the mandibular-cranial joint from the middle ear enabled mammals to amplify, compared to reptiles, not only the range of mandibular movements but also the spectrum of perceptible frequencies (Heffner & Heffner, 2018).

In tupaiids (once classified among primates), lemurs, tarsiers, and anthropoid apes, the tympanic ring, upon which the eardrum membrane is stretched, has an evolutionary origin from the reptilian angular bone of the mandible. In more primitive forms, it is represented by an incomplete ring, which later forms and deepens to create the external acoustic meatus.

The evolutionary pathway and embryological development of the anatomical segments that separate the TMJ from the middle ear make it clearer how connections exist between these two regions in anthropoid primates and humans. It is precisely this adjacency that is responsible for the presence of border pathologies that can involve both structures (Isberg et al., 1986; Myers, 1988; Loughner, 1989; Ren & Isberg, 1995; Williamson, 1990; Riga et al., 2010) (Fig. 2.29).

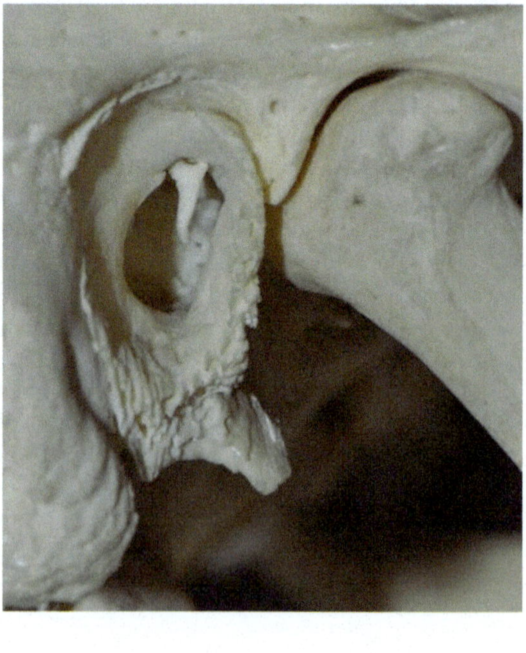

Fig. 2.29 Acoustic meatus and mandibular condyle: inside the acoustic meatus, note the malleus with the handle and the lateral process. Specimen from the Museum of Anthropology of the University of Florence

From a clinical perspective, many issues in this area can be attributed to the posterior displacement of the condyle in the fossa, resulting in inevitable compression of the retrodiscal volumes. These retrodiscal volumes are not anatomically or histologically well-suited to withstand compressive forces. This situation is often caused by a reduction in the vertical occlusal dimension, which can result from factors such as the loss of posterior teeth, pronounced Class II malocclusions, or an anterior functional obstruction, like a deep bite. In cases of a dysfunctional overbite, characterized by excessive overlap of the upper incisors over the lower ones, causing a disruption of functional pathways, it is advisable nowadays to consider the following: the increasing use of aesthetic veneers for anterior sector restorations should always include a prior assessment of proper horizontal mandibular movements and how these movements may be affected by aesthetic restorations.

2.3.3 Formation of the Bony Palate

The palate is a structure in vertebrates that tends to separate, to varying degrees, the nasal cavities from the oral cavity. It is divided into a primary palate, composed of the premaxilla located in front of the incisors and originating from the frontonasal process, and a secondary palate that extends from the incisive foramen to the uvula. In all vertebrates, the secondary palate presents as bilateral outgrowths originating from the maxillary and palatine processes.

In bony fish, the palatine bone is merely a plate positioned on the inner edge of the maxillary bone. In many reptiles and birds, it originates from two longitudinal lateral folds that do not fuse, resulting in a physiological palatal cleft and a vault that varies in development. It is less developed in lizards and snakes, extends in turtles, and in most amphibians, it only forms a bar between the vomer and the maxilla. It reaches its maximum development in crocodiles, where it extends all the way to the pharynx (Ferguson, 1988). It is noteworthy that among the various orders of reptiles, it is in crocodiles that the secondary palate reaches its maximum development. Interestingly, crocodiles are among the very few non-mammalian vertebrates to possess dental roots, which are unmistakable structural adaptations to withstand intense biting forces. Indeed, the presence of the palatal vault and dental roots—anatomical features that characterize mammals and have evolved in this class to better absorb masticatory forces—might seem incongruous, given that crocodiles do not chew their food. However, these reptiles employ particularly violent predatory techniques during which extremely powerful forces are generated, necessitating effective dissipation within their cranial and dental structures. In crocodiles, the key to understanding these two anatomical structures lies in their need to withstand the tremendous stresses of their potent bites. The secondary palate significantly contributes to increasing rigidity and resistance to the stresses generated during chewing (Thomason & Russell, 1986). It first appeared in the ancestors of mammals approximately 250 million years ago, strengthening the skull and allowing for the dissipation of the greater twisting forces exerted by powerful adductor muscles

(Thomason & Russell, 1986; Wall & Smith, 2001; Liem, 2001). Additionally, it divides the mouth from the nasal cavity, which improved airflow, the olfactory system, milk suction in neonates, and the structural development of dentition (Maier et al., 1996; Hildebrand et al., 1995). Furthermore, the secondary palate allowed air to enter the lungs without necessarily passing through the oral cavity, enabling simultaneous chewing and breathing. This separation of air and food passages during chewing prevented choking and facilitated the initial pharyngeal depression during swallowing (Ungar, 2017). In mammals, the primary palate, or premaxilla, extends into a vault formed by two horizontal plates, the maxillary and palatine bones, which fuse together posteriorly to delineate the choanae. The soft palate, unsupported by bone, extends posteriorly from the hard palate, further separating the oral cavity from the nasal cavity and facilitating the passage of air during chewing.

The bones that make up the palate contribute to the formation of the facial structure, and their homology, or phylogenetic correspondence, is evident across different taxonomic categories of animals. In most amniotes, the premaxilla is always the most distal bone, the maxilla is in an intermediate position, and the palatine bone is the most proximal. These bones are connected by sutures that immobilize the upper jaw block (Richman et al., 2006) (Fig. 2.30).

In vertebrates, the phylogenetic evolution of the maxillae preserves, albeit with modifications, constant developmental patterns. Even in modern humans, there

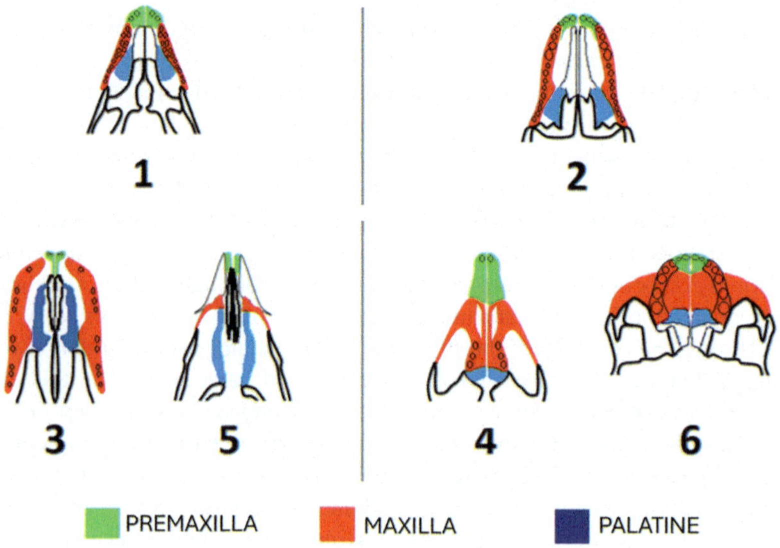

Fig. 2.30 Bone modules composing the palate in amniote vertebrates. 1: Sauropsid, ancestor of the reptile evolutionary lineage; 2: Synapsid, ancestor of the mammalian evolutionary lineage; 3: Python; 4: Mouse; 5: Rooster; 6: Human. The subdivision of the bony components of the upper jawblock shows a high degree of homology: note how the premaxilla is also in humans a distinct bony component, whose origin and developmental mechanisms are rooted in pre-mammalian vertebrates

have been no significant evolutionary changes in the integration of these structures into the facial complex. The central region of the face, as typically defined by anthropologists, is almost entirely composed of the bones of the upper maxilla. This is reflected in the composition of the palate, which consists of three distinct bony elements: the premaxilla in the anterior position, the maxillary bones in the central position, and the palatine bones posteriorly.

In hominids, the variation in the premaxilla has long been identified as a distinctive feature of mid-facial anatomy, with the premaxilla always formed by an independent bony module. In australopithecines, the modularity of the palate appears to have been a key element in phenotypic variation, facilitating various dietary adaptations imposed by environmental pressures. The independent development of palatal modules may have indeed facilitated these dietary adaptations, contributing to the "early radiation" of hominids. Specifically, the modularity of the palate may have played a role in the retraction of the upper facial block relative to the rest of the skull (Villmoare et al., 2014).

The entire facial skeleton originates from neural crest cells (Jugessur et al., 2009), and in *Homo sapiens*, during embryonic development, the maxillae originate from distinct elements. Around the fourth week of intrauterine life, between the cranial swelling that will give rise to the head and the cardiac primordium, the four branchial arches appear. The first arch consists of two lateral swellings, and later these swellings split into two tubercles, one upper and one lower. The upper tubercle gives rise to the lateral processes of the maxilla, while the lower tubercle gives rise to the mandibular processes, which rapidly grow toward the midline, where they meet to form the mandibular arch. The upper maxilla thus forms from the fusion of three distinct components: the right maxillary tubercle, the left maxillary tubercle, and a fronto-nasal process that converge at the level of the canines (Fig. 2.31).

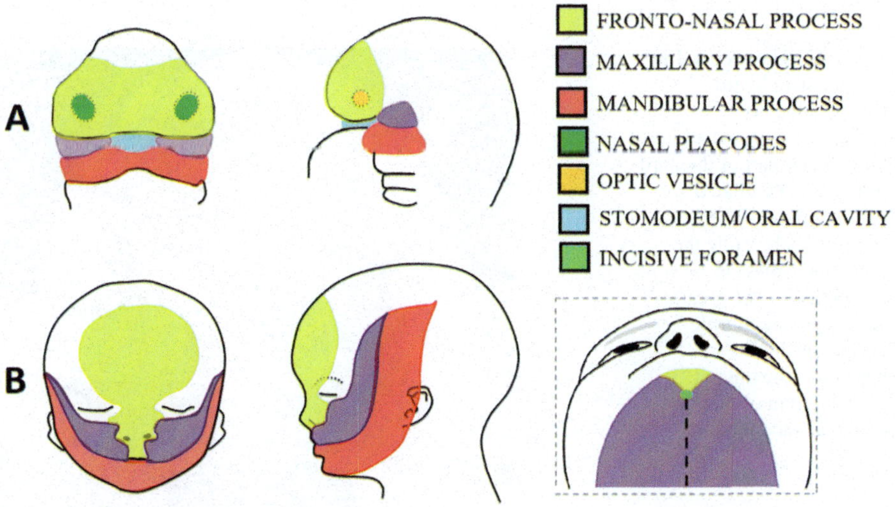

Fig. 2.31 Embryonic origin of the maxillae; frontal and lateral perspectives of the fetal head. (**a**): 4th week; (**b**): 14th week

From an evolutionary perspective, the sizes of the anterior teeth can be seen as an important factor that has influenced the dimensions of the premaxilla. This influence has led to adaptations favoring the selection of a smaller pre-maxillary module to accommodate a reduced anterior dentition. It's worth noting that not only are the upper maxillae composed of distinct bony components with separate embryonic origins, but the mandible is also formed by two equal and separate parts, both in terms of phylogeny and ontogeny. These parts develop individually and converge at the symphysis (Figs. 2.32 and 2.33).

Additionally, the mentioned bony components of the maxillae have their own innervation and vascularization, allowing each of them to independently modulate their development in response to the forces generated during mastication

The regulation of bone homeostasis is coordinated by various factors, both local and systemic, with mechanical stimulation playing a crucial role as a key regulator. Bone tissue not only serves as a mechanical transmitter but also acts as a receptor for the neural system under mechanical load (Qiao et al., 2019). The skeleton is well-innervated, but only recently have the functions of this complex bone network begun to be understood. The peripheral nervous system actively participates in osteogenesis and bone remodeling, locally exerting an indispensable influence on bone metabolism through numerous neurotransmitters. This influence is evident in the regulation of both anabolic and catabolic processes in bone, both during embryonic development and throughout adult life (Grässel, 2014; Heffner et al., 2017). Similar to other parts of the body, sensory innervation in bone serves to detect and respond to various stimuli, including mechanical load, thereby regulating the tissue's autonomous functions, including homeostatic remodeling.

Nerve fibers are more commonly found in metabolically active bone. In fact, innervation is higher in the periosteum and marrow, while mineralized bone contains relatively few fibers. The number of fibers increases peripherally on the

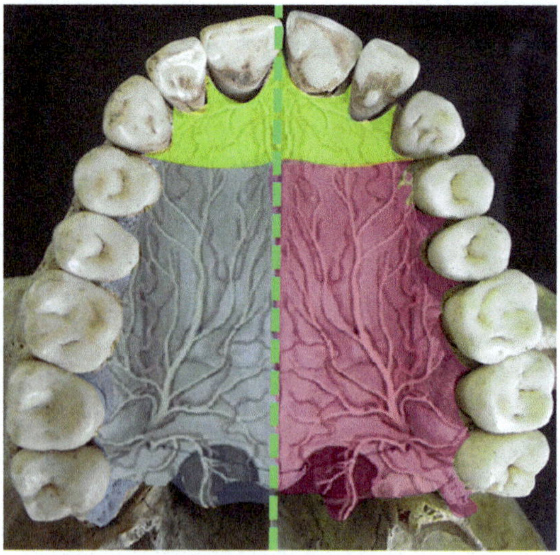

Fig. 2.32 The premaxilla is innervated by the nasopalatine nerve (yellow), which innervates the anterior part of the hard palate and the lingual periodontium of the anterior teeth; while the remaining palate is innervated by the greater palatine nerve (purple, gray), which also innervates the lingual periodontium of the premolars and molars

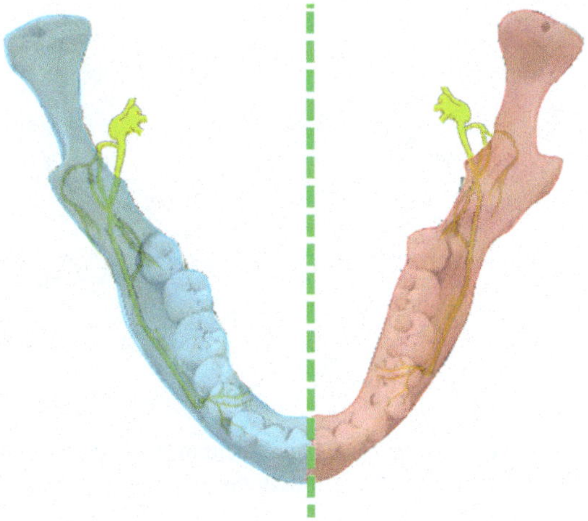

Fig. 2.33 The two hemimandibles—separated in most mammals—develop embryonically from two distinct processes, which in humans fuse at the symphysis level

surfaces undergoing active remodeling. Both sensory and autonomous nerve fibers have been identified around periosteal vessels, in Volkmann's canals, within the bone marrow, at the osteochondral junction of the growth plate, and at the attachment of the synovial membrane. On the bone's surface, periosteal nerves are more abundant at the epiphysis than in the central region (García-Castellano et al., 2000).

In the processes of skeletal growth, innervation of developing bone structures occurs concurrently with their increasing mineralization (Sisask et al., 2013). During embryonic development, bone innervation occurs almost simultaneously with endochondral ossification. The sympathetic nervous system stimulates bone resorption, negatively impacting new bone formation, while parasympathetic nervous system activity inhibits resorption, leading to an accumulation of bone mass. After birth, the density of nerve fibers in bone continues to increase, coinciding with developmental remodeling. Aging reduces the presence of nerve fibers and their organization (Tomlinson et al., 2020). During osteogenesis, the shape of the bone is determined by multiple interacting factors, and in the case of the maxillary bones, one of these factors is the force exerted by the masticatory muscles (Richman et al., 2006). The metabolic activity of the maxillary bones is closely tied to biomechanical stimulation from chewing and, as we've seen, is mediated by innervation. This combination of factors would explain the topographic differentiation of growth in the maxillaries described by Pedro Planas in his developmental laws (Planas, 1994) (Fig. 2.34). For example, unilateral, non-alternating chewing leads to an asymmetry in growth stimulation between the right and left sides due to differential development of the various components that make up the bony structure of the stomatognathic apparatus.

The functional or adaptive nature of a structure or system is evident through its phenotypic plasticity, which is a significant source of morphological variation within populations. Among mammals, plasticity concerning the differential load on masticatory elements, particularly the mandibular body and the temporomandibular joint, is well-documented.

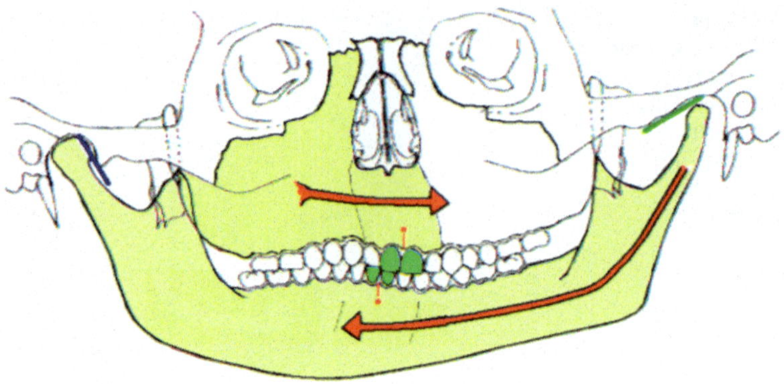

Fig. 2.34 Asymmetric development of the maxillary bones linked to unilateral chewing. Drawing from "Evolution of the functional dynamic relationship between occlusion and TMJ", Dettori et al., 2001

Throughout an individual's life, the external dimensions of the maxillary bones, the thickness of the cortical bone, trabecular density, and the composition of non-dental mandibular cartilage elements are significantly influenced by variations in peak masticatory forces and/or cyclic loads. Bone modules exhibit patterns of plasticity similar to those observed in postcranial tissues and structures.

The morphology of the bony palate (which, situated between the rows of maxillary teeth, separates the oral cavity from the nasal cavity) is also influenced by its resistance to bite forces and post-canine chewing. Consequently, its shape reflects the functional requirements for withstanding mechanical forces (Menegaz et al., 2009) (Fig. 2.35).

2.3.4 Development of the Zygomatic Arch

The division of changes in various components of the maxillofacial region can be useful from a descriptive standpoint, but it's essential to always bear in mind that structural modifications within this region did not occur independently of one another; rather, they mutually influenced each other. The alteration of one characteristic subsequently affects the change in all others functionally connected to it, which, in turn, influence the first. The evolution of different characteristics within an anatomical region occurs collectively under the cross-influence of these mutual dependencies.

Even before the advent of mammals, in some reptilian evolutionary lineages, a temporal fenestration developed in the skull, associated with the modification of adductor muscles (Liem, 2001). The earliest reptiles, the anapsids, had a compact skull without fenestrations. Subsequently, in the skulls of certain reptilian lineages, relatively large temporal fenestrations appeared (one in synapsids and two in diapsids) (Fig. 2.36). These fenestrations served as attachment points for larger and

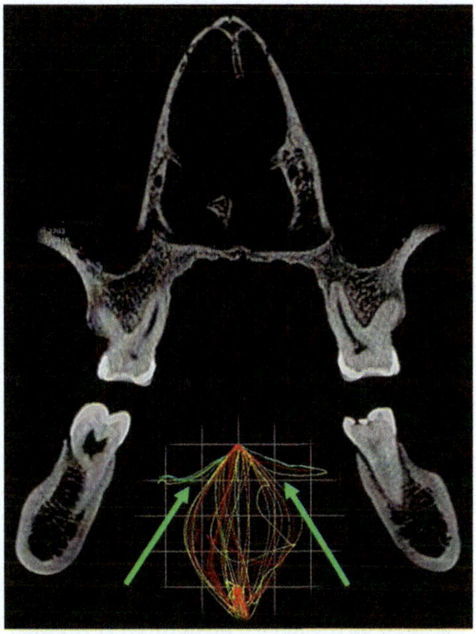

Fig. 2.35 Cone beam CT, coronal section at M1 level: the bony palate strengthens the upper jaw, making it more capable of withstanding and dissipating masticatory stresses. In the center and at the bottom of the figure, the two lateral pathways guiding the entry and exit phases of masticatory cycles. The green arrows indicate the entry vectors of the cycles corresponding to the inclinations of the alveolar processes of the mandible and upper jaw

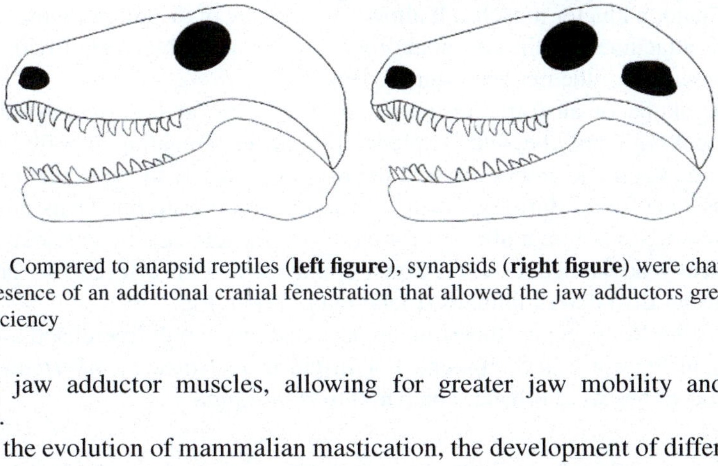

Fig. 2.36 Compared to anapsid reptiles (**left figure**), synapsids (**right figure**) were characterized by the presence of an additional cranial fenestration that allowed the jaw adductors greater functional efficiency

stronger jaw adductor muscles, allowing for greater jaw mobility and mouth opening.

With the evolution of mammalian mastication, the development of differentiated musculature for new movements necessitated a different shape for its anchoring points, the bones.

In mammals, the largest and most powerful jaw adductor differentiated into the temporal and masseter muscles, which in turn split into superficial and deep heads (Liem, 2001; Ungar, 2017). As the external adductor muscles increased and differentiated, the bony bar that defined the temporal fenestra expanded significantly, fully structuring itself as the zygomatic arch (Liem, 2001; Berkovitz & Shellis, 2018). In mammals, the design of the zygomatic arch enhances jaw closure efficiency and strength, provides more space for the insertion of the masseter muscle,

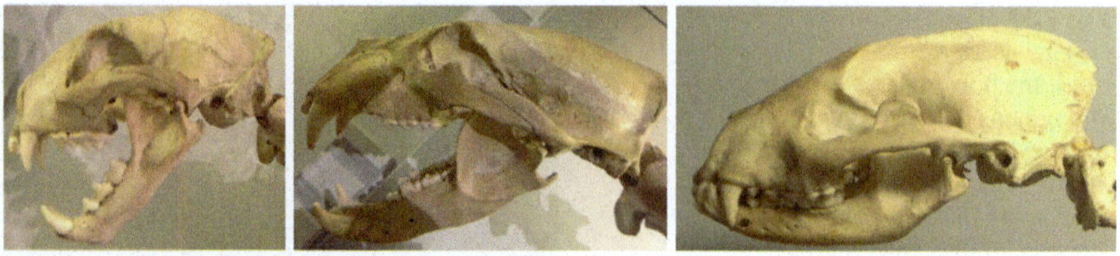

Fig. 2.37 Highly developed zygomatic arches in carnivorous mammals; from top to bottom: cougar (Puma concolor); polar bear (Ursus maritimus); badger (Meles meles). Note how in this order the surface area of insertion of the temporal muscle is very extensive and the coronoid process of the mandible is equally developed. Specimens from the Museum of Zoology of the University of Florence

and allows room for the temporal muscle to pass beneath its arch. This enables the expansion of muscle size and, consequently, an increase in their strength.

The varying development of the temporal and masseter muscles among carnivores, herbivores, and rodents influences the shape and extent of their zygomatic arches. For instance, in carnivores, both the temporal muscle and the coronoid process are highly developed, leading to a broad zygomatic arch to accommodate the passage of these two structures (Fig. 2.37).

This bony bridge is associated with the action and, consequently, the volume of the two major adductor muscles. It allows the passage of the temporal muscle, facilitating its attachment to the coronoid process of the mandible, and provides support for the insertion of the masseter muscle (Fig. 2.38).

In the skulls of australopithecines, the zygomatic arches were notably long, extending prominently beyond the plane of the face, elongating laterally. Wide and protruding zygomatic arches allowed the masseter muscle to apply intense vertical and transverse forces during chewing. *Paranthropus boisei*, due to its unique anatomical configuration that allowed for extensive muscle attachment areas and biomechanically advantageous lever arms, could chew with a force two and a half times greater than that of humans (Lieberman, 2011) (Fig. 2.39).

Within the *Homo* genus, this structure also exhibits variability related to different cranial and muscular development. For instance, in *Homo neanderthalensis*, the zygomatic arches were broader compared to *Homo sapiens*.

2.3.5 The Development of the Mandible with its Processes

The appearance of the mandible marked a turning point in vertebrate evolution because it enabled the capture and processing of large, mobile prey (Cerny et al., 2010). Throughout their evolutionary journey, vertebrates became capable of seeking and even hunting for their food, becoming predators. The first organ to evolve for this purpose was the mandible (Le Douarin, 2004). The emergence of the mandible was a fundamental event that "freed" the agnathans (jawless aquatic

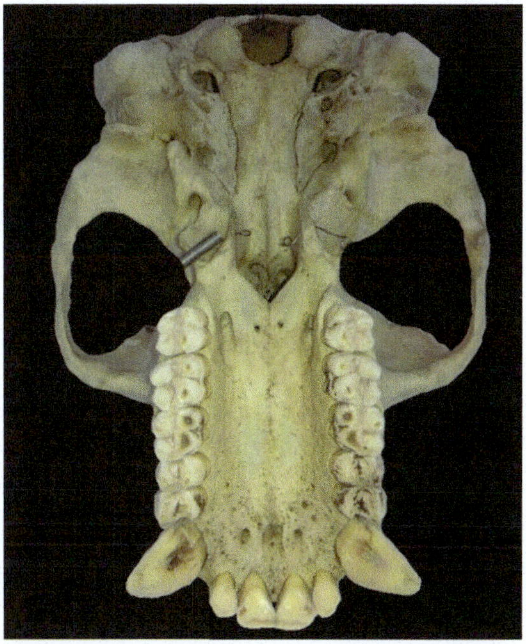

Fig. 2.38 Gorilla (*Gorilla gorilla*). Note the extensive zygomatic arches for the passage of powerful temporal muscles and for the insertion of equally strong masseter muscles. Specimen from the Museum of Zoology of the University of Florence

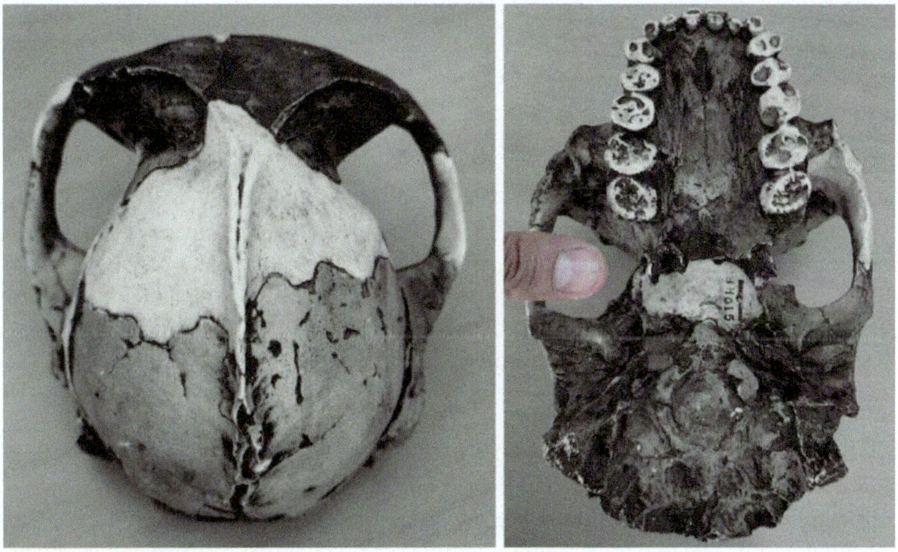

Fig. 2.39 *Paranthropus boisei* skull. Note the wide zygomatic arches allowing the passage of powerful temporal muscles; the power of this musculature also results in a characteristic median sagittal bony crest. In this species, there was a molarization of the teeth that increased in size to meet increased masticatory demands: the surface of a molar was approximately as large as the thumbnail of a human

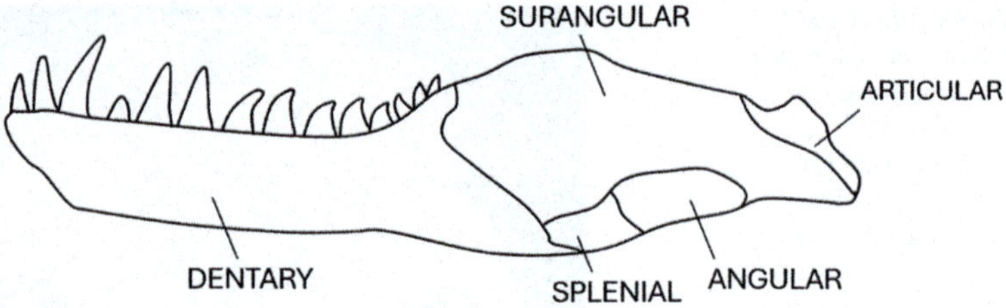

Fig. 2.40 Different bony components of the reptilian mandible

Fig. 2.41 Reptile mandible (**above**) and mammal mandible (**below**)

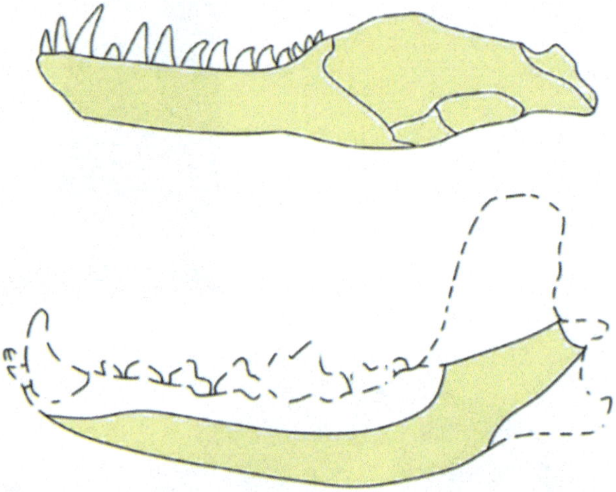

vertebrates) from a filter-feeding or scavenging lifestyle, allowing them to feed in various ways and on a wide range of foods. This offered numerous opportunities for evolutionary diversification.

In the evolutionary adaptation to these different lifestyles, skeletal elements associated with feeding, such as the maxillary bones, also changed in shape and size (Anthwal et al., 2015). In pre-mammalian vertebrates, there are several bones that make up the lower jaw (Fig. 2.40), while in mammals, the mandible consists of two symmetrical portions, right and left, which unite in the area of the symphysis.

The pre-mammalian mandible is often divided into two hemimandibles that do not unite in the midline (Fig. 2.41) and can be connected by an elastic ligament, allowing the right portion to move away from the left, thus increasing the mouth's opening width. The two lower hemimandibles can also move forward and backward independently of each other, which aids in prey ingestion.

The profile of reptilian mandibles is elongated and often lacks pronounced features, except occasionally for a small coronoidean protrusion of the supra-angular bone. The absence of bony processes makes the reptilian mandibular body resemble the appearance of the basal bone of mammalian mandibles (Fig. 2.41).

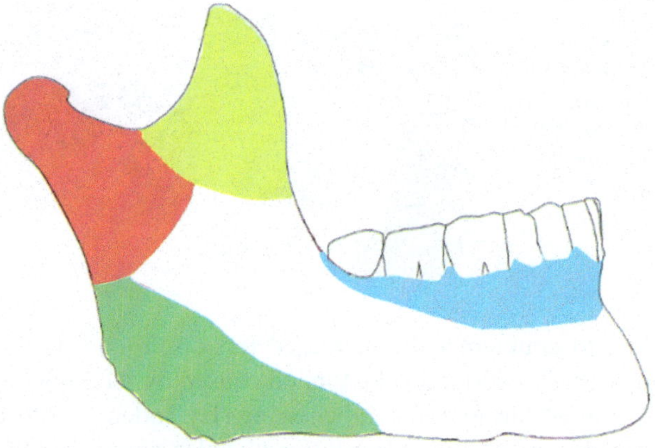

Fig. 2.42 Mandibular processes are specialized areas developed in mammals to fulfill mastication: for the insertions of masticatory muscles (yellow: coronoid process for the temporal muscle; green: angular process for the masseter and internal pterygoid muscles); for the articulation with the skull (red: condylar process); to support dental roots (light blue: alveolar process)

Mammals adapted to a variety of ecological niches and a wide range of diets, from purely carnivorous to exclusively herbivorous. Unlike other gnathostomes, the mammalian mandible reflects dietary specializations and comprises a single bone (the dentary) that underwent increased complexity and modularity. In mammals, all regions of the mandible, except the basal body, have adapted in response to the mechanical stimuli of chewing. The mandible always features, although with variable proportions, four processes: the alveolar process for supporting dental roots; the angular process, which serves as the attachment site for powerful adductor muscles such as the masseter and external pterygoid; the condylar process, which is the mandibular component of the new joint; and the coronoid process, which serves as the attachment area for another important elevating muscle, the temporal muscle (Fig. 2.42).

The mandible, aided by the degrees of freedom granted by the TMJ, is the most mobile and independent bone in the human body. Its functional complexity subjects it to continuous muscular action, which instant by instant controls its various functions. Chewing places the mandible under multiple stresses, and the osseous tissue has had to differentiate in its cortical-to-medullary ratio in different regions: the bony processes that support muscle attachments and the mandibular branch that supports them are highly cortical and possess a lower percentage of spongy bone. Meanwhile, the mandibular body, while also having a thick cortical layer, has a significant internal area with a more sparse trabecular structure, allowing for an exceptional biomechanical synthesis of strength, elasticity, and lightweight properties (Fig. 2.43).

Due to its unique configuration, the mandible is among the most resistant bones in the human body to post-mortem decomposition and, therefore, more likely to undergo fossilization (Villmoare et al., 2015). The mandible is transversely thick in

Fig. 2.43 Mandibular body seen from above in the absence of the alveolar process bony ridge. Note the thickness of the cortical bone compared to a spongy one with very rarefied trabeculae

the molar region to counteract the increased torsion around the long axis of the mandibular body on the working side, torsion caused by the high intensity of masticatory muscle forces during unilateral chewing (Hylander, 1979a, 1979b, 1979c).

Anatomical components like the inferior alveolar nerve, which longitudinally traverses the mandibular body, or the dental roots inserted within it, are encased and supported by a thin layer of bone tissue (Fig. 2.44) connected to the cortical bone through trabeculae. These trabeculae act like struts, suspending these structures within the mandibular body (Fig. 2.45).

This arrangement essentially creates a tensile structure in which stability is provided by the combination of precompression and tension forces in its individual components. When force is applied to a single element, it directly and indirectly affects all other components, causing a reversible deformation of the entire structure through the temporary reorientation of its individual parts. The advantages of tensile structures include:

- Overall Strength: These structures are much stronger than the sum of the strengths of their individual components.
- Lightweight: Tensile structures, while mechanically strong, are lightweight compared to compression-based structures of similar strength.
- Flexibility: They offer a high degree of reversible dynamic adaptability, with local deformations modulated by the entire structure.
- Mechanical and Functional Interconnection: All constituent elements are mechanically and functionally interconnected, allowing continuous and constant communication.

Nature has harnessed this fundamental construction principle in many ways and at all scales, creating increasingly complex multi-modular and hierarchical structures that have led to the emergence and evolution of living cells and organisms (Ingber et al., 2014).

Morphogenesis, through which cells self-assemble into tissues and complex organs with specialized forms and functions, is mediated by the action of many genes and chemical signals. However, the mechanical environment plays a central role in both the formation of embryonic tissues, developmental stages, and during the homeostatic control in adult life (Mammoto et al., 2013). Although genetics and biochemistry have dominated the field of biology and medicine in the last century, the emergence of "mechanobiology" has highlighted that physical forces play an

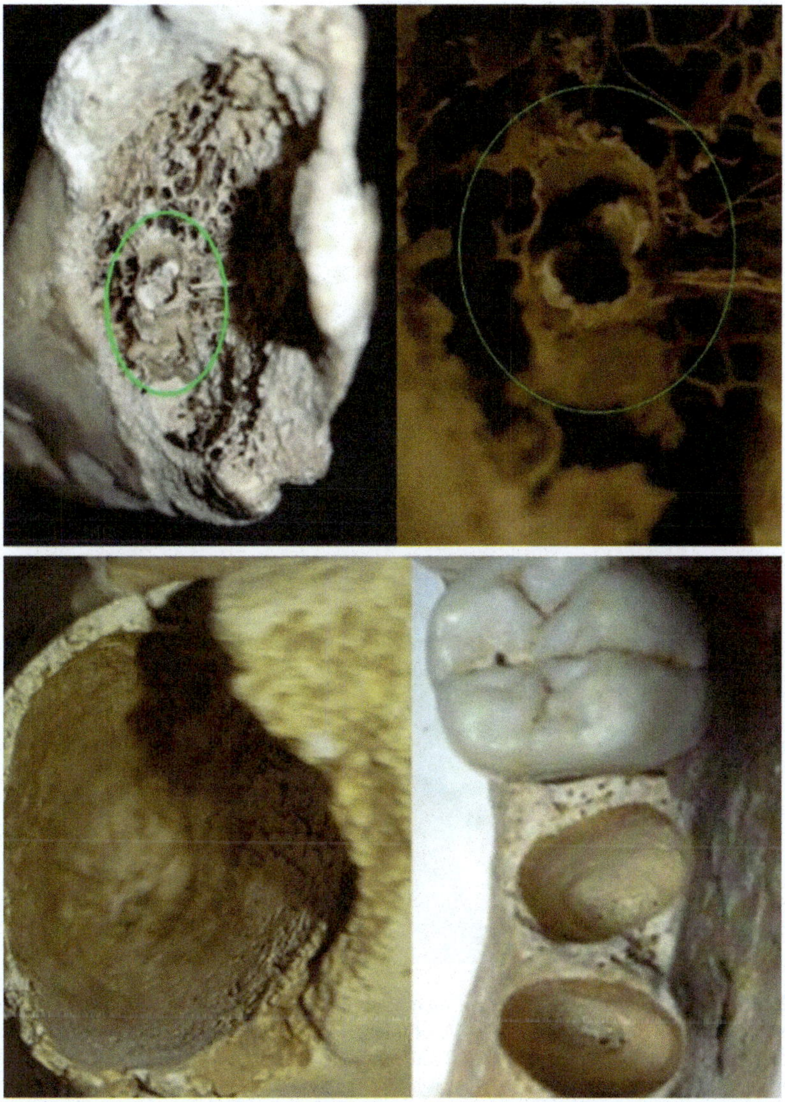

Fig. 2.44 In the first two images at the top, canal of the inferior alveolar nerve: posterior view through fracture of the ascending branch and its magnification; note how the nerve canal, highlighted by the circle, is delimited by a thin bony tube suspended in the mandibular body by a framework of trabeculae. Skull 3112; Museum of Anthropology of the University of Florence. In the two images at the bottom, the compact bone lamina forming the dental alveoli. On the left: upper canine alveolus of cave bear (*Ursus spelaeus*); on the right, mandibular alveoli of human (*Homo sapiens*)

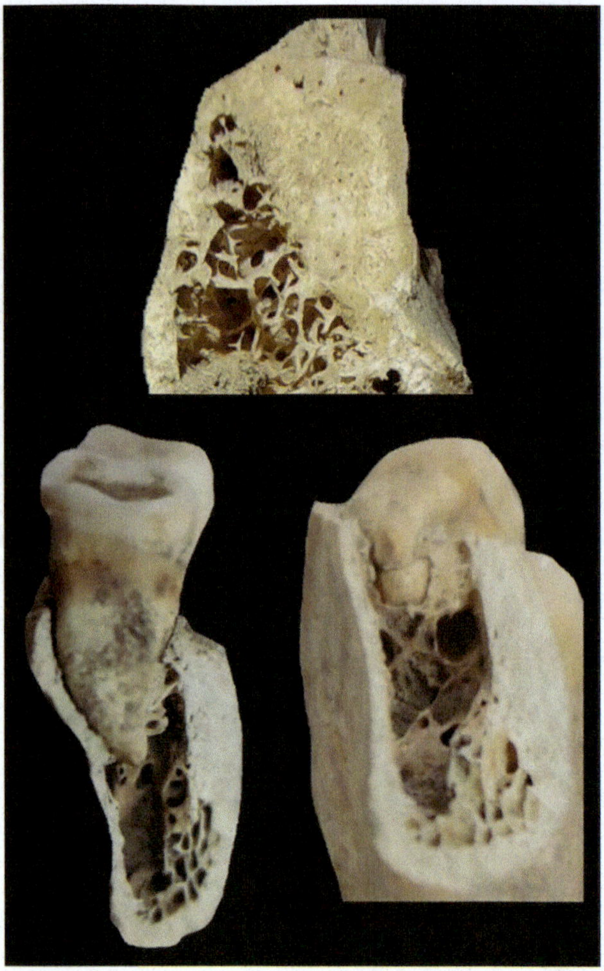

Fig. 2.45 Post mortem fractured human mandible: through the fracture, the internal bone structure is highlighted. Top: lateral perspective of a fractured alveolus; on the right side of the specimen, note the alveolar wall, and on the left, the complex of trabeculae supporting the hard lamina. Bottom right: bottom view of the roots of a lower molar; the trabeculae support the hard lamina surrounding the molar roots. Bottom left: lateral perspective of a molar root without the alveolar bone layer; the root appears suspended in an apparently empty space between the two cortices. The hard lamina can be seen as a thin cradle surrounding the roots, made of non-Haversian compact bone and suspended to the cortices through a trabecular framework; this organization, essentially comparable to a tensile structure, provides elasticity and strength, allowing effective dissipation of intense masticatory forces. Specimen from the Museum of Anthropology of the University of Florence

equally important role in controlling the development of cells and tissues as chemicals and genes (Ingber et al., 2014).

The bone that surrounds dental roots is characterized by a large number of blood vessels that pass through the so-called *Volkmann's perforating canals*. These canals give the structure a characteristic perforated appearance, from which it also derives the name *Cribriform lamina* or *Cribriform plate* (from the Latin *cribrum* meaning sieve) (Mjor, 1980) (Figs. 2.46 and 2.47). This intense vascularization is necessary to support the high metabolic activity of the periodontium.

Any pressure exerted on the teeth during chewing is directly transmitted to the trabecular bone, which continues to remodel and change throughout life in response to masticatory stimuli, tooth movement, and loss (Moon et al., 2004).

The outer cortical and inner spongy bone are held together by the inherent cohesive forces of the material. External forces that subject the bone structure to bending produce internal compressions and expansions, which are strongly contained by the resistance created by the trabecular arrangement (Montemurro et al., 2019). The varied relationship between cortical and spongy bone makes the mandible an extremely efficient tensile structure composed of rigid and elastic elements capable of withstanding and dissipating the intense masticatory forces. As previously noted, the overall strength of this structure increases significantly, surpassing the sum of the strengths of its individual components, while maintaining a significantly reduced weight and flexibility that allows local deformations to be dissipated globally.

Fig. 2.46 **Left**: fragment of a dental alveolus with the alveolar bone lamina perforated by numerous vascular canals; note the scaffold of trabeculae supporting the alveolus by suspending it within the alveolar bone. **Right**: fragment of alveolar bone in which, through the space left vacant by the dental root, the internal wall of the alveolus is visible; note, through the interruption of the external surface of the maxillary bone, the outer wall of the bone lamina surrounding the alveolus, also obviously perforated by Volkmann's canals and inserted in a substantially 'empty' bone: this bony cradle is like suspended and harnessed by the trabecular network. Specimen from the Museum of Anthropology of the University of Florence

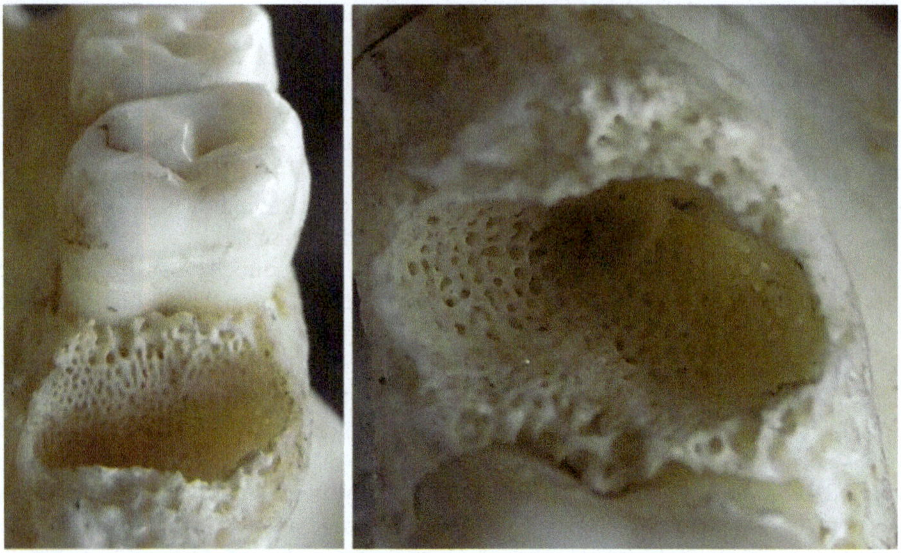

Fig. 2.47 The alveolar cortical bone is perforated by a dense network of vascular channels for the nourishment of the periodontal ligament, evidence of the intense activity of this structure; this reticular aspect also gives it greater elasticity and resistance to stresses. Skull 120. Specimen from the Museum of Anthropology of the University of Florence

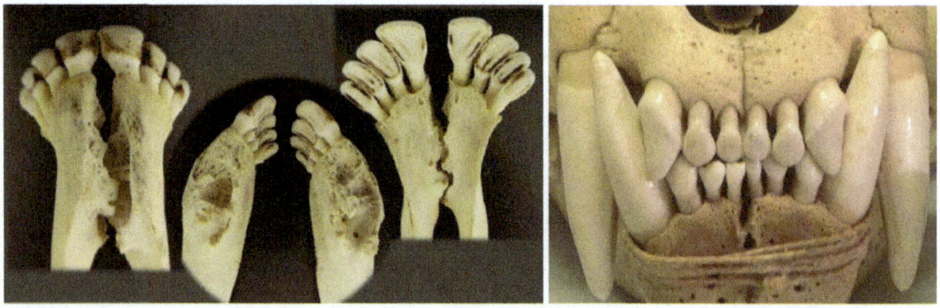

Fig. 2.48 Mandibular symphyses constituted, in life, by fibrocartilaginous tissue. **Left**: roe deer (Capreolus capreolus); **right**: lion (Panthera leo). Note: the irregularity of the symphyseal surface due to ligamentous insertions; and the string, which can be seen in the specimen below, necessary to keep the otherwise separable hemimandibles together

Even the symphyseal area of the mandible in mammals exhibits variations. It can be composed of fibrocartilage that loosely joins the two hemimandibles (Fig. 2.48). Alternatively, in various taxa of mammals, including anthropoid primates, this area can be formed by bone tissue so dense that it is less vascularized compared to other regions of the mandibular spongiosa.

The symphysis, consisting of rigid fibrous tissue, allows for mutual mobility between the two parts as it primarily functions as a mobile joint that enables each

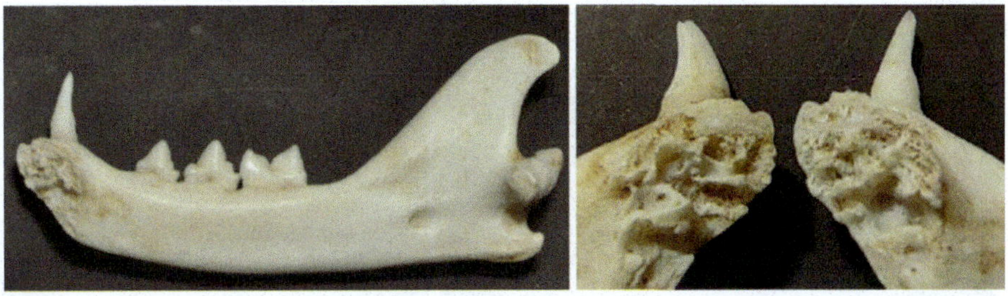

Fig. 2.49 Hemimandible of a cat (Felis catus). The symphysis area presents an irregular surface due to the insertion of connective fibers that unite the two hemimandibles

hemimandibular body to rotate around its own longitudinal axis. In most mammals, the mobility of the mandibular symphysis, in relation to the dynamics of the TMJ, allows for a wide range of movements. This is achieved through a set of transverse and crossed ligaments that effectively transfer vertically oriented forces from the working side to the balancing side by rotating the mandibular branches around their longitudinal axes. Rotation and derotation are induced by the masticatory power stroke, and in some mammals, this mechanism represents the primary means for medial translation of the lower arch relative to the upper arch (Crompton, 1989).

Fused symphyses have independently evolved in numerous mammal taxa to recruit the closing muscles on both sides of the jaw (Popowics & Herring, 2006; Wall & Smith, 2001) and are stronger and more rigid than fibrocartilaginous symphyses (Hylander, 1979a, 1979b, 1979c). Among mammals, symphyseal fusion significantly correlates with occlusal orientation, as species with more transversely oriented occlusal planes tend to have fused symphyses (Fig. 2.49).

Symphyseal fusion is very likely an adaptation to increase the efficiency of transferring transversely oriented occlusal forces. Among primates, true monkeys—unlike prosimians—possess a fused mandibular symphysis to effectively withstand the different forces acting on the two halves of the mandible during unilateral chewing. This fusion allows for the transfer of muscle force from the balancing side to the chewing side, utilizing an additional and distinct muscular force, thereby making unilateral chewing more efficient than bilateral chewing (Dean, 1990).

The maximum stress on the mandibular symphysis occurs during the latter part of the power stroke. At this moment, the force on the working side teeth opposes the force of the muscles on the balancing side, causing the mandibular symphysis to experience a shearing force with the two halves of the mandible flexed in opposite directions (Hylander, 1984). The concentration of masticatory forces at the symphysis is responsible for its high bone density (Fig. 2.50). A highly compact trabecular bone tends to have a lower vascular component, making the symphysis a more delicate area for dental implant osteointegration.

Compared to phylogenetically older vertebrates, the mandible of mammals experienced increased muscular stresses due to chewing, leading to the formation and remodeling of bony portions to accommodate the powerful bundles of masticatory muscles. The mammalian mandible is composed of different sectors that respond to

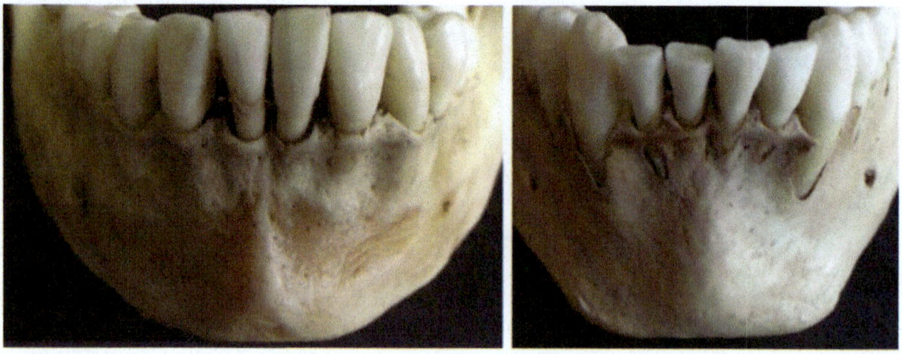

Fig. 2.50 Symphyseal areas of human mandibles (*Homo sapiens*): skull 327 (**left**); skull 5586 (**right**). Museum of Anthropology of the University of Florence

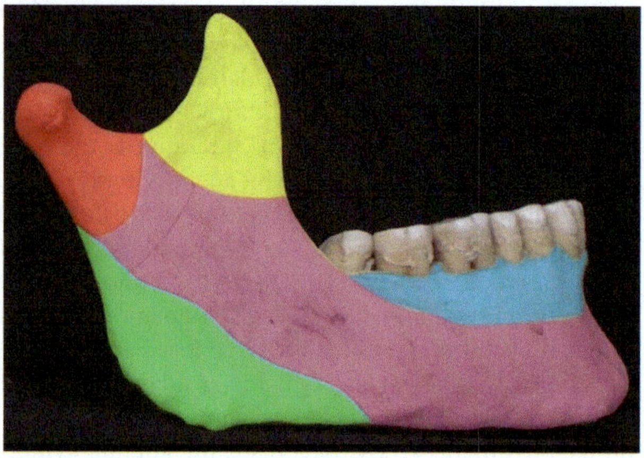

Fig. 2.51 The different anatomical components of the mandible (specifically addressed in the following subsections). Specimen from the Museum of Anthropology of the University of Florence

various functional needs: a very stable bone base (basal bone) to which several bony processes attach; the alveolar process, which contains and supports dental roots; the condylar process, which forms the joint part of the mandible; the coronoid process, where the temporal muscle attaches; and the angular process, where the masseter and internal pterygoid muscles attach. These are a mosaic of anatomically, morphologically, and sometimes embryologically distinct components characterized by the action different muscles exert on them (Fig. 2.51). Each of these portions, except for the basal bone, undergoes ontogenetic functional remodeling (Popowics & Herring, 2006).

2.3.5.1 Basal Bone

The portion of the mandible that maintains its volume nearly unchanged throughout an individual's life is the basal bone (Figs. 2.52, 2.53 and 2.54). In addition to serving as a platform for various processes of the mandible that support muscle

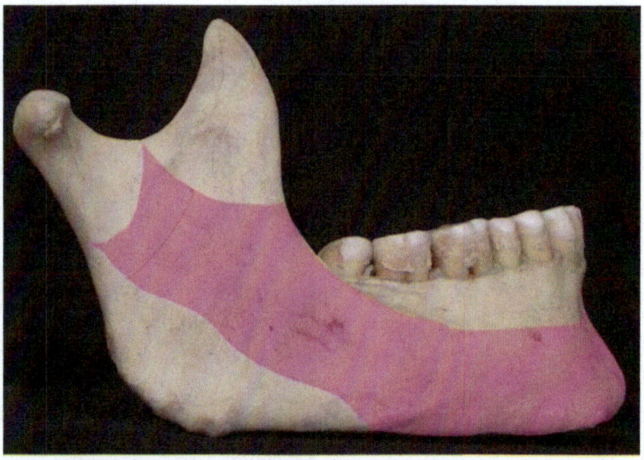

Fig. 2.52 Basal bone of the mandible. Specimen from the Museum of Anthropology of the University of Florence

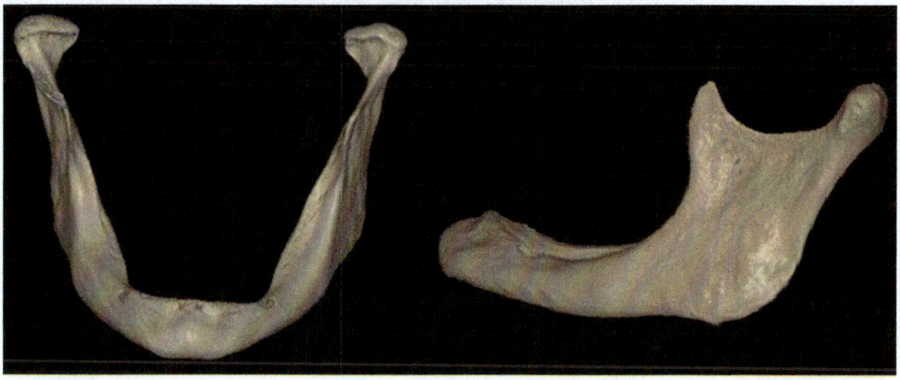

Fig. 2.53 Basal bone of the mandible. Specimen from the Museum of Anthropology of the University of Florence

insertions, it protects the course of the inferior alveolar nerve from its entry point at the Spyx spine to the mental foramen (Figs. 2.52 and 2.53).

The stability of the basal bone of the mandible safeguards the course of the inferior alveolar nerve and provides essential support to maintain airway patency through the insertions of the muscles that support the tongue, hyoid bone, and pharynx: specifically, the mylohyoid, digastric, and superior pharyngeal constrictor muscles (Fig. 2.54).

Even in an edentulous mandible with significant resorption of the alveolar process, there are no significant changes in the basal bone, both in the anteroposterior and lateral dimensions (Fig. 2.55).

The impression of centrifugal resorption of the lower alveolar bone is accentuated by the clinical observation of the edentulous ridge, which, due to the anteroposterior inclination of the alveolar process, lowers and moves forward in relation to the upper ridge (Fig. 2.56).

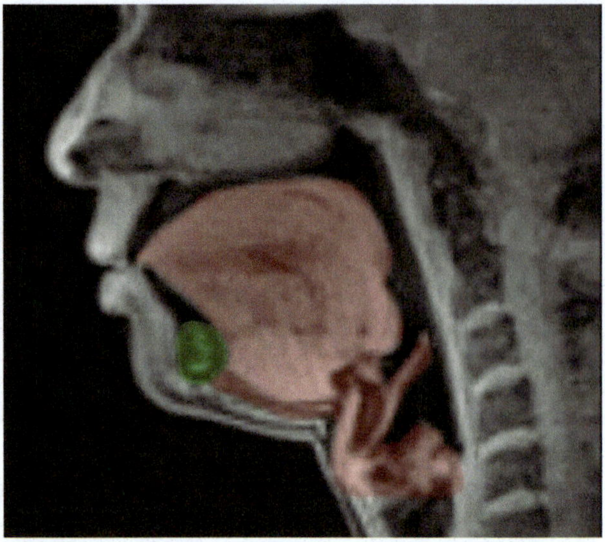

Fig. 2.54 Relationship between the basal bone of the mandible (green) and the lingual, hyoid, and laryngeal musculature (red)

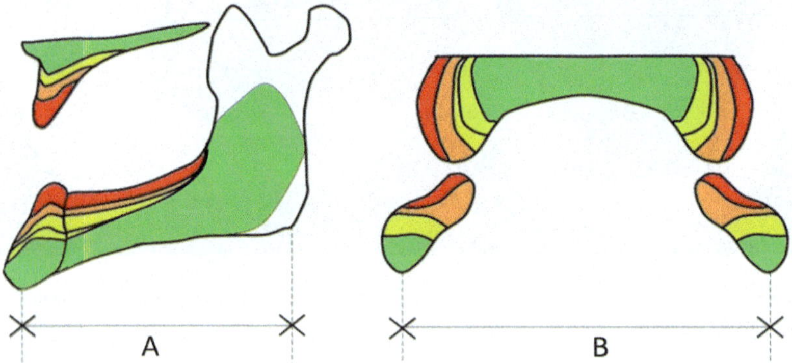

Fig. 2.55 Even in cases of severe atrophy of the alveolar bone, the basal bone maintains both its anteroposterior (**a**) and lateral (**b**) dimensions, both of which are indispensable to prevent collapse of the upper airways

In cases of severe resorption of the upper alveolar bone, the retraction stops at the anterior nasal spine and the cortical boundary of the maxillary sinuses, preserving the structure of the upper airways (Fig. 2.57).

In the mandible, the resorption of the alveolar bone is limited by muscle insertions, and the demarcation line between alveolar bone and basal bone can be identified by the path of the internal mylohyoid line, the insertion area of the mylohyoid muscle, and the trajectory that ideally connects it to the genial tubercles, the insertion points of the geniohyoid and genioglossus muscles (Figs. 2.58 and 2.59). Resorption ceases when critical bone components that are vital for survival could be compromised. It's important to note that in areas where shape or protection is of primary importance, the architecture of bone tissue remains under direct genetic control, remaining stable even under functional stimuli (Lanyon, 1987).

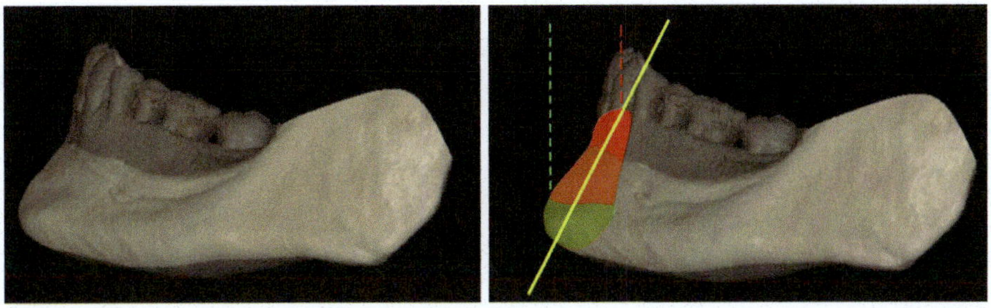

Fig. 2.56 Superimposition of the anterior profile of a mandible with a complete arch (dark gray) with an edentulous mandible (light gray): the apparent centrifugal resorption (red and green dashed lines) is due to the posterior inclination of the alveolar process in the symphyseal area (yellow line). In red, alveolar bone supporting dental roots; in green, alveolar bone undergoing resorption. Cone beam CT scan of a specimen from the Museum of Anthropology of the University of Florence

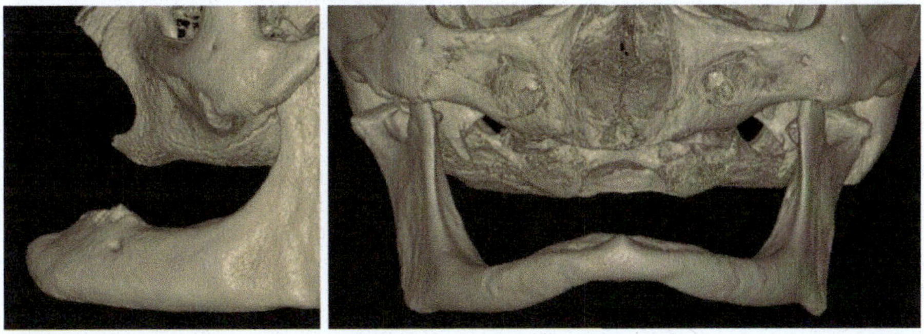

Fig. 2.57 Severe resorption of the maxillae: in the upper maxilla, retraction stops at the cortical of the anterior nasal spine, the anterior floor of the nasal cavities, and the maxillary sinus. Cone beam CT scan of a specimen from the Museum of Anthropology of the University of Florence

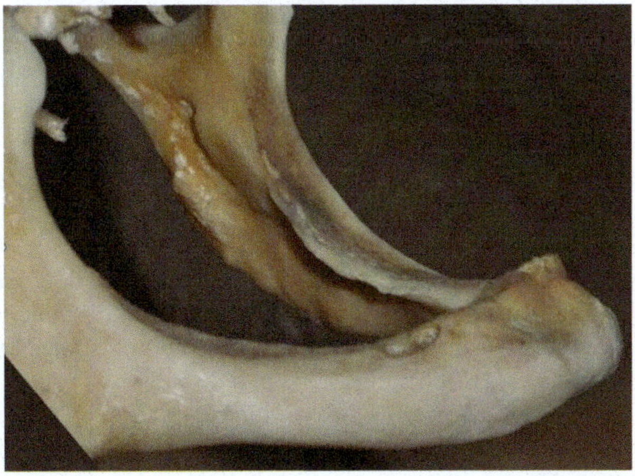

Fig. 2.58 Resorption of the alveolar bone, which stops at the mylohyoid line, insertion of the muscle fundamental for lingual support and stability. Note how the mental foramen, following the strong resorption of the alveolar bone, is located crestally. Specimen 1425 from the Museum of Anthropology of the University of Florence

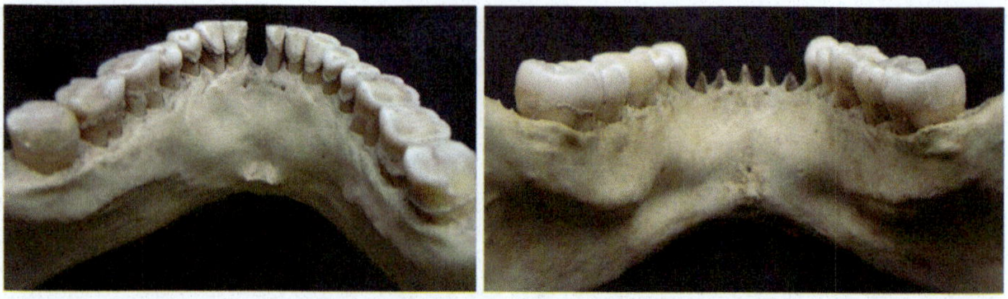

Fig. 2.59 Mylohyoid line and genial tubercle. **Left**: mandible 4781; **right**: mandible 3781. Museum of Anthropology of the University of Florence

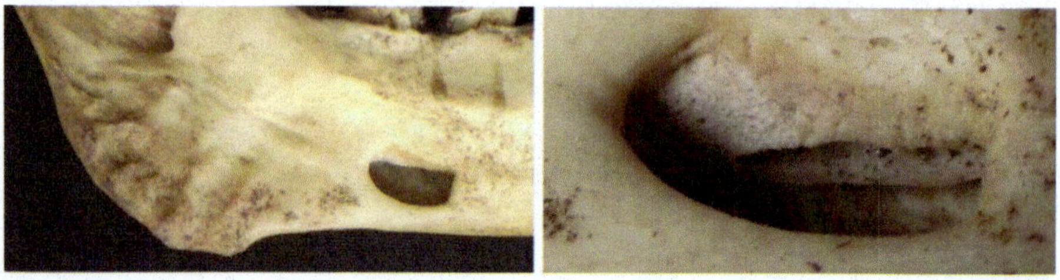

Fig. 2.60 On the inner surface of the mandibular body, an elliptical cavity attributed to Stafne's defect is evident, a local osseous adaptation due to the inclusion of an aberrant lobe of a salivary gland; in the enlarged detail (right image), note the course of the mandibular canal which, at the defect, has its lower part resorbed. Skull 4229 of the Museum of Anthropology of the University of Florence

Although subject to minimal resorption, the basal bone remains an active tissue capable of adapting to local stimuli (Fig. 2.60).

The stability of the maxillary basal bone is due to its supportive function for important adjacent structures. The basal bone of the mandible, as previously observed, serves as the insertion area for the muscle structure that forms the framework of the upper airways (Fig. 2.61).

In this regard, it should be noted that the prominent chin is a defining characteristic of *modern Homo sapiens*, distinguishing them from all other primates, including more archaic *Homo* species and Neanderthals (Daegling, 1993). The insertions of the mentioned musculature are necessary to support and maintain the patency of the upper airways. This is likely why *Homo sapiens* is the only species within its genus that possesses the chin prominence on the anterior mandible; all other *Homo* species had a receding mandibular profile (Fig. 2.62).

The human chin is often cited as an example of a non-adaptive trait, and in modern humans, its development is not linked to the functional demands placed on the mandible during chewing, which evolved in response to other biomechanical needs (Ichim et al., 2006) (Figs. 2.63 and 2.64).

Throughout various evolutionary processes at the cranial level, a series of changes in the relationships between various components have accumulated. If we consider the process of encephalization in *Homo sapiens*, characterized by a

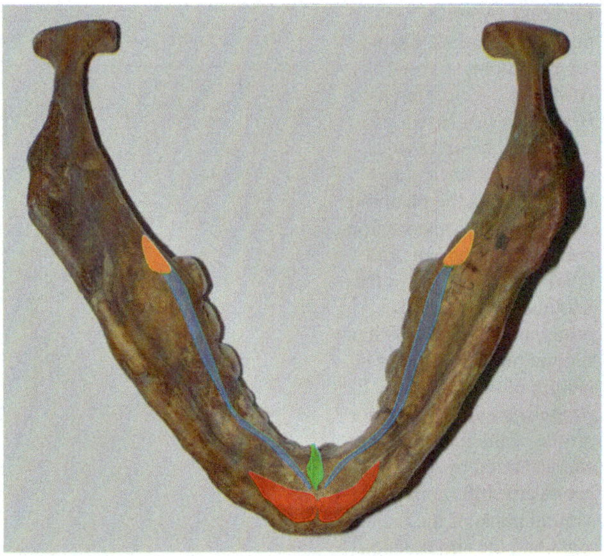

Fig. 2.61 Muscle insertion areas on the inner and lower part of the mandibular body. Red: insertion area of the anterior belly of the digastric muscle; green: insertion area of the geniohyoid and genioglossus muscles; light blue: insertion area of the mylohyoid muscle; orange: insertion area of the superior pharyngeal constrictor muscle. Specimen from the Museum of Anthropology of the University of Florence

globular development of the neurocranium, it led to a decrease in its anteroposterior diameter associated with a backward movement of the maxillae (Fig. 2.65). However, the basal bone of the mandible did not follow the backward movement of the other maxillary structures, thus preventing a posterior shift of the tongue, suprahyoid musculature, and pharyngeal structures that would have compromised airway patency (Figs. 2.66 and 2.67). Therefore, the chin prominence can be seen as a mandibular bony component of anteroposterior stability, functionally distinct from other regions of the maxillae involved in a series of morphological adaptations related to the unique globular encephalization in *H. sapiens*.

The posterior shift in the sagittal plane of the alveolar bone in *H. sapiens* compared to the basal bone, along with the resulting formation of the chin, has led to a change in the relationship between the major axis of the incisors and the basal portion of the mandibular body (Fig. 2.68).

In the anterior region of an edentulous mandible with pronounced alveolar bone resorption, prior to performing dental implant rehabilitation, a careful evaluation of bone morphology is necessary. When placing an implant, attention must be paid to the angulation of the drill, aligning its axis in relation to the anatomy of the basal bone (Fig. 2.69).

As mentioned earlier, the basal bone of the mandible is the origin and development site for various bony processes related to dental roots, the mandibular joint, and the insertion of masticatory muscles:

- The alveolar process, which supports dental roots.
- The condylar process, forming the joint portion.
- The coronoid process, serving as the insertion point for the temporal muscle.
- The angular process, where the masseter and internal pterygoid muscles attach.

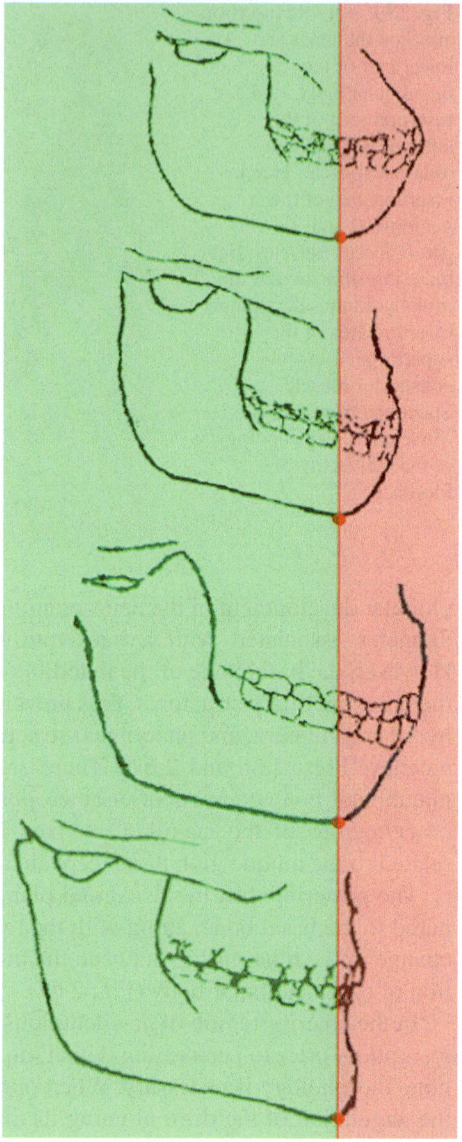

Fig. 2.62 Anterior maxillary profiles. From top to bottom: Australopithecus afarensis, *Homo erectus, Homo neanderthalensis, Homo sapiens*. The vertical red line intersects the gnathion (also known as 'mentonion point'): the craniometric point corresponding to the maximum downward prominence, on the median sagittal plane, of the lower margin of the mandible. In *Homo sapiens*, the anterior profile of the maxillae is characterized by a backward shift on the sagittal plane of the facial mass and the simultaneous development of chin prominence. Drawing by Marco Boggioni and the author

2.3.5.2 Alveolar Process

The alveolar bone (Fig. 2.70) is closely related both phylogenetically and ontogenetically to mechanical stresses associated with chewing. It is almost always considered as the foundational support for dental roots, but in reality, this supportive function is reciprocal. Just think about the extensive resorption that occurs in the alveolar bone after the loss of a tooth.

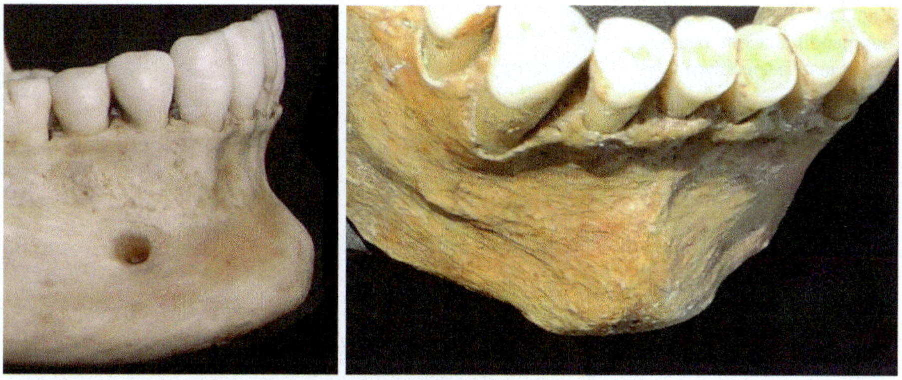

Fig. 2.63 Mental protuberance. Specimen from the Museum of Anthropology of the University of Florence

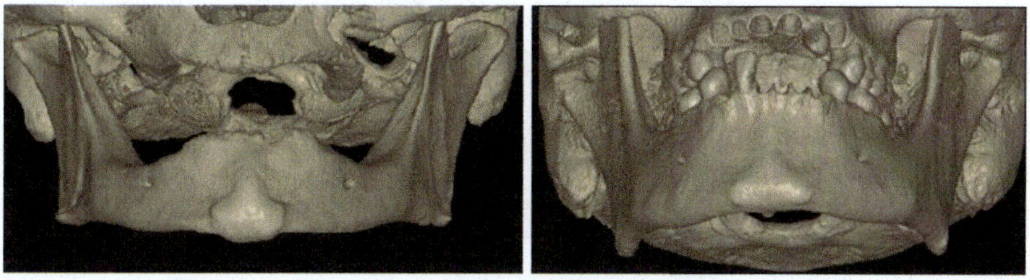

Fig. 2.64 Mental protuberance. Cone beam CT scan of specimens from the Museum of Anthropology of the University of Florence

Fig. 2.65 Comparison of the cranial profile in *H. sapiens* (black) with the profile in *Early Homo* (red): retraction of the upper maxillary base, upper and lower alveolar processes, but not of the mandibular basal bone, resulting in the formation of the chin. Drawing by Marco Boggioni

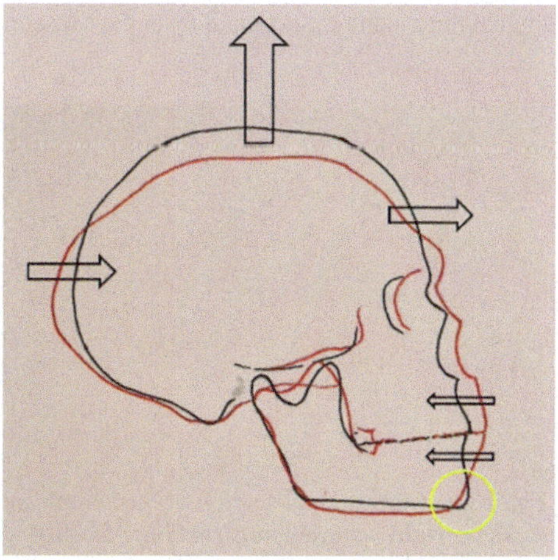

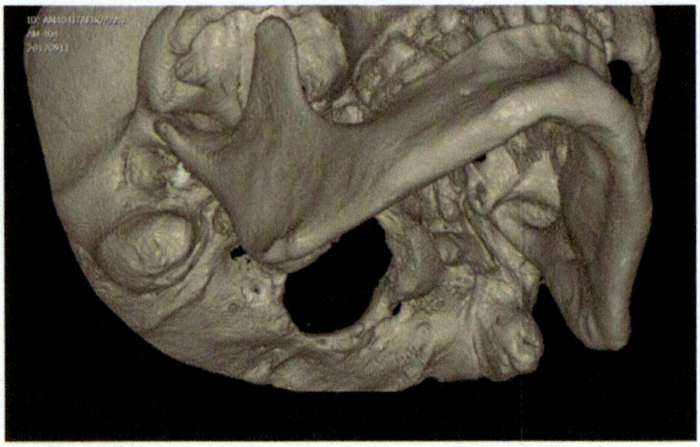

Fig. 2.66 Basal bone of the mandible. Cone beam CT scan of a specimen from the Museum of Anthropology of the University of Florence

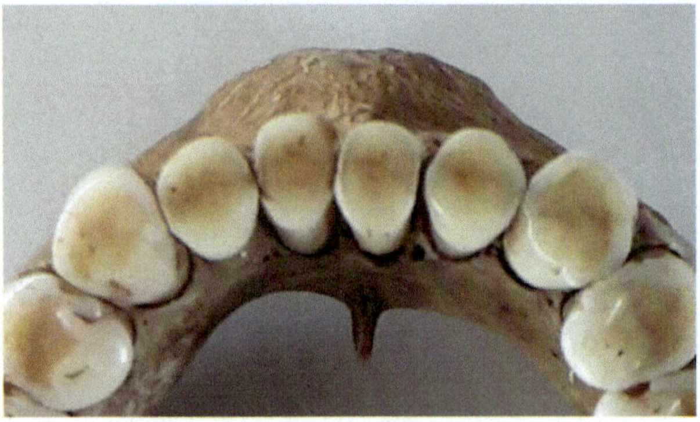

Fig. 2.67 Note the three structures that, due to mid-facial retrusion in *Homo sapiens*, had to find a new balance: the anterior margin of the mandible forming the chin prominence, the teeth that followed the retrusion of the maxillae, and the genial tubercle, the insertion area of part of the lingual musculature. Specimen from the Museum of Anthropology of the University of Florence

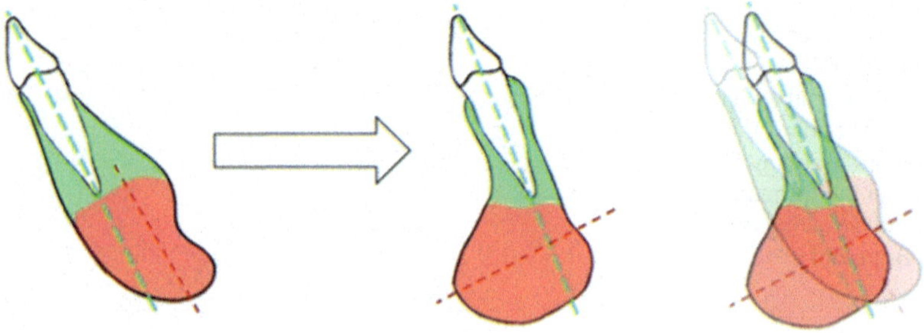

Fig. 2.68 **Left**: sagittal section of the symphysis in archaic *Homo*; center: same section in *Homo sapiens*; **right**: overlapping sections. Dashed green lines: dental axis; dashed red lines: basal bone axis

Fig. 2.69 Change of direction in the symphyseal region of the axis of the surgical implant drill (outlined in red), relative to the dental axis. Note, at the level of the internal surface of the symphysis, the depression of the profile caused by the entrance of a branch of the lingual artery: the point at which this artery enters the mandibular body should be carefully evaluated during surgery in this area

Fig. 2.70 Alveolar process of the mandible. Specimen from the Museum of Anthropology of the University of Florence

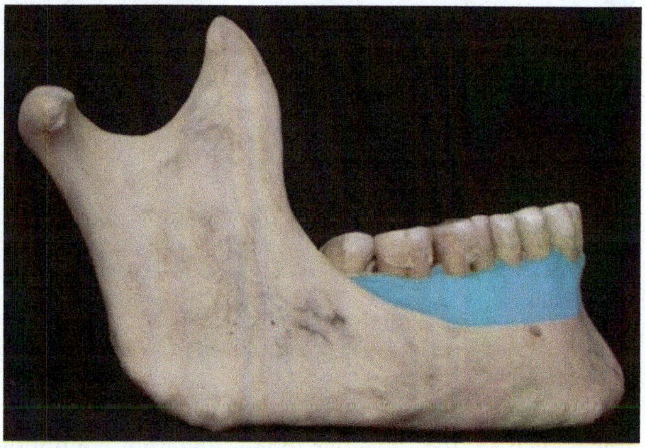

The alveolar process is common to both the mandible and the three bony components of the upper jaw (the premaxilla and the maxillary bones) (Fig. 2.71).

Its height is correlated with the length of the tooth roots (Atchley, 1993). The high vascularity of the bone surrounding the dental roots is necessary to support the intense periodontal metabolism and the continuous functional responsiveness of this bone component. The dento-alveolar connection is typical, though not exclusive, to mammals. All other types of connections between dental roots and bone are phylogenetically older and are limited to a more or less rigid union without any adaptive capacity. The periodontium is an extraordinary tissue, a masterpiece of nature that has evolved finely to facilitate chewing (Diekwisch, 2016). The

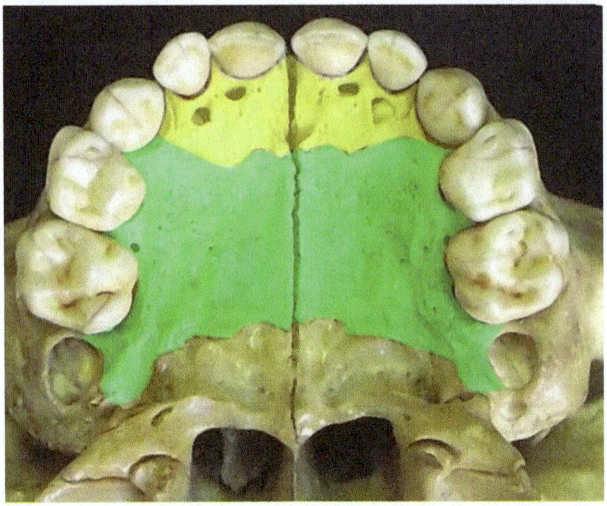

Fig. 2.71 Premaxilla (yellow) and maxillary bone (green). Skull 2857 from the Museum of Anthropology of the University of Florence

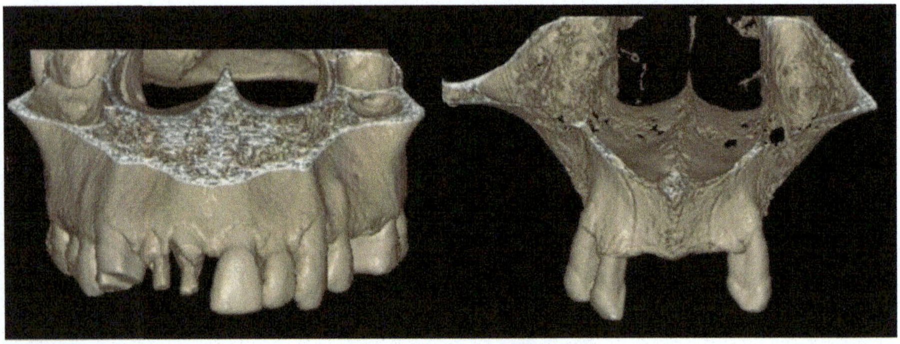

Fig. 2.72 Different volume of the alveolar bone in relation to the presence or absence of dental elements. Cone beam CT scan of a specimen from the Museum of Anthropology of the University of Florence

trabeculae of the alveolar bone undergo continuous remodeling to optimize responses to the mechanical loads of chewing, dissipate forces, and prevent damage (Bodic et al., 2005). This continuous remodeling is crucial to adapt to the inevitable changes in occlusion that occur during an individual's lifetime, whether they are physiological, such as enamel wear, or pathological, such as coronal damage or tooth loss.

Periodontitis weakens and leads to resorption of this structural support, and it is the primary cause of tooth loss (Watson et al., 2018), often followed by the irreversible process of alveolar crest bone resorption (Hansson & Halldin, 2012) (Fig. 2.72).

From an evolutionary, embryological, anatomical, and biomechanical perspective, the presence of alveolar bone in mammals is closely related to the root that inserts into it (Fig. 2.73).

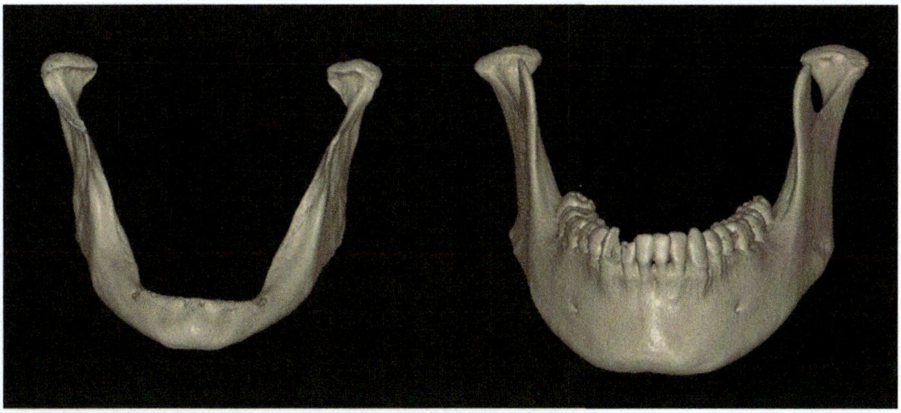

Fig. 2.73 Different volume of the alveolar bone in relation to the presence or absence of dental elements. Cone beam CT scan of a specimen from the Museum of Anthropology of the University of Florence

The teeth become firmly anchored in bone cavities, in a condition called 'thecodont,' from the greek *theca* = sheath and *odous, -dontos* = tooth (Liem, 2001), which is achieved through a 'gomphosis,' a type of joint formed by a conical surface (the root) that fits into a bone cavity, acting like a mold. The gomphosis develops from the dental follicle with the simultaneous formation of cementum, periodontal ligament, and alveolar bone; the alveolar bone begins to form early during the bell stage of embryological development and continues to remodel in response to tooth growth (Cho & Garant, 2000). The alveolar bone, in conjunction with the dental roots and the periodontal ligament, constitutes a complex that has developed as a functional response to allow teeth to withstand the intense forces imposed by chewing.

2.3.5.3 Condylar Process

In mammals, there was a drastic transformation of the mandibulo-cranial joint compared to that of reptiles: the concave part of the joint transfers to the cranial surface, while the joint pivot moves to the lower portion, structuring itself in the mandibular condyle (Fig. 2.74). (The description of this process will continue a little later in the section dedicated to the TMJ—temporomandibular joint.)

2.3.5.4 Coronoid Process

In all primates and, more generally, in all mammals, the mandibular branch terminates superiorly in two processes: the condylar process, which constitutes the mandibular component of the TMJ (temporomandibular joint), and the coronoid process, where the common tendon of the temporal muscle inserts. Between these two processes lies the U-shaped sigmoid notch (Terhune et al., 2014) (Fig. 2.75).

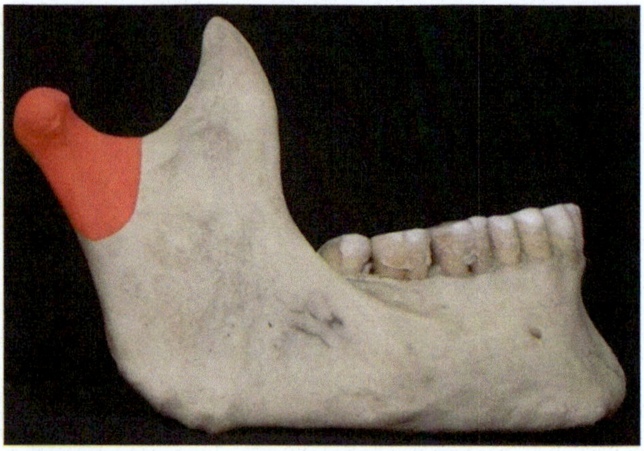

Fig. 2.74 Condylar process of the mandible. Specimen from the Museum of Anthropology of the University of Florence

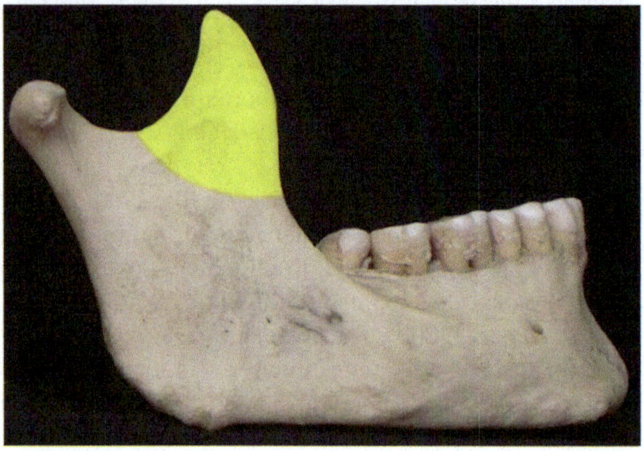

Fig. 2.75 Coronoid process. Specimen from the Museum of Anthropology of the University of Florence

The coronoid process in mammals evolves from the earlier small coronoid eminence in reptiles, and the temporal muscle tendon attaches to this new area of insertion (Liem, 2001). The shape of the coronoid process varies between species depending on the form and power of the temporal muscle (Popowics & Herring, 2006). Carnivorous mammals tend to have a much more developed coronoid process compared to herbivores, while in humans, it exhibits a highly variable shape in relation to the movements through which the dental arches functionally interact with each other. Within hominids, some authors suggest that this morphological variation may reflect phylogenetic development patterns (Rak et al., 2007).

In mammals, the coronoid process (like the secondary palate) is an evolutionary novelty and is one of the morphological units that provide valuable insights into adaptations that have occurred during evolution. In the fossil records of pre-mammalian reptiles, the emergence of the coronoid process precedes that of the TMJ (temporomandibular joint). Existing carnivorous, omnivorous, and insectivorous mammals have a large coronoid process, similar to that in the ancestral mammal, whereas herbivores have a reduced coronoid process. This feature is a significant aspect of the mandibular skeleton in mammals and is evolutionarily associated with a shift in the insertion of the precursor muscle of the temporal muscle. This shift allowed for the development of increased bite force and played an important role in the evolution of mandibular adductor muscles.

The formation of the temporal muscle and its insertion on the coronoid process are essential for the development and maintenance of the coronoid. The extrinsic effect of the muscle directly influences its size, while its formation occurs through genetic growth regulation (Anthwal et al., 2015).

2.3.5.5 Angular Process

The angular process of the mandible (Fig. 2.76) is associated with the action of the masseter muscle (Moss & Simon, 1968) and the internal pterygoid muscle. Its significant variability in shape and size in mammals reflects its close relationship with dietary adaptations.

Herbivores, with well-developed masseters and medial pterygoids, tend to have a pronounced angular process, whereas it is generally reduced in small insectivorous marsupials (Sánchez-Villagra & Smith, 1997).

In humans, the angular process of the mandible undergoes age-related changes to provide necessary support for the insertion of these muscles, adapting on the external side to changes in masseter muscle action (Moss & Simon, 1968) and on the internal side to the biomechanical influence of the internal pterygoid. The traction exerted by these muscle insertions tends to create roughness on the bone surface (Figs. 2.77 and 2.78).

In individuals with a deep antegonial notch, a depression on the lower border of the mandible located in front of the mandibular angle, increased muscular activity of the masseter and internal pterygoid muscles has been observed. However, no statistically significant correlation has been observed between vertical craniofacial morphology and the depth of the antegonial notch (Tomer & Kishnani, 2011).

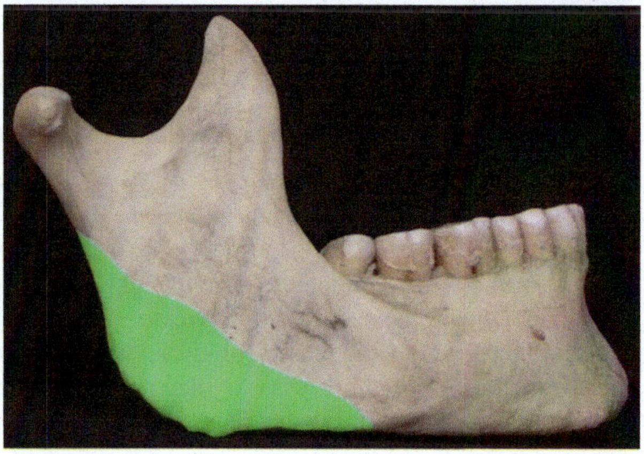

Fig. 2.76 Angular process. Specimen from the Museum of Anthropology of the University of Florence

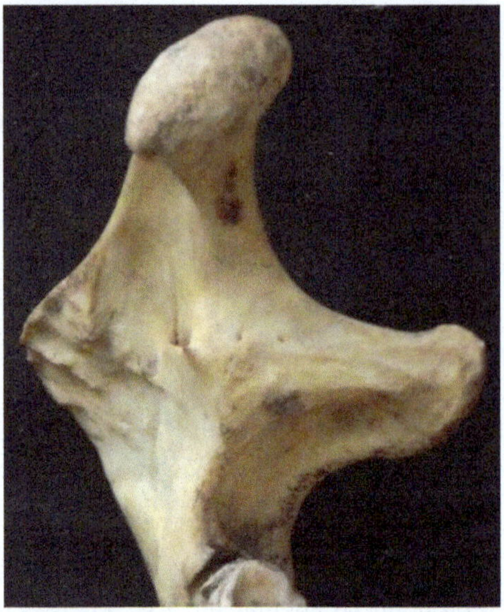

Fig. 2.77 Angular process of the mandible: 'imprints' of the insertions of the internal pterygoid muscle. Note on the medial side of the coronoid process, the reliefs of the cortical bone shaped by the insertion of the temporal muscle. Skull 4234 from the Museum of Anthropology of the University of Florence

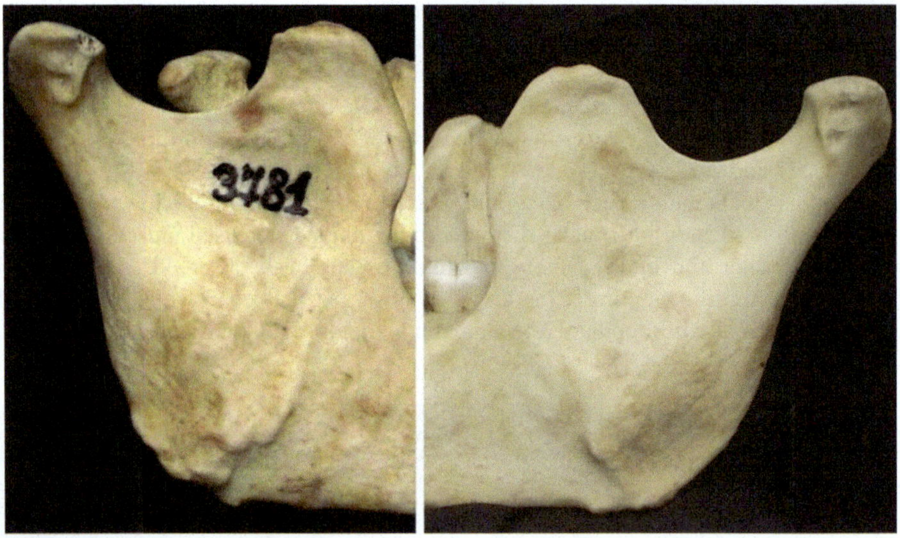

Fig. 2.78 Angular process of the mandible: 'imprints' of the insertions of the masseter muscle. Skull 3781 from the Museum of Anthropology of the University of Florence

Therefore, the bony components of the angular and coronoid processes are highly sensitive to the biomechanical stimuli exerted by the muscles that insert onto them: the internal pterygoid and masseter muscles on the angular process and the temporal muscle on the coronoid process (Atchley, 1993) (Figs. 2.79, 2.80, 2.81).

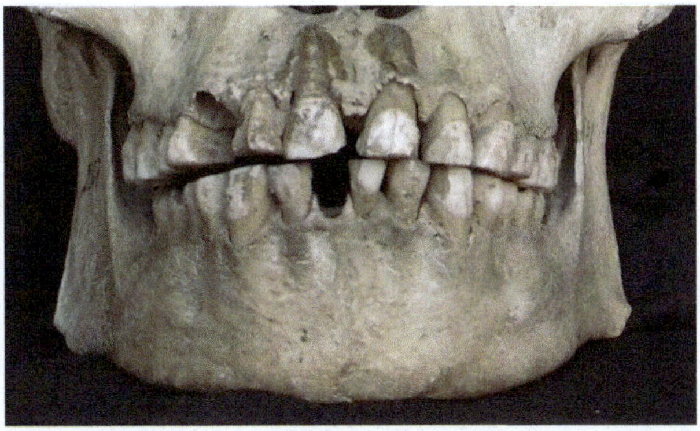

Fig. 2.79 Angular processes particularly developed in an individual who lived in a pre-industrial era: this specific development is the adaptive response of this region of the mandible to strong muscular insertions; together with dental wear, this bone morphology fits into a harmonious framework of intense activity of the masticatory apparatus. Skull 2507 from the Museum of Anthropology of the University of Florence

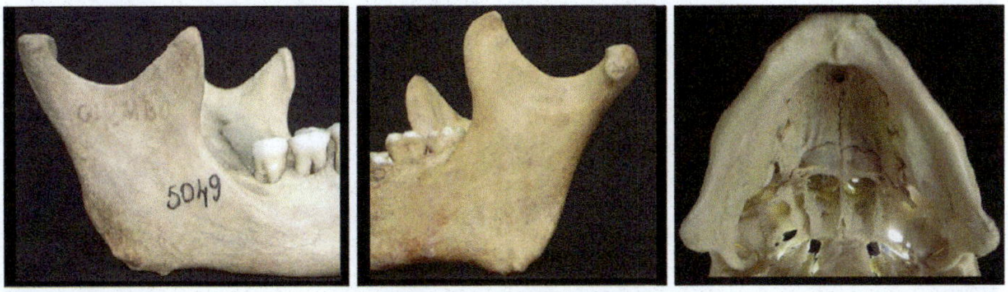

Fig. 2.80 Angular process of the mandible. Top: skull 5049; middle: skull 6694; bottom: skull 311. Museum of Anthropology of the University of Florence

Regarding the mandibular angle, there is an interesting anatomical feature in this region that is peculiar to a species of humans, *Homo neanderthalensis*. The Neanderthal man was our closest relative, a genetic sibling, and a generational cousin. It was the last extinct human species, leaving limited biomolecular traces within us, with only a few bones remaining. Neanderthals were present until relatively recently, approximately 40,000 years ago. Despite the aura of ruggedness that has surrounded them as the archetypal prehistoric humans, *Homo neanderthalensis* has always evoked sympathy and even a bit of affection in the collective imagination.

Their unique characteristics, similar yet distinct from ours, have always generated considerable interest, affecting the entire skeleton and especially the craniocervical region. The anatomical features of the head specific to Neanderthals have long been a subject of debate within the scientific community. The Neanderthal mandible has a distinctive morphology that sets it apart from that of *Homo sapiens*. For example, the ascending branch of the mandibular ramus is wider in relation to its height compared to *H. sapiens* (Fig. 2.82).

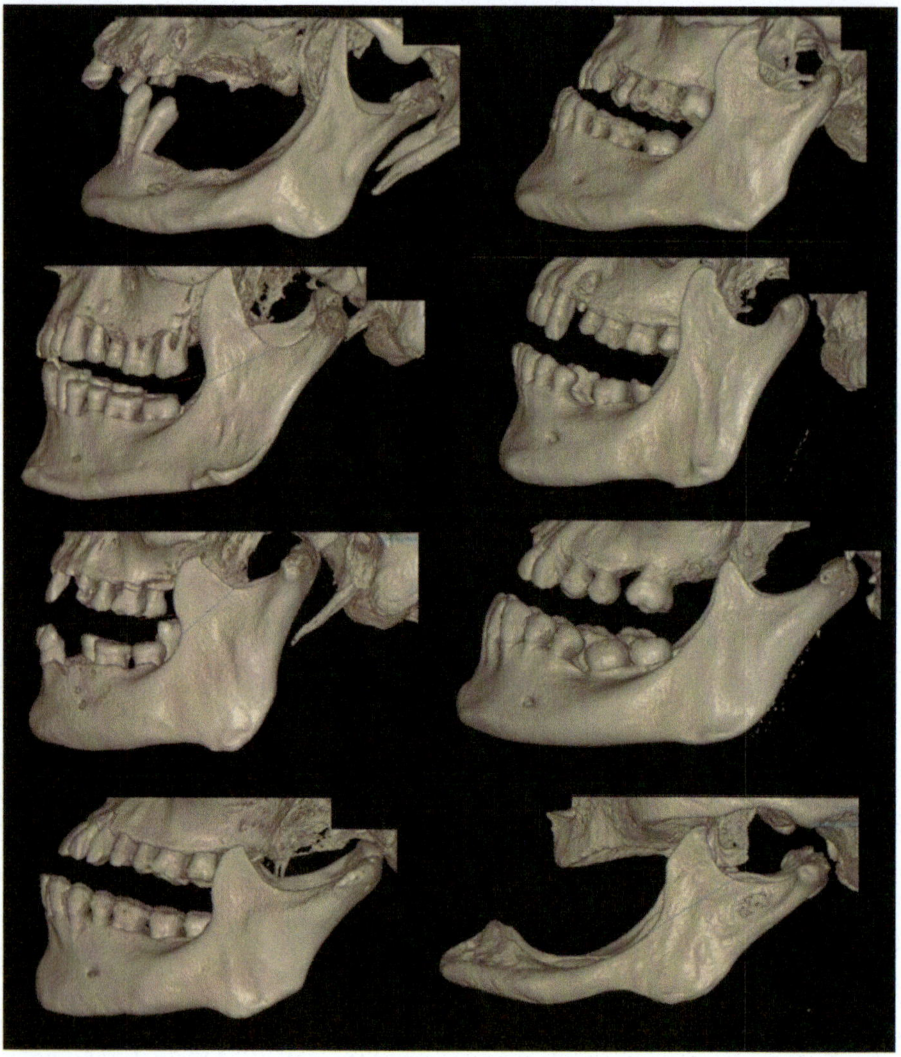

Fig. 2.81 Morphological variability of different mandibular processes. Cone beam CT scan in sagittal section of specimens from the Museum of Anthropology of the University of Florence

The width of the ascending branch was measured from the point of maximum anterior concavity to the point of maximum posterior concavity, while the height of the ascending branch was measured from the deepest point of the antegonial notch to the deepest point of the sigmoid notch (Papini et al., 2012) (Fig. 2.83).

In a study conducted by the author and Dr. Marco Boggioni, a greater width of the ascending branch of the Neanderthal mandible was highlighted. In fact, the ratio of height to width in Neanderthal mandibles from Amud, Guattari III, Regourdou 1, Tabun C1, and Ehringsdorf is 1.4262, while in mandibles belonging to the

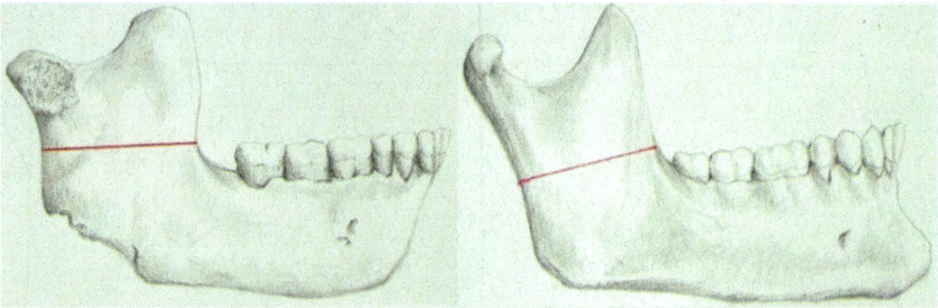

Fig. 2.82 The ascending branch of the Neanderthal mandible (on the left) is wider compared to that present in *Homo sapiens*. Drawing by Marco Boggioni

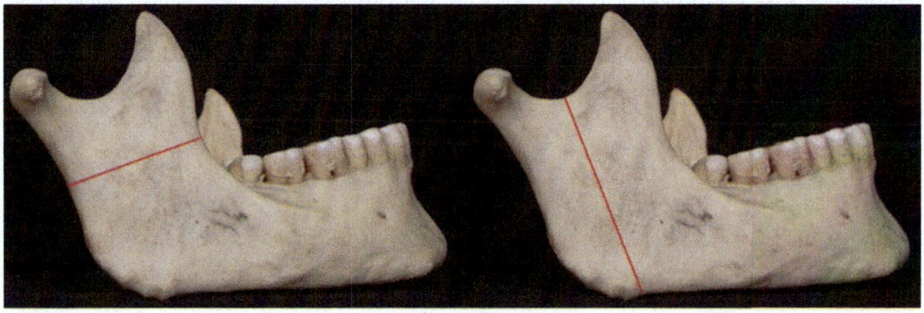

Fig. 2.83 Width and height of the mandibular ramus. In the Neanderthal mandible, the ratio between these measurements of the ascending branch is different compared to that present in *Homo sapiens*. Specimen from the Museum of Anthropology of the University of Florence

collection of human skeletal remains at the Section of Anthropology and Ethnology of the Museum of Natural History in Florence, the ratio is 1.5837. This configuration of the Neanderthal mandibular ramus tends to change the relationships with the structures in the retro-mandibular space, reducing it (Fig. 2.84).

From the same anatomical study, it also emerges that the Neanderthal gonial angle is smaller compared to that in *H. sapiens*. The average gonial angle of Neanderthal mandibles from Amud, Guattari III, Regourdou 1, Tabun C1 is 107.48 degrees, while that of the mandibles studied in the cranioteque of the Anthropological Museum at the University of Florence is 123.78 degrees. This Neanderthal characteristic affects the geometry of maximum opening movement, tending to increase the posteriorization of the mandibular angle (Fig. 2.85).

In the maximum opening movement, the smaller gonial angle in *Homo neanderthalensis* would induce greater posteriorization, compared to *H. sapiens*, of the mandibular angle and all the musculature that inserts there (including the floor of the tongue), with a tendency to compress the airways and vascular bundles of the neck. However, the specific blunt shape of the Neanderthal mandibular angle geometrically reduces this backward movement, allowing, at the same mouth opening, the maintenance of airway patency (Fig. 2.86).

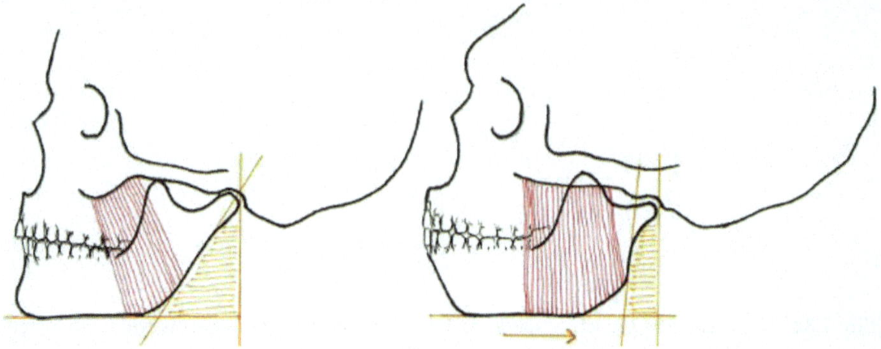

Fig. 2.84 In Neanderthals, the mandibular ramus, wider than in *Homo sapiens*, tends to reduce the retro-mandibular space delimited posteriorly by the cervical spine; the blunted angle of the Neanderthal mandible, during maximum opening, helps reduce the posteriorization of the mandibular angle and reduce compression of the airways and large vascular bundles of the neck. Drawing by Marco Boggioni

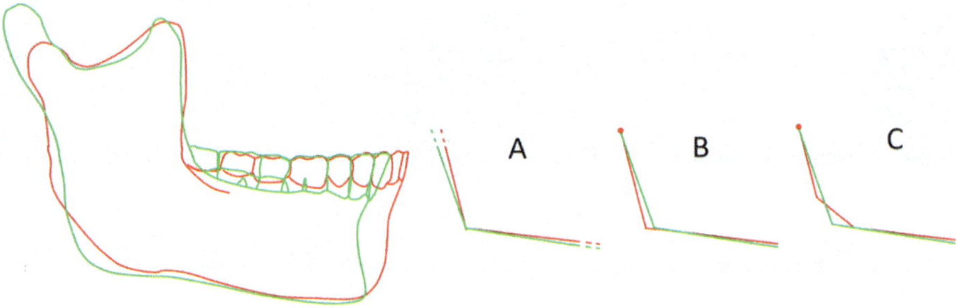

Fig. 2.85 In the left figure, overlapping mandibular profiles: red, Neanderthal mandible; green, mandible in *Homo sapiens*. On the right, stylized, their respective gonial angles: (**a**), their overlap by coinciding angles; (**b**), superimposing a hypothetical joint; (**c**), highlighting the blunting of the angle that reduces its posterior bulk. Drawing by Marco Boggioni and the author

It should be emphasized that the musculo-articular structure of Neanderthals was thicker, more massive, and powerful compared to that of *H. sapiens*. This characteristic was also evident in the cervical region, with more powerful muscles and lower, wider vertebral bodies (Gómez-Olivencia et al., 2007, 2013). This suggests reduced agility and extension in flexor-extensor movements of the head (Gómez-Olivencia et al., 2013). With such a massive neck, it is reasonable to assume significantly reduced flexo-extension movements. This functional condition could be considered another autapomorphic character (an evolutionary novelty, a unique and exclusive trait) of Neanderthals.

The authors suggest that the reduced ability to extend the head could have been disadvantageous in the maximum mouth-opening movement because it decreased the portion of movement resulting from cranial extension (Fig. 2.87).

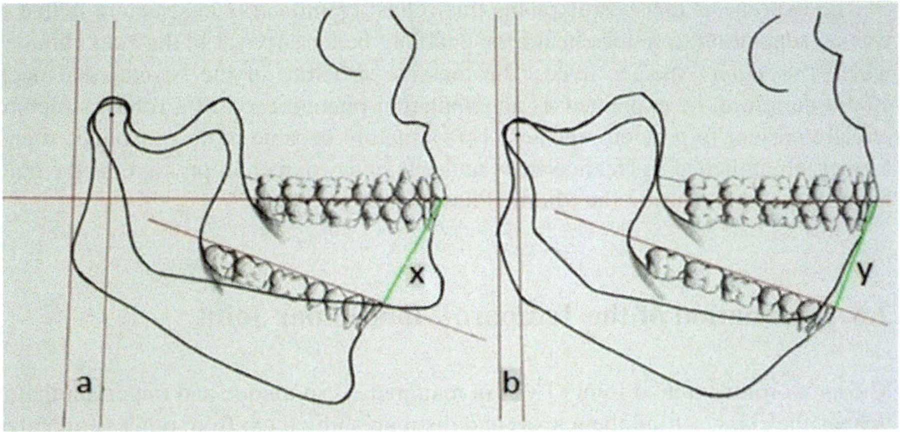

Fig. 2.86 With the mouth opened to the same extent, in Neanderthals, the blunted angle of the mandible reduces the posterior-inferior retraction of the mandibular portion: x = y; a > b. Drawing by Marco Boggioni

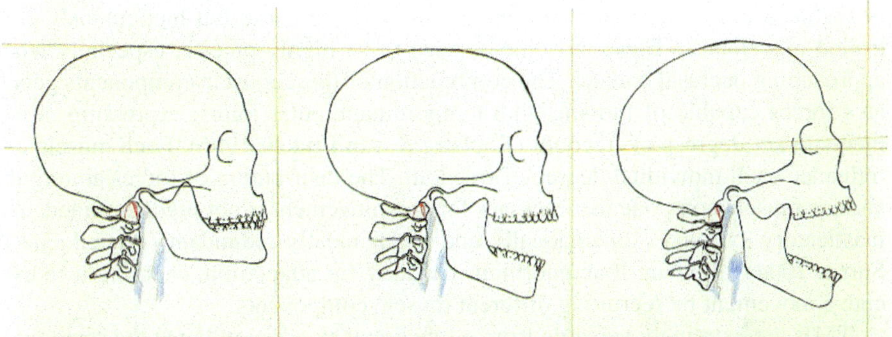

Fig. 2.87 At the top, from left to right, the movement of maximum opening in *Homo sapiens*; at the bottom, from left to right, the hypothesized movement of maximum opening in *Homo neanderthalensis*: note the lack of cranial extension in the Neanderthal. The powerful musculo-articular structure of the Neanderthal neck reduced flexo-extensor movements of the head, penalizing the maximum opening movement of the mouth, which probably occurred with minimal cranial extension, exclusively with mandibular movement; for this reason, the blunted mandibular angle was valuable for the patency of the airways and vascular trunks of the neck. Drawing by Marco Boggioni

The hypothesis that best explains this robust cranio-cervical structure is that it was an adaptation to reduce heat loss, enabling better survival in the cold climates where this *Homo* species lived. The massive structure of the Neanderthal neck could, therefore, be explained as an exaptation phenomenon of a robust structure already present in previous species. This structure became more consolidated and further strengthened in Neanderthals under the environmental pressure of the cold European climate during the Middle Pleistocene.

2.4 Formation of the Temporo-Mandibular Joint

The temporomandibular joint (TMJ) in mammals is so unique and important that it defines the class, setting them apart and distinguishing them from other vertebrates (Irish & Scott, 2015). Within this class, the TMJ exhibits remarkable morphological and functional variations in different species. This reflects not only the great adaptive radiation of mammals in feeding mechanisms but also a freedom from constraints, such as body weight (Herring, 2003).

The TMJ serves as the pivot point for mandibular movements, guided by as many as eighteen muscle groups that generate often asymmetric but meticulously controlled movements. These movements need to be highly precise, especially when approaching occlusal contact. The combination of these muscle components generates forces capable of moving both temporomandibular joints, expressing movements in six degrees of freedom (Koolstra & van Eijden, 1996). Each muscle can influence each individual degree of freedom. The distribution of forces among the various muscle components necessary for any movement is not predetermined. The masticatory system is cinematically and mechanically redundant (Pileicikiene & Surna, 2004), allowing it an enormous capacity for adaptation, enabling it to execute a movement by recruiting different muscle components.

TMJs are extremely versatile from a functional standpoint, but at the same time, they are anatomically ill-defined. The reason for this morphology can be expressed with a quote from Oscar Wilde's "The Picture of Dorian Gray": "To define is to limit." In humans, the significant freedom of movement in the masticatory system is reflected in the morphological indeterminacy of the TMJs, where the mandibular condyle articulates anatomically incongruously with the articular surface of the temporal bone (Koolstra & van Eijden, 1999). The limited anatomical reciprocity between the condylar articular surface and the mandibular fossa, excluding the anterior-superior articular surface, allowed for a wide range of mandibular movements. This made humans one of the most flexible species in exploiting diverse food sources, characteristic of an omnivorous diet (Pileicikiene & Surna, 2004; Fiorenza et al., 2020). It also enabled all the movements required by the various functions of the stomatognathic apparatus.

In mammals, the articulation between the skull and the mandible had to adapt to handle movements different from simple rotation because chewing required an articular structure that was flexible enough to develop both horizontal and vertical movements while also being stable enough to withstand and dissipate the forces needed for food fracturing. Almost all mammals are capable of chewing and often have excellent occlusion between their upper and lower teeth, allowing for food processing. This is achieved through the mandible's ability to move laterally (Tucker, 2017). The development of chewing allowed the mandible to achieve horizontal movements by enabling the condyles to translate in various directions within the mandibular fossa, both laterally and anteroposteriorly. In the human TMJ, the articular eminence should ideally have the same inclination on both sides, characterized by a symmetric growth pattern with no statistically significant difference between the development of the right and left sides (Katsavrias, 2002).

Before mammals, reptiles had a simple rotating cranio-mandibular joint that allowed only basic rotational movements. The quadrate bone on the lower part of the skull projected downward, articulating with the concave and socket-like articular bone on the back of the mandible (Modesto et al., 2001; Luo et al., 2004; Kemp, 2005). The ability of the adductor muscles to close the jaws quickly was greater than opening them, which allowed them to capture the prey and prevent escape more effectively. In mammals, the first step in forming a distinct ear and mandible was the development of two separate joints. Having two joints provided greater resistance against the forces generated during chewing, which introduced a twisting motion to the mandible (Kermack, 1972; Hylander & Crompton, 1986; Kemp, 2005). The mandibular condyle gained the ability to translate within the fossa. Mediolateral condylar translation is a characteristic of all mammals, even in species with limited or no anteroposterior translation (Hylander, 2006). With mammals came complex and differentiated dental forms, which corresponded to equally complex muscle-joint movements. Most mammals have an intra-articular division created by the fibrocartilaginous disc, dividing each articular cavity into an upper and lower compartment. The disc is functional in allowing complex movements, following translatory movements by attaching to the medial and lateral poles of the condyle. The translatory part of the movement mainly involves the upper compartment, while the rotational part occurs within the lower joint between the condyle and the disc (Popowics & Herring, 2006). The advent of a double joint was directly linked to changes in feeding habits and alterations in tooth form (Anthwal et al., 2013).

Although the components of the TMJ are constant among mammals, there is significant diversity in the morphology of the joint, as well as the size and orientation of the articular surfaces (Popowics & Herring, 2006). The TMJ exhibits considerable morphological and functional variation across different species (Dean, 1990). For example, rodents experience minimal joint loads during chewing, while carnivores support higher loads than primates (Herring et al., 2001). This substantial variability reflects not only the remarkable adaptation of mammals to feeding mechanisms but also their different biomechanical behaviors (Naples, 1999). Extreme evolutionary variants in adaptation are also evident, such as the loss of the synovial cavity in some whales, the loss of the disc in certain marsupials, and in some edentates like anteaters and sloths, or variations in the orientation of the articular

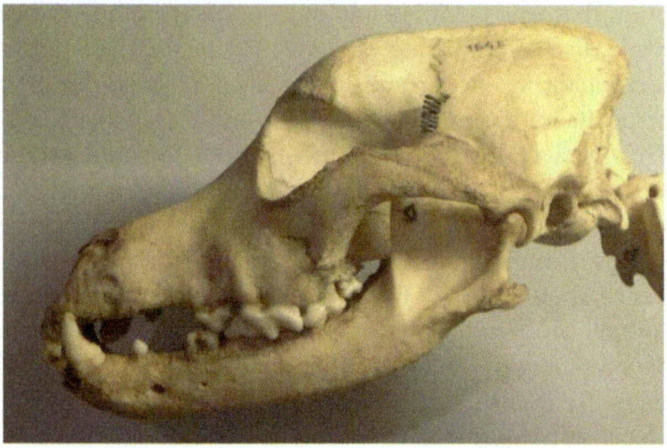

Fig. 2.88 Skull of a dog (*Canis lupus familiaris*). Museum of Zoology, University of Florence. Note how the entire lateral surface of the skull is occupied by the temporal fossa, where a particularly extensive temporal muscle originates and inserts onto a robust coronoid process; the high bite force is also indicated by the pronounced development of the angular process of the mandible, on whose outer surface the superficial head of the masseter muscle inserts: the conformation of this bone process, along with a powerful zygomatic arch (the other insertion area of this muscle), indicates a very strong masseter muscle. A bite supported by such powerful musculature requires a specific conformation of the TMJ, capable of resisting intense dislocating forces

cavity—sagittal in many rodents or transverse in many carnivores, where the condyle is wrapped by the mandibular fossa like a hinge on a door. The TMJ structure typical of carnivores has limited movement primarily to rotation to ensure maximum condylar stability and prevent joint dislocation due to the unpredictability of forces applied when these animals subdue struggling prey (Fig. 2.88).

The TMJ in herbivorous or omnivorous mammals is a complex joint that combines the characteristics of both a hinge joint (with a hinge-like geometry) and a gliding joint (with flat, sliding surfaces). This type of joint contrasts with the TMJ of carnivorous mammals, which functions primarily as a hinge joint, allowing only a rotational hinge-like movement (opening and closing) of the cylindrical condyle (Vallittu et al., 2021).

One way to prevent mandibular dislocation is to have a tightly fitting TMJ that cannot translate anteroposteriorly due to its bony and ligamentous configuration (Hylander, 2006) (Fig. 2.89).

In the badger for instance, the condyle is so tightly wrapped by the mandibular fossa that in a dry skull, the mandible cannot be disengaged except by laterally sliding the two hemimandibles apart, thanks to the unfused symphysis (Fig. 2.90).

In herbivores, there is often an inversion of the usual convexity of the craniomandibular joint surfaces: the condylar surface becomes the concave element, as seen in many ungulates such as sheep and cattle (Angelo et al., 2016) (Fig. 2.91).

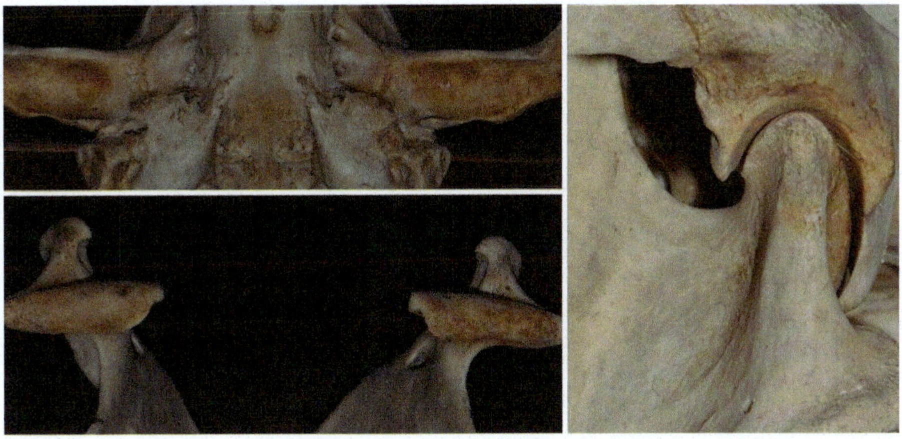

Fig. 2.89 Condyles and TMJ of a lion (*Panthera leo*). Museum of Zoology of the University of Florence

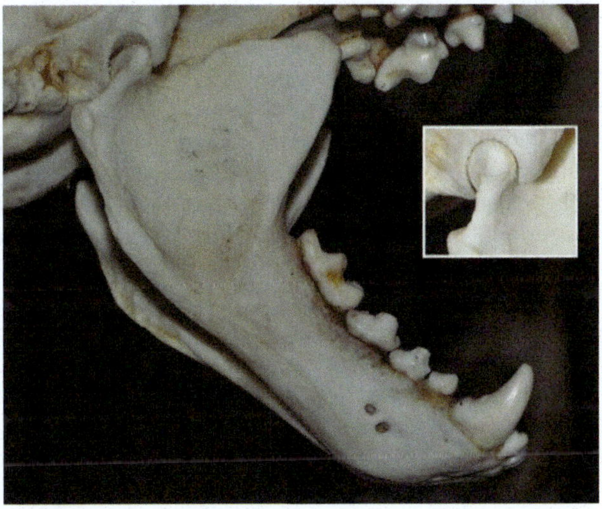

Fig. 2.90 Mandible and temporomandibular joint of a badger (*Meles meles*). Note, in addition to the extreme reciprocal shape between the condyle and mandibular fossa, the highly developed coronoid process. Museum of Zoology of the University of Florence

2.4.1 Mediolateral Translation

Early vertebrates applied maximum bite force simultaneously on both sides of their jaws, while mammals were and are the only animals to develop the ability to chew with maximum force concentrated on one side of the jaw only (although there are exceptions, such as primates that feed on soft foods and may exhibit simultaneous bilateral chewing movements).

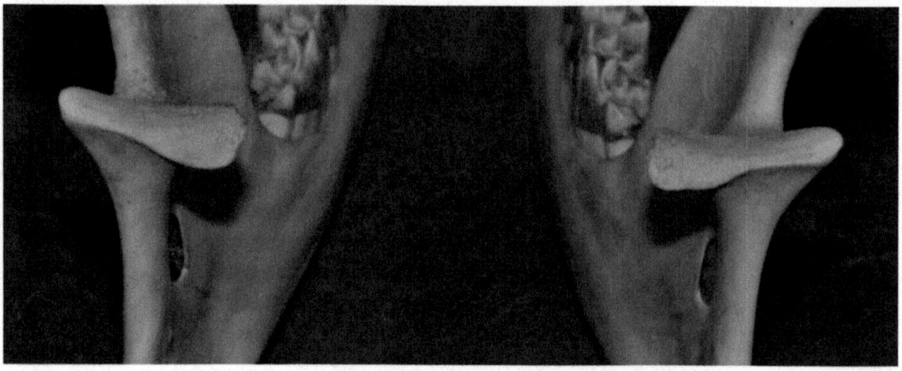

Fig. 2.91 Condyles of a waterbuck (*Kobus ellipsiprymnus*). Museum of Zoology of the University of Florence

Two parameters influence the geometry of chewing cycles: the mechanical characteristics of food (consistency and hardness) and the anatomy of the stomatognathic apparatus. Although the typical cycles of monolateral chewing draw orbits that can have different shapes depending on the type of diet, the drop-like movement (see Fig. 2.97) is not species-specific, as it is found in many mammal species (Hiiemae, 1978; Fish & Mendal, 1982).

Mediolateral condylar movement also occurs in species that have limited or no anteroposterior translation, which is hindered by highly restrictive bony processes. It is the only movement their jaws can make horizontally thanks to the action of the posterior fibers of the temporalis muscle, which is essential for their chewing to take place. Often in carnivores like felids, the lateral pterygoid muscle is absent or very small (Hartstone-Rose et al., 2012; Herring, 2007), and the translatory movement of the condyles is essentially produced by the contraction of the temporalis muscle fibers, which is particularly developed and powerful in this order of mammals.

As previously mentioned, carnivores do not exhibit anteroposterior translation of the condyle because their TMJs are designed to ensure maximum condylar stability. This is to prevent possible dislocation of the mandible caused by the unpredictable intensity of forces in the field when carnivorous animals need to subdue their prey. Their way of preventing mandibular dislocation involves having a well-locked TMJ that cannot slide anteroposteriorly due to its bony and ligamentous configuration. This joint behavior characterized the earliest members of this order (Hylander, 2006).

Unlike humans, many mammals are unable to bring opposing dental arches into contact without mediolateral translation because the upper dental arch is much wider mediolaterally than the lower one (Hylander, 2006). Due to the smaller radius of the lower dental arch compared to the upper one, mediolateral movement of the mandible is necessary to allow chewing on one side at a time. The purpose of mediolateral condylar translation is to improve masticatory function, facilitating transverse movements of the mandible's grinding teeth in relation to their antagonists. Lateral movements of the mandible are necessary to engage the lower teeth

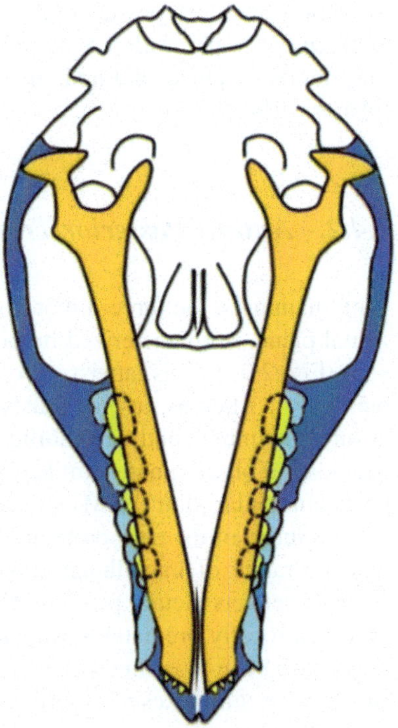

Fig. 2.92 Skull of a mammal, basal view. Note the narrower mandibular arch compared to the maxillary arch, so that the lower teeth, moving laterally during chewing, interface with the upper teeth

with the upper ones for effective food crushing within the mandibular arch that always fits within the maxillary arch (Fig. 2.92).

From this perspective, we could consider the lateral translational movement of the jaw, which in early mammals allowed the lower molars on the working side to engage with the upper ones, as the primitive form of jaw lateral movement: a lateral shift of the condyle that is also present in humans and described as 'immediate Bennett movement,' which can be seen as a residue of the primordial lateral movement in ancient mammals. In humans, lateral movements are continuously influenced by the external pterygoid muscle, which, with its lower part, attaches directly to the condylar fossa, playing a remodeling role on the condyle during the early years of life (De Toffol et al., 2004).

The frequency of alternating the chewing side is usually regular in most mammals, including humans; in some species, the alternation pattern decreases significantly, as seen in many ungulates and rabbits, which chew on the same side for extended periods and sometimes for days. Only a few animals, such as camels and some rodents, chew with perfect alternation on their right and left sides (Weijs, 1994).

Most mammals, in addition to chewing on one side at a time (Popowics & Herring, 2006), have the lower molar cusps offset anteriorly compared to their homologous counterparts in the opposing molars. Acting like a pestle in a mortar, the supporting cusp of the upper molar aligns with the central fossa of the antagonistic molar to naturally create precise occlusion between the various elements of the dental arches, defined in humans as the 'Angle Class I occlusion' (Osborn, 1982).

2.4.2 Antero-Posterior Translation

With mammals, the three-dimensional chewing cycle emerges, which, on the horizontal plane, is characterized by both a lateral-lateral and antero-posterior component (Fig. 2.93). This latter type of movement is found in rodents, herbivores, and omnivores, whereas, as previously mentioned, it does not occur in carnivores to ensure maximum condylar stability during predation. In herbivores, where grinding mastication predominates unlike carnivores or omnivores, the surfaces of the temporomandibular joint (ATM) are flat.

The shape of the temporomandibular joint (TMJ) reflects the masticatory forces and the range of mandibular movements, much in the same way that dentition strongly reflects dietary profiles. The morphology of the TMJ and dental form vary in a reciprocally predictable way, and both are related to diet. Significant relationships have been identified between molars, the skull, the mandible, and the glenoid fossa: these complexes of different shapes reflect diverse feeding strategies (Terhune et al., 2015).

The lateral movement of the mandible is primarily driven by the contraction of the controlateral lateral pterygoid muscle (LPM). Concerning human TMJ mobility,

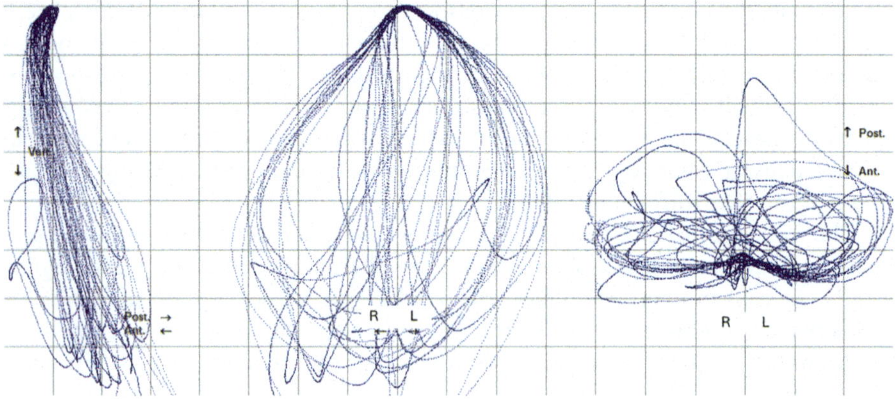

Fig. 2.93 Graphical traces of masticatory axes: the masticatory cycles are drawn in the three planes of space from the movement of the interincisal point. Note how on the frontal and horizontal planes, the 'bundle' formed by the masticatory cycles is symmetric on both sides, indicating a physiological alternating unilateral chewing

it is important to specify the decisive role of the LPM: this muscle has a fan-shaped orientation of its fibers and is capable of generating a finely graded range of force vectors on the condyle and articular disc, thus precisely controlling mandibular movements, as required by mastication and other functions (Murray et al., 2001).

2.5 Formation of the Middle Ear

The formation of the middle ear is a topic that cannot be separated from the development of the temporomandibular joint (TMJ) because these two structures are interconnected from an evolutionary, embryological, anatomical, physiological, and pathological perspective. Approximately 200 million years ago, the first mammals appeared (Jenkins, 1969), small, furry animals capable of maintaining a higher body temperature than contemporary species. They possessed sweat glands, some of which, in females, specialized in milk production to nourish their young. However, the most distinguishing feature setting mammals apart from reptiles and all other vertebrates is the presence of two small bones in the middle ear: the evolution of the auditory ossicles of the middle ear from the jawbones of reptiles, and their change in position and function, are an excellent example of exaptation (Hogervorst et al., 2009). As sometimes occurs in evolutionary pathways, a pre-existing structural element becomes a fundamental feature for a new function through a process of 'reuse,' in which nature employs what is available in descendants from ancestors. However, it can also happen that evolution, while reshaping the available material, can create products that are far from perfect (Pievani, 2019).

Reptiles and pre-mammalian synapsids have a jaw composed of multiple bones but only a single bone (homologous to the stirrup) in the middle ear. In contrast, all mammals have a single bone (the primitive dentary bone) in the jaw and multiple bones in the middle ear, namely the hammer, anvil, and stirrup (Urban et al., 2017). Fossil data, comparative anatomy, and developmental biology demonstrate that the hammer and anvil, the two new bones in the middle ear of mammals, are homologous to the two bones in the reptilian jaw that form the joint between the upper and lower jaws: the quadrate bone (which evolves to become the anvil) and the articular bone (which transforms into the hammer). The incorporation of two bony portions of the reptilian jaw into the middle ear of mammals (Fig. 2.94) was only possible with the evolution of a new method of jaw articulation (Anthwal et al., 2013; Maier & Ruf, 2016).

Before mammals, the middle ear had only one tiny bone called the 'columella,' which, due to its large size, primarily allowed low-frequency vibrations to be transmitted through the jaw-cranium joint. However, this type of joint was (and is in present-day reptiles) decidedly too massive and therefore not very sensitive for transmitting high-frequency air-borne vibrations (Liem, 2001).

The hammer, anvil, and stirrup of the middle ear are among the defining features of mammals and are closely linked to the evolution of a new temporomandibular joint (TMJ). This TMJ evolved alongside the ossicular chain, effectively disconnecting the jaw from the columella, thereby separating the apparatus for feeding

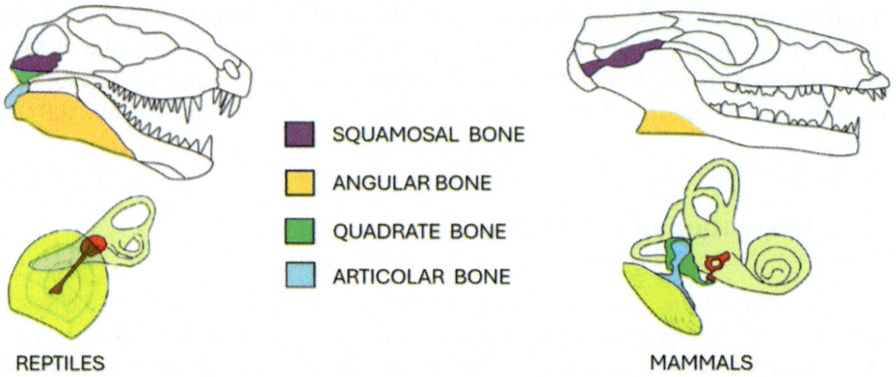

Fig. 2.94 Incorporation of the bones of the reptilian cranio-mandibular joint (quadrate bone and articular bone of reptiles) into the middle ear of mammals

from that for hearing. In mammals, the quadrate and articular bones of reptiles moved upward and backward to form the hammer and anvil. This modification, along with a reduction in the size of these bones, led to a better perception of higher-frequency sound (Benton, 2000).

The connections between the temporomandibular joint (TMJ) and the ossicular chain extend beyond the realm of phylogeny; they also have embryological, anatomical, and functional aspects. It's interesting to note that the innervation of the tensor tympani muscle, which is associated with the tensor veli palatini muscle (Aristeguieta et al., 2010), is provided by a branch of the mandibular nerve. Contraction of this muscle increases tension in the tympanic membrane, limiting its vibrational amplitude and reducing sound transmission to the inner ear. This is an evolutionary adaptation to protect the structures of the inner ear from loud noises, and it also aids in coordinating verbal expression.

In non-mammalian vertebrates, the embryonic origin of the mandible involves several bones that surround the Meckel's cartilage (CdM), which traverses the lateral body of the mandible. In its proximal part, it divides to form the two skeletal elements (the quadrate and articular bones) that will form the characteristic joint of non-mammalian vertebrates (Woronowicz & Schneider, 2019).

The first crucial step in the embryological formation of the middle ear in mammals was the splitting of CdM into two portions, allowing the functional separation of the apparatuses for feeding and hearing, thereby removing (almost entirely) the connection between these two structures (Anthwal et al., 2013). The separation between the masticatory apparatus and the auditory apparatus, along with the evolution of the hammer and anvil and their disconnection from the jaws, occurred independently multiple times, at different times, and in distinct evolutionary lineages. In mammals, during the development of the cartilaginous body from CdM, there is a separation of its posterior portion (Urban et al., 2017), which later divides into the hammer and anvil—initially united as a single cartilaginous condensation (Fig. 2.95).

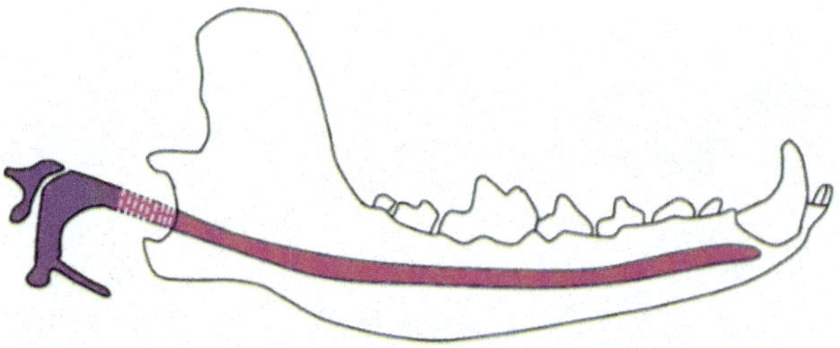

Fig. 2.95 The TMJ is represented in color; on the left in darker color: malleus, incus. The dashed white line indicates the separation of the more proximal portion of the cartilage, which allows the separation of the middle ear from the mandible

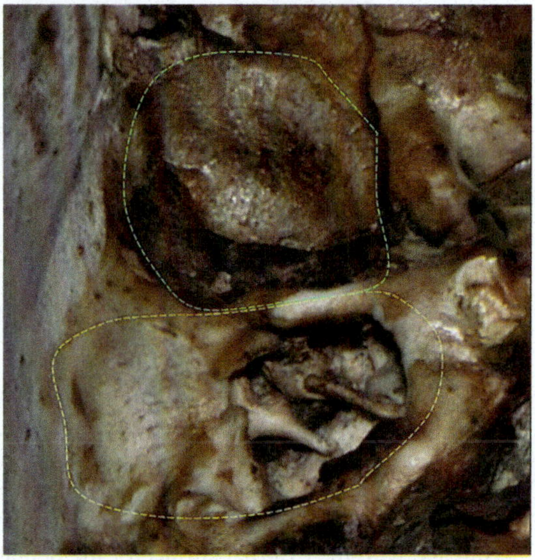

Fig. 2.96 Skull of a newborn, bottom view. Note in the lower part (yellow hatching) the already developed sizes of the incus and malleus, in relation to the still immature area of the mandibular fossa (green hatching). Specimen from the Museum of Anthropology of the University of Florence

In humans and anthropoid apes, these structures are anatomically connected even in adulthood, through ligaments that persist after growth is completed. These connections remain anatomically evident, such as the anterior malleolar ligament formed from a modification of the posterior part of CdM or the disc-malleolar ligament, which is the embryological continuation of the sheath of the lateral pterygoid muscle. Both of these ligaments traverse the Civinini canal (Öğütcen-Toller, 1995; Cheynet et al., 2003).

The hammer and anvil ossify relatively early in development, fixing their sizes early on in contrast to the still-growing skull and mandible (Figs. 2.96 and 2.97). This keeps the ossicles stable as the head grows, facilitating their posterior movement to separate the middle ear from the temporomandibular joint (Anthwal et al., 2013).

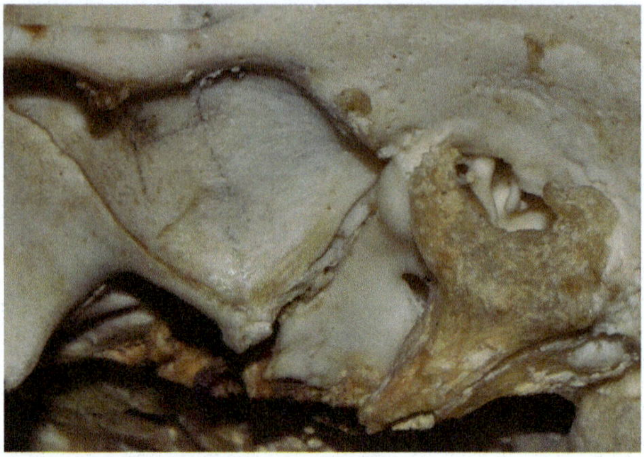

Fig. 2.97 Skull of a 16-month-old child. Lateral-inferior view. Malleus, incus, and stapes are partially visible and still relatively large; note the slight slope of the glenoid roof of the TMJ, the petrotympanic fissure separating the anterior joint portion from the posterior, and on the latter the not yet closed Huschke's foramen. Skull 1379 from the Museum of Anthropology of the University of Florence

2.6 Formation of the Periodontium

Among vertebrates, teeth can connect to the bone through three types of unions (Fig. 2.98):

- ACRODONT, in which the roots are absent, and the teeth are superficially attached to the apex of the jawbones, with their basal portion fused to the surface of the underlying jawbone (seen in snakes, fish, frogs, and lizards) (Fig. 2.98a).
- PLEURODONT, in which the roots are absent, and the teeth are superficially attached to the inner side of the jawbones, connecting to the bone through direct linkage with the external surface of the dental base via elastic and flexible fibrous tissue. This arrangement facilitates the sliding of prey toward the back of the mouth, preventing it from escaping (seen in monitor lizards and iguanas) (Fig. 2.98b).
- TECODONT, in which dental roots appear, inserted into deep bony sockets, enabling them to withstand significant forces (found in mammals, crocodiles, barracuda, and some types of bluefish) (Fig. 2.98c).

In the majority of fish, reptiles, and amphibians, teeth are rigidly connected to the jawbones. This fundamental difference between mammals and other vertebrates is precisely what distinguishes them. In non-mammalian vertebrates, the periodontal ligament is more of an exception than the rule (Diekwisch, 2016) (Fig. 2.99).

Non-mammalian vertebrates that have dental attachment to the bone similar to mammals are exceptions and are characterized by voracious predation (Klein-MacPhee, 2002) on large, strong, and hard-to-catch prey (Bemis et al., 2005). These feeding habits have led these non-mammalian vertebrates to develop dental support,

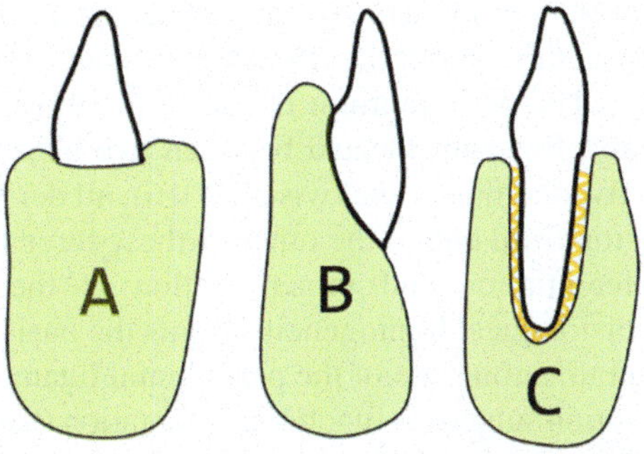

Fig. 2.98 Scheme of the different types of tooth attachment to the bone. (**a**): acrodont; (**b**): pleurodont; (**c**): thecodont

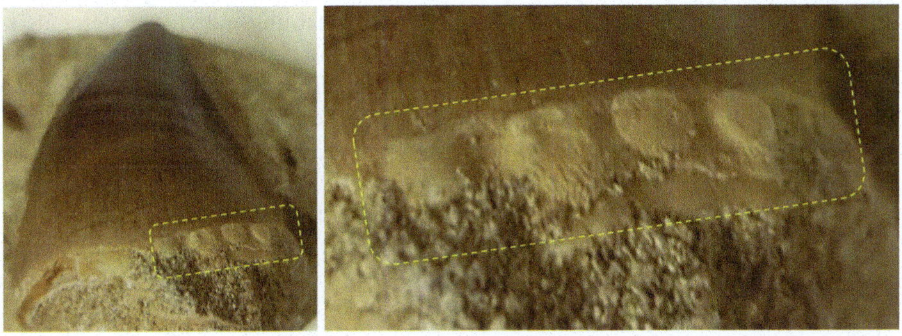

Fig. 2.99 Fossil tooth of Mosasaurus: note the attachment area of the tooth beyond the enamel edge

akin to that of mammals, capable of withstanding strong stresses and dissipating the intense biting forces.

Crocodiles, for example, are one of the few reptile species that possess a root structure necessary for their unique way of feeding. They hunt large prey in the water, from which they need to tear and disarticulate large portions. Additionally, in the aquatic environment, there are no stable anchor points to exert the necessary force to remove pieces of meat from the prey. These reptiles have adopted a technique in which, once they have bitten into the prey, they spin rapidly on themselves to tear away what is firmly held between their teeth. During this phase, the teeth are subjected to extremely high forces, both in the bite and especially in the twisting force due to the unique technique. In these animals, dental elements require a deeper attachment to the bony bases than a simple fusion, which is more effective in dissipating stress. This anchoring is provided by root-like structures similar to those of mammals.

In mammals, the periodontium connects the teeth to the jawbones through fibrous joints that provide them with an elastic suspensory apparatus capable of withstanding the forces generated during normal chewing functions. This type of connection also allows dental elements to move within the bone, modifying their position

(Mjor, 1980). This remarkable adaptability enables each tooth to optimize its position throughout life, ensuring it functions at its best.

The periodontium is one of the most unique structures in the human body—a soft ligament located between two hard mineralized tissues, root cementum, and alveolar bone (Diekwisch, 2016), all derived from mesenchymal tissue. The transition from one to the other of these three layers occurs with varying degrees of differentiation and mineralization of the same original connective tissue. This histological homogeneity forms the basis of its biomechanical reactivity. The mineralization state of the periodontal ligament is a dynamic feature in vertebrate evolution subject to functional adaptation (Luan et al., 2009). The periodontal ligament, root cementum, and alveolar bone should be seen as a structural (Berkovitz, 2004) and functional (Ivanovski et al., 2006) unit that allows teeth to dissipate mechanical stresses generated during chewing, as well as enabling individual elements to undergo adaptive movements within the alveolar bone (Bosshardt et al., 2008). Within certain limits, mammalian teeth can move within the bone as independent units. This continuous repositioning, made possible by periodontal remodeling, occurs during jaw growth and following tooth wear, as a result of occlusal forces (Fig. 2.100).

The composition and organization of the periodontal structures in mammals emerged with the advent of the new masticatory function, which is the need for the supporting tissues of the tooth to dissipate intense biomechanical stresses with appropriate structural responses. Not only does the structure of the dental crown change, but its supporting structures also develop or undergo modification: alveolar bone, periodontal ligament, and dental roots.

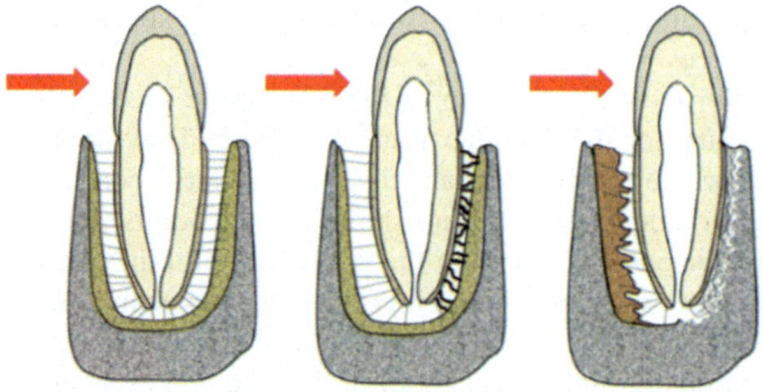

Fig. 2.100 Movement of the tooth within the alveolar bone. On the left, a constant, continuous, or repeated force begins to act on the tooth; in the center, the consequent deformation of the periodontal fibers which stretch on one side while compressing on the other; on the right, there is the apposition of new bone on the traction side and the resorption of the alveolar wall on the compression side. This mechanism allows mammalian teeth to adapt their position within the bone to meet changing functional needs

2.6.1 Alveolar Bone

For information about the alveolar bone, please refer to Sect. 2.3.5.2 regarding maxillofacial bones.

2.6.2 Periodontal Ligament

The periodontal ligament is a non-mineralized tissue that allows dental roots to have a robust yet resilient anchorage (capable of returning to an equilibrium state after structural alteration) within the alveolar bone. It ensures minimal trauma to dental elements during the approximately 3000 masticatory cycles that humans perform each day (Diekwisch, 2016). The periodontal ligament controls the dissipation of mechanical stresses generated during chewing and traumatic impacts (Bosshardt et al., 2008) and prevents root resorption or ankylosis to the surrounding alveolar bone (Bertsen et al., 1997).

In mammals, teeth "move" within the bony component as independent units. This is made possible by the continuous remodeling of the periodontium, which is necessary to adapt occlusion to the bone growth of the jaws, tooth wear (Dhakray et al., 2012), or dental injuries. The insertions of the periodontal fibers on the hard lamina, comprising about 75% in volume, can be thought of as the structure of a hammock supporting the tooth. It essentially transforms compressive forces from occlusion directed towards the base of the alveolus into tensile forces (McCormack, 2016).

From an evolutionary perspective, the action imposed by masticatory dynamics led to histological changes in the hard lamina surrounding dental roots, where periodontal ligament fibers attach. It adapted to the type of stress imposed by the anatomical and functional conformation and is composed of non-Haversian compact bone. The Haversian structure is better suited to withstand compressive stresses rather than tensile forces, making it biomechanically less effective (Mayya et al., 2013).

2.6.3 Cementum

In mammals, cementum is a mineralized tissue that surrounds dental roots and contributes to the tooth's supporting apparatus, in conjunction with periodontal fibers and alveolar bone. In herbivorous mammals, cementum is also found on the dental crown, filling the space between cusps and enamel lamellae (Berkovitz & Shellis, 2018). Cementum is often histologically described as similar to bone tissue; however, it is avascular and does not undergo any dynamic remodeling, only increasing in thickness throughout life. Cementum is classified as acellular cementum without cementocytes and with extrinsic fibers, which thinly covers the root in the cervical area, and cellular cementum with intrinsic fibers and cementocytes, which more frequently covers the root in the apical area.

The surface and volume of the roots are correlated with bite force, both in molars and canines of carnivores and some primates (Berkovitz & Shellis, 2018). In molars, in particular, cementum is generally thicker because its quantity is proportional to the masticatory stress transmitted to the tooth (Yamamoto et al., 2016).

Acellular cementum is better suited to provide tooth support (Bosshardt & Selvig, 1997), while cellular cementum appears to be more involved in adapting and remodeling the root surface during tooth movement, partly compensating for crown wear and potentially filling the resorbed root surface area (Jones, 1981; Schroeder, 1986).

As previously mentioned, in most vertebrates, including all amphibians and most reptiles, teeth are rigidly attached to the bone through fusion of the two tissues (Diekwisch, 2016). Crocodiles are the only reptiles that have teeth attached to the jaws with roots covered by a bone-like tissue (cementum) and inserted into bony sockets (alveoli). The roots are as long as the dental crowns and are held in bony sockets by a ligament that is an intermediate between the non-mineralized ligament in mammals and the mineralized reptilian attachment similar to ankylosis (McIntosh et al., 2002). Additionally, in crocodiles, both the replacement and deciduous teeth develop and erupt within the same socket (tecodontic pleurodonty), whereas in mammals, the replacement tooth occupies its own socket, and the periodontal ligament and deciduous socket disappear during growth remodeling (Berkovitz & Shellis, 2016).

Dental roots, while not exclusive anatomical features distinguishing the class of mammals, can be considered anatomical parts that have developed with a degree of complexity previously unknown in this class. The development of a root and its related anatomical structures, although present in a more rudimentary form in some pre-mammalian vertebrates, has evolved in this class thanks to the evolution of mastication. This anatomical characteristic should be considered an evolutionary response to the intense stresses imposed by chewing: the function of the roots is to transmit the chewing forces from the root cementum to the alveolar bone.

In mammals, all tooth-supporting structures (alveolar bone, tooth root, cementum, and periodontal ligament) originate, forming together with the tooth socket, from the dental follicle (Cho & Garant, 2000). The evolutionary divergence, which occurred in mammals with the development of tooth sockets and the consequent insertion of teeth into the jawbones, is due to a modification of the structure of Hertwig's epithelial root sheath (GdH), an essential change for root development: in crocodiles and mammals, as compared to other vertebrates, the GdH determines the formation of a periodontal structure with all its specialized tissues, rather than a rigid fusion between the tooth and the bone (Luan et al., 2006).

A gradual evolutionary transition in the connection between the tooth and bone occurred at the histological level: from the coronal enamel organ cap in cartilaginous and bony fish, to the continuous cervical epithelial collar in amphibians, which facilitates the fusion of the root's terminal part, finally reaching the transient epithelial network in crocodiles and mammals. This network allows mesenchymal cells from the dental follicle to breach the epithelial barrier and secrete cementum on the dentin surface of the root (Luan et al., 2006).

Hertwig's Epithelial Root Sheath (GdH) was discovered in 1874 by Oscar Hertwig, a German zoologist and biologist. It consists of a double cell layer that, like a sheath, surrounds the dental roots of amphibians. GdH is one of the few tissues discovered for the first time in amphibians and not in mammals. The reason for its discovery is related to its longevity in amphibian dentition. From an evolutionary perspective, it appears that GdH evolved first to provide elasticity in anchoring amphibian teeth, which may have facilitated the formation and maintenance of a non-mineralized periodontal ligament in crocodiles and mammals (Luan et al., 2006). Among various vertebrate species, the morphology, evolution, and development of GdH are closely linked to the type of dental attachment, from the simple, anchylothic-like attachment characteristic of pre-mammalian vertebrates to the complex mammalian tooth sockets.

While in reptiles, GdH remains continuous and limited to the base of the dental crown, in crocodiles, it is notably different: perforated and covering the entire length of the root, allowing for the rigid fusion of the root's terminal part to the bone (McIntosh et al., 2002). Mammalian tooth sockets develop from the dental follicle with the alveolar bone beginning to form early during the bell stage and continually remodeling in response to dental development, eruption, and jaw growth.

During root development, Hertwig's sheath begins to move away from the crown, recruiting mesenchymal cells from the follicle. The development of the root begins when crown formation has reached the future enamel-cementum junction. The epithelial diaphragm is formed with a plane that remains relatively fixed during root development and growth, proliferating coronally and shaping the root's form while creating intermediate cementum. During the pre-eruptive phase, once the crown has formed at the cervical margin of the enamel organ, a ring is structured by a double layer of epithelial cells that forms the GdH. This induces the differentiation and proliferation of mesenchymal cells in the dental papilla adjacent to the diaphragm.

The epithelial cells on the inner part of the sheath promote the recruitment and cytodifferentiation of pulp progenitor cells. Some of these cells migrate to the inner surface of the GdH and become odontoblasts, which are organized in a palisade-like structure and will form the root dentin (Matalová et al., 2015). In mammals, after the deposition of the first layer of dentin, the GdH loses its continuity and becomes similar to a network of epithelial filaments that allow, through its meshwork, the penetration of undifferentiated mesenchymal cells surrounding the tooth germ. These mesenchymal cells from the dental follicle, entering into contact with the surface of the root dentin through the gaps in the sheath, differentiate into cementoblasts (Huang et al., 2009).

The differentiation of the outermost cells will give rise to populations of fibroblasts (which, by secreting collagen and other proteins, will form Sharpey's fibers, later incorporated into developing cementum) as well as populations of osteoblasts contributing to the formation of alveolar bone, in a coordinated development of tooth-supporting tissues (Fig. 2.101).

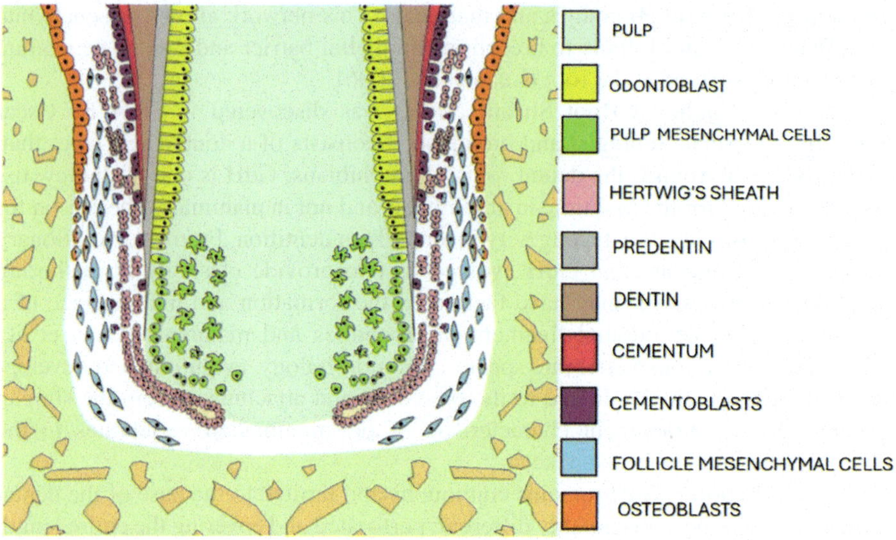

Fig. 2.101 Root development: Hertwig's sheath. Drawing freely adapted from "Stem Cell Biology and Tissue Engineering in Dental Sciences," Vishwakarma et al. (2014)

The GdH (Gingival Epithelial Downgrowth) is involved in recruiting cells that will give rise to the periodontal ligament, in the construction of the alveolus, and in establishing the shape of the root (Matalová et al., 2015). All of this occurs in a contextual development with the alveolar bone, which undergoes active remodeling as the tooth element grows. This mutual balance is maintained throughout the functional life of the tooth (Vishwakarma et al., 2014).

In multi-rooted teeth, the epithelial diaphragm also proliferates horizontally, producing tongue-like extensions (2 extensions in bi-rooted teeth and 3 in tri-rooted teeth). The free ends of these epithelial extensions grow toward each other and merge, dividing the root opening into two or three partitions. There is an intimate association between the penetration of the epithelial barrier by the connective tissue cells of the dental follicle and the subsequent formation of a periodontal ligament. This ligament, replacing the GdH, occupies the surface of the root (Fig. 2.102).

The alveolar bone, periodontal ligament, cementum, and root dentin all form simultaneously in the same embryogenic process (Fig. 2.103). In mammals, the entire periodontium matures and functions as a single anatomo-functional unit, with most of the periodontal tissues developing in conjunction with root formation and originating from the dental follicle, derived from the neural crest (Cho & Garant, 2000).

Just as the positioning of teeth within the oral cavity is influenced by factors like cheeks, tongue, and occlusion, the eruption of teeth from the bone is also facilitated by a remodeling process driven by the functional forces acting on the jawbones. The tooth's eruption naturally leads to the remodeling of the bony crypt that accommodates the tooth. Eventually, this remodeled crypt merges with the cortical bone of

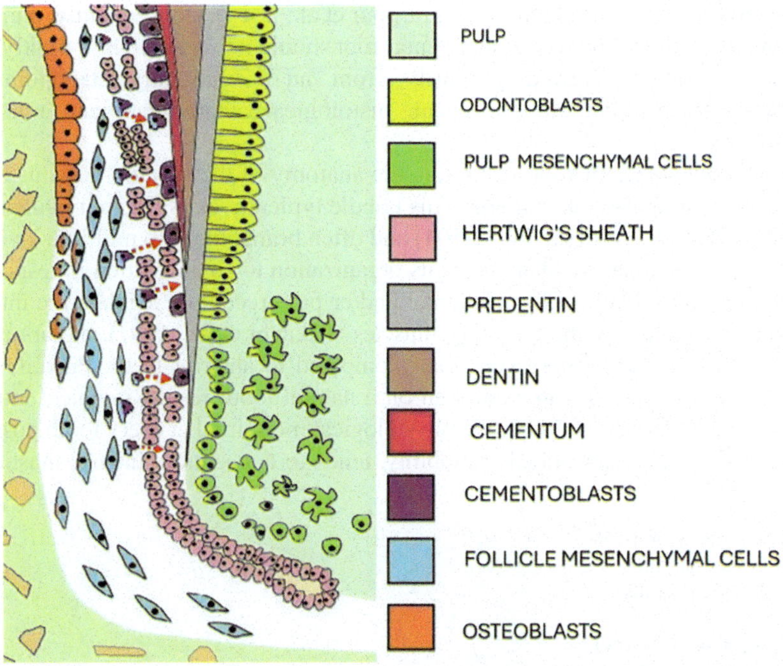

Fig. 2.102 Loss of continuity of Hertwig's sheath allowing the passage and differentiation of mesenchymal cells from the follicle into the different cell types of the tissues that will form the periodontium. Drawing freely adapted from "Stem Cell Biology and Tissue Engineering in Dental Sciences," Vishwakarma et al. (2014)

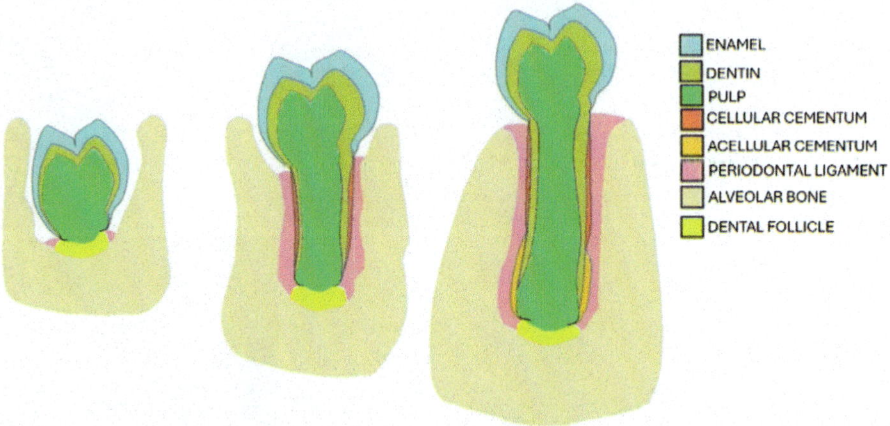

Fig. 2.103 Simultaneous development of the root tissues of the periodontal ligament and the alveolar bone. Drawing freely adapted from "Stem Cell Biology and Tissue Engineering in Dental Sciences," Vishwakarma et al. (2014)

the jaw to form the 'hard lamina' (Sarrafpour et al., 2013). Consequently, the alveolar bone, periodontal ligament, and dental root should be regarded as a tightly interconnected system, viewed primarily from an evolutionary standpoint and subsequently from an embryological, histological, anatomical, and functional perspective.

In the final stages of root formation, the anatomy of the vasculo-nervous bundle in the apical area also takes shape. This bundle typically exits the root while avoiding the geometric apex (see Fig. 2.104) and often branching into multiple divisions.

This specific anatomy likely owes its organization to an adaptation in response to masticatory loads. Although periodontal and/or pulp receptors regulate the intensity of forces generated by the elevating muscles (Kikuchi et al., 1997), molars consistently endure higher compressive loads compared to anterior teeth (Pileicikiene & Surna, 2004). Given that an axial load on a natural tooth results in apical intrusion (Kim et al., 2005; Gross, 2008), the physiological mobility that every tooth possesses in its socket, especially intrusive mobility, tends to concentrate intense masticatory

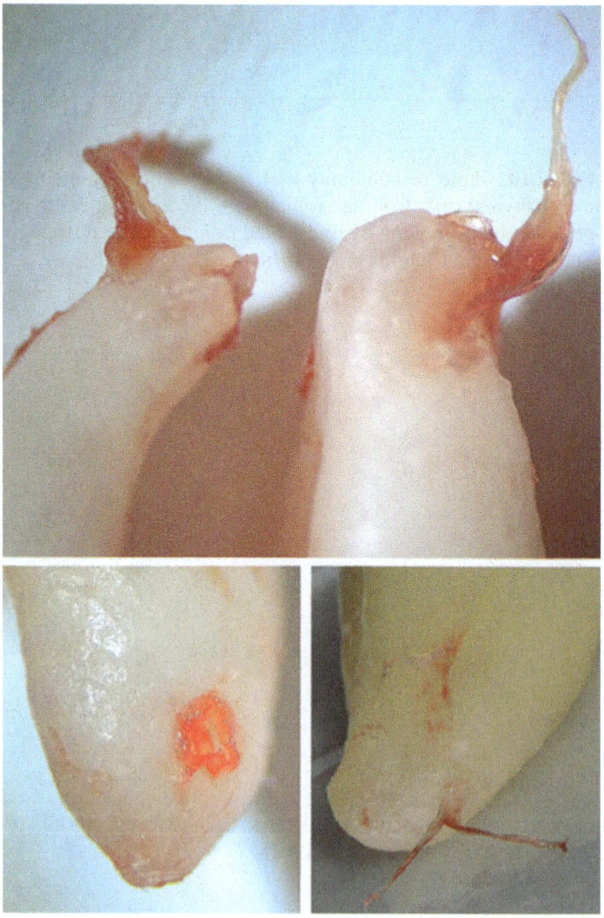

Fig. 2.104 Entry of the main vascular-nervous bundle located laterally and lower than the geometric apex of the root

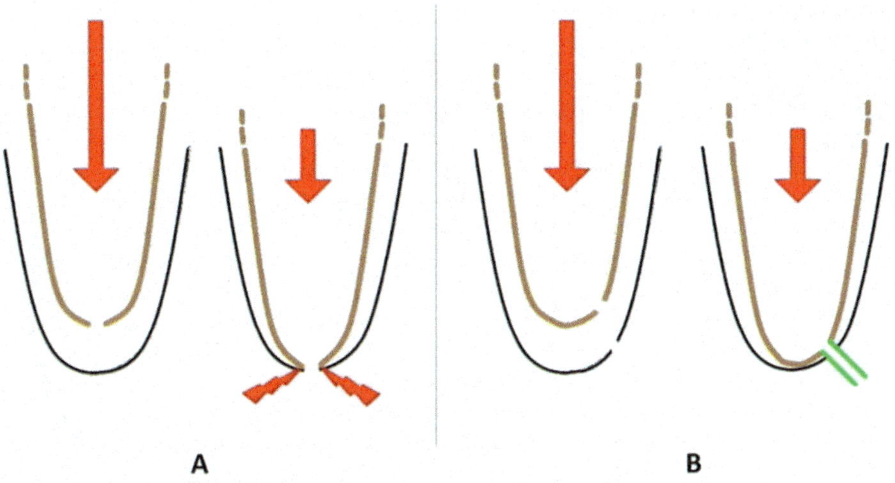

Fig. 2.105 Intrusion of the geometric apex of the root into the alveolar bone: (**a**), compression of the vascular-nervous bundle located on the geometric apex; (**b**), vascular-nervous bundle located laterally to the geometric apex, reducing the possibility of harmful compressions

forces at the geometric apex of the root tip. If the main foramen were to be located at the geometric apex, there would be a higher likelihood of damage to the vasculo-nervous bundle due to a phenomenon of ischemic compression associated with tooth intrusion into its socket. The more frequent positioning of the main foramen laterally to the apex, in a biomechanically less stressed area, tends to ensure both the vascularization and sensitivity of the dental pulp are more secure (see Fig. 2.105).

The subdivision of the vasculo-nervous bundle into multiple branches with different outlets likely has its origin in risk partitioning. A compressive biomechanical event that is non-axial and concentrates its load at a precise point on the apical curvature, even if it causes limited damage to the involved vasculo-nervous bundle, is less likely to compromise the overall pulp vitality, thanks to the presence of other vasculo-nervous bundles (Fig. 2.106). Suggestive is the higher frequency in the delta apex of molars, which are subject to more intense loads compared to anterior teeth (Gao et al., 2016).

The loss of periodontal tissues, either due to tooth avulsion or periodontal disease, stimulates and exacerbates the course of pathological processes. This disorganizes the proprioceptive system and feedback on mandibular mobility due to a decrease in the number of receptors in the periodontal ligaments. Consequently, the management of chewing forces becomes less precise, leading to less controlled and more traumatic force distribution.

The restoration of functional integrity in the dental arch is one of the most significant challenges in clinical dentistry. This is because the functional balance between dental arches ensures harmonious functional stimulation of the entire masticatory apparatus, contributing to its proper development and maintenance (Pileicikiene & Surna, 2004). Each tooth is a mechanically inert element that

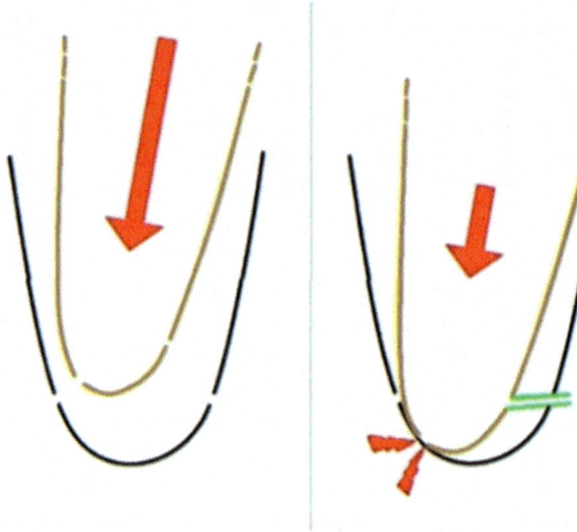

Fig. 2.106 Reduction in the likelihood of compromising pulp vitality due to the presence of multiple vascular-nervous bundles

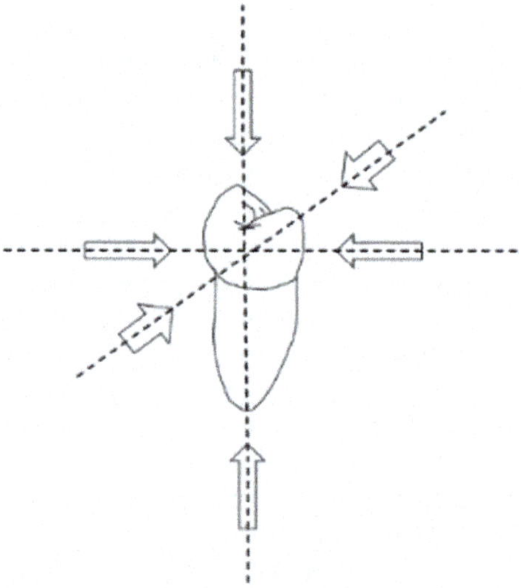

Fig. 2.107 Each tooth is subject to movement within the bone when the forces converging on it have a resultant

positions and stabilizes itself within the bone over time when the resultant forces applied to it are balanced (see Fig. 2.107).

If the sum of the mechanical stresses on the tooth cannot be dissipated, the remarkable adaptability of the periodontium allows for compensatory adjustments of the dental element. These adjustments help maintain masticatory and sensory effectiveness for as long as possible.

The functional wear of occlusal surfaces, as well as interproximal surfaces, is compensated for by vertical eruption and transverse displacement, movements that represent an adaptive response of the periodontium. This ensures the correct functional reciprocity of the surfaces, which is essential for maintaining the highest possible masticatory efficiency. The concept of lifelong active tooth eruption was first proposed by Bernhard Gottlieb in 1927, and several metric studies conducted in the last four decades have confirmed its presence in populations with heavily worn dentitions (Kaifu et al., 2003). In light of these considerations, we should begin to view non-canonical tooth positions or enamel wear as the best possible strategy that each individual tooth employs to adapt to changes in its biomechanical environment.

2.7 Modifications of Dental Tissues

Even, and perhaps especially, hard dental tissues have had to adapt their structure to meet the demands of a new biomechanical environment imposed by their new function. We will now discuss enamel and dentin, as cementum, a tissue related to the periodontium, has already been covered in the section concerning the periodontium.

2.7.1 *Enamel*

One of the almost constant premises when reading about dental enamel is that it is the hardest and most mineralized substance in our body. Enamel is, indeed, the most highly mineralized tissue derivative (approximately 97% by weight). Its composition and structure define its ability to withstand stress and resist fractures (Popowics & Herring, 2006). However, among its remarkable characteristics is its ability to progressively adapt to its primary function, mastication, through a targeted process: wear. Its biochemical, embryological, structural, and ultrastructural aspects result from a gradual evolutionary development that has acted over millions of years, starting from deep time. Like every tissue and organ in our body, dental enamel is the product of how the external environment (forces, molecules, temperatures) has influenced our genes throughout evolution (Lieberman, 2014). In reptiles, enamel typically tends to be a thin, uniform layer (see Fig. 2.108), with thicknesses usually not exceeding a few tens of microns (Sahni, 1987; LeBlanc et al., 2020), and its structure appears devoid of prisms.

The evolution towards a prismatic structure occurred gradually from reptiles similar to mammals (Stem & Crompton, 1995). The enamel of some of these reptiles, known as pre-prismatic enamel, exhibited discontinuities in its crystalline structure, representing a transitional stage between aprismatic and prismatic enamel (Legasa et al., 1994).

In mammals, enamel undergoes a microstructural change, becoming prismatic, thereby increasing its resistance to the thousands of daily masticatory cycles. Prismatic enamel is considered a distinctive feature of mammalian teeth, closely associated with their masticatory function (Irish & Scott, 2015).

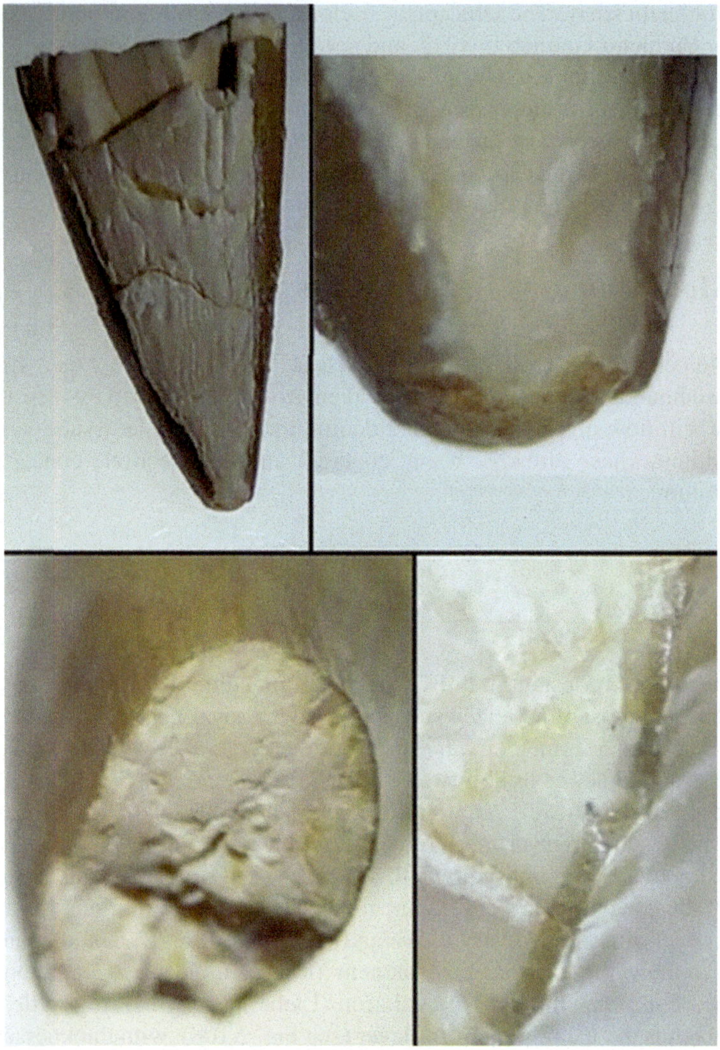

Fig. 2.108 Fossil tooth of Mosasaurus. Note the thin layer of enamel, fossilized

Certain adaptive solutions appear repeatedly in the evolutionary history of phylogenetically distant animals, but only in some they become distinctive characteristics. In mammals, enamel exhibits significant variation in form, size, structure, and properties. Given its primarily mineral nature (it lacks cells), once formed, enamel cannot be remodeled except through ion exchange within the mouth. This represents a unique example of the conservation and permanent record of ontogenesis (Maas & Dumont, 1999).

The evolutionary process of mammalian enamel is not solely due to an anatomical rearrangement of tissue configuration but, most importantly, to a modification of the histological structure of the ameloblast. This 'small' change in the cells responsible for enamel production, which subsequently die after completing their task, has led to profound structural changes and a multitude of biomechanical variations in this substance, all essential for the efficiency of mammalian masticatory function. Mammalian enamel adjusts quantitatively and qualitatively to withstand the stresses induced by food triturition, becoming both the hardest substance in the body and one capable of progressively shaping itself through the functional wear of its surfaces.

The modifications to this substance in mammals extend beyond its microscopic and submicroscopic structure and also affect its macroscopic appearance, with varied thicknesses depending on the dietary habits of individual species. For instance, humans possess thicker enamel than pongids (Molnar & Gantt, 1977). Among primates, compared to modern humans, significant variations in the thickness of masticatory teeth exist. In big African apes, our closest living relatives, enamel is considerably thinner, while in Asian orangutans, it is only slightly thinner. Even in modern humans, enamel thickness can vary significantly between adjacent teeth or even within the crown of a single tooth.

Increased enamel thickness enhances resistance to stress and can be considered an adaptation to diets that include hard components, demanding high fracture resistance from teeth triturition, A greater thickness prolongs the life of enamel, particularly in diets rich in abrasive foods, which require prolonged wear resistance (Berkowitz & Shellis, 2018).

The hard objects most commonly consumed by both large and small mammals are likely seeds or fruits protected by tough shells. These shells are designed to withstand fracture and often have a dual-layer structure, with a hard outer layer enclosing a soft kernel. From a biomechanical perspective, there are structural similarities between the constitution of human teeth and the shells of seeds, which exhibit adaptations similar to those of enamel. Some shells have radial fibers, others have fibers that intersect throughout their thickness, and some have crossed fibers limited to the inner part of the shell with radial fibers on the outside (Lucas et al., 1995, 2008; Jennings & Macmillan, 1986; Williamson & Lucas, 1995). In the face of similar biomechanical demands, nature tends to propose similar solutions.

The nature of enamel and its evolution will be further explored in Chap. 3, dedicated to wear.

2.7.2 Dentin

As we have seen, enamel, once a hard and homogenous "cuticle" covering a replaceable and essentially inert tooth, evolved into a substance with varied thicknesses in response to diet. Similarly, other dental tissues, under the influence of the biomechanical environment, adapted to their new functional roles. In non-mammalian vertebrates, several types of dentin exist, but by far the most widely distributed type is *orthodentin*. It forms a dense layer that surrounds the pulp, permeated by a vast number of minute tubules containing cytoplasmic extensions of odontoblastic processes. In other varieties of dentin, the dentinal layer may exhibit folds (*plicidentin*), be subdivided (osteodentin), or be rich in capillaries (*vasodentin*) (Berkovitz & Shellis, 2016).

Within the class of mammals, there can be variations in the microstructure of dentin, such as differences in the degree of peritubular dentin mineralization. These variances can have a significant impact on the strength and elasticity of dental elements. In mammals, dentin is a vital and highly innervated tissue, sensitive, and capable of responding to both mechanical and chemical stimuli from the environment. It is so interconnected with the pulp that it is considered an integral part of the pulpo-dentinal complex (Berkovitz & Shellis, 2016). The dental formation process begins with the deposition of primary dentin. In mammals, the tooth's shape is determined by the folding and growth of the interface between enamel epithelium and odontoblasts forming dentin, which acts like a diaphragm traversing the interior of the dental follicle and serves as a precursor to the external enamel surface morphology (see Fig. 2.109) (Guy et al., 2015; Hakkinen et al., 2019).

During the eruption phase, once the teeth come into occlusion or after the closure of the root apices, dentin formation slows down while continuing to form regularly as secondary dentin. Over the course of life, its continuous deposition tends to fill the pulp chamber. Tertiary dentin is a reactive tissue in response to injuries, cavities, or wear. With enamel wear, this type of dentin forms to protect the pulp, blocking the tubules running from the exposed surface to the pulp chamber, preventing the intratubular penetration of bacteria and substances that could irreversibly damage it (Mrör, 2001; Ungar, 2010).

Dentin, originally serving as a supportive structure for enamel, evolved into a biologically active tissue capable of reacting to physical, chemical, and biological injuries (Tjäderhane, 2019). With the ability to deposit new layers of internal tertiary dentin (see Fig. 2.110), a process was initiated to limit damage associated with enamel wear, thereby preventing excessive and harmful pulp chamber exposure.

As wear progresses due to chewing, dentin becomes more exposed and actively contributes to the food grinding process. The ability to form tertiary dentin serves as protection for the pulp organ, responding to external insults that may be physical, chemical, or bacterial. The reaction tends to be proportional to the duration of the stimulus. The formation of tertiary dentin plays a role similar to enamel, helping

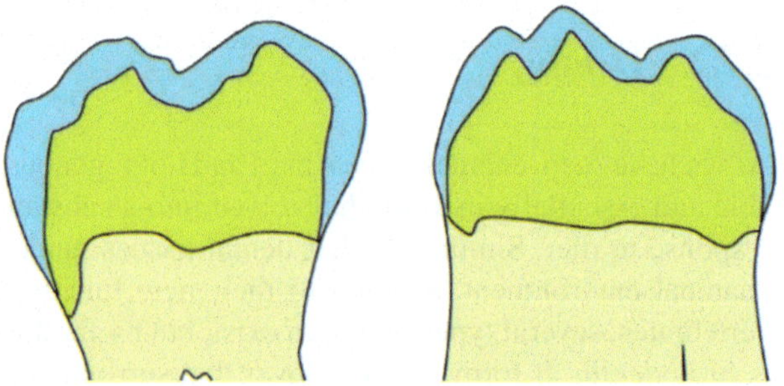

Fig. 2.109 The dentin mass supports the enamel to make it as uniform as possible

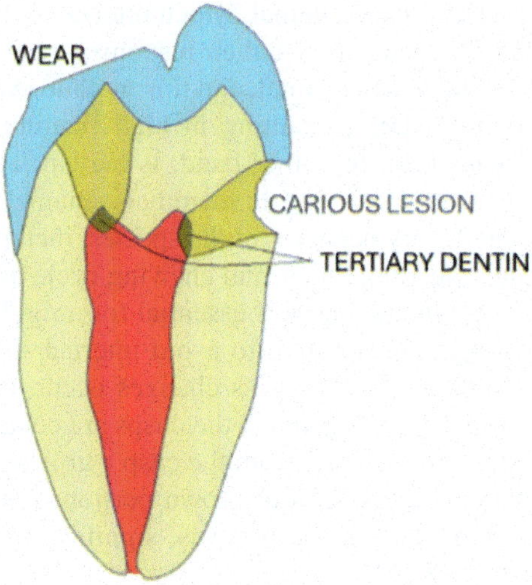

Fig. 2.110 Apposition of tertiary dentin resulting from caries and wear

maintain cutting function and extending a tooth's lifespan by preventing loss of functionality (Towle, 2019). Rapid or severe inflammations lead to the release of cytokines that can stimulate undifferentiated pulp cells to differentiate into new odontoblasts, which deposit tertiary dentin to repair the damaged site (Berkovitz & Shellis, 2018).

The formation of tertiary dentin plays a particularly crucial role in the dentition of grazing mammals, which possess dental pulp capable of compensating for significant wear through active and continuous production of new reactive dentin. Some species appear predisposed to produce tertiary dentin earlier, more rapidly, or more frequently than others. The presence of tertiary dentin in three hominid species (*Australopithecus africanus*, *Homo naledi*, and *Paranthropus robustus*) ranges from 12% to 16.13%, while in two species of large apes, the gorilla (*Gorilla gorilla*) exhibits 90% and the chimpanzee (*Pan troglodytes*) 47.21%.

In humans, the maximum capacity to produce new odontoblasts and thus maintain pulp health is observed up to around 25 years. This data can be viewed as a "histological indicator of obsolescence" as it highlights a temporal limit to the pulp's ability to most effectively cope with environmental pressures. The reactivity of undifferentiated pulp cells and dental wear appear to be factors contributing to the programmed obsolescence of the tooth and an animal's life in its natural state. As histological reactivity diminishes over time, dental pulp becomes less adept at fulfilling its functions, effectively weakening the tooth, and consequently, the entire organism.

It's important to consider that a quarter of a century is not very far from the average lifespan of the *Homo sapiens* species until relatively recent times. Until a "short" time ago, various factors, including food scarcity, diseases and the absence of treatments, intense competition for survival, greatly reduced what we now consider a reasonable life expectancy.

Dentin and enamel, which interface with a reciprocal relationship, exhibit different characteristics in their response to mechanical stress. Dentin has a higher elastic modulus than enamel, and this modulus varies only slightly depending on the direction of stress, resulting in high fracture resistance (Berkovitz & Shellis, 2018). Enamel, on the other hand, is harder, less isotropic, and more brittle than dentin (Chun et al., 2014). In mammals, enamel thickness depends on body size, lifespan, and dietary behavior of the species, including the physical properties of their foods, daily feeding times, and chewing cycle duration (Guy et al., 2015).

While thickness is essential for a rigid material's fracture resistance, uniformity is equally important to avoid internal stress concentrations that would localize at points where thickness changes occur, potentially initiating fracture lines. At the onset of amelogenesis, the cusps are centers for the early maturation of enamel epithelium cells, and from the cusp's tip, this process extends cervically. This differentiated start in various crown portions results in the dental tip's growth continuing longer than in the grooves, significantly increasing cusp enamel thickness (Guy et al., 2015).

Enamel thickness in mammals must meet two opposing needs: it should be as uniform as possible while having differentiated and thicker sections in areas subjected to more wear, such as occlusal surfaces in general and cusps and edges in particular. Cusps are the occlusal areas most involved in chewing and the regions most susceptible to wear. Their greater enamel quantity compensates for increased abrasion while also enduring stresses where food contacts enamel (Lee et al., 2010), thereby maintaining tooth effectiveness for as long as possible.

Even though hominids (chimpanzees, bonobos, and humans) have evolved enamel thickness modifications on their molars multiple times, we must consider the close relationship that this rigid thickness establishes with the underlying, more elastic dentin. There is a structural need to avoid excessive enamel thickness variations to reduce internal stresses that would result from it (Schwartz et al., 2020)

In a way, nature has anticipated the techniques used by dental technicians in the construction of metal-ceramic prostheses by several million years. The structure of the more elastic and resistant metal substrate must be designed to support a generally uniform ceramic layer to prevent fractures triggered by internal stresses that concentrate in areas with varying thickness (Fig. 2.111).

2.7.3 Pulp

Knowledge about the dental pulp of non-mammalian vertebrates (aside from dentin formation) is fragmentary, unlike what we know about the pulp of mammals, whose functions have been the subject of numerous studies. In non-mammalian vertebrates, the function of dental pulp remains largely unclear, as its nutritive function is very limited or absent due to repeated tooth replacement.

In mammals, dental pulp, in addition to dentin formation, took on the role of regulating tooth homeostasis, becoming an active element in protecting its integrity,

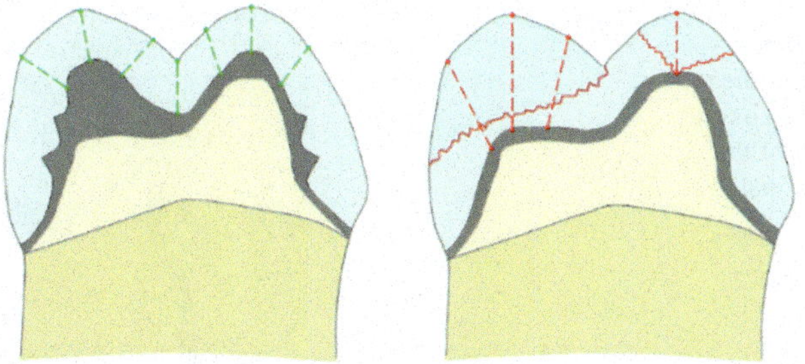

Fig. 2.111 Section of a metal-ceramic crown: on the left, modeling of the metal (gray) able to support a uniform layer of ceramic material (light blue); on the right, the modeling of the metallic framework, which does not compensate for irregularities in the preparation of the dental stump, results in a differentiated thickness of the overlying rigid layer, leading to the initiation of fractures. Drawing by Antonio Meoni

with the ability to react to potentially harmful events. Compared to non-mammalian vertebrates, the pulp-dentin complex became essential for the maintenance and nourishment of odontoblasts, capable of organizing reparative responses to wear and infections. This role is crucial because mammals have only two sets of teeth, and the dental element must remain efficient throughout the animal's life (Berkovitz & Shellis, 2016).

Anders Adolph Retzius described the course of nerves inside the dental pulp in various classes of vertebrates. In fish and amphibians, pulpal nerve fibers branch but do not extend to the apex of the pulp. In reptiles, thin nerve fibers from the main trunks run toward the sides and pass beyond the layer of odontoblasts, terminating convolutedly on the inner surface of the pulp chamber. However, in both cases, the pattern differs from that of mammals (Tuisku & Hildebrand, 1995; Stewart, 1927).

In mammals, the dental pulp is composed of odontoblasts, undifferentiated mesenchymal cells, fibroblasts, macrophages, leukocytes, lymphocytes, blood vessels, nerves, collagen fibers, and the extracellular matrix. The nerve fibers within it run almost undivided to the top of the pulp chamber, where, in the area beneath the occlusal surface, they branch into numerous ramifications, forming the Raschkow Plexus. From the Raschkow Plexus, thin nerve endings extend into the dentinal tubules, running inside them. The concentration of innervated tubules varies depending on the portion of the dental element considered: pulp horns 40%, lateral parts of the pulp chamber 4.8%, cervical regions 1%, while in radicular dentin, there are only sporadic nerve endings (Fig. 2.112) (Byers, 1984). This significant decrease in nerve endings in the corono-apical direction makes it challenging to explain the sensitivity that is often experienced with the consumption of cold beverages.

The greater innervation of dentinal tubules in the pulp horns and roof of the pulp chamber appears instead to be motivated by the greater deformation that occurs in these areas under load (Costa et al., 2014). Teeth are not mechanically inert structures but can tilt and twist more than one might think, and they possess an internal sensory system capable of detecting even the slightest compression or torsion (Dawson, 2006). Therefore, the tooth can be considered as a mechanoreceptor

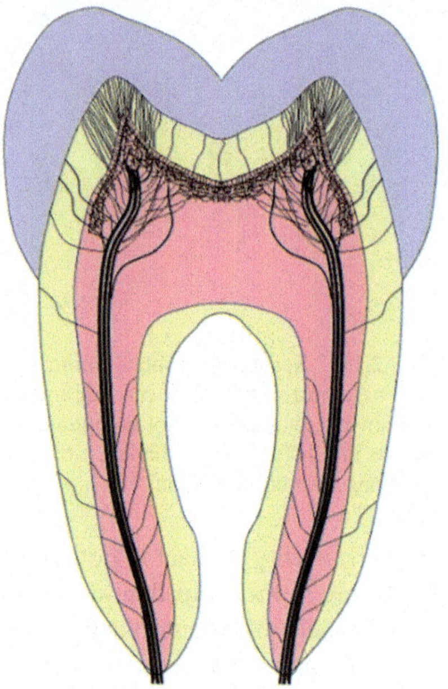

Fig. 2.112 Diagram of the course of nerve fibers within the pulp; at the top, the Raschkow plexus and the continuation of the fibers in the dentinal tubules

Fig. 2.113 Star-nosed mole (Condylura cristata): on the right, an enlargement of the characteristic nose. Drawing by Enrico Ricci

station whose internal sensitivity, linked to a neurosensory network that responds to the slightest distortions of odontoblastic tubules and that can provide rapid feedback to the musculature, is amplified by a multi-layered structure with different elasticity modules and different deformation under load.

In humans, along the interface between pulp and dentin, there are approximately 30,000 dentinal tubules per square millimeter, and in the area of the pulp horns, half of the tubules are innervated. This density of nerve endings rivals only the one found in the highly sensitive touch organ represented by the star-nosed mole's snout (Fig. 2.113), whose refined sensitivity is comparable to the density of receptor units in the human hand.

The innervation of the pulp and pulp horns varies greatly depending on species, age, and types of teeth but is structured in line with the idea that dental elements can be considered proprioceptive organs capable of detecting vibrations (Byers & Cornel, 2018).

2.8 Modification of Cervical Vertebrae

Although it is not part of the stomatognathic system, the cervical region is included in this discussion because, as we will see, these two anatomical regions are closely related. The neck of mammals adopts a wide variety of postures during daily life and regulates numerous head movements. Despite the considerable morphological and functional variability among species, regardless of neck length and body size, the number of cervical vertebrae is almost always limited to seven (Galis et al., 2006). In fact, in this class of vertebrates (with a few exceptions, such as manatees and sloths), the cervical portion of the spine consists of seven vertebrae: the first two (C1 atlas and C2 axis) further specialize, morphologically differing from the rest of the cervical vertebrae, allowing exceptional freedom of movement for the head (Liem, 2001).

The extremely low interspecific variation in the number of cervical vertebrae in mammals has perplexed biologists for over a century and a half: in birds, reptiles, and amphibians, this number varies considerably, and even in mammals, the number of vertebrae in other regions of the spine is variable. Therefore, due to this constancy in the number of vertebrae, the cervical region of the spine seems to be an evolutionary constraint resisting potential variations; the number of cervical vertebrae is determined during the early stage of organogenesis and strongly conserved in mammals because changes in the number of cervical vertebrae are associated with alterations that could severely impact an individual's survival (Galis et al., 2006). While maintaining this body plan, the increasing functional specialization of neck modules, however, did not evolve all at once but began much earlier in the upper part compared to the lower part.

The concept of modularity corresponds to the recognition that the phenotype can be broken down into modules or parts that are closely interconnected by numerous strong interactions. These relationships manifest at various levels of integration, including genetic, anatomical, and functional interactions (Esteve-Altava, 2017).

While vertebral proportions are relatively uniform in most mammals, the general pattern is altered by increasing the proportion of one of the three modules (C1/C2, C3/C5, C6/C7) under specific loading conditions on the cervical spine (such as having large heads or an aquatic lifestyle). For instance, in antelopes and camelids, long necks are formed by increasing the proportions of vertebrae from C3 to C6. Conversely, relatively short necks have the portion from C2 to C7 shortened, but not C1, which thus increases its proportion and maintains head support at the craniocervical junction. Therefore, it is not surprising to find that the neck exhibits modularity divided into a kinematic chain consisting of three sectors, with the upper and

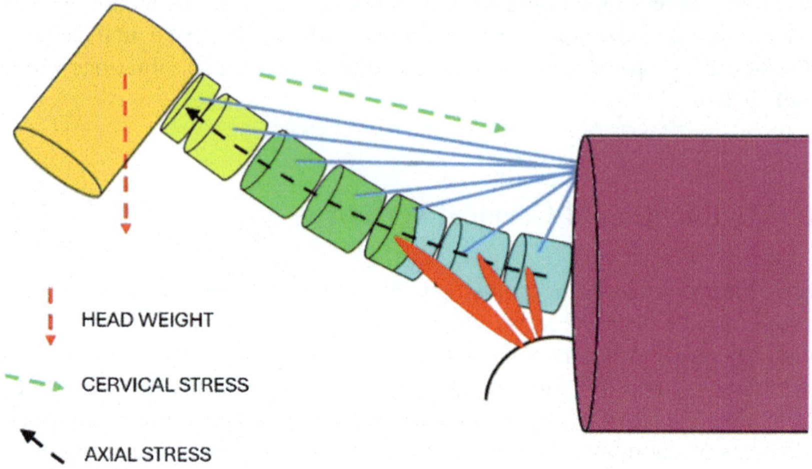

Fig. 2.114 The weight of the head is supported by the neck muscles and ligaments. This stress stabilizes the terminal part of the vertebral column, which functionally connects with the forelimb. Drawing freely adapted from: "The physiological linkage between molar inclination and dental macrowear pattern," Arnold, 2020

lower elements tending to change in relation to their transitional function (cranio-cervical and cervico-thoracic transitions). The cervical vertebrae in the middle section do not show functional specializations precisely because they do not have such a close functional connection to other body parts (Arnold, 2020) (Fig. 2.114).

In adult mammals, the atlanto-axial complex typically consists of only two bones, whereas in living reptiles, there can be at least eight separate bones. The complex formed by these two vertebrae is an innovation that allows mammals precise movements of the head (Van der Linden et al., 2019). The atlas and axis are functionally and anatomically very different from the rest of the cervical spine, and their unique ability to support the fine adjustments required for head movements is well-documented. Although both the joints between the skull and C1 and between C1 and C2 support horizontal rotational and sagittal flexion-extension movements, the rotation movement is more pronounced between C1 and C2, while cranial-atlas articulation is more involved in flexion-extension (Fig. 2.115).

The odontoid process (or odontoid peg, or simply odontoid) is a bony projection on the upper surface of the second cervical vertebra that articulates with the anterior arch of the atlas. The odontoid process evolved when, with mammals, the C1-C2 joint specialized for head rotation (Jenkins Jr., 1969). The odontoid is present in all mammals except cetaceans (Jenkins, 1969). These mammals have followed an apparently regressive evolutionary path, characterized by the abandonment of heterodonty for homodonty, the reacquisition of a conical shape for the individual dental element, and a return to aprismatic enamel with reduced and uniform thickness. This apparent "regression" must instead be considered a series of adaptations that responded to the drastic environmental changes induced by evolution in these mammals.

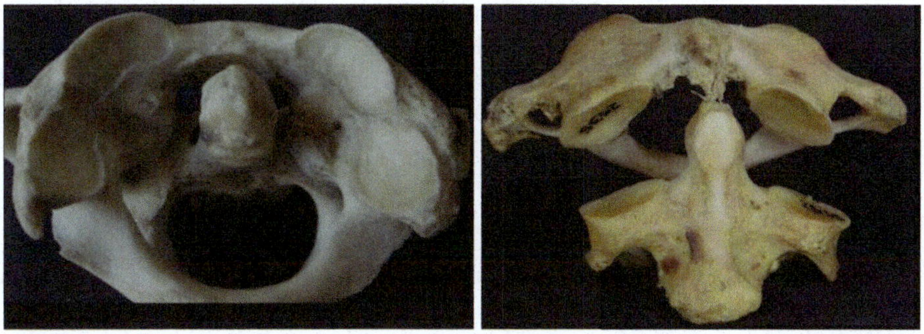

Fig. 2.115 Articulation between the odontoid process and the anterior arch of the atlas. **Left**: specimen 2538, top view; **right**: specimen 2535, front view with C1 slightly lifted from its relationship with C2 to highlight the articular surface. Museum of Anthropology, University of Florence

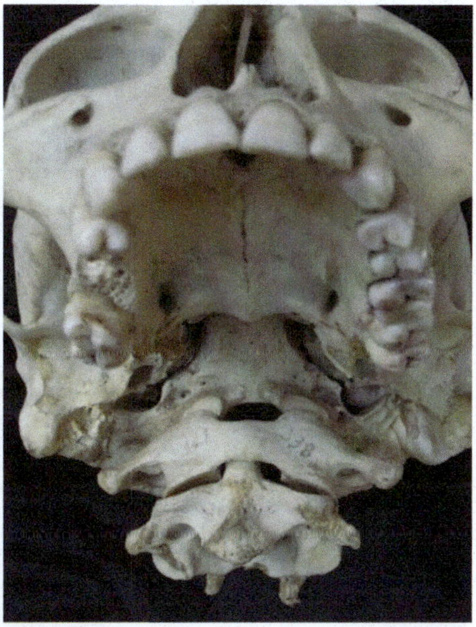

Fig. 2.116 Relationships between the atlas, axis and skull. Skull 2358 from the Museum of Anthropology, University of Florence

The dimensions of the odontoid process are relatively small when considering the stresses to which it is subjected, including those related to the weight of the head. In addition to the physiologically transmitted stresses, we must also consider the possibility of violent impacts transmitted by rapidly changing dynamic forces, which could put the structural strength of the odontoid bone to the test.

In the upright posture typical of humans, the weight of the head tends to induce a significant amount of stress, especially on the C1 and C2 joints, more so than on the other cervical joints (Fig. 2.116).

It's noteworthy that in mammals, there has been an evolutionary convergence between the appearance of the odontoid peg, the expansion of horizontal cranial movements on C1/C2, and the development of jaw movements on the horizontal plane. This suggests that the connection between cervical musculoskeletal structures and the jaw (Galli et al., 2018) is not only anatomical and functional but also evolutionary (Fig. 2.117).

Any change in mandibular position involves an adaptive movement of the skull, which in turn modifies its position on C1. Therefore, we should consider the temporomandibular joint and the cranio-cervical joint as a single musculoskeletal unit (Fig. 2.118).

With the acquisition of an upright posture and encephalization, along with a forward shift of the occipital foramen, there was also a remodeling of the articular surfaces of the occiput (Fig. 2.119), atlas, and axis, which can be considered atypical cervical vertebrae (Aiello & Dean, 1990).

The shape and curvature of the articular facets in the joint between the occiput and C1 determine the range of head movements in terms of cranial flexion-extension, while in the joint between C1 and C2, they are mainly responsible for the lateral rotation of the head. Morphological analysis of these features can provide information about the range of cranial movements.

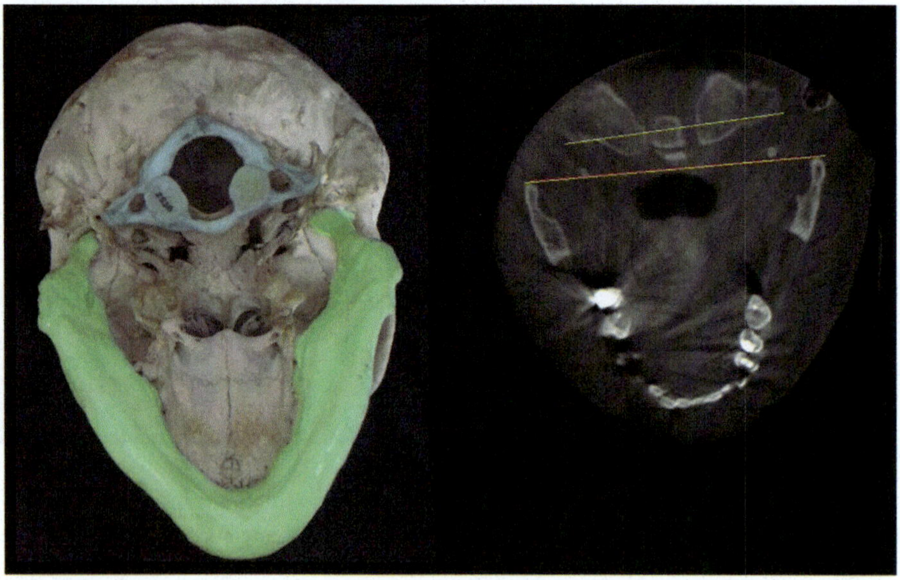

Fig. 2.117 A mandibular lateral deviation in the maxillary-mandibular relationship of maximum intercuspation often coincides with a homolateral rotation adaptation of C1 (Galli et al., 2018). **Left**: author's photo of a specimen from the Museum of Anthropology, University of Florence, modified to create a deviation of the mandible and C1 relative to the skull; **right**: horizontal section of a CT scan with two colored segments highlighting the rotation in the same direction of the mandible and the lateral masses of the atlas

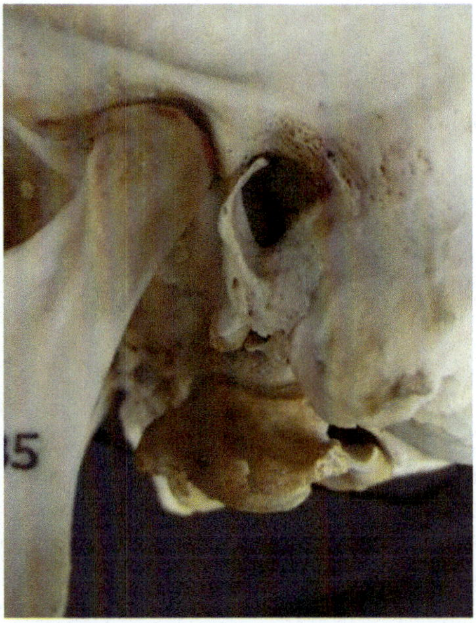

Fig. 2.118 Temporomandibular joint and cranio-atlas joint. Skull 2535 from the Museum of Anthropology, University of Florence

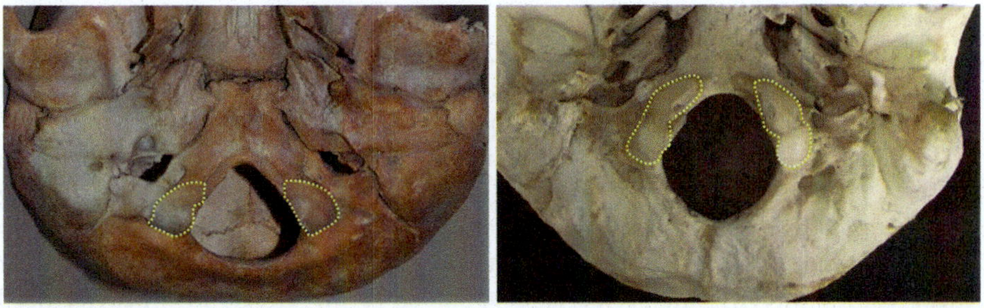

Fig. 2.119 With an upright posture, not only did the foramen magnum change, but also the shape, extension, and inclination of the articular surfaces of the occipital condyles. Left: chimpanzee (*Pan troglodytes*), bottom view. Museum of Zoology, University of Florence. Right: human (*Homo sapiens*), bottom view. Specimen from the Museum of Anthropology, University of Florence. Note the larger extension of the articular surfaces of the occipital condyles in humans

The articular surfaces of chimpanzees, gorillas, and orangutans are more inclined than in humans, whereas in australopithecines, they have an intermediate posterior-anterior inclination between non-human and human comparative taxa. They are less inclined than in anthropoid apes but more inclined than in *Homo* species (Beaudet et al., 2020). In humans, the articular facets of the first two vertebrae also expanded their surface while simultaneously reducing their inclination (Fig. 2.120).

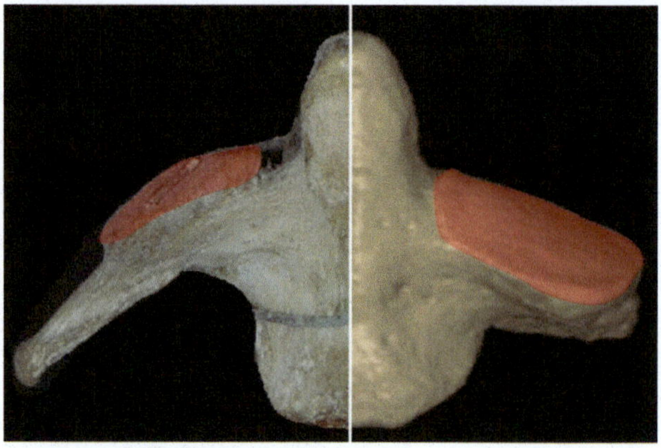

Fig. 2.120 **Left**: portion of the axis of a gorilla (*Gorilla savagei*). Museum of Zoology, University of Florence. **Right**: portion of the axis in *Homo sapiens*: note the greater extension and lesser verticality of the articular surfaces in humans. Cone beam CT of a specimen from the skull collection of the Museum of Anthropology, Florence

In apes and monkeys, both anthropoid and non-anthropoid, the vertebral arteries follow an ascending course, and at the level of the first two cervical vertebrae, their path is not tortuous. In these primates, the transverse foramen of C2 guides the vertebral artery more as an invitation to change direction rather than a conduit with a near 90-degree bend, as seen in humans. In humans, the arteries are forced to make a sharp deviation due to the overlying articular surfaces of the axis (C2) and atlas (C1) (Figs. 2.121, 2.122, 2.123).

Inside the human axis (C2), the course of the transverse foramen deviates almost at a right angle, forcing the vertebral artery to move outward. With a sudden change in direction, it heads upward to enter the foramen of C1 (Fig. 2.124).

In humans, the transverse foramen of C2 obliges the vertebral artery to exhibit its characteristic curvature to avoid obstruction by the broad and extensive articular surfaces of the atlas and axis (Fig. 2.125).

In humans, the articular surfaces between C1 and C2 are flatter compared to those in apes and monkeys. While it's important to be cautious about adopting explanations that assert a one-to-one relationship between form and function, it's undeniable that cervical vertebrae bear the weight of the head, and changes in their shape can be directly interpreted as adaptations or the effects of a heavier head (Manfreda et al., 2006). In humans, the greater extension and lesser inclination of the articular surfaces of the first two cervical vertebrae were crucial not only for stabilizing an increasingly heavy and bulky skull on a vertically oriented spine but also for accommodating the refined rotational and flexion-extension mobility of the head. As previously discussed, these adaptations of the atlas and axis induced a change in the course of the vertebral arteries within C2 to bypass the obstruction caused by their more extensive articular surfaces. During head movements, the odontoid process, or dens, plays a pivotal role as both a pivot point for head rotation (Pang & Li, 2004) and support for flexion-extension movements (Jenkins, 1969; Shellock, 1997; Akobo et al., 2015).

During these movements, the dens, with its posterior inclination, supports the movement of the anterior arch of the atlas (Fig. 2.126).

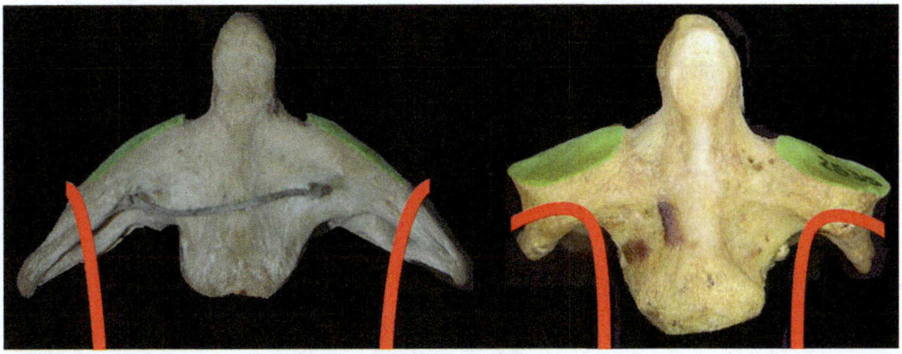

Fig. 2.121 **Left**: axis of a gorilla (*Gorilla savagei*), specimen from the Museum of Zoology, University of Florence; **right**: human axis (*Homo sapiens*), specimen from the Museum of Anthropology, University of Florence: in red, the trajectory of the vertebral arteries; in green, the articular surfaces of C2 on which the atlas rests

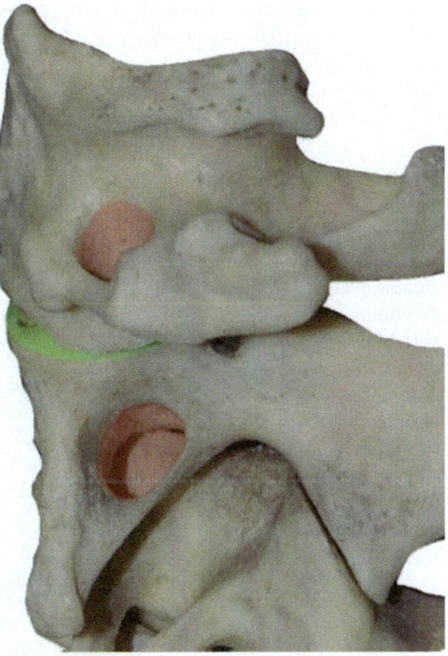

Fig. 2.122 Atlas and axis of a chimpanzee (*Pan troglodytes).* Note how the extension and inclination of the articular surface between C1 and C2 determine only an outward inclination of the transverse foramen of the axis, slightly deviating the ascending course of the vertebral artery which then engages in the foramen of C1. Specimen from the Museum of Zoology, University of Florence

The odontoid process experiences significant stresses, both torsional and flexion-extension, around its axis. It exhibits a very specific functional adaptation in both its external shape and internal structure (Koebke, 1979). At the level of the articulation with the anterior arch of the atlas, it condenses into a characteristic trabecular structure (Fig. 2.127).

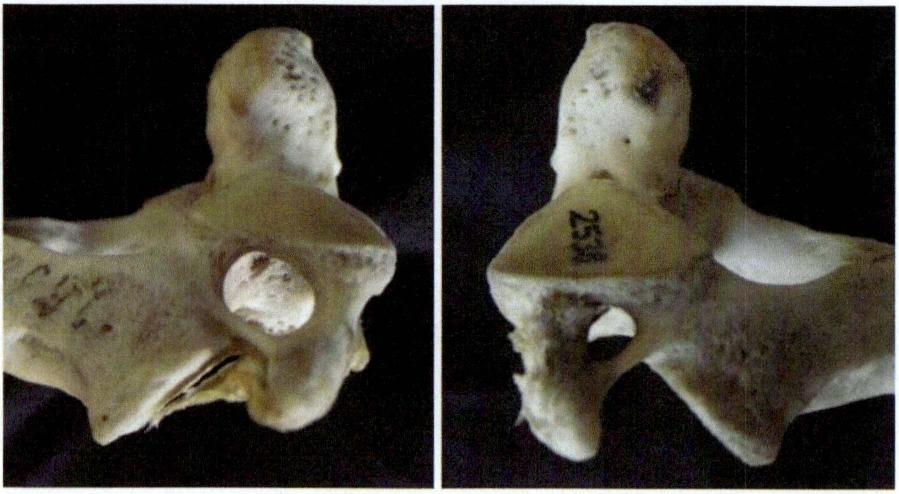

Fig. 2.123 Human axis vertebra, lateral view. Note the exit of the transverse foramen with an almost horizontal direction; the ascending course of the vertebral artery is hindered by an extensive articular surface. The vertebral artery is forced in its ascending course into a sharp latero-posterior curvature within the body of C2. Skull 2358 from the Museum of Anthropology at the University of Florence

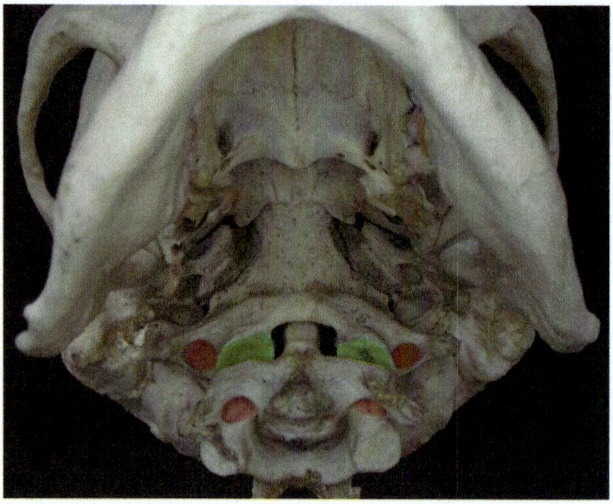

Fig. 2.124 Relationships between C2, C1, and the cranial base: the articular surface of C1 (in green) creates a barrier to the ascending course of the vertebral artery, which changes direction sharply within the body of C2, moving outward before resuming its upward course and entering the transverse foramen of C1 (transverse foramina in red)

The internal structure of the upper third of the odontoid process can be a significant example of how trabecular bone structures in spongy tissue are optimally arranged to dissipate the stresses they undergo, distributing them to make the structure more resistant (Fig. 2.128).

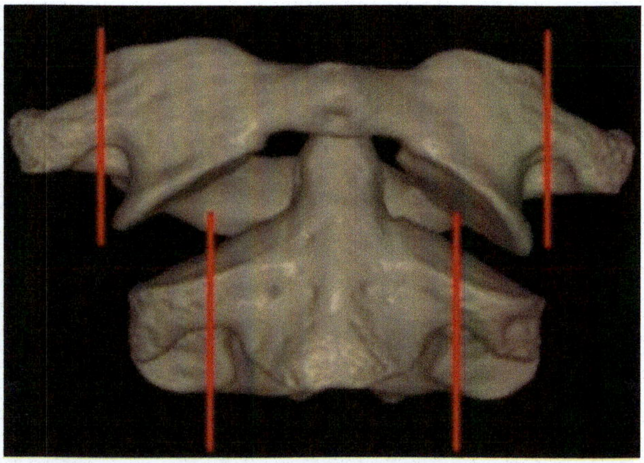

Fig. 2.125 The different axes of the transverse foramina of C1 and C2. Cone beam CT of a specimen from the Museum of Anthropology, University of Florence

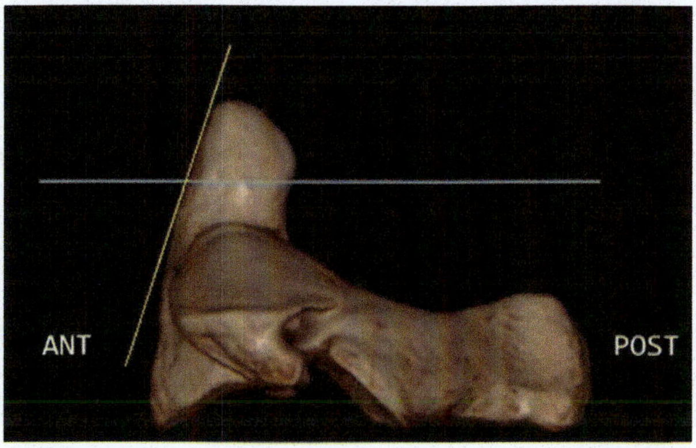

Fig. 2.126 *Homo sapiens:* axis in lateral view. The yellow line indicates the posterior inclination of the odontoid process relative to the horizontal plane (light blue line). This characteristic inclination accommodates the sliding of the anterior arch of the atlas, facilitating flexion-extension movements of the skull. Cone beam CT of a specimen from the Museum of Anthropology, University of Florence

The spongy part of the odontoid process often exhibits a characteristic trabecular pattern resembling a '**Y**' when viewed horizontally, with the center coinciding with the odontoid axis and located at the joint with the anterior arch of the atlas (Montemurro et al., 2019). To resist twisting moments, this reinforcement geometry is most effective through the numerous bone cells that fill the intertrabecular space. The structure, forming a three-pointed star, divides the odontoid circumference into three sections of approximately equivalent circles (Figs. 2.129 and 2.130). This represents the most efficient architectural design with the least volume to withstand flexion-extension stresses according to Roux's law (Lee & Taylor, 1999).

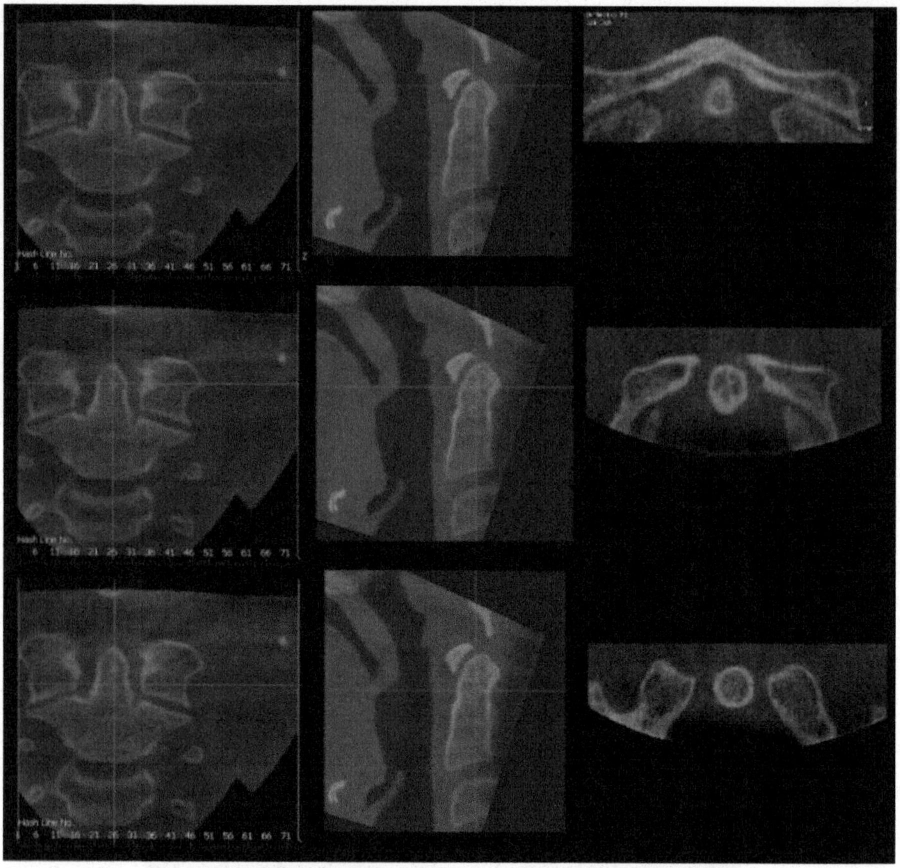

Fig. 2.127 CT scan of the cervical area. Odontoid process: coronal (**left**), sagittal (**center**), horizontal (**right**) sections at three different planes of section. Note the internal structure in the second horizontal section

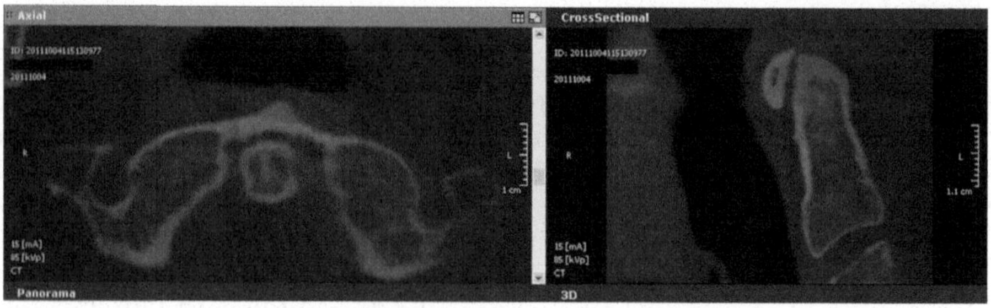

Fig. 2.128 CT scan of C1 and C2. **Left**: horizontal section showing the lateral masses of the atlas body, its anterior arch, and the odontoid process of the axis; note the three characteristic reinforcement trabeculae within the odontoid section. **Right**: sagittal section of the odontoid and anterior arch of the atlas; internally to the odontoid process, at the area of the articular surface with the anterior arch of the atlas, note the bone densification given by the sagittal reinforcement lamina

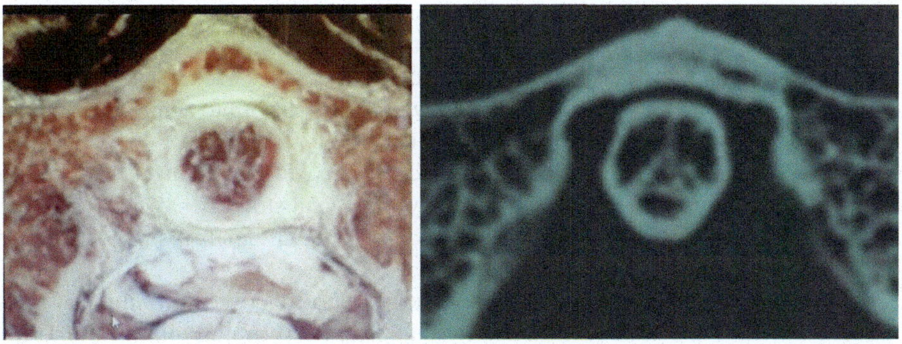

Fig. 2.129 Note the internal arrangement of the trabeculae, which tend to organize into three segments converging at the center of the bone section on the horizontal plane, dividing the structure into three equivalent portions

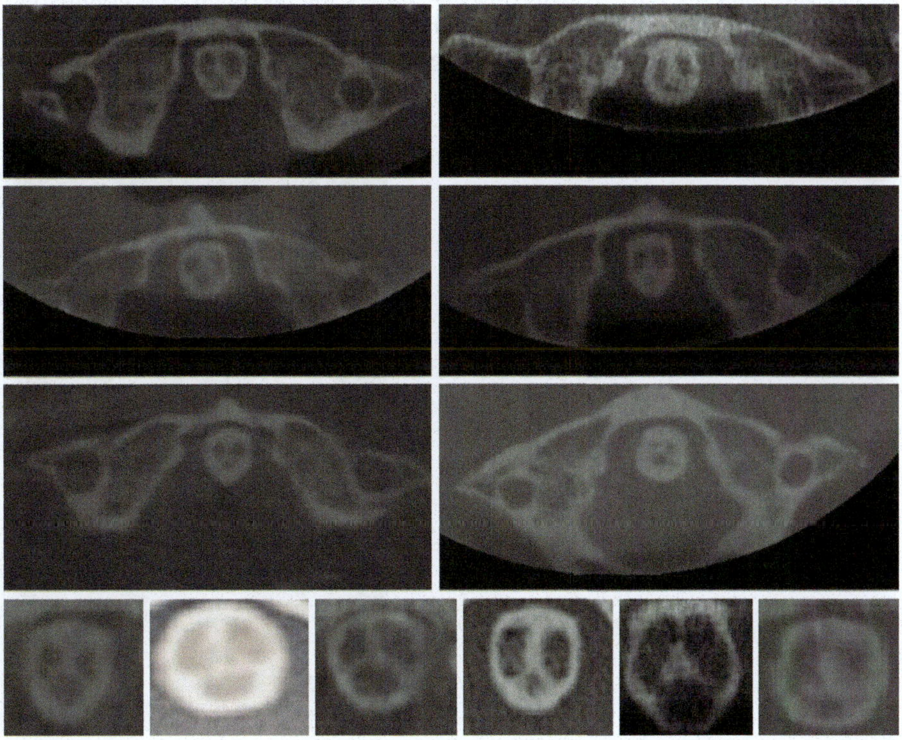

Fig. 2.130 Images of horizontal sections of the odontoid process. Photos courtesy of Massimo Galli

To resist flexural stresses, the reinforcement geometry is most effective with this rib structure formed by three bony lamellae that converge at the center, dividing the odontoid process into three segments of about 120 degrees each. This is the ideal reinforcement geometry that allows a hollow tubular structure to achieve maximum resistance to stresses with the least possible mass (Fig. 2.131).

This type of structural organization aimed at increasing resistance to bending in a tubular shape can be observed within many biological structures. Once again, we notice how nature, faced with similar biomechanical needs, tends to propose similar solutions (Fig. 2.132).

In this chapter, we have tried to outline the morpho-functional evolutionary continuum that links and unites the human stomatognathic apparatus with the life forms with which we share a common origin: from non-anthropomorphic primates to anthropomorphic primates and encompassing the entire class of mammals.

There is a clear demarcation, unfortunately often ignored, between the wisdom of recognizing the anatomical and functional evolutionary continuity of every part of our body and the folly of believing that we are the inevitable end product of a predestined design for "magnificent and progressive destinies." The human species is merely a point on a short trajectory, traveled by a small spark, within the vast pyrotechnic fire of life.

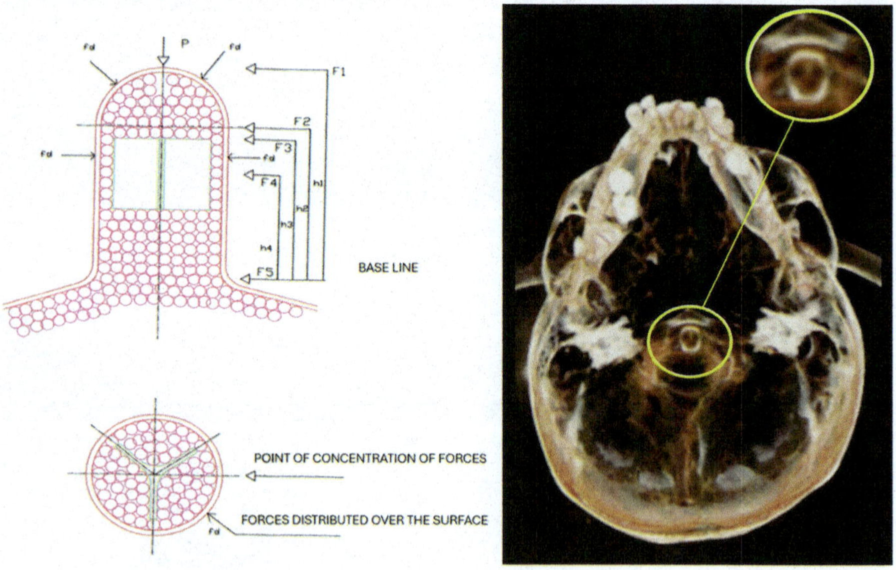

Fig. 2.131 **Left**: diagram of the internal structure of the odontoid process. Drawing by engineer Vittoriano Mangini. **Right**: CT section

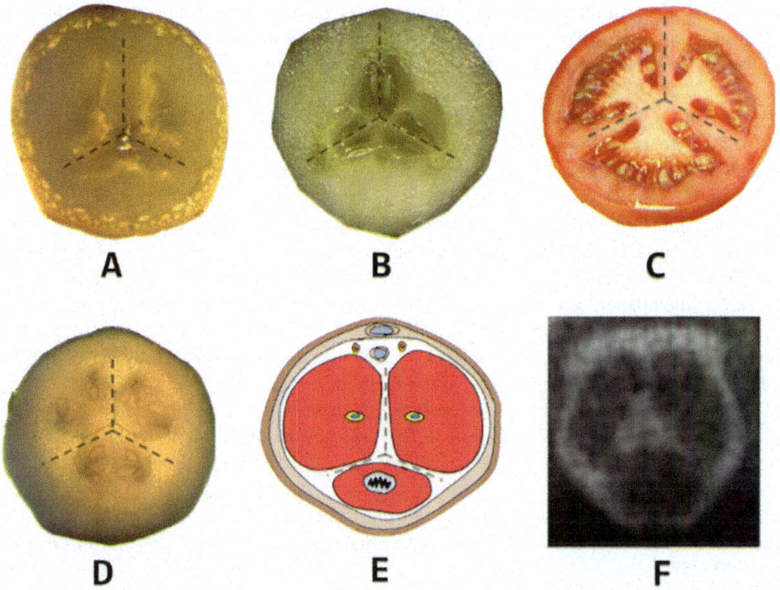

Fig. 2.132 Transverse sections of tubular organic structures found in nature. (**a**): Banana, (**b**): Cucumber, (**c**): Tomato variety S. Marzano, D: Zucchini, E: Penis, F: Odontoid process; all these structures share an internal ribbing organized in a 'Y' shape, dividing the circular section into three equal parts of approximately 120 degrees

References

Aiello, L., & Dean, C. (1990). *An introduction to human evolutionary anatomy*. Academic Press.

Akobo, S., Rizk, E., Loukas, M., Chapman, J. R., Oskouian, R. J., & Tubbs, R. S. (2015). The odontoid process: A comprehensive review of its anatomy, embryology, and variations. *Child's Nervous System, 31*(11), 2025–2034.

Angelo, D. F., Morouço, P., Alves, N., Viana, T., Santos, F., González, R., ... & Pinho, M. (2016). Choosing sheep (*Ovis aries*) as animal model for temporomandibular joint research: Morphological, histological and biomechanical characterization of the joint disc. *Morphologie, 100*(331), 223–233.

Angle, E. H. (1981). *Treatment of malocclusion of the teeth* (The classics of dentistry library) (pp. 198–262). Birmingham Alabama Division of Gryphon Editions.

Anthwal, N., Joshi, L., & Tucker, A. S. (2013). Evolution of the mammalian middle ear and jaw: Adaptations and novel structures. *Journal of Anatomy, 222*(1), 147–160.

Anthwal, N., Peters, H., & Tucker, A. S. (2015). Species-specific modifications of mandible shape reveal independent mechanisms for growth and initiation of the coronoid. *EvoDevo, 6*(1), 1–14.

Anthwal, N., et al. (2017). Meckel's cartilage breakdown offers clues to mammalian middle ear evolution. *Nature Ecology & Evolution, 1*(4), 1–6.

Aristeguieta, L. M., Ramirez, L. E., Acuña, B., & Ortiz, G. P. S. (2010). Tensor veli palatini and tensor tympani muscles: Anatomical, functional and symptomatic links. *Acta Otorrinolaringologica (English Edition), 61*(1), 26–33.

Arnold, P. (2020). Evolution of the mammalian neck from developmental, morpho-functional, and paleontological perspectives. *Journal of Mammalian Evolution, 28*, 173–183.
Atchley, W. R. (1993). Genetic and developmental aspects of variability in the mammalian mandible. *The Skull, 1*, 207–247.
Aung, P. T., Kato, C., Fujita, A., Abe, Y., Ogawa, T., Ishidori, H., Okihara, H., Kokai, S., & Ono, T. (2021). Effects of low occlusal loading on the neuromuscular behavioral development of cortically-elicited jaw movements in growing rats. *Scientific Reports, 11*(1), 1–12.
Baumel, J. J. (1974). Trigeminal-facial nerve communications: Their function in facial muscle innervation and reinnervation. *Archives of Otolaryngology, 99*(1), 34–44.
Beaudet, A., Clarke, R. J., Heaton, J. L., Pickering, T. R., Carlson, K. J., Crompton, R. H., Jashashvili, T., Bruxelles, L., Jakata, K., Bam, L., Van Hoorebeke, L., Kuman, K., & Stratford, D. (2020). The atlas of StW 573 and the late emergence of human-like head mobility and brain metabolism. *Scientific Reports, 10*(1), 1–13.
Beecher, R. M., & Corruccini, R. S. (1981). Effects of dietary consistency on craniofacial and occlusal development in the rat. *The Angle Orthodontist, 51*, 61–69.
Bels, V. (2019). *Feeding in vertebrates: Evolution, morphology, behavior, biomechanics*. Springer.
Bemis, W. E., Giuliano, A., & McGuire, B. (2005). Structure, attachment, replacement and growth of teeth in bluefish, *Pomatomus saltatrix* (Linnaeus, 1776), a teleost with deeply socketed teeth. *Zoology, 108*(4), 317–327.
Benton, M. J. (2000). *Paleontologia dei vertebrati*. Lucisano.
Bergqvist, L. P. (2003). The role of teeth in mammal history. *Brazilian Journal of Oral Sciences, 2*(6), 249–257.
Berkovitz, B. K. B. (2004). Periodontal ligament: Structural and clinical correlates. *Dental Update, 31*(1), 46–54.
Berkovitz, B. K. B., & Shellis, R. P. (2016). *The teeth of non-mammalian vertebrates*. Academic Press.
Berkovitz, B., & Shellis, R. P. (2018). *The teeth of mammalian vertebrates* (1st ed.). Academic Press. eBook ISBN: 9780128028193, Hardcover ISBN: 9780128028186, Published Date: 17th August 2018.
Bertsen, W., McCulloch, C. A. G., & Sodek, J. (1997). The periodontal ligament: A unique, multifunctional connective tissue. *Periodontology 2000, 13*(1), 20–40.
Biewener, A. A. (1991). Musculoskeletal design in relation to body size. *Journal of Biomechanics, 24*, 19–29.
Bodic, F., et al. (2005). Bone loss and teeth. *Joint Bone Spine, 72*(3), 215–221.
Bosshardt, D. D., & Selvig, K. A. (1997). Dental cementum: The dynamic tissue covering of the root. *Periodontology 2000, 13*(1), 41–75.
Bosshardt, D. D., Bergomi, M., Vaglio, G., & Wiskott, A. (2008). Regional structural characteristics of bovine periodontal ligament samples and their suitability for biomechanical tests. *Journal of Anatomy, 212*(3), 319–329.
Bresin, A., Johansson, C., & Kiliaridis, S. (1994). Effects of occlusal strain on the development of the dentoalveolar process in the growing rat. A morphometric study. *European Journal of Musculoskeletal Diseases, 3*, 112–122.
Burrows, A. M. (2008). The facial expression musculature in primates and its evolutionary significance. *Bioessays, 30*(3), 212–225.
Butler, P. M. (1952). The milk-molars of *Perissodactyla*, with remarks on molar occlusion. *Proceedings of the Zoological Society of London, 121*(4), 777–817. Blackwell.
Butler, P. M. (1972). Some functional aspects of molar evolution. *Evolution, 26*(3), 474–483.
Byers, M. R. (1984). Dental sensory receptors. *International Review of Neurobiology, 25*, 39–94. Academic Press.
Byers, M. R., & Cornel, L. M. (2018). Multiple complex somatosensory systems in mature rat molars defined by immunohistochemistry. *Archives of Oral Biology, 85*, 84–97.
Carlson, D. S. (1985). Craniofacial biology as normal science. In *New vistas in orthodontics* (pp. 12–37). Lea & Febiger.

Carlson, D. S. (1999). Growth modification: From molecules to mandibles. *Craniofacial Growth Series, 35*, 17–62.
Cawood, J. I., & Howell, R. A. (1988). A classification of the edentulous jaws. *International Journal of Oral and Maxillofacial Surgery, 17*(4), 232–236.
Cerny, R., Cattell, M., Sauka-Spengler, T., Bronner-Fraser, M., Yu, F., & Medeiros, D. M. (2010). Evidence for the prepattern/cooption model of vertebrate jaw evolution. *Proceedings of the National Academy of Sciences, 107*(40), 17262–17267.
Cheynet, F., Guyot, L., Richard, O., Layoun, W., & Gola, R. (2003). Discomallear and malleomandibular ligaments: Anatomical study and clinical applications. *Surgical and Radiologic Anatomy, 25*(2), 152–157.
Cho, M.-I., & Garant, P. R. (2000). Development and general structure of the periodontium. *Periodontology, 200024*(1), 9–27.
Choi, D. S., Cha, B. K., Jang, I., Kang, K. H., & Kim, S. C. (2012). Three-dimensional finite element analysis of occlusal stress distribution in the human skull with premolar extraction. *The Angle Orthodontist, 83*(2), 204–211.
Chun, K. J., Choi, H. H., & Lee, J. Y. (2014). Comparison of mechanical property and role between enamel and dentin in the human teeth. *Journal of Dental Biomechanics, 5*.
Constantino, P. J., Bush, M. B., Barani, A., & Lawn, B. R. (2016). On the evolutionary advantage of multi-cusped teeth. *Journal of the Royal Society Interface, 13*(121), 20160374.
Corruccini, R. S. (1984). An epidemiologic transition in dental occlusion in world populations. *American Journal of Orthodontics, 86*(5), 419–426.
Costa, A. K., Xavier, T. A., Paes-Junior, T. J., Andreatta-Filho, O. D., & Borges, A. L. (2014). Influence of occlusal contact area on cusp defection and stress distribution. *The Journal of Contemporary Dental Practice, 15*(6), 699–704.
Cowin, C. S. (2007). The significance of bone microstructure in mechanotransduction. *Journal of Biomechanic, 40*(Suppl 1), S105–S109.
Cowin, S. C. (1986). Wolff's law of trabecular architecture at remodeling equilibrium. *Journal of Biomechanical Engineering, 108*(1), 83–88. https://doi.org/10.1115/1.3138584
Crompton, A. W. (1963). The evolution of the mammalian jaw. *Evolution*, 431–439.
Crompton, A. W. (1989). The evolution of mammalian mastication. In D. B. Wake & G. Roth (Eds.), *Complex organismal functions: Integration and evolution in vertebrates* (pp. 23–40). Wiley.
Cuccia, A. M., Caradonna, C., & Caradonna, D. (2011). Manual therapy of the mandibular accessory ligaments for the management of temporomandibular joint disorders. *The Journal of the American Osteopathic Association, 111*(2), 102–112.
Currey, J. D. (2006). *Bones: Structure and mechanics*. Princeton University Press.
Daegling, D. J. (1993). Functional morphology of the human chin. *Evolutionary Anthropology: Issues, News, and Reviews, 1*(5), 170–177.
Davis, B. M. (2011). Evolution of the tribosphenic molar pattern in early mammals, with comments on the "dual-origin" hypothesis. *Journal of Mammalian Evolution, 18*(4), 227.
Dawson, P. E. (2006). *Functional occlusion-E-Book: From TMJ to smile design*. Elsevier Health Sciences.
Dean, C. (1990). *An introduction to human evolutionary anatomy*. Academic Press.
De Toffol, L., Cozza, P., di Girolamo, M., & Cirillo, F. (2004). Il condilio mandibolare. *Virtual Journal of Orthodontics [Serial Online], 6*(2), 1–8.
Dettori, S., Confaloni, A., & Pazzola, T. (2001). *RNO: evoluzione del rapporto dinamico funzionale tra occlusione ed ATM*. Mercurio editore.
Dhakray, V., Mittal, M., Khanna, P., Jain, M., & Yadav, B. (2012). Evolution and history of the periodontal ligament-A. *Internet Journal of Medical Technology, 6*(1).
Diekwisch, T. G. H. (2016). Our periodontal tissue: A masterpiece of evolution. *Journal of Clinical Periodontology, 43*(4), 320–322.
Diogo, R., & Powell, V. (2019). The origin and evolution of mammalian head muscles with special emphasis on the facial myology of primates and modern humans. In *Heads, jaws, and muscles* (pp. 253–293). Springer.

Diogo, R., Wood, B. A., Aziz, M. A., & Burrows, A. (2009). On the origin, homologies and evolution of primate facial muscles, with a particular focus on hominoids and a suggested unifying nomenclature for the facial muscles of the Mammalia. *Journal of Anatomy, 215*(3), 300–319.

Di Vincenzo, F. (2011). Evoluzione: da *Australopithecus* a *homo*.

Du Brul, E. L. (1977). Early hominid feeding mechanisms. *American Journal of Physical Anthropology, 47*(2), 305–320.

Duncan, R. L., & Turner, C. H. (1995). Mechanotransduction and the functional response of bone to mechanical strain. *Calcified Tissue International, 57*(5), 344–358.

Dupont, S., Morsut, L., Aragona, M., Enzo, E., Giulitti, S., Cordenonsi, M., Zanconato, F., Le Digabel, J., Forcato, M., Bicciato, S., Elvassore, N., & Piccolo, S. (2011). Role of YAP/TAZ in mechanotransduction. *Nature, 474*(7350), 179–183.

Esteve-Altava, B. (2017). In search of morphological modules: A systematic review. *Biological Reviews, 92*(3), 1332–1347.

Evans, A. R., & Sanson, G. D. (2003). The tooth of perfection: Functional and spatial constraints on mammalian tooth shape. *Biological Journal of the Linnean Society, 78*(2), 173–191.

Ferguson, M. W. J. (1988). Palate development. *Development, 103*(Suppl), 41–60.

Fiorenza, L., Benazzi, S., Estalrrich, A., & Kullmer, O. (2020). Diet and cultural diversity in Neanderthals and modern humans from dental macrowear analyses. In C. W. Schmidt & J. T. Watson (Eds.), *Dental wear in evolutionary and biocultural contexts* (pp. 39–72). Academic Press.

Fish, D., & Mendel, F. (1982). Mandibular movement patterns relative to food types in common tree shrews (*Tupaia glis*). *American Journal of Physical Anthropology, 58*, 255–269.

Galis, F., Van Dooren, T. J. M., Feuth, J. D., Metzt, J. A. J., Witkam, A., Ruinard, S., Steigenga, M. J., & Wijnaendts, L. C. D. (2006). Extreme selection in humans against homeotic transformations of cervical vertebrae. *Evolution, 60*, 2643–2654.

Galli, M., Papini, A., Buffoli, B., Ferrari, M., Labanca, M., Rezzani, R., Migliario, M., Lucchina, A. G., & Rodella, L. F. (2018). Assessment of atlanto-axial and mandibular rotation by cone beam computed tomography. *Journal of Craniofacial Surgery, 29*(8), 2237–2240.

Galton, P. M. (1973). The cheeks of ornithischian dinosaurs. *Lethaia, 6*(1), 67–89.

Gao, X., Tay, F. R., Gutmann, J. L., Fan, W., Xu, T., & Fan, B. (2016). Micro-CT evaluation of apical delta morphologies in human teeth. *Scientific Reports, 6*(1), 36501.

García-Castellano, J. M., Díaz-Herrera, P., & Morcuende, J. A. (2000). Is bone a target-tissue for the nervous system?: New advances on the understanding of their interactions. *The Iowa Orthopaedic Journal, 20*, 49.

Ghodrati, M., Mosallanezhad, Z., Shati, M., Noroozi, M., Moghadam, A. N., Rostami, M., & Nourbakhsh, M. R. (2020). Adding temporomandibular joint treatments to routine physiotherapy for patients with non-specific chronic neck pain: A randomized clinical study. *Journal of Bodywork and Movement Therapies, 24*(2), 202–212.

Gilbert, R. J., Napadow, V. J., Gaige, T. A., & Wedeen, V. J. (2007). Anatomical basis of lingual hydrostatic deformation. *Journal of Experimental Biology, 210*(23), 4069–4082.

Gómez-Olivencia, A., Been, E., Arsuaga, J. L., & Stock, J. T. (2013). The Neandertal vertebral column 1: The cervical spine. *Journal of Human Evolution, 64*(6), 608–630.

Gómez-Olivencia, A., Carretero, J. M., Arsuaga, J. L., Rodríguez-García, L., García-González, R., & Martínez, I. (2007). Metric and morphological study of the upper cervical spine from the Sima de los Huesos site (Sierra de Atapuerca, Burgos, Spain). *Journal of Human Evolution, 53*(1), 6–25.

Gotthard, K., & Nylin, S. (1995). Adaptive plasticity and plasticity as an adaptation: A selective review of plasticity in animal morphology and life history. *Oikos, 74*, 3–17.

Gould, S. J., & Vrba, E. S. (1982). Exaptation—A missing term in the science of form. *Paleobiology, 8*(1), 4–15.

Gow, B. S. (1980). Circulatory correlates: Vascular impedance, resistance, and capacity. *Comprehensive Physiology*. https://doi.org/10.1002/cphy.cp020214

Grässel, S. (2014). The role of peripheral nerve fibers and their neurotransmitters in cartilage and bone physiology and pathophysiology. *Arthritis Research & Therapy, 16*(6), 1–13.

Gross, M. D. (2008). Occlusion in implant dentistry. A review of the literature of prosthetic determinants and current concepts. *Australian Dental Journal, 53*(Suppl 1), S60–S68.

Guy, F., Lazzari, V., Gilissen, E., & Thiery, G. (2015). To what extent is primate second molar enamel occlusal morphology shaped by the enamel-dentine junction? *PLoS One, 10*(9), e0138802.

Häkkinen, T. J., Sova, S. S., Corfe, I. J., Tjäderhane, L., Hannukainen, A., & Jernvall, J. (2019). Modeling enamel matrix secretion in mammalian teeth. *PLoS Computational Biology, 15*(5), e1007058.

Hansson, S., & Halldin, A. (2012). Alveolar ridge resorption after tooth extraction: A consequence of a fundamental principle of bone physiology. *Journal of Dental Biomechanics, 3*.

Hartstone-Rose, A., Perry, J. M. G., & Morrow, C. J. (2012). Bite force estimation and the fiber architecture of felid masticatory muscles. *The Anatomical Record: Advances in Integrative Anatomy and Evolutionary Biology, 295*(8), 1336–1351.

He, T. (2004). Craniofacial morphology and growth in the ferret: Effects from alteration of masticatory function. *Swedish Dental Journal, 165*(suppl), 1–72.

Heffner, H. E., & Heffner, R. S. (2018). The evolution of mammalian hearing. *AIP Conference Proceedings, 1965*(1) AIP.

Heffner, M. A., Genetos, D. C., & Christiansen, B. A. (2017). Bone adaptation to mechanical loading in a mouse model of reduced peripheral sensory nerve function. *PLoS One, 12*(10), e0187354.

Herring, S. W. (2003). TMJ anatomy and animal models. *Journal of Musculoskeletal Neuronal Interactions, 3*, 391–394.

Herring, S. W. (2007). Masticatory muscles and the skull: A comparative perspective. *Archives of Oral Biology, 52*(4), 296–299.

Herring, S. W., & Lakars, T. C. (1982). Craniofacial development in the absence of muscle. *Journal of Craniofacial Genetics and Development Biology, 1*(4), 341–357.

Herring, S. W., & Liu, Z. J. (2001). Loading of the temporomandibular joint: Anatomical and in vivo evidence from the bones. *Cells Tissues Organs, 169*, 193–200.

Herring, S. W., & Teng, S. (2000). Strain in the braincase and its sutures during function. *American Journal of Physical Anthropology: The Official Publication of the American Association of Physical Anthropologists, 112*(4), 575–593.

Herring, S. W., Teng, S., Huang, X., Mucci, R. J., & Freeman, J. (1996). Patterns of bone strain in the zygomatic arch. *The Anatomical Record: An Official Publication of the American Association of Anatomists, 246*(4), 446–457.

Herring, S. W., Rafferty, K. L., Liu, Z. J., & Marshall, C. D. (2001). Jaw muscles and the skull in mammals: The biomechanics of mastication. *Comparative Biochemistry and Physiology Part A: Molecular & Integrative Physiology, 131*(1), 207–219.

Hiiemae, K. M. (1978). Mammalian mastication: A review of the activity of the jaw muscles and the movements they produce in chewing. In P. M. Butler & K. A. Joysey (Eds.), *Development, function and evolution of teeth* (pp. 359–398). Academic Press.

Hiiemae, K. (1984). Functional aspects of primate jaw morphology. In D. J. Chivers, B. A. Wood, & A. Bilsborough (Eds.), *Food acquisition and processing in primates* (pp. 257–281). Springer.

Hiiemae, K. M., & Palmer, J. B. (2003). Tongue movements in feeding and speech. *Critical Reviews in Oral Biology & Medicine, 14*(6), 413–429.

Hildebrand, M., Goslow, G. E., & Hildebrand, V. (1995). *Analysis of vertebrate structure*. Wiley.

Hogervorst, T., Bouma, H. W., & De Vos, J. (2009). Evolution of the hip and pelvis. *Acta orthopaedica, 80*(suppl 336), 1–39.

Huang, X., Bringas Jr, P., Slavkin, H. C., & Chai, Y. (2009). Fate of HERS during tooth root development. *Developmental Biology, 334*(1), 22–30.

Hylander, W. L. (1979a). The functional significance of primate mandibular form. *Journal of Morphology, 160*(2), 223–239.

Hylander, W. L. (1979b). Mandibular function in *Galago crassicaudatus* and *Macaca fascicularis*: An in vivo approach to stress analysis of the mandible. *Journal of Morphology, 159*(2), 253–296.
Hylander, W. L. (1979c). An experimental analysis of temporomandibular joint reaction force in macaques. *American Journal of Physical Anthropology, 51*(3), 433–456.
Hylander, W. L. (1984). Stress and strain in the mandibular symphysis of primates: A test of competing hypotheses. *American Journal of Physical Anthropology, 64*(1), 1–46.
Hylander, W. L. (2006). Functional anatomy and biomechanics of the masticatory apparatus. In *Temporomandibular disorders: An evidenced approach to diagnosis and treatment*. Quintessence.
Hylander, W. L., & Crompton, A. W. (1986). Jaw movements and patterns of mandibular bone strain during mastication in the monkey *Macaca fascicularis*. *Archives of Oral Biology, 31*(12), 841–848.
Ichim, I., Swain, M. V., & Kieser, J. A. (2006). Mandibular stiffness in humans: Numerical predictions. *Journal of Biomechanics, 39*(10), 1903–1913.
Ingber, D. E., Wang, N., & Stamenović, D. (2014). Tensegrity, cellular biophysics, and the mechanics of living systems. *Reports on Progress in Physics, 77*(4), 046603.
Iordansky, N. N. (2010). Pterygoideus muscles and other jaw adductors in amphibians and reptiles. *Biology Bulletin, 37*(9), 905–914.
Irish, J. D., & Scott, G. R. (2015). *A companion to dental anthropology*. Wiley.
Isberg, A., Isacsson, G., Johansson, A. S., & Larson, O. (1986). Hyperplastic soft-tissue formation in the temporomandibular joint associated with internal derangement: A radiographic and histologic study. *Oral Surgery, Oral Medicine, Oral Pathology, 61*(1), 32–38.
Ivanovski, S., Gronthos, S., Shi, S., & Bartold, P. M. (2006). Stem cells in the periodontal ligament. *Oral Diseases, 12*(4), 358–363.
Iwasaki, S.-i. (2002). Evolution of the structure and function of the vertebrate tongue. *Journal of Anatomy, 201*(1), 1–13.
Jenkins, F. A. (1969). The evolution and development of the dens of the mammalian axis. *The Anatomical Record, 164*(2), 173–184.
Jennings, J. S., & Macmillan, N. H. (1986). A tough nut to crack. *Journal of Materials Science, 21*(5), 1517–1524.
Johnsen, S. E., & Trulsson, M. (2005). Encoding of amplitude and rate of tooth loads by human periodontal afferents from premolar and molar teeth. *Journal of Neurophysiology, 93*(4), 1889–1897.
Jones, H. H., Priest, J. D., Hayes, W. C., Tichenor, C. C., & Nagel, D. A. (1977). Humeral hypertrophy in response to exercise. *Journal of Bone and Joint Surgery, 59*(2), 204–208.
Jones, S. J. (1981). Dental tissues. In J. W. Osborn (Ed.), *Dental anatomy and embryology* (pp. 66–209). Blackwell.
Jugessur, A., Farlie, P. G., & Kilpatrick, N. (2009). The genetics of isolated orofacial clefts: From genotypes to subphenotypes. *Oral Diseases, 15*(7), 437–453.
Kaifu, Y., Kasai, K., Townsend, G. C., & Richards, L. C. (2003). Tooth wear and the "design" of the human dentition: A perspective from evolutionary medicine. *American Journal of Physical Anthropology: The Official Publication of the American Association of Physical Anthropologists, 122*(S37), 47–61.
Kardong, K. V. (1995). Vertebrates: Comparative anatomy, function, evolution. No. QL 805. K35 1995.
Katsavrias, E. G. (2002). Changes in articular eminence inclination during the craniofacial growth period. *The Angle Orthodontist, 72*(3), 258–264.
Kemp, T. S. (2005). *The origin and evolution of mammals*. Oxford University Press on Demand.
Kermack, K. A. (1972). Evolution of mammalian dental structures: The origin of mammals and the evolution of the temporomandibular joint. *Proceedings of the Royal Society of Medicine, 65*(4), 389–392.

Kikuchi, M., Korioth, T. W. P., & Hannam, A. G. (1997). The association among occlusal contacts, clenching effort and bite force distribution in man. *Journal of Dental Research, 76*, 1316–1312.

Kiliaridis, S., Bresin, A., Holm, J., et al. (1996). Effects of masticatory muscle function on bone mass in the mandible of the growing rat. *Acta Anatomica, 155*, 200–205.

Kiliaridis, S., Engström, C., & Thilander, B. (1988). Histochemical analysis of masticatory muscle in the growing rat after prolonged alteration in the consistency of the diet. *Archives of Oral Biology, 33*(3), 187–193.

Kim, Y., Oh, T. J., Misch, C. E., & Wang, H. L. (2005). Occlusal considerations in implant therapy: Clinical guidelines with biomechanical rationale. *Clinical Oral Implants Research, 16*, 26–35.

Klein-MacPhee, G. (2002). Bluefish, family Pomatomidae. In *Bigelow and Schroeder's fishes of the Gulf of Maine* (3rd ed., pp. 400–406). Smithsonian Institution Press.

Klemetti, E. (1996). A review of residual ridge resorption and bone density. *The Journal of Prosthetic Dentistry, 75*(5), 512–514.

Koebke, J. (1979). Morphological and functional studies on the odontoid process of the human axis. *Anatomy and Embryology, 155*(2), 197–208.

Koolstra, J. H., & van Eijden, T. M. (1996). Influence of the dynamical properties of the human masticatory muscles on jaw closing movements. *European Journal of Morphology, 34*, 11–18.

Koolstra, J. H., & van Eijden, T. M. (1999). Three-dimensional dynamical capabilities of the human masticatory muscles. *Journal of Biomechanics, 32*, 145–152.

Krahl, H., Michaelis, U., Pieper, H. G., Quack, G., & Montag, M. (1994, November–December). Stimulation of bone growth through sports. A radiologic investigation of the upper extremities in professional tennis players. *American Journal of Sports Medicine, 22*(6), 751–757.

Lanyon, L. E. (1987). Functional strain in bone tissue as an objective, and controlling stimulus for adaptive bone remodeling. *Journal of Biomechanics, 20*(11–12), 1083–1093.

Lautenschlager, S. P., Gill, P., Luo, Z. X., Fagan, M. J., & Rayfield, E. J. (2016). Morphological evolution of the mammalian jaw adductor complex. *Biological Reviews of the Cambridge Philosophical Society*. https://doi.org/10.1111/brv.12314

LeBlanc, A. R., Apesteguía, S., Larsson, H. C., & Caldwell, M. W. (2020). Unique tooth morphology and prismatic enamel in Late Cretaceous sphenodontians from Argentina. *Current Biology, 30*(9), 1755–1761.

Le Douarin, N. M. (2004). The avian embryo as a model to study the development of the neural crest: A long and still ongoing story. *Mechanisms of Development, 121*(9), 1089–1102.

Lee, J. J. W., Morris, D., Constantino, P. J., Lucas, P. W., Smith, T. M., & Lawn, B. R. (2010). Properties of tooth enamel in great apes. *Acta Biomaterialia, 6*(12), 4560–4565.

Lee, T. C., & Taylor, D. (1999). Bone remodelling: Should we cry Wolff? *Irish Journal of Medical Science, 168*(2), 102.

Legasa, O., Buscalioni, A. D., & Gasparini, Z. (1994). The serrated teeth of Sebecus and the Iberoccitanian crocodile, a morphological and ultrastructural comparison. *Studia Geologica Salmanticensia, 29*, 127–144.

Lieberman, D. (2011). *The evolution of the human head*. Harvard University Press.

Lieberman, D. (2014). *The story of the human body: Evolution, health, and disease*. Vintage.

Liem, K. F., et al. (2001). *Functional anatomy of the vertebrates: An evolutionary perspective*. No. 596 FUN.

Limme, M. (2002). Conduites alimentaires et croissance des arcades dentaires. *Revue d'Orthopédie Dento-Faciale, 36*(3), 289–309.

Lorenzon, G. (2010). *Implantologia funzionale*. Ed. Martina. ISBN 978-88-7572-094-0.

Loughner, B. A., Larkin, L. H., & Mahan, P. E. (1989). Discomalleolar and anterior malleolar ligaments: Possible causes of middle ear damage during temporomandibular joint surgery. *Oral Surgery, Oral Medicine, Oral Pathology, 68*(1), 14–22.

Luan, X., Ito, Y., & Diekwisch, T. G. H. (2006). Evolution and development of Hertwig's epithelial root sheath. *Developmental Dynamics: An Official Publication of the American Association of Anatomists, 235*(5), 1167–1180.

Luan, X., Walker, C., Dangaria, S., Ito, Y., Druzinsky, R., Jarosius, K., Lesot, H., & Rieppel, O. (2009). The mosasaur tooth attachment apparatus as paradigm for the evolution of the gnathostome periodontium. *Evolution & Development, 11*(3), 247–259.

Lucas, P. W. (1979). The dental-dietary adaptations of mammals. *Neues Jahrbuch für Geologie und Paläontologie, Monatshefte, 8*, 486–512.

Lucas, P. W. (1982). Basic principles of tooth design. In B. Kurtén (Ed.), *Teeth: Form, function and evolution*. Columbia University Press 15416.

Lucas, P. W. (2004). *Dental functional morphology: How teeth work*. Cambridge University Press.

Lucas, P., Constantino, P., Wood, B., & Lawn, B. (2008). Dental enamel as a dietary indicator in mammals. *BioEssays, 30*(4), 374–385.

Lucas, P. W., Darvell, B. W., Lee, P. K. D., Yuen, T. D. B., & Choong, M. F. (1995). The toughness of plant cell walls. *Philosophical Transactions of the Royal Society of London. Series B: Biological Sciences, 348*(1325), 363–372.

Luddleston, A. L., Rockwell, D., Kulund, D. N., & Harrison, R. B. (1980, September 5). Bone mass in lifetime tennis athletes. *JAMA, 244*(10), 1107–1109. https://doi.org/10.1001/jama.1980.03310100025022

Luo, Z.-X., Kielan-Jaworowska, Z., & Cifelli, R. L. (2004). Evolution of dental replacement in mammals. *Bulletin of Carnegie Museum of Natural History, 36*, 159–175.

Maas, M. C., & Dumont, E. R. (1999). Built to last: The structure, function, and evolution of primate dental enamel. *Evolutionary Anthropology: Issues, News, and Reviews: Issues, News, and Reviews, 8*(4), 133–152.

Maier, W., & Ruf, I. (2016). Evolution of the mammalian middle ear: A historical review. *Journal of Anatomy, 228*(2), 270–283.

Maier, W., Van den Heever, J., & Durand, F. (1996). New therapsid specimens and the origin of the secondary hard and soft palate of mammals. *Journal of Zoological Systematics and Evolutionary Research, 34*(1), 9–19.

Mammoto, T., Mammoto, A., & Ingber, D. E. (2013). Mechanobiology and developmental control. *Annual Review of Cell and Developmental Biology, 29*, 27–61.

Manfreda, E., Mitteroecker, P., Bookstein, F. L., & Schaefer, K. (2006). Functional morphology of the first cervical vertebra in humans and nonhuman primates. *The Anatomical Record. Part B, The New Anatomist, 289*(5), 184–194.

Manzi, G. (2013). *Il grande racconto dell'evoluzione umana*. Società Editrice il Mulino.

Marchi, D. (2016). *Il mistero di Homo naledi*. Ed Mondadori.

Matalová, E., Lungová, V., & Sharpe, P. (2015). Development of tooth and associated structures. In A. Vishwakarma, P. Sharpe, S. Shi, & M. Ramalingam (Eds.), *Stem cell biology and tissue engineering in dental sciences* (pp. 335–346). Academic Press.

Mayya, A., Banerjee, A., & Rajesh, R. (2013). Mammalian cortical bone in tension is non-Haversian. *Scientific Reports, 3*, 2533.

McCormack, S. W. (2016). *Biomechanical function of the periodontal ligament in biting and orthodontic tooth movement*. Diss. University of Hull.

McIntosh, J. E., Anderton, X., Flores-De-Jacoby, L., Carlson, D. S., Shuler, C. F., & Diekwisch, T. G. (2002). Caiman periodontium as an intermediate between basal vertebrate ankylosis-type attachment and mammalian "true" periodontium. *Microscopy Research and Technique, 59*(5), 449–459.

Menegaz, R. A., Sublett, S. V., Figueroa, S. D., Hoffman, T. J., & Ravosa, M. J. (2009). Phenotypic plasticity and function of the hard palate in growing rabbits. *The Anatomical Record: Advances in Integrative Anatomy and Evolutionary Biology: Advances in Integrative Anatomy and Evolutionary Biology, 292*(2), 277–284.

Mjor, I. A. (1980). *Histology of the human tooth*. Munksgaard.

Modesto, S., Sidor, C. A., Rubidge, B. S., & Welman, J. (2001). A second varanopseid skull from the Upper Permian of South Africa: Implications for Late Permian 'pelycosaur' evolution. *Lethaia, 34*(4), 249–259.

Molnar, S., & Gantt, D. G. (1977). Functional implications of primate enamel thickness. *American Journal of Physical Anthropology, 46*(3), 447–454.

Montemurro, N., Perrini, P., Mangini, V., Galli, M., & Papini, A. (2019). The Y-shaped trabecular bone structure in the odontoid process of the axis: A CT scan study in 54 healthy subjects and biomechanical considerations. *Journal of Neurosurgery: Spine, 30*(5), 585–592.

Moon, H. S., Won, Y. Y., Kim, K. D., Ruprecht, A., Kim, H. J., Kook, H. K., & Chung, M. K. (2004). The three-dimensional microstructure of the trabecular bone in the mandible. *Surgical and Radiologic Anatomy, 26*(6), 466–473.

Moss, M. L. (1962). *The functional matrix. Vistas in orthodontics*. Lea & Febiger.

Moss, M. L., & Simon, M. R. (1968). Growth of the human mandibular angular process: A functional cranial analysis. *American Journal of Physical Anthropology, 28*(2), 127–138.

Mrör, I. A. (2001). Pulp-dentin biology in restorative dentistry. Part 5: Clinical management and tissue changes associated with wear and trauma. *Quintessence International, 32*(10).

Murray, G. M., et al. (2001). The role of the human lateral pterygoid muscle in the control of horizontal jaw movements. *Journal of Orofacial Pain, 15*(4), 279–292.

Myers, L. J. (1988). Possible inflammatory pathways relating temporomandibular joint dysfunction to otic symptoms: An anatomical hypothesis. *CRANIO®, 6*(1), 64–70.

Naples, V. L. (1999). Morphology, evolution and function of feeding in the giant anteater (*Myrmecophaga tridactyla*). *Journal of Zoology, 249*, 19–41. [13].

Öğütcen-Toller, M. (1995). The morphogenesis of the human discomalleolar and sphenomandibular ligaments. *Journal of Cranio-Maxillofacial Surgery, 23*(1), 42–46.

Osborn, J. W. (1982). Helicoidal plane of dental occlusion. *American Journal of Physical Anthropology, 57*(3), 273–281.

Ota, M. S., et al. (2009). Patterning of molar tooth roots in mammals. *Journal of Oral Biosciences, 51*(4), 193–198.

Oxilia, G., Bortolini, E., Martini, S., Papini, A., Boggioni, M., Buti, L., Figus, C., Sorrentino, R., Townsend, G., Kaidonis, J., Fiorenza, L., Cristiani, E., Kullmer, O., Moggi-Cecchi, J., & Benazzi, S. (2018). The physiological linkage between molar inclination and dental macrowear pattern. *American Journal of Physical Anthropology, 166*(4), 941–951.

Pang, D., & Li, V. (2004). Atlantoaxial rotatory fixation: Part 1—Biomechanics of normal rotation at the atlantoaxial joint in children. *Neurosurgery, 55*(3), 614–626.

Papini, A., Galli, M., Di Martino, N., Zavattaro, M., & Cecchi, J. M. (2012). Studio delle simmetrie e asimmetrie ossee mascellari in collezioni italiane presenti nella cranioteca del Museo di Storia Naturale dell'Università di Firenze tramite CBCT.

Parfitt, A. M. (1997). Genetic effects on bone mass and turnover-relevance to black/white differences. *Journal of the American College of Nutrition, 16*(4), 325–333.

Pearson, O. M., & Lieberman, D. E. (2004). The aging of Wolff's "law": Ontogeny and responses to mechanical loading in cortical bone. *American Journal of Physical Anthropology, 125*(S39), 63–99.

Pievani, T. (2019). *Imperfezione: una storia naturale*. Raffaello Cortina Editore.

Pileicikiene, G., & Surna, A. (2004). The human masticatory system from a biomechanical perspective: A review. *Stomatologija, 6*(3), 81–84.

Planas, P. (1994). *Rehabilitación neuro-oclusal*. Editorial Salvat.

Popowics, T. E., & Herring, S. W. (2006). Teeth, jaws and muscles in mammalian mastication. In V. L. Bels (Ed.), *Feeding in domestic vertebrates: From structure to behaviour* (pp. 61–83). CABI.

Porto, G. G., Vasconcelos, B. C. D. E., Andrade, E. S. D. S., & Silva-Junior, V. A. (2010). Comparison between human and rat TMJ: Anatomic and histopathologic features. *Acta Cirurgica Brasileira, 25*(3), 290–293.

Qiao, Y., Wang, Y., Zhou, Y., Jiang, F., Huang, T., Chen, L., Lan, J., Yang, C., Guo, Y., Yan, S., Wei, Z., & Li, J. (2019). The role of nervous system in adaptive response of bone to mechanical loading. *Journal of Cellular Physiology, 234*(6), 7771–7780.

Rak, Y., Ginzburg, A., & Geffen, E. (2007). Gorilla-like anatomy on Australopithecus afarensis mandibles suggests Au. afarensis link to robust australopiths. *Proceedings of the National Academy of Sciences, 104*(16), 6568–6572.

Rasch, L. J., Martin, K. J., Cooper, R. L., Metscher, B. D., Underwood, C. J., & Fraser, G. J. (2016). An ancient dental gene set governs development and continuous regeneration of teeth in sharks. *Developmental Biology, 415*(2), 347–370.

Rashed, F. (2015). A comparative study of the dentition and temporomandibular joint anatomy and histology adult dogs. *Biological Systems: Open Access, 4*(147), 2.

Rathee, M., & Jain, P. (2020). *Anatomy, head and neck, lateral pterygoid muscle*. StatPearls [Internet].

Ravosa, M. J., Lopez, E. K., Menegaz, R. A., Stock, S. R., Stack, M. S., & Hamrick, M. W. (2008). Adaptive plasticity in the mammalian masticatory complex: You are what, and how, you eat. In C. Vinyard, M. J. Ravosa, & C. Wall (Eds.), *Primate craniofacial function and biology* (pp. 293–328). Springer.

Reif, W. E. (1984). Pattern regulation in shark dentition. In G. M. Malacinski & W. B. Bryant (Eds.), *Pattern formation: A primer in developmental biology* (pp. 603–621). Macmillan.

Ren, Y.-F., & Isberg, A. (1995). Tinnitus in patients with temporomandibular joint internal derangement. *CRANIO®, 13*(2), 75–80.

Richman, J. M., Buchtová, M., & Boughner, J. C. (2006). Comparative ontogeny and phylogeny of the upper jaw skeleton in amniotes. *Developmental Dynamics: An Official Publication of the American Association of Anatomists, 235*(5), 1230–1243.

Riga, M., Xenellis, J., Peraki, E., Ferekidou, E., & Korres, S. (2010). Aural symptoms in patients with temporomandibular joint disorders: Multiple frequency tympanometry provides objective evidence of changes in middle ear impedance. *Otology & Neurotology, 31*(9), 1359–1364.

Ross, C. F., Eckhardt, A., Herrel, A., Hylander, W. L., Metzger, K. A., Schaerlaeken, V., Washington, R. L., & Williams, S. H. (2007). Modulation of intra-oral processing in mammals and lepidosaurs. *Integrative and Comparative Biology, 47*(1), 118–136.

Rubidge, B. S., & Sidor, C. A. (2001). Evolutionary patterns among Permo-Triassic therapsids. *Annual Review of Ecology and Systematics, 32*(1), 449–480.

Rubin, C. T., & Lanyon, L. E. (1987). Osteoregulatory nature of mechanical stimuli: Function as a determinant for adaptive remodeling in bone. *Journal of Orthopaedic, 5*(2), 300–310.

Ruff, C. B., Walker, A., & Trinkaus, E. (1994). Postcranial robusticity in Homo. III: Ontogeny. *American Journal of Physical Anthropology, 93*(1), 35–54.

Sahni, A. (1987). Evolutionary aspects of reptilian and mammalian enamel structure. *Scanning Microscopy, 1*(4), 1903–1912.

Sánchez-Villagra, M. R., & Smith, K. K. (1997). Diversity and evolution of the marsupial mandibular angular process. *Journal of Mammalian Evolution, 4*(2), 119–144.

Sarrafpour, B., Swain, M., Li, Q., & Zoellner, H. (2013). Tooth eruption results from bone remodelling driven by bite forces sensed by soft tissue dental follicles: A finite element analysis. *PLoS One, 8*(3), e58803.

Schroeder, H. E. (1986). Development, structure, and function of periodontal tissues. In *The periodontium* (pp. 23–323). Springer.

Schwartz, G. T., McGrosky, A., & Strait, D. S. (2020). Fracture mechanics, enamel thickness and the evolution of molar form in hominins. *Biology Letters, 16*(1), 20190671.

Schwenk, K. (Ed.). (2000). *Feeding: Form, function and evolution in tetrapod vertebrates*. Elsevier.

Seo, H., Kim, J., Hwang, J. J., Jeong, H. G., Han, S. S., Park, W., ... & Cho, S. W. (2017). Regulation of root patterns in mammalian teeth. *Scientific Reports, 7*(1), 12714.

Shellock, F. G. (1997). Kinematic MRI of the joints. *Seminars in Musculoskeletal Radiology, 1*(01) © 1997 by Thieme Medical Publishers.

Sisask, G., Silfverswärd, C. J., Bjurholm, A., & Nilsson, O. (2013). Ontogeny of sensory and autonomic nerves in the developing mouse skeleton. *Autonomic Neuroscience, 177*(2), 237–243.

Smith, P. (1972). Diet and attrition in the Natufians. *American Journal of Physical Anthropology, 37*(2), 233–238.

Spencer, M. A. (2003). Tooth-root form and function in platyrrhine seed-eaters. *American Journal of Physical Anthropology: The Official Publication of the American Association of Physical Anthropologists, 122*(4), 325–335.

Stem, D. N., & Crompton, A. W. (1995). *A study of enamel organization, from reptiles to mammals*. International Institute for the Study of Man.

Stewart, D. (1927). The problem of the innervation of the dentine. *Journal of Anatomy, 61*(Pt 4), 439.

Ström, D., Holm, S., Clemensson, E., Haraldson, T., & Carlsson, G. E. (1988). Gross anatomy of the craniomandibular joint and masticatory muscles of the dog. *Archives of Oral Biology, 33*(8), 597–604.

Terhune, C. E., Cooke, S. B., & Otárola-Castillo, E. (2015). Form and function in the platyrrhine skull: A three-dimensional analysis of dental and TMJ morphology. *The Anatomical Record, 298*(1), 29–47.

Terhune, C. E., Robinson, C. A., & Ritzman, T. B. (2014). Ontogenetic variation in the mandibular ramus of great apes and humans. *Journal of Morphology, 275*(6), 661–677.

Thomason, J. J., & Russell, A. P. (1986). Mechanical factors in the evolution of the mammalian secondary palate: A theoretical analysis. *Journal of Morphology, 189*(2), 199–213.

Tjäderhane, L. (2019). Dentin basic structure, composition, and function. In M. A. Versiani, B. Basrani, & M. D. Sousa-Neto (Eds.), *The root canal anatomy in permanent dentition* (pp. 17–27). Springer.

Tomer, G., & Kishnani, R. (2011). Correlation of antegonial notch depth with craniofacial morphology—A cephalometric and electromyographic study. *Journal of Pierre Fauchard Academy (India Section), 25*(3), 163–171.

Tomlinson, R. E., Christiansen, B. A., Giannone, A. A., & Genetos, D. C. (2020). The role of nerves during skeletal development, mechano-adaptation, fracture, and aging. *Frontiers in Endocrinology, 11*, 646.

Towle, I. (2019). Tertiary dentine frequencies in extant great apes and fossil hominins. *Open Quaternary, 5*(1).

Trinkaus, E., Churchill, S. E., & Ruff, C. B. (1994). Postcranial robusticity in Homo. II: Humeral bilateral asymmetry and bone plasticity. *American Journal of Physical Anthropology, 93*(1), 1–34.

Tucker, A. S. (2017). Major evolutionary transitions and innovations: The tympanic middle ear. *Philosophical Transactions of the Royal Society B: Biological Sciences, 372*(1713), 20150483.

Tuisku, F., & Hildebrand, C. (1995). Immunohistochemical and electron microscopic demonstration of nerve fibres in relation to gingiva, tooth germs and functional teeth in the lower jaw of the cichlid *Tilapia mariae*. *Archives of Oral Biology, 40*(6), 513–520.

Turner, C. H., Anne, V., & Pidaparti, R. M. V. (1997). A uniform strain criterion for trabecular bone adaptation: Do continuum-level strain gradients drive adaptation? *Journal of Biomechanics, 30*(6), 555–563.

Uhthoff, H. K., & Jaworski, Z. F. (1978). Bone loss in response to long-term immobilisation. *The Journal of Bone and Joint Surgery. British Volume, 60*(3), 420–429.

Uhthoff, H. K., Sékaly, G., & Jaworski, Z. F. (1985). Effect of long-term nontraumatic immobilization on metaphyseal spongiosa in young adult and old beagle dogs. *Clinical Orthopaedics and Related Research, 192*, 278–283.

Ungar, P. S. (2010). *Mammal teeth: Origin, evolution, and diversity*. JHU Press.

Ungar, P. S. (2014). *Teeth: A very short introduction*. OUP Oxford.

Ungar, P. S. (2015). Mammalian dental function and wear: A review. *Biosurface and Biotribology, 1*(1), 25–41.

Ungar, P. S. (2017). *Evolution's bite: A story of teeth, diet, and human origins*. Princeton University Press.

Urban, D. J., Anthwal, N., Luo, Z. X., Maier, J. A., Sadier, A., Tucker, A. S., & Sears, K. E. (2017). A new developmental mechanism for the separation of the mammalian middle ear ossicles from the jaw. *Proceedings of the Royal Society B: Biological Sciences, 284*(1848), 20162416.

Vallittu, P. K., Varrela, J., Salo, J., Rengui, L., Shanshan, L., Shan, H., Zhang, H., & Niemelä, P. (2021). Temporomandibular joint and Giant Panda's (*Ailuropoda melanoleuca*) adaptation to bamboo diet. *Scientific Reports, 11*(1), 1–6.

Vander Linden, A., Campbell, K. M., Bryar, E. K., & Santana, S. E. (2019). Head-turning morphologies: Evolution of shape diversity in the mammalian atlas–axis complex. *Evolution, 73*(10), 2060–2071.

Villmoare, B., Kimbel, W. H., Seyoum, C., Campisano, C. J., DiMaggio, E. N., Rowan, J., Braun, D. R., Arrowsmith, J. R., & Reed, K. E. (2015). Early Homo at 2.8 Ma from Ledi-Geraru, Afar, Ethiopia. *Science, 347*(6228), 1352–1355.

Villmoare, B. A., et al. (2014). Craniofacial modularity, character analysis, and the evolution of the premaxilla in early African hominins. *Journal of Human Evolution*. https://doi.org/10.1016/j.jhevol.2014.06.014

Vishwakarma, A., Sharpe, P., Shi, S., & Ramalingam, M. (Eds.). (2014). *Stem cell biology and tissue engineering in dental sciences*. Academic Press.

Walczyńska-Dragon, K., Baron, S., Nitecka-Buchta, A., & Tkacz, E. (2014). Correlation between TMD and cervical spine pain and mobility: Is the whole body balance TMJ related? *BioMed Research International*. https://doi.org/10.1155/2014/582414

Wall, C. E., & Smith, K. K. (2001). Ingestion in mammals. *e LS*. https://doi.org/10.1038/npg.els.0001837

Watson, P. J., Fitton, L. C., Meloro, C., Fagan, M. J., & Gröning, F. (2018). Mechanical adaptation of trabecular bone morphology in the mammalian mandible. *Scientific Reports, 8*(1), 1–12.

Weijs, W. A. (1994). Evolutionary approach of masticatory motor patterns in mammals. In V. L. Bels, M. Chardon, & P. Vandewalle (Eds.), *Biomechanics of feeding in vertebrates* (pp. 281–320). Springer.

Wilken, A. T., Middleton, K. M., Sellers, K. C., Cost, I. N., & Holliday, C. M. (2019). The roles of joint tissues and jaw muscles in palatal biomechanics of the Savannah monitor (*Varanus exanthematicus*) and their significance for cranial kinesis. *Journal of Experimental Biology, 222*(18), jeb201459.

Williamson, E. H. (1990). The interrelationship of internal derangements of the temporomandibular joint, headache, vertigo, and tinnitus: A survey of 25 patients. *CRANIO®, 8*(4), 301–306.

Williamson, L., & Lucas, P. (1995). The effect of moisture content on the mechanical properties of a seed shell. *Journal of Materials Science, 30*(1), 162–166.

Woo, S. L., Kuei, S. C., Amiel, D., Gomez, M. A., Hayes, W. C., White, F. C., & Akeson, W. H. (1981). The effect of prolonged physical training on the properties of long bone: A study of Wolff's Law. *The Journal of Bone and Joint Surgery. American Volume, 63*(5), 780–787.

Woronowicz, K. C., & Schneider, R. A. (2019). Molecular and cellular mechanisms underlying the evolution of form and function in the amniote jaw. *EvoDevo, 10*(1), 1–21.

Yamamoto, T., et al. (2016). Histology of human cementum: Its structure, function, and development. *Japanese Dental Science Review, 52*(3), 63–74.

Chapter 3
Tooth Wear

Abstract Chapter provides an in-depth analysis of tooth wear, presenting it as a fundamental physiological and functional adaptation in the evolution of mammals—particularly humans. Rather than being viewed as mere tissue loss, dental wear is described as a necessary process for occlusal maturation, stomatognathic function, and efficient mastication.

The chapter explains how wear allows teeth to progressively adapt to masticatory forces, improving force distribution, shaping functional occlusal surfaces, and supporting the biomechanical development of the entire oral system. It distinguishes physiological from pathological wear and discusses the dietary changes in human history, from prehistoric to industrial times. In the past, the natural abrasiveness of food ensured consistent wear, which prevented many modern occlusal problems such as crowding and malocclusion.

The text also explores the structural levels of dental enamel—millimetric, micrometric, and nanometric—demonstrating how its hierarchical organization provides fracture resistance, adaptability, and long-term functionality. The evolution of prismatic enamel in mammals is highlighted as a biomechanical breakthrough.

Finally, the importance of both interproximal and occlusal wear is emphasized in shaping human facial structure and preventing modern dental pathologies, supported by paleoanthropological evidence and current clinical data. Dental wear, rather than being a destructive process, is reframed as an evolutionarily advantageous phenomenon that has shaped the function and form of human dentition.

By nature's law it is decreed,
All forms do fade and change with speed,
Each thing, though strong, in time concedes,
To time's relentless, steady needs.
Slowly shifts each firmest frame,
Why then, dost thou harbor shame?
Anonymous Poet, from Prato, Tuscany. Twenty-First Century

Exploring the fascinating and fundamental process of dental wear, it should be noted that this morphological adaptation allows a newly erupted tooth to coordinate with all other dental elements. It contributes its part in achieving and maintaining maximum efficiency in species-specific mastication.

A. Papini, *On the Origin of Human Masticatory Function*,
https://doi.org/10.1007/978-3-031-98664-2_3

Imagine a tooth erupting in the dental arch as a beautiful and pristine instrument not yet tuned to fit harmoniously into the masticatory apparatus. The inevitable wear experienced by any enamel surface that rubs against its counterpart allows for a process of occlusal relationship maturation. In mammals, enamel wear is the primary force that individually shapes the surfaces of each tooth, according to characteristic motor patterns of each species. For instance, throughout human evolutionary history, the eruption of canines initially regulated chewing cycles. However, the abrasive contact concentrated on the canine guides quickly wears down their steep and disclusive surfaces, leading to the involvement of other functional support areas in the posterior teeth. This progressive involvement marks the transition from canine guidance to a group function, supported on the opposite side by a gradual engagement of balancing support surfaces.

Physiological wear of the chewing surfaces inevitably leads to group function with contralateral balancing contacts. From the beginning of human evolutionary history, this occlusal maturation has been the functional objective that characterizes our species.

Although we tend to think of teeth as tools for chewing, it's only in mammals that dental elements truly began to engage in the act of chewing. The earliest mammals added mastication to the simple ingestion performed by reptiles. To understand how teeth should function and, consequently, how they should wear, it is essential to grasp the transition from teeth used for obtaining food to those used for processing it (Ungar, 2017).

Wear is an inexorable physical process between two opposing surfaces, and dental enamel is no exception. Enamel wear is a physiological process that modern dentistry often too hastily categorizes as pathological, without considering that the refined ultrastructural processes governing enamel wear are the expression of an evolutionary adaptation that began over 250 million years ago.

Reptiles typically have simple, conical teeth that are replaced multiple times during the animal's life, and their enamel does not need to last for their entire existence. In addition to microstructural differences compared to mammalian enamel (chemical composition, lack of prisms), reptilian enamel also differs in thickness, generally being thin and consistent across the tooth, except for a thickening at the conical tip (Ungar, 2017). Until the advent of mammals, enamel commonly had a structure devoid of prisms but was still effective in withstanding the functional stresses to which it was subjected. Reptiles lack a complex dental occlusion, are devoid of horizontal movement, and do not have precise and corresponding cutting surfaces between opposing teeth.

While the prismatic structure of enamel is considered a feature of mammals, we must always keep in mind that complex morphologies, a reduced number of replacements, heterodonty, and various other specializations have repeatedly evolved within the reptilian taxonomic group. Prismatic enamel has also been found in some cynodont dinosaurs, the Mississippi alligator, and certain Jurassic and Cretaceous theropod dinosaurs (Carlson, 1990), as well as in the lizard *Uromastyx nigriventris* (Cooper & Poole, 1973), which inhabits northern Africa and India—one of the very few reptiles to have developed a form of dental occlusion. In primitive mammals, the selection of prismatic enamel was favored by increased biomechanical stress imposed by chewing, heterodonty, and diphyodont dentition (Kemp, 1982).

Mammalian enamel evolved from a hypermineralized enameloid tissue that covered/covers the teeth of cartilaginous fish. Although enamel and enameloid have a similar composition, they form through distinct developmental mechanisms. Enameloid originates from mesenchymal tissue with long (over one micrometer) fluorapatite crystals (>1 μm) and a matrix primarily composed of collagen fibers. In contrast, mammalian enamel originates from epithelial tissue with smaller hydroxyapatite crystals and an organic matrix that does not contain collagen.

Since the early vertebrates, such as conodonts, enamel was composed of calcium hydroxyapatite crystals that were flat and oblong, exceptionally large, and generally parallel to each other, perpendicular to the tooth's surface, with no evidence of prisms (Kemp, 2002). In most amphibians and reptiles, enamel has remained aprismatic, with crystals approximately parallel to each other and perpendicular to the enamel surface (Line & Novaes, 2017). This unique arrangement of crystals has been preserved even in the more organized enamel of mammals, maintaining, as we will see, a crucial ultrastructural characteristic for regulating mastication wear—an ancestral structural arrangement later repurposed for a new function.

Differences in enamel thickness among various species provide insights into taxonomy, their life history, and the functional properties of their teeth (Hakkinen et al., 2019). Mammals that consume hard foods typically have thicker enamel compared to closely related species that feed on soft foods (Kay, 1981; Dumont, 1995).

A very illustrative example of the relationship between enamel and masticatory function can be observed in odontocetes, a suborder of cetaceans (sperm whales, dolphins, orcas, etc.), which, unlike mysticetes (baleen whales), have teeth instead of baleen plates. The terrestrial ancestors of these mammals, during their evolutionary journey, returned to the aquatic environment, leading to significant adaptations related to their new way of life. The only significant radiation of mammals that lost a precise dental occlusion occurred in these animals. As they became aquatic, they adapted to feeding without a precise dental occlusion, altering the developmental mechanisms that regulate tooth class and number (Armfield et al., 2013). Since chewing in an aquatic environment poses numerous challenges, modern odontocetes are the only mammals that have lost the reciprocity of occlusal contacts. They use their teeth solely for grasping and holding onto food without the ability to chew it (Werth, 2000; Armfield et al., 2013). This necessity led to a series of functional modifications, accompanied by changes in the number and shape of their teeth, enamel structure, and a thinning of enamel thickness (Fig. 3.1), resembling that typically found in pre-mammalian vertebrates (Loch et al., 2014, 2015; Schwenk, 2000).

In odontocetes, the absence of precise dental occlusion is associated with the absence of chewing (Ishiyama, 1987), which also resulted in a shift towards homodont dentition. In all odontocetes, feeding occurs in four cyclical phases, combining a predatory, sudden, and violent moment that incapacitates the prey with a subsequent phase where the prey is aspirated into the oral cavity through the action of the hyoid and lingual muscles. These muscles contract and retract, creating a

Fig. 3.1 Fossil tooth of Mosasaurus. Note the wear on the coronal apex revealing the thin enamel thickness

depression that draws water and prey into the mouth. Some of the aspirated water may enter the stomach, but a significant portion accumulates in specialized pockets that expand within the oral cavity, thanks to a depression in its floor. Subsequently, the accumulated water is expelled from the sides of the mouth (Berkovitz & Shellis, 2018).

In the Miocene odontocete whales, enamel still retained the prismatic appearance of their terrestrial ancestors. However, in most living odontocetes such as dolphins, orcas, and sperm whales, pseudoprismatic and aprismatic enamel has become predominant (Sahni, 1981).

The remarkable changes in occlusion, dental form, and enamel ultrastructure in modern odontocetes may appear, at first glance, as a regression, as if evolution had taken a wrong turn instead of following a more "direct" path toward a particular goal, almost as if it wasted time, generations, and energy. However, this is an example of how evolution is not a predetermined destiny. Through natural selection, the evolutionary product is a design without a creator (Silvertown, 2018). Evolution is not a preconceived plan and is not necessarily linear because it can manifest through independent pathways. It does not progress from "worse" forms to "better" forms but continuously adapts to environmental pressures.

Evolution does not plan, think, or desire anything. It can only build and adapt existing bodies with materials from the past and the DNA of parents (Hogervorst et al., 2009). Evolution proceeds through the branching of trajectories, not by transforming one form into another, as if the older form dissolves into the triumph of the newer one to transcend its previous and flawed self. As Stephen Jay Gould humorously and scientifically precise once said: "Novelties don't begin as butterfly wings, but as buds on old branches, and they don't transform Joe Shmoe into Michael Jordan" (Gould, 2002).

For most mammals, including humans, the formation of dental enamel is a finite process. However, this substance must last a lifetime, resisting wear and breakage from millions of chewing cycles and contact pressures (Zhou et al., 2013). Mammalian teeth face the formidable challenge of being both hard and resistant to strong stresses, and mammals achieve this with complex dental mineralization structures and compositions that have evolved to withstand fractures (Lucas, 2004;

Popowics & Herring, 2006; Ungar, 2008). Fracture resistance is a critical aspect of dental form and function because mammals have only two dental generations, and the fracture of a tooth can be catastrophic for the animal's life, often equating to a death sentence (Lucas, 2004).

The shape and structure of each tooth are designed to efficiently break down food into particles suitable for extracting the contained energy. Any loss of efficiency in chewing can significantly impact an individual's ability to survive (Lieberman, 2011).

Prismatic enamel is a hard, acellular substance with high mechanical and chemical resistance, capable of withstanding the forces of mammalian chewing. Due to the oblique contacts with opposing teeth and the food bolus during chewing, enamel is exposed to cutting forces that could easily trigger a tooth fracture, unlike calcified skeletal structures, which are not repairable (He & Swain, 2008).

In contemporary human populations, dental wear is occurring with different patterns compared to those of the past. This lack of occlusal surface maturation, combined with the substantial phenotypic variation within the human species, has made it challenging for various dental schools to establish a common and definitive occlusal rule. This condition of apparent indeterminacy can be summarized by the following words: "*In rehabilitative procedures, an optimal occlusion should be achieved based on the patient's needs. Masticatory efficiency can exist across a wide range of dental shapes and occlusal patterns, so no fixed rule can be applied to all patients. Occlusal rehabilitation should be performed by the dentist based on their knowledge of the various philosophies followed and their clinical skills*" (Tiwari et al., 2014).

Faced with a multitude of perspectives and equally numerous therapeutic approaches, seemingly without any fixed rules to follow, we must broaden our perspective. This involves looking both in space, considering the anatomical and physiological patterns that unite the various populations of *Homo sapiens*, and in time, observing how changes in dietary conditions have induced phenotypic changes in the stomatognathic system that are now considered the norm.

By making forms and their relationships mutable, the process of dental wear may appear to complicate the proper interpretation of occlusion. However, it actually points the way to better assess the shapes of occlusal surfaces and their interactions in relation to a person's age and biotype.

In dental practice, dental wear is often considered a negative phenomenon and, therefore, undesirable in itself. It seemingly weakens dental structure due to substance loss, potentially causing damage to the crown's integrity. Proper dental wear, on the other hand, is a physiological adaptive phenomenon related to chewing (Benazzi, Kullmer, et al., 2013). It is necessary for the correct maturation of occlusal surfaces, which is essential for developing a proper function (Fortelius, 1985; Ungar, 2015).

Occlusal wear is a biomechanical cornerstone around which the entire masticatory process revolves. It enhances the grinding effectiveness of occlusal surfaces by literally sculpting the teeth, continually adapting them to achieve the best possible shape for transforming food into a mush that allows optimal nutrient intake. Enamel responds to stress based on its micro and nanometric structure. The enamel prisms biomechanically react differently, transitioning from point contacts to areas of reciprocal friction necessary for cutting and crushing food bolus.

From its inception in humans, enamel wear has remained virtually unchanged. Only recently, in a very short span of time, has it undergone a radical reduction. In recent generations, dental wear has significantly decreased, and a potentially pathological pattern has replaced the physiological abrasive pattern of the past. Research on large prehistoric skeletal collections has demonstrated that past populations possessed nearly universally well-aligned dental arches with correct margins (Brace, 1977). This pattern was predominant in human populations until around 1900. Consequently, due to changing dietary habits in industrial societies, a new human phenotype has emerged over the last 150 years. In this phenotype, teeth no longer exhibit abrasions caused by a diet based on local and unprocessed foods. The cusps maintain a more complex morphology, necessitating an adaptation in the functional control of the masticatory organ (Slavicek, 2002).

The achievement of musculoskeletal maturity, also obtained through occlusal wear caused by the action of correct masticatory forces, tends toward a species-specific phenotypic model—a functional and morphological model linked to the typical masticatory dynamics of our species.Disruption of this model, due to the lack of morphological and functional maturation, can occur at various stages of individual development and for different reasons: this 'variability of interferences' leads to adaptive functional strategies necessary to 'push the mouth beyond the obstacle.' This altered growth model significantly increases the phenotypic variability of our patients' mouths.

Over the past century, populations that have adopted diets consisting of industrially processed foods have experienced a notable increase in certain dental conditions, including cavities, which had a sudden emergence previously with the advent of agriculture (Price, 1934, 1936). Other dental issues include misalignments, crowding, periodontal diseases, and non-carious cervical lesions with the loss of dental substance (enamel and dentin) in the buccal cervical region of the tooth. These diseases, now ubiquitous and widespread worldwide, were very sporadic in modern humans from pre-industrial societies. In contemporary populations that consume processed foods from the food industry, older individuals exhibit much less dental wear than in the past. Teeth are exposed much longer and more intensively to stresses concentrated in the buccal cervical region, also favoring the activity of other destructive processes such as biocorrosion. Due to the lack of dental abrasion, our masticatory system faces new challenges that can only be understood from an evolutionary perspective (Benazzi, Kullmer, et al., 2013; Benazzi, Nguyen, Kullmer, et al., 2013; Benazzi, Nguyen, Schulz, et al., 2013).

Even though dietary wear has decreased in contemporary populations, this process remains a natural adaptation that should not be stopped, except in pathological cases such as parafunctions or eating disorders (D'Incau & Saulue, 2012).

The wear of occlusal surfaces is related to the individual's lifespan and can be useful for estimating their approximate age (Molnar et al., 1983; Brown & Chapman, 1991; Riga et al., 2020). It continually remodels the occlusal surfaces, achieving the necessary adaptation that allows our hardest and most resistant substance to harmonize with the changes due to age, diseases, traumas, and postural adjustments that the stomatognathic system undergoes.

Wear enables a better distribution of occlusal forces by transitioning from a "point" area with high stress concentration to a more extensive area. Surface modeling related to abrasive wear and the resulting reduction in occlusal load have had a crucial influence on the evolutionary adaptation of teeth, increasing their specific resistance as life progresses (Lucas et al., 2013; Benazzi, Kullmer, et al., 2013; Benazzi, Nguyen, Kullmer, et al., 2013; Benazzi, Nguyen, Schulz, et al., 2013).

The movement of the mandible during chewing generates contacts and facets from consistent tooth-to-tooth friction (Evans & Fortelius, 2008). Enamel adapts morphologically to stress thanks to its fine nanometric structure. The biomechanical response of enamel prisms, as we have seen, is differentiated and programmed to make possible those areas of reciprocal friction necessary for processing the food bolus (Zhou et al., 2013). Wear should be understood as a necessary physiological process for the morphological and functional maturation not only of individual teeth or the entire occlusal surface but of the entire stomatognathic system.

The loss of dental substance due to wear is compensated by an alveolar-parodontal adaptive response that ensures the maintenance of proper reciprocity of enamel surfaces that come into contact. This parodontal response counteracts the loss of vertical dimension in occlusal wear and also contributes to the mesialization of dental elements associated with interproximal wear.

The idea that human teeth undergo continuous eruption throughout life was first proposed by Bernhard Gottlieb (in 1927) and later developed by other authors (Herzberg & Holic, 1943; Begg, 1954; Sicher, 1953; Sarnäs, 1957). The wear of occlusal surfaces is compensated by continuous and active dental growth without an increase in vertical dimension (Berry & Poole, 1976; Molnar & Gantt, 1977). This eruption is two-thirds related to bone apposition in the alveoli and one-third to cementum apposition along the roots (D'Incau & Saulue, 2012).

In most mammals, the occlusal surfaces of newly erupted teeth do not have the precise micrometric complementarity required for efficient chewing, and this fine interlocking of occlusal surfaces can only be achieved through proper wear (Osborn, 1982). Dental wear is so important that some mammals grind their teeth even in utero so that they are functionally adapted and ready for use from birth (Teaford & Walker, 1983).

As mammals developed increasingly complex motor patterns characteristic of mastication, well-defined wear facets with sharp and cutting edges developed on the occlusal enamel, both on the teeth responsible for guiding chewing (anterior dental group) (Fig. 3.2) and those necessary for food fragmentation, such as tribosphenic molars (Fig. 3.3).

When two flat enamel surfaces come into contact, such as the wear facets on two antagonistic teeth, the edges of the wear areas maintain their cutting efficiency only if they remain sharp; if the edges become dull, there will be a reduction in cutting efficiency (Saito et al., 2013).

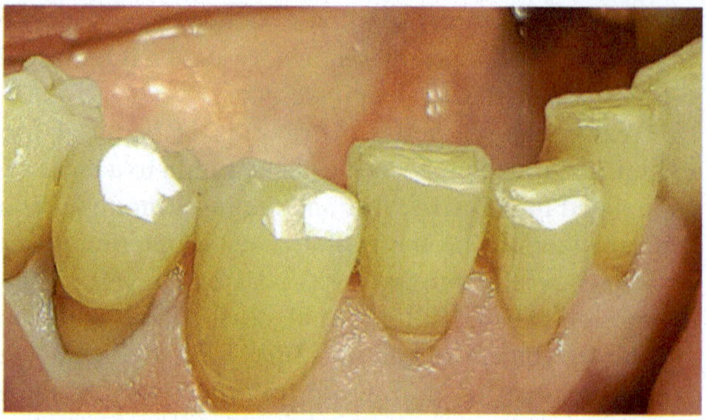

Fig. 3.2 Occlusal wear facets on the incisor group and premolars teeth

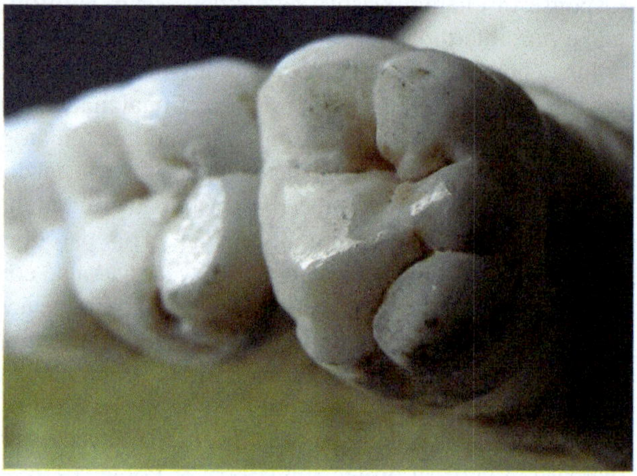

Fig. 3.3 Wear facets on the molars. Skull 5050 from the Museum of Anthropology, University of Florence. Occlusal

During food breakdown, abrasion occurs on the occlusal surface of teeth due to the interposed food (three-body interactions), with compressive forces distributed across all involved surfaces in a complex and instantaneous combination that makes it practically impossible to measure occlusal load during chewing (Katona, 2009). As the food bolus reduces to a thin layer, direct occlusal contacts occur between antagonistic teeth (Benazzi et al., 2011), establishing a two-body interaction.

The wear process and the resulting reduction in occlusal stress concentration may have had a crucial influence on the evolutionary adaptation of teeth, increasing their resistance to specific stresses in later stages of an individual's life. As wear progresses, facets and load directions change from an oblique direction (Fig. 3.4) to one that is nearly perpendicular to the tooth axis (Fig. 3.5).

The forces at play on the enamel surface are significant. The maximum bite force typically averages around 1000 newtons (Waltimo & Könönen, 1995), although this can vary from individual to individual and depending on the tooth in question:

- MEN: Incisal region ~380 N, Molar region ~900 N.
- WOMEN: Incisal region ~330 N, Molar region ~780 N.

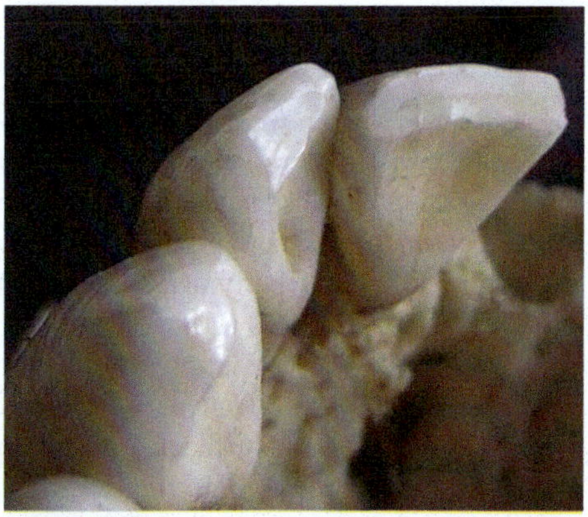

Fig. 3.4 Initial wear with an oblique pattern on the incisal edge. Skull 120 from the Museum of Anthropology, University of Florence

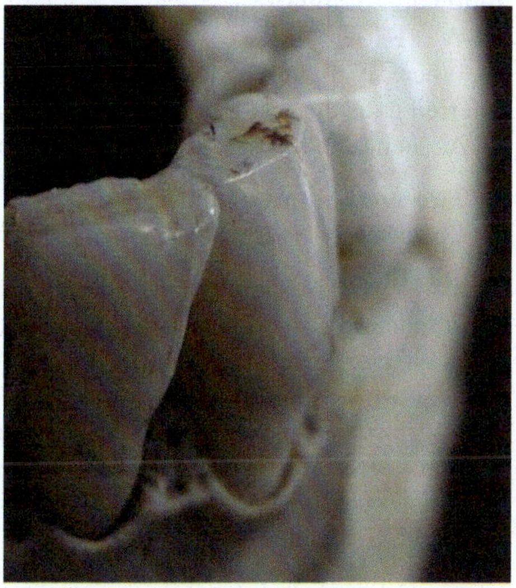

Fig. 3.5 Advanced incisal wear with a flat pattern. Skull 5049 from the Museum of Anthropology, University of Florence

Enamel, harder but also more fragile than dentin, serves a protective function for the underlying layers. When wear exceeds the thickness of enamel, the tooth has reactive capabilities within certain limits, such as dentin sclerosis and the deposition of tertiary dentin, which allow the pulp to survive biological insults. Dentin is more elastic and less hard and resistant than enamel, and its biomechanical role is to absorb the forces of biting while supporting the protective layer (Chun et al., 2014).

From a physical perspective, wear is a gradual removal of material from the surface of a solid body, which is in relative motion with respect to another solid or a fluid. Several types of wear have been classified (adhesive, abrasive, corrosive, erosive, fatigue), but the one we will consider is abrasive wear, which occurs during chewing.

3.1 Abrasive Wear

Abrasive wear is the wear that occurs through contact between teeth and food bolus (three-body wear) and between teeth themselves (two-body wear). In this type of physical phenomenon, saliva serves as a biotribological lubricant (tribology—from the Greek "tríbein" meaning "to rub"—is the discipline that studies friction, wear, and lubrication between surfaces in relative motion), reducing friction and wear between dental surfaces (Ungar, 2015).

Wear is a complex phenomenon for which it is essential to consider the microscopic characteristics of surfaces. Abrasive wear is directly proportional to the compressive load between surfaces, inversely proportional to their hardness, and related to a set of conditions, including the morphological features of the surfaces, their extent, and the presence of a lubricating element. Wear can be summarized by the following formula:

$$\mathbf{W = FxC/H}$$

Where **W** quantifies wear, **F** is the total load, **H** is hardness, and **C** is a constant, referred to as the coefficient of wear (Harris et al., 2002).

As mentioned earlier, there are two types of abrasive wear:

- **Two-body wear**, which occurs with direct tooth-to-tooth contact.
- **Three-body wear**, which occurs with tooth-food-tooth contact.

Abrasive wear is related to the degree of surface roughness. No matter how smooth and regular a surface may appear to the naked eye, at increasing magnifications, it constantly exhibits irregularities that define its roughness, which is the irregular deviation from flatness. It is precisely the nature of these micro-asperities, along with surface hardness, that potentially influences the type and degree of wear. Under a microscope, no surface appears completely smooth. When two surfaces, even if apparently smooth, come into contact, it is their asperities that make contact and act as abrasive particles (D'Incau et al., 2012).

In the physical world, there is no point load (Benazzi et al., 2011); hence, the actual contact area is the sum of all the small contact areas and depends on deformation, surface roughness, and the intensity of the load. In a situation of reduced force and high rigidity, the contact area can be reduced to one-thousandth of the apparent one, and even if the overall pressure exerted between the two materials is weak, the local pressure is concentrated on micro-contacts so intensely that it can cause deformation or breakage (Fig. 3.6).

When two surfaces with different hardness levels come into contact, the asperities of the harder substance plow the surface of the more ductile one, creating a groove that repeats with each pass. These grooves continue to accumulate, weakening the surface by locally deforming it through their contiguity and removing micrometric amounts of substance in various ways: either through microgrooves without plastic deformation or through the propagation of small fractures that, when joined, result in material detachment. The cumulative effect of the repetition of these steps causes the sum of these infinitesimal losses to become macroscopic, resulting in visible wear (Zum Gahr, 1987) (Fig. 3.7).

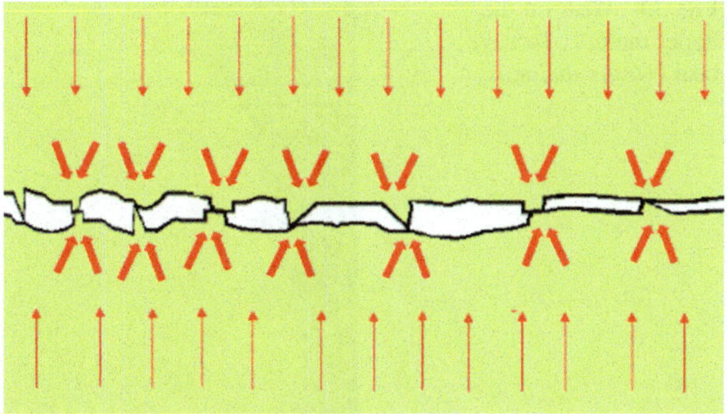

Fig. 3.6 Micro-contacts create a contact surface much smaller than what appears macroscopically: forces concentrate on these reduced areas

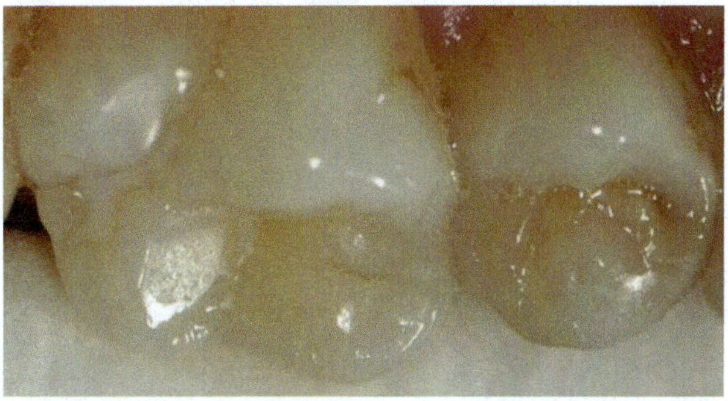

Fig. 3.7 Wear on the cutting/guiding surfaces. The pronounced facet on the first upper molar indicates the work done by this surface in regulating the entry phase of the masticatory cycle, protecting against the retrusion of the working condyle

Microscopic investigation of enamel wear can provide numerous insights into dental function, allowing the determination of chewing motion patterns, different types of diets, and non-food use of teeth: the depth, length, and width of these marks reflect the type of diet consumed (Kaidonis, 2008). Species that consume leaves and other non-reproductive plant parts appear to accumulate enamel surfaces with aligned scratches, while animals that feed on harder foods like seeds and nuts tend to show more irregular marks and pitted surfaces (Ungar, 2010; Lucas et al., 2013). These marks represent both dietary habits and the stresses associated with non-food use, they overlap and fade within a few days (Ungar, 2008), representing a trace of the last meals consumed. The evolution of mammalian enamel has been guided by a combination of evolutionary and geometric constraints, functional influences, and phylogenetic history (Maas & Dumont, 1999). The evolutionary path led mammals to drastically reduce the number of dental replacements, reducing them to two, in order not to nullify the finely tuned dental gears necessary for chewing through frequent tooth renewal: in mammals, each individual tooth must function for a considerably long period of time, which is why it needs to be supported by more complex structural patterns and extremely precise occlusal dynamics compared to other vertebrates.

Each tooth evolved into a highly precise tool for cutting and crushing food (Figs. 3.8, 3.9), which dental wear perfected by sharpening its edges and continually adapting the complementary nature of opposing surfaces.

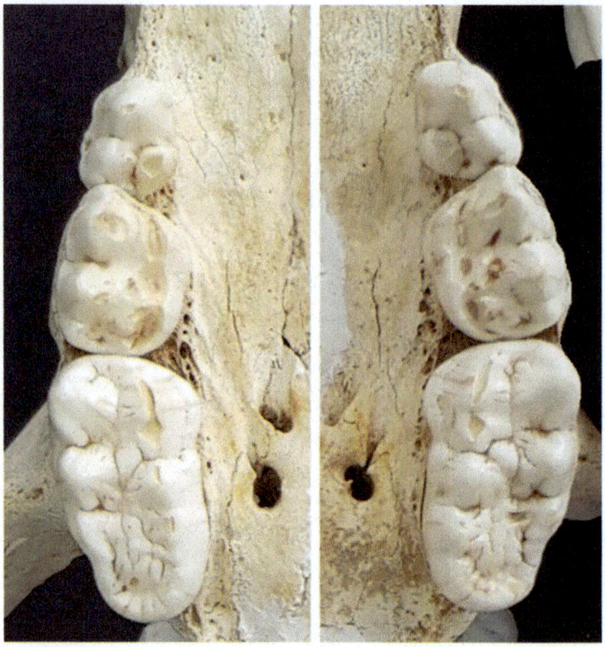

Fig. 3.8 Wear on the upper molars of a cave bear (*Ursus spelaeus*)

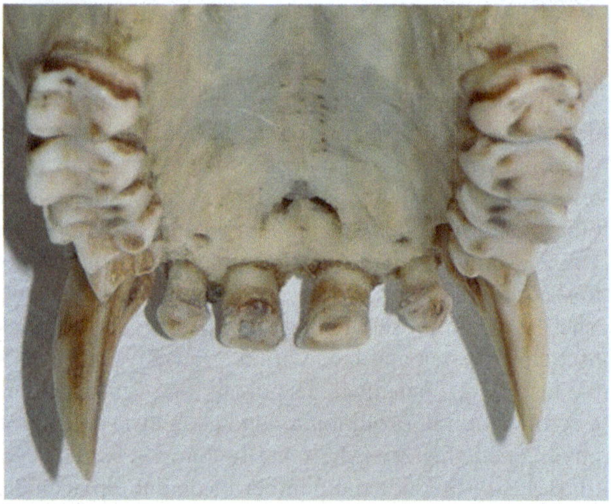

Fig. 3.9 Wear on the upper teeth of a gibbon (*Bunopithecus hoolock*). Note the widespread wear on the entire occlusal surface. Specimen 3216, collection of the Museum of Zoology, University of Florence

The prismatic organization of mammalian enamel enabled their teeth to adapt to the biomechanical stresses imposed by chewing, simultaneously meeting two opposing needs: high resistance to fracture and the ability to morphologically adapt to function.

Most non-mammalian vertebrates can replace their teeth when they break, while in mammals, permanent teeth are not replaced. Generally, in this class of vertebrates, enamel growth is a finite process, and once deposited in genetically predetermined thicknesses, this acellular substance is unable to self-regenerate. This makes it more susceptible to a "programmed obsolescence" compared to cellular tissues. Once mature, dental enamel must withstand millions of chewing cycles, with forces ranging from 28 to 1200 N (Yilmaz et al., 2015), lasting a lifetime and resisting both breakage and excessive wear (Zhou et al., 2013).

The Homo genus began to differentiate from the hominid family around two million years ago, and our species, *Homo sapiens*, appeared on the scene around 220,000 years ago. Although genetic data are increasingly important (considering that there is over 98% DNA in common between humans and chimpanzees), diversity is still better represented by the phenotype, i.e., how the organism appears, reflecting the evolutionary and adaptive history of its species (Manzi 2013).

Throughout the history of *Homo sapiens*, for a very long time, until the development of agriculture (around 10/12,000 years ago), sustenance was primarily linked to hunting and gathering activities, with a diet consisting of hard, fibrous, and substantial foods, certainly tougher than those consumed in later epochs. During this evolutionary phase, almost all disorders of the masticatory apparatus were very rare.

Until fairly recent times, human populations shared common occlusal patterns and were characterized by well-aligned dentitions. Epidemiological data show that the decline in dental wear and the increase in malocclusion occurred simultaneously with the adoption of a 'modern' lifestyle, which led to a decrease in masticatory function. Paleopathological research shows that the reduction in dental wear and the increase in occlusal variability occurred almost universally and concurrently among human populations (Corruccini, 1984). Oral health began to deteriorate with the dietary shift due to the introduction of agricultural techniques. The substantial change in diet during the transition from a subsistence based on hunting to one based on agriculture, with a significant increase in carbohydrate consumption, also brought about a marked reduction in food variety and consistency. This, in turn, led to a decrease in masticatory function and a widespread decline in oral health (Hunt, 1961; Larsen, 2002).

Environmental pressures on the human genome dramatically increased following two events: the advent of agriculture about 12,000 years ago, marking the beginning of the Neolithic period, and the onset of the industrial era around 200 years ago, characterized by the widespread availability of soft, sugar-rich foods.

Today, these rapid environmental changes are believed to be responsible for creating a closely interconnected group of oral health problems, including dental caries, periodontal disease, crowding, and malocclusions. The prevalence of dental caries has been estimated to be less than 2% in fossilized human teeth from the Paleolithic era. However, the first significant increase in caries incidence (9–10%) occurred at the beginning of the Neolithic period with the spread of agriculture. The second and more dramatic increase took place between the 19th and 20th centuries when caries became widespread, affecting between 50% and 90% of individuals (Saratti et al., Part 1, 2021). Therefore, dental caries, which was quite rare in anthropoid apes and among hunter-gatherers, began to spread significantly after the practice of agriculture became established, reaching its peak in the late nineteenth and twentieth centuries (Lieberman, 2014). A similar incidence was observed in malocclusions caused by cross-bites, posteriorization, Class II or III sagittal molar relationships, dental misalignments, crossbites, and anomalies in oral segments (Corruccini et al., 1983; Larsen, 2002). The more pronounced and widespread deterioration on a global scale occurred around the turn of the 19th and 20th centuries with the second 'food revolution,' marked by the global dissemination of processed and refined foods produced by the food industry (Price, 1945).

Today, in our dental practices, we are confronted daily with these changes, to which our stomatognathic system has not yet had the necessary evolutionary time to adapt. The first farmers lived only about 500/600 generations ago, and in most parts of the world, agriculture has been practiced for fewer than 300 generations. Starting around 10,000 years ago, there has been a 'cranial gracilization' and 'contraction' of the jaws, with a significant reduction in facial measurements resulting from dietary changes (Mew, 2004). We have modified our foods, and the changed foods, in turn, are modifying us.

Over the past 100–150 years, all human populations have undergone significant changes in dentition and occlusion, with a significant increase in the prevalence of malocclusions (Varrela, 1990; Kaul & Corruccini, 1984). This increase is highlighted by the absence of malocclusions in previous populations (Corruccini, 1984). Some dental problems in contemporary societies reflect the disparity between the original design of our dentition and our current environment. The human dentition is fundamentally 'designed' with the expectation of extensive wear, a reasonable assumption when considering that humans evolved in relatively recent times in environments with highly abrasive foods and a significantly shorter life expectancy compared to today (Kaifu et al., 2003) (Fig. 3.10).

The processes of food industrialization dramatically increased the consumption of processed, less fibrous, and often softer foods, which are much less abrasive (Molnar, 1971; Kaifu, 1999). Additionally, these foods are characterized by being loaded with refined sugars, saturated fats, and super-fine salt to better stimulate taste buds (Moss, 2013).

Evolution has encoded our passion for these three appealing and high-energy-density substances into ancient brain circuits. All three substances are quite challenging to find in nature, and due to their infrequent intake, the central nervous system processes a rewarding response by 'activating' brain areas associated with pleasure and stimulated by the release of neurotransmitters like dopamine in the diencephalon or endogenous opioids within the nucleus accumbens. The nucleus accumbens is a site in the forebrain crucial for regulating reward-related behavior and plays a significant role in appetite control (Will et al., 2003; Zhang & Kelley, 2000). This made it more rewarding to take advantage of the infrequent and enticing opportunities to stockpile rare and precious substances, making these hyper-nutrient foods like sugars and fats, or salt, necessary for proper electrolyte balance, an irresistible temptation.

The food industry quickly recognized this human 'food weakness' and skillfully exploited it on a global scale. In many countries, the consumption of these foods has become a social problem with indiscriminate intake of sweet, fatty, and salty foods, leading to serious diseases such as diabetes, obesity, and hypertension (Misra et al., 2011).

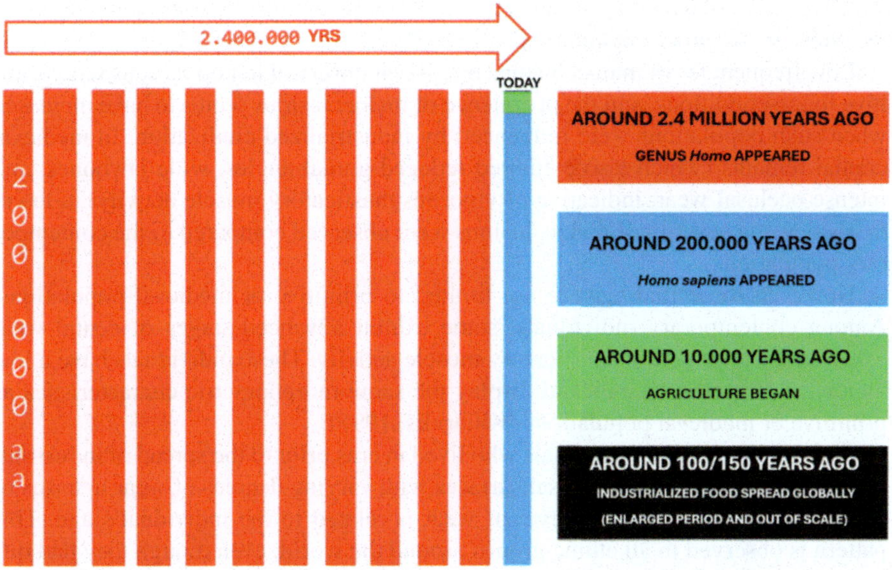

Fig. 3.10 Timeline of the presence of humans and *Homo sapiens* on Earth. Note the out-of-scale enlargement of the last century, during which enamel wear has drastically reduced (see Fig. 144) and simultaneously there has been a spread of malocclusions due to underdevelopment of the jaws. It is remarkable that modern gnathology is based on anatomical-functional studies conducted within the very thin black bar of the graph; in this last and very brief period of our history (about 150 years), a new phenotype of man has essentially emerged, driven by the changed dietary habits imposed by industrialized food, first appearing in Western societies and then rapidly spreading globally (Proffit et al., 2006)

Returning to the effects of the inconsistency of refined foods, dental wear, which has remained unchanged since the dawn of humanity, has recently undergone profound changes in a very short period. In past populations, wear was omnipresent, intense, abrasive, and physiological (Proença et al., 2014) because it was related to poorly processed foods before consumption. However, in current populations, enamel wear has been dramatically reduced.

The reduced masticatory function resulting from foods with low mechanical consistency also leads to reduced development of the jawbones. This relationship has been well-documented in experimental studies comparing animals fed on hard diets with those fed on soft diets. Consistently, animals fed on soft diets exhibited a significant reduction in facial and mandibular dimensions compared to animals on a hard diet (Corruccini & Beecher, 1982).

Chewing forces, in addition to wear, are crucial for the proper development of jawbone structures. For instance, Aboriginal children raised on a consistent and fibrous diet tend to have generally wide and low dental arches, characteristics also found in late Pleistocene European populations (Hunt, 1961).

From a biological perspective, in response to a continuous reduction in biomechanical pressure on the masticatory system, our bodies have reacted with a volumetric reduction of the system itself, leading to greater phenotypic variability and the emergence of asymmetries during the development, growth, and remodeling of the entire masticatory apparatus (Oxilia et al., 2018).

Low frequencies of malocclusions have been observed in populations with primitive living conditions, and the prevalence of malocclusions seems to have increased when such populations were influenced by industrialized civilization. In medieval cranial remains (which also exhibited reduced gonial angles, wide jawbones, and intense occlusal wear, indicative of vigorous masticatory muscle activity), a lower frequency and severity of malocclusions were observed compared to the contemporary population.

However, these findings are not limited to primitive or medieval populations. Among contemporary individuals, some exhibit advanced stages of dental wear associated with vigorous masticatory muscle activity. These individuals have a low incidence of malocclusions and display the same morphometric characteristics as primitive or medieval populations (Kiliaridis, 1995).

Skulls belonging to individuals who lived in eras prior to the spread of industrialized food typically show occlusal surfaces with varying degrees of wear in terms of intensity and extent, but the level of wear is related to the individual's age. This pattern is observed in all ethnic groups around the world, even though they had different diets, but were united by minimal food processing

Caries and most of the aforementioned malocclusions are diseases resulting from a phenomenon called evolutionary mismatch (Lieberman, 2014). This means that these conditions arise from traits that evolved in response to certain environmental pressures but become disadvantageous due to changes in the environment itself. In the modern industrialized world, malocclusion is endemic (Moore et al., 1968; Dickson, 1970), but it is still less frequent in populations living in primitive conditions. For example, in India, a sample of children living in urbanized areas exhibited deep overbites at a rate nine times higher than their peers living in rural areas (Corruccini, 1984).

Abrasive wear that occurs on dental surfaces can be divided into two categories:

- *Interproximal wear*
- *Occlusal wear*

3.1.1 Interproximal Wear

Interproximal wear (UI) is a gradual phenomenon primarily linked to the cumulative effect of interdental contacts and micro-frictions that occur at the points (which later become areas) of direct contact between interproximal surfaces

UI happens due to dental movements during both chewing cycles and non-masticatory activities. During chewing, forces with multiple directions are exerted on the teeth, leading to a physiological movement of dental elements in relation to different dynamic components. This movement induces a friction between the proximal contact surfaces of the enamel against each other (Pileicikiene & Surna, 2004).

UI occurs both with a downward movement into the alveolus (about 30 micrometers) that the teeth experience during maximum intercuspation or chewing, and through lateral motion between adjacent teeth, caused by the viscoelasticity of the periodontal ligament (Parfitt, 1960), resulting in a movement of 50 to 110 micrometers (50–110 μm) (Kim et al., 2005; Gross, 2008).

During chewing, there is also a posterior-anterior movement of each dental element that distributes the mesial component of occlusal forces throughout the entire dental arch, pushing each tooth mesially. This movement is significantly intensified by a decrease in the von Spee curve, typically associated with considerable occlusal wear (Wolpoff, 1971; Gaspard, 1979). While occlusal surface wear indicates both chewing force and the abrasive properties of food, interproximal wear depends more on the amount of chewing force applied to the teeth, as it is the lateral component of this force that causes wear (Wolpoff, 1971).

UI and the resulting mesial migration of teeth reduce the frequency of malocclusion and crowding of anterior teeth, allowing sufficient space for the eruption of the third molar (Begg, 1954; Pereira & Evans, 1975). The size, position, and shape of interproximal contact areas depend on the individual anatomical characteristics of interproximal surfaces (Pileicikiene & Surna, 2004; Saito et al., 2013) (Fig. 3.11).

Before the adoption of an 'industrialized' diet, UI and occlusal wear began early, characterizing even the deciduous dentition. Once the transition occurred, even permanent teeth adapted to function by wearing out quickly (Figs. 3.12, 3.13, 3.14, 3.15, 3.16, 3.17).

Consequently, due to the nature of the masticatory cycle, the movement associated with UI leads to a difference between the highly worn and concave mesial surfaces and the distal surfaces, which tend to remain convex (Gaspard, 1979) (Figs. 3.18, 3.19).

UI leads to the formation of actual contact surfaces with a progressive mesial drift of teeth and a reduction in arch length (Begg, 1954) (Figs. 3.20, 3.21, 3.22).

Occlusal wear associated with interproximal wear resulted in a natural maturation of the mandibular-cranial relationship. In past populations, incisal overbite was

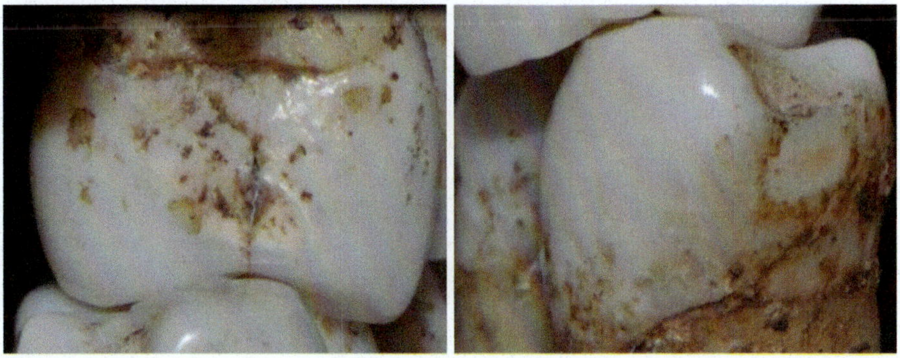

Fig. 3.11 Interproximal wear: note the extent of the contact areas. Skull 318 from the Museum of Anthropology, University of Florence

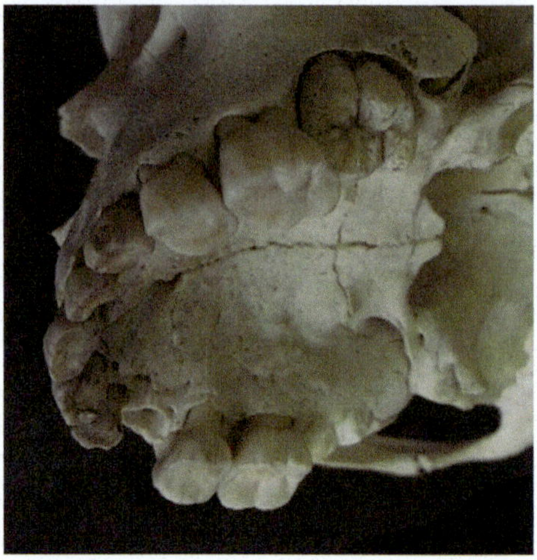

Fig. 3.12 Dental wear in deciduous dentition. Skull 3603 from the Museum of Anthropology, University of Florence

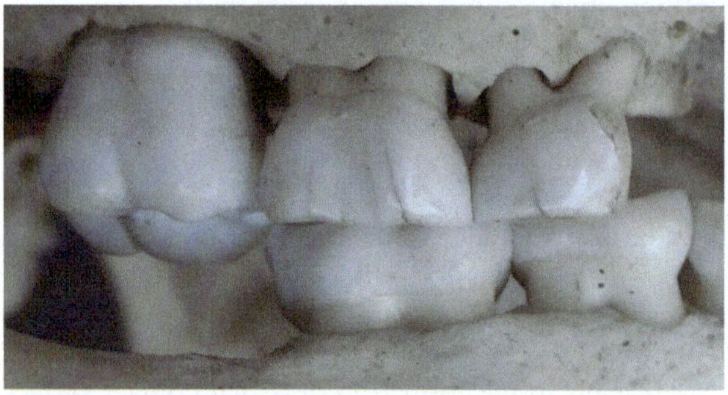

Fig. 3.13 Dental wear in deciduous teeth in mixed dentition. Skull 6153 from the Museum of Anthropology, University of Florence

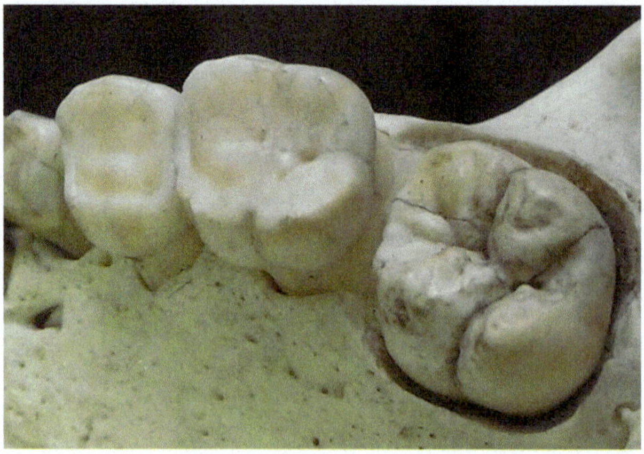

Fig. 3.14 Dental wear in deciduous teeth in mixed dentition. Note the difference between the surface of the deciduous teeth, sculpted by chewing, and that of the first permanent molar, still intact as a 'tool' to be adjusted to masticatory function. Skull 3063 from the Museum of Anthropology, University of Florence

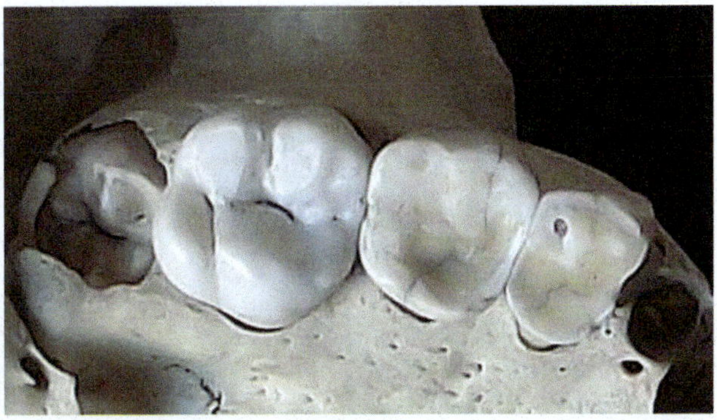

Fig. 3.15 Dental wear in deciduous teeth in mixed dentition. Note the strong wear of deciduous molars and the early formation of wear facets on the occlusal surface of the first permanent molar. Skull 6153 from the Museum of Anthropology, University of Florence

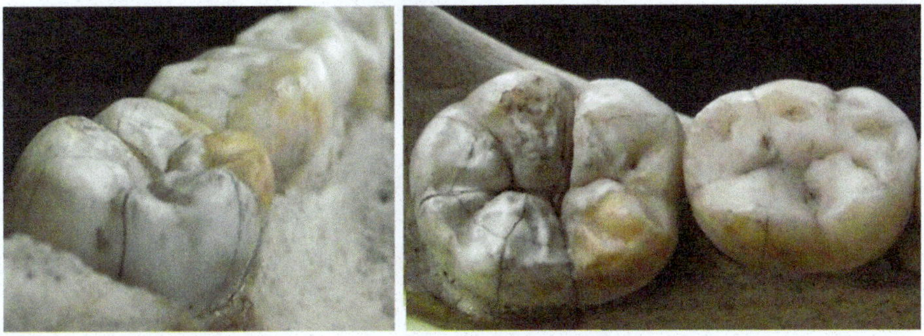

Fig. 3.16 Different degrees of wear between deciduous and permanent dentition. Note the strong wear of the supporting cusps of the deciduous molar and the early formation of wear facets on the lingual surface of the buccal cusps of the first permanent molar. Skull 3603 from the Museum of Anthropology, University of Florence

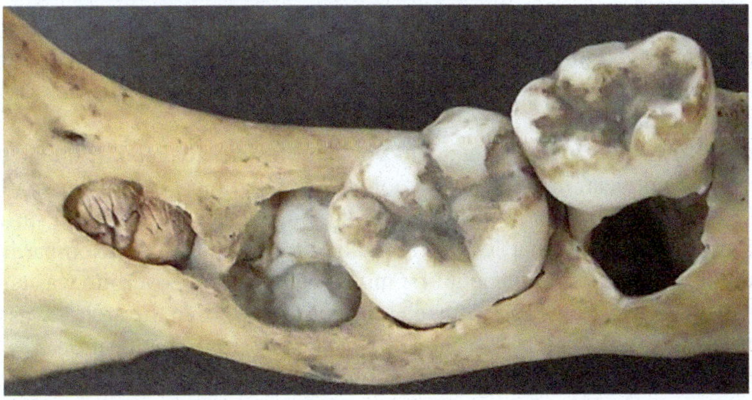

Fig. 3.17 Wear in mixed dentition. Note the presence of wear facets on the inner slopes of the buccal cusps of the first molar despite its recent eruption in the arch. Skull 4230 from the Museum of Anthropology, University of Florence

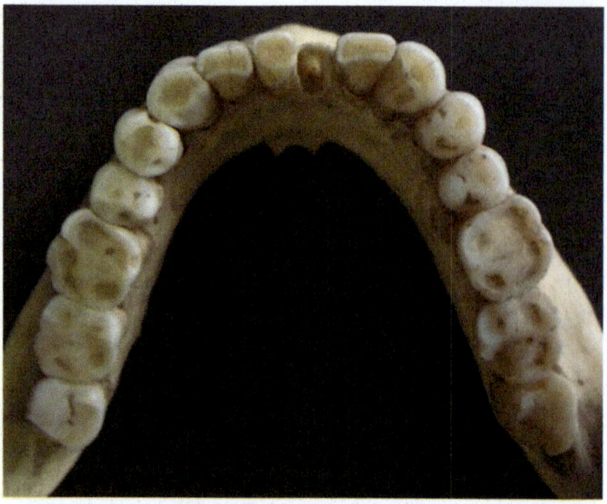

Fig. 3.18 Occlusal and interproximal wear. Skull 4781 from the Museum of Anthropology, University of Florence

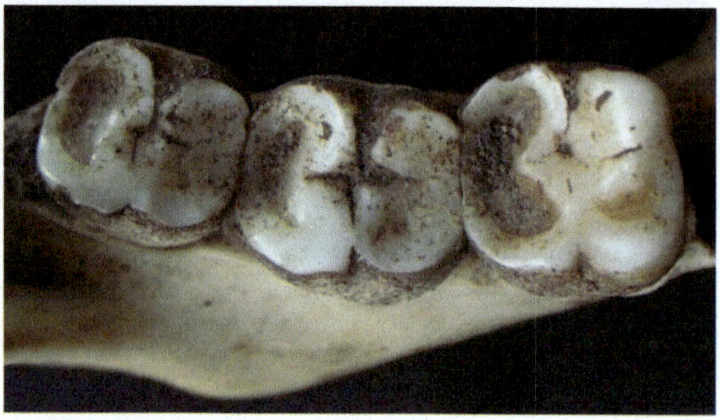

Fig. 3.19 Occlusal and interproximal wear in permanent dentition. Skull 4234 from the Museum of Anthropology, University of Florence

transient. Already at the age of 5, the mesial movement caused by UI and the wear of the incisal edge led to a head-to-head position of the deciduous incisors (Hunt, 1961) (Fig. 3.23).

Dental crowding was rare in ancient populations, whereas today, due to the lack of interproximal wear, it is endemic among all populations with an "industrialized" diet. The mesial migration of posterior teeth compensates for the tooth surface lost due to attrition from a diet that does not stimulate proper dental wear (Lombardi, 1982).

With the eruption of permanent first molars and incisors, overbite was temporarily regained. However, by maturity, the relationship between the arches was already heading towards a second head-to-head incisal occlusion. The attrition was usually so pronounced that dental crowding was only a transient phase during the eruption of permanent teeth (Figs. 3.24, 3.25, 3.26) (Hunt, 1961).

In contemporary industrialized populations, the same adaptation occurs but at a significantly reduced rate (D'Incau et al., 2012).

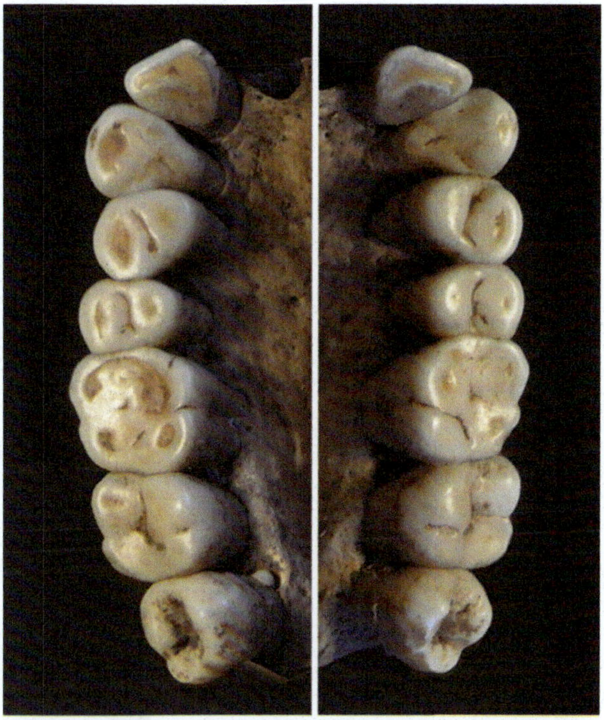

Fig. 3.20 Occlusal and interproximal wear. Note the concave profile of the mesial interproximal surfaces and the convex profile of the distal interproximal surfaces. Skull 3049 from the Museum of Anthropology, University of Florence

Fig. 3.21 Occlusal and interproximal wear in permanent dentition. Skull 4290 from the Museum of Anthropology, University of Florence

The movement associated with physiological intrusion translates into proximal enamel wear over time, transforming from 'point-like' areas to more extensive contact zones, which must be respected and reproduced in the case of conservative or prosthetic reconstructions (Fig. 3.27).

A brief digression is necessary regarding Non-Carious Cervical Lesions (NCCL), a phenomenon seemingly related to abrasive wear but also linked to reduced masticatory intensity, which in turn results from a diet dominated by 'processed' foods.

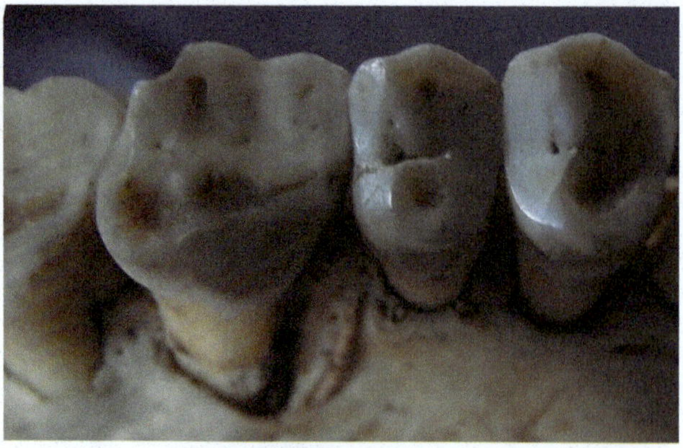

Fig. 3.22 Occlusal and interproximal wear in permanent dentition. Skull 4325 from the Museum of Anthropology, University of Florence

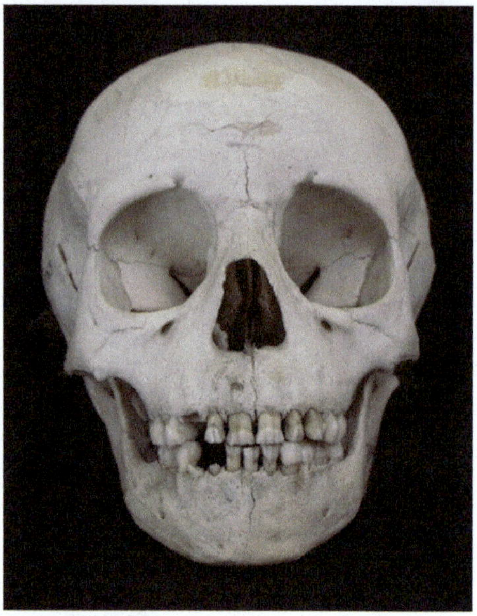

Fig. 3.23 Wear in deciduous dentition. Skull 3603 from the Museum of Anthropology, University of Florence

Over the last century, individuals in industrialized societies have witnessed a significant increase in certain dental diseases. A critical issue involves the loss of dental substance (enamel and dentin) in the buccal and cervical region of the teeth. NCCLs are currently widespread and prevalent among human populations worldwide, exposed to tensile stresses in the buccal cervical region for decades.

It has been observed that reducing cusp height decreases the tensile stresses exerted on the tooth (Chang et al., 2009). However, these considerations do not suggest that having a completely flat occlusal surface is preferable, as there is no doubt that occlusal ridges play a vital role in food processing. However, a reduction in cusp slope due to wear reduces tensile stresses on the crown. This reduction could be beneficial when crests and ridges, grooves, enamel thickness, and crown height

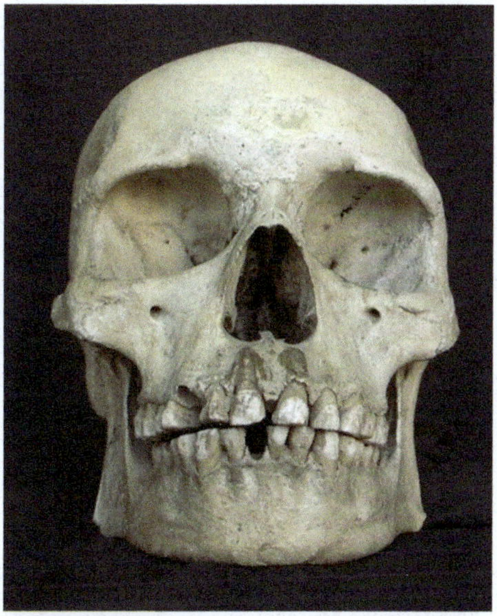

Fig. 3.24 Wear in permanent dentition. Skull 2507 from the Museum of Anthropology, University of Florence

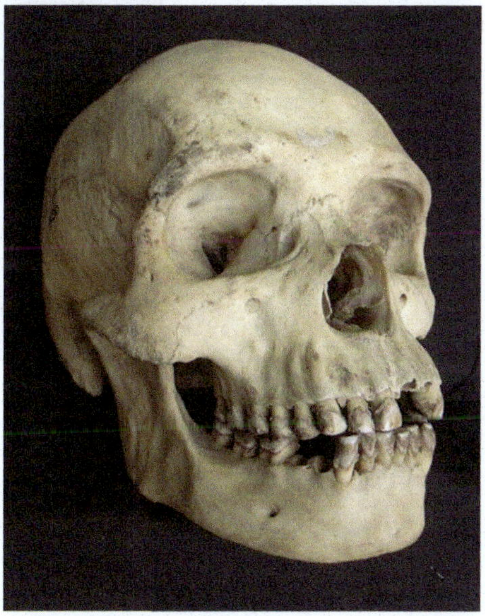

Fig. 3.25 Wear in permanent dentition. Skull 4290 from the Museum of Anthropology, University of Florence

have been removed or significantly reduced due to tooth wear (Benazzi, Kullmer, et al., 2013).

In the modern industrialized world, elderly individuals exhibit much milder dental wear compared to the past, and their teeth are exposed to strong tensile stresses in the buccal cervical region for many more decades, given the increase in average life expectancy. This is the most likely mechanism explaining enamel loss in the

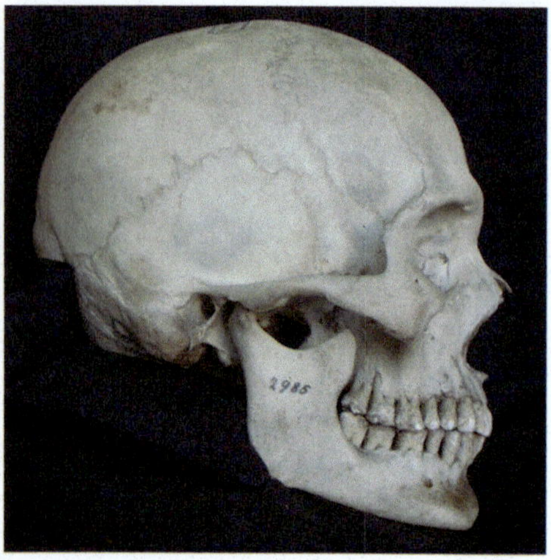

Fig. 3.26 Wear in permanent dentition. Skull 2985 from the Museum of Anthropology, University of Florence

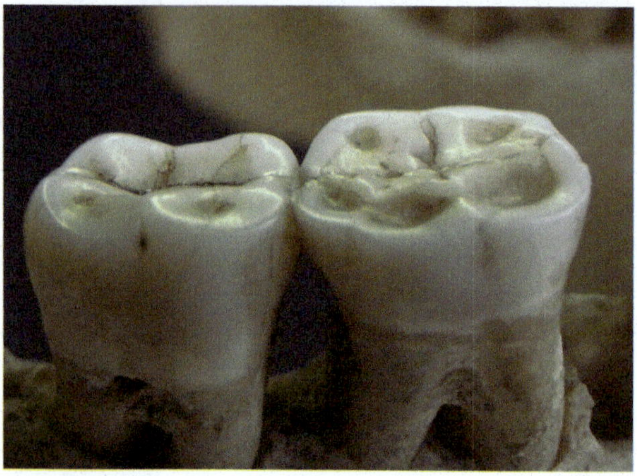

Fig. 3.27 Wear in permanent dentition. Skull 2985 from the Museum of Anthropology, University of Florence

cervical region, promoting the activity of other destructive processes like biocorrosion (Benazzi, Kullmer, et al., 2013).

During the masticatory force, the occlusal load on molars exerts high forces transmitted to the tooth neck, stressing both the cortical bone surrounding the tooth (Benazzi et al., 2012) and the dental neck region (Figs. 3.28, 3.29), causing structural weakening that can pave the way for subsequent biocorrosive phenomena and even fracture (Fig. 3.30).

A precise and individualized reconstruction of the occlusal surface of dental arches is crucial for a functional and balanced distribution of stresses, as well as optimal masticatory efficiency. This approach is essential in clinical dentistry because the individual evaluation and positioning of occlusal forces are important factors for estimating and achieving long-term success in orthodontic, conservative, implant, and prosthetic treatments (Benazzi et al., 2011).

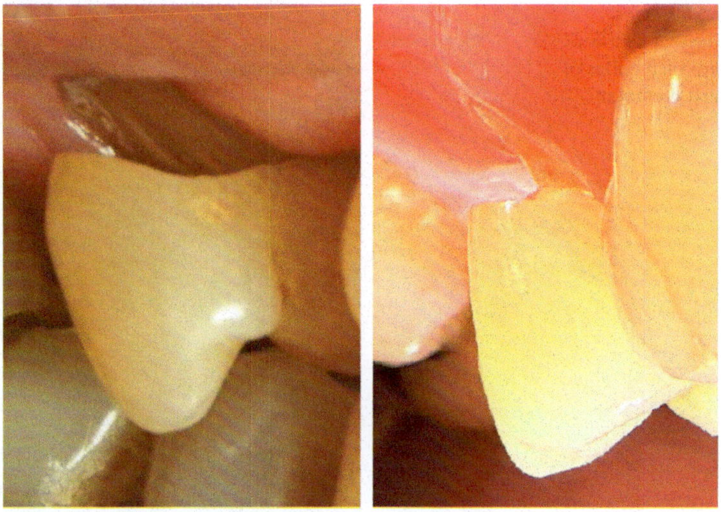

Fig. 3.28 Non-carious cervical lesion: note the marked height of the vestibular cusp of the first upper premolar and the inclination of its crown towards the palate, which increases the depth of the occlusal constraint in lateral movement. The rotation of the canine along its axis prevents its guiding function in lateral movement, favoring the accumulation of transverse stress on the premolar

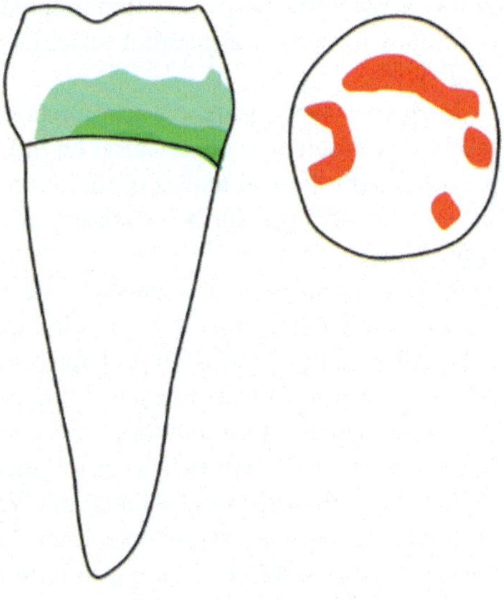

Fig. 3.29 Occlusal compressive stress (red); cervical tensile stress (green). Drawing freely adapted from: "The evolutionary paradox of tooth wear: simply destruction or inevitable adaptation?", Benazzi, Kullmer, et al., 2013

3.1.2 Occlusal Wear

Whenever opposing teeth come into contact, forces are exerted, leading to wear. During chewing (and also during functional and parafunctional contacts outside of it), there is abrasion of the occlusal surface. Initially, due to the interposed food, this

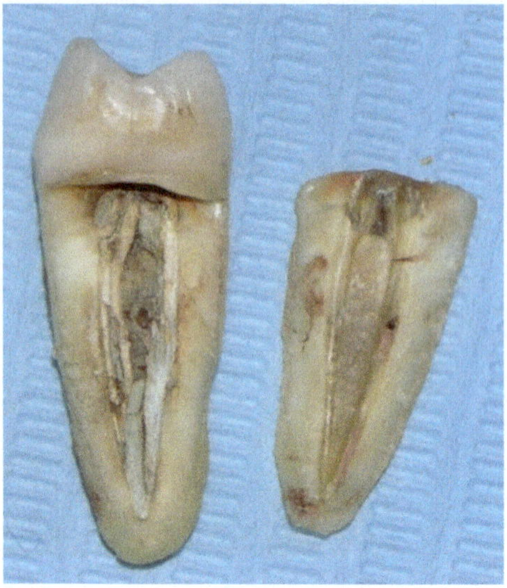

Fig. 3.30 Singular fracture line that follows the interproximal collar of a devitalized premolar

interaction involves three bodies. However, as the food bolus reduces to a thin layer, occlusal contacts between antagonistic teeth cause abrasive interactions between two bodies (Benazzi et al., 2011). The resulting wear creates wear facets—flat, glossy areas with sharp and often pointed edges (Kaidonis, 2008). These facets are covered with tiny sub-parallel striations that reflect the direction of movement (Fig. 3.31).

Each wear facet is always accompanied by a corresponding one on the opposing tooth. If we focus our observation on individual teeth, it's not possible to assess the potential pathological nature of the worn surfaces. This judgment requires a global view of the occlusal surfaces, taking into account the subject's age simultaneously (Fig. 3.32).

Physiologically, wear facets should be evenly distributed, with their relative extent mainly determined by the individual's age. Limited wear on individual teeth or small isolated groups should raise suspicion of parafunction or dysfunction. Abrasive wear is considered physiological when it is a result of aging, specifically the consequences of masticatory function influenced by an individual's diet and age. In the posterior regions, two types of facets are created, working facets, and balancing facets, both outcomes of function (Woda et al., 1987).

In our patients, we rarely encounter the extensive dental wear that characterized past populations. It's interesting to note that wear also serves as a protective phenomenon for the integrity of the stomatognathic apparatus. By increasing the contact areas in number and extent, it progressively improves the dispersion of occlusal forces, reducing the occlusal force per unit area (Benazzi, Kullmer, et al., 2013; Benazzi, Nguyen, Kullmer, et al., 2013; Benazzi, Nguyen, Schulz, et al., 2013). Wear facets have a crucial influence on stress distribution across the entire tooth, whereas grooves and fissures on the occlusal surface are critical points as they concentrate tensile stresses (Benazzi et al., 2011).

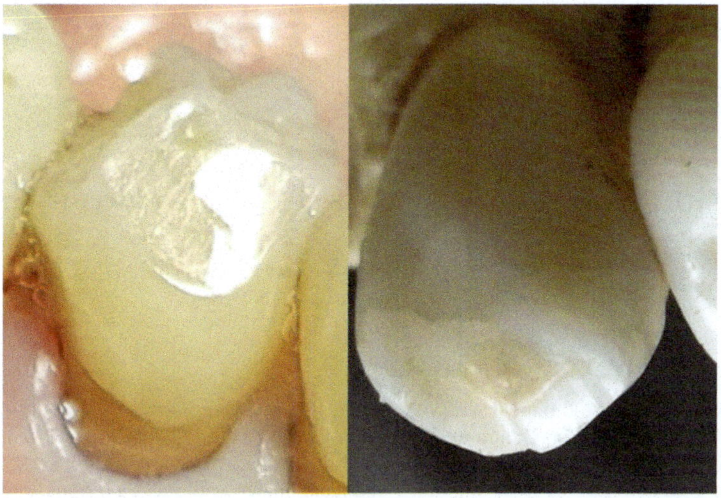

Fig. 3.31 Wear facets. Skull 322 (on the right). Museum of Anthropology, University of Florence

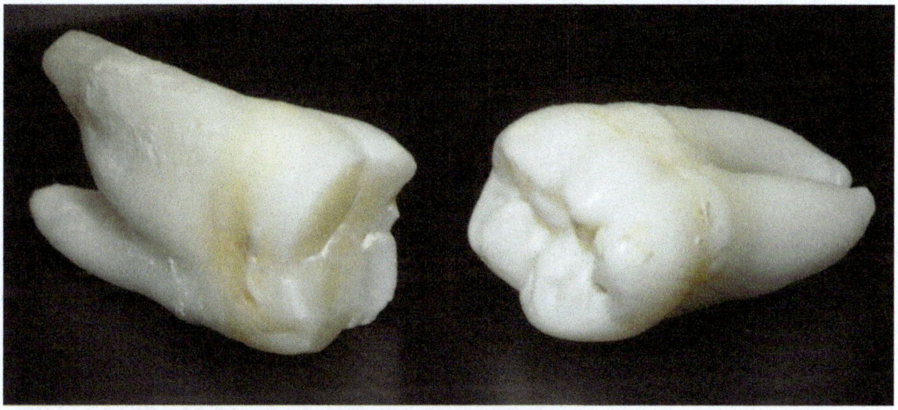

Fig. 3.32 With age, wear shapes occlusal surfaces: on the left, upper first molar of a 70-year-old subject. On the right, lower third molar in an 18-year-old subject. The wear on the left tooth, physiological at 70 years, would be all the more pathological the earlier it appears

In patients with periodontal disease, tissue damage often accompanies a lack of occlusal wear. The occlusal inclines of opposing cusps, especially if highly inclined (remember that cusp height is one of the determinants of occlusion), can create significant dynamic obstructions. These interlocking contacts generate internal stresses within the coronal structure that also affect the cervical region. The absence of functional remodeling of two opposing occlusal surfaces results in a concentration of forces, which is greater when the contact area is smaller. A pronounced cusp height with steep inclines and deep interlocking is undoubtedly a biomechanical environment in which mandibular movements generate transverse forces that exert

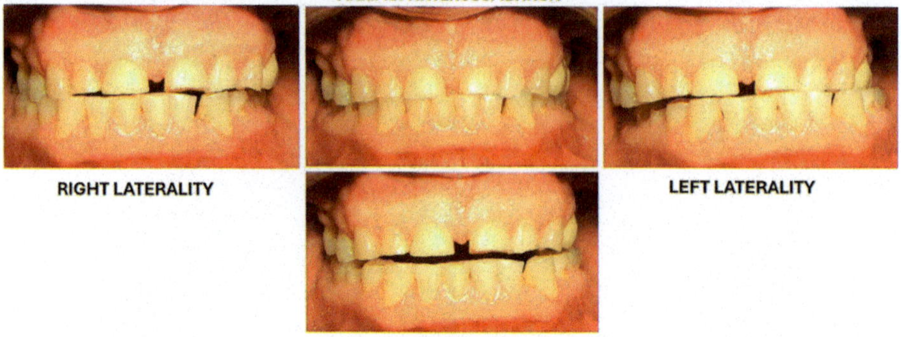

Fig. 3.33 An 81-year-old patient with physiological dental wear that has matured occlusal surfaces: loss of overbite in maximum intercuspation, laterality with group guidance, and contralateral balancing. Obviously, the physiological condition of these wears is related to the age of the subject: the same degree of wear in a much younger subject would take on frankly pathological connotations

Fig. 3.34 Occlusal wear, working and balancing inclined planes. Skull 4325 from the Museum of Anthropology, University of Florence

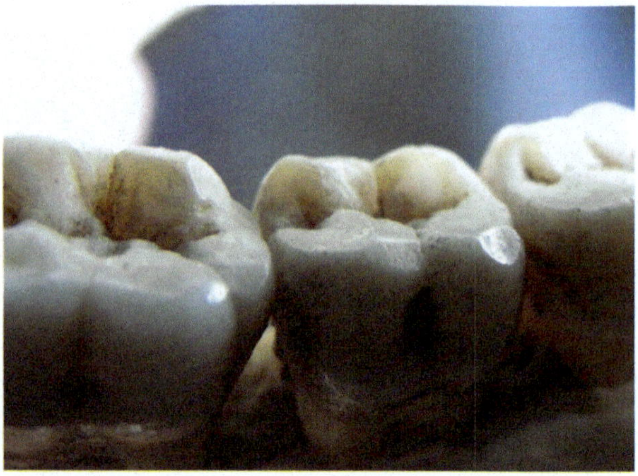

significant stress on the periodontium. Nevertheless, a reduction in cusp height tends to reduce occlusal stress, and individuals in whom wear has led to the maturation of occlusal contacts are more commonly free from significant periodontal diseases (Fig. 3.33). An extraction in these patients is often a challenging experience for both the patient and the dentist.

The chewing of robust foods (unprocessed foods tends to be more abrasive and requires more prolonged and vigorous chewing) creates wear facets on both working and non-working cusps. These facets reciprocally guide each other during function (Kim et al., 2001) (Figs. 3.34, 3.35).

In populations with a diet of robust foods, deciduous dentition was subject to both interproximal and occlusal wear, which, coupled with a reduction in overbite, led to the mesialization of the lower arch, facilitating the proper eruption of first molars (Figs. 3.36, 3.37).

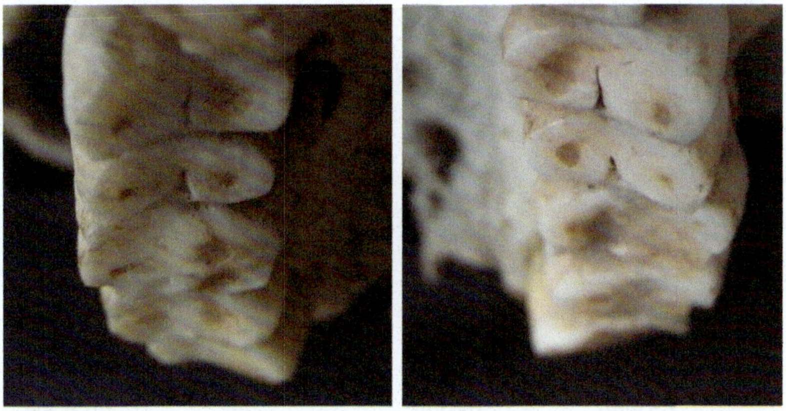

Fig. 3.35 Occlusal wear, working and balancing inclined planes. Skull 5050 from the Museum of Anthropology, University of Florence

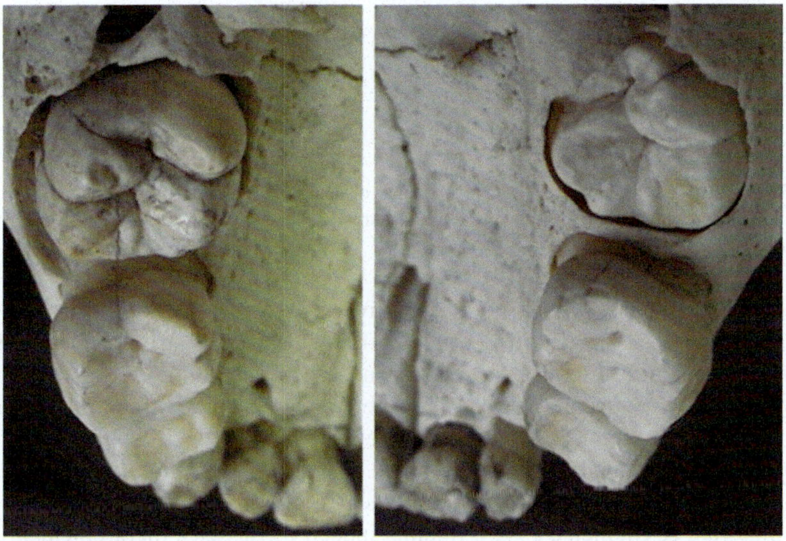

Fig. 3.36 Occlusal wear and wear facets of deciduous molars. The first permanent molars, not yet in occlusion, retain their occlusal morphology intact. Skull 3063 from the Museum of Anthropology, University of Florence

3.2 Structural Fundamentals of Enamel Wear

The surface of a newly erupted tooth is like a blank canvas on which wear has not yet left its functional marks. Enamel wear is a fascinating mechanism that allows the tooth to adapt to changing functional situations while reconciling its distinctive traits—extreme resistance and remarkable hardness—with the need for occlusal

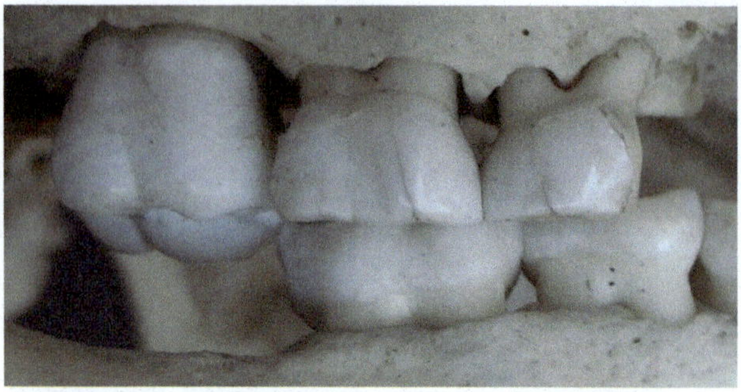

Fig. 3.37 Occlusal wear and wear facets of deciduous molars with their mesialization, setting the correct class I relationship in the first permanent molars. Skull 6153 from the Museum of Anthropology, University of Florence

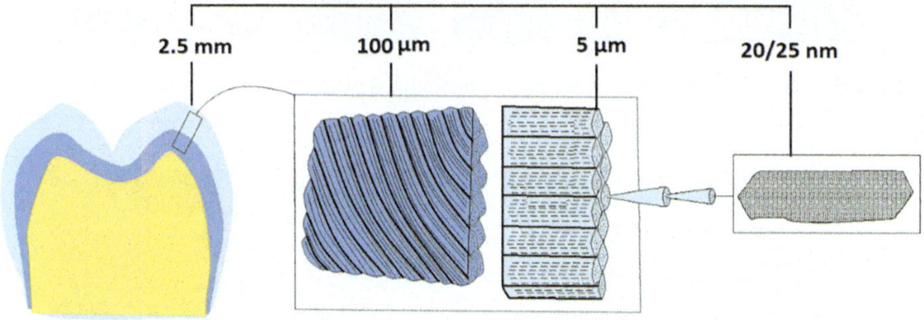

Fig. 3.38 The different hierarchical levels of enamel structure: from the millimetric dimensions of its thickness, to the organization of layers, prisms, hydroxyapatite nanofibrils, to the single crystallite

shape alterations. The origins of this adaptive capacity are found in various dimensional levels (Benazzi, Kullmer, et al., 2013). Despite the strong functional constraints on enamel structure, its evolutionary differentiation reflects phylogeny. Many evolutionary groups can be characterized by specific structural features of their enamel, corresponding to varying levels of complexity (Koenigswald, 2020).

Describing and analyzing the complex structure of enamel can be facilitated through the application of a hierarchical system of structural complexity (Koenigswald & Clemens, 1992). Enamel indeed possesses a hierarchical structure organized across multiple levels, and its mechanical behavior is influenced by structural aspects at the millimeter, micrometer, and nanometer levels (Hakkinen et al., 2019; Cui & Ge, 2007).

The hierarchical organization of enamel ranges from millimeter-scale to smaller dimensions, covering a range from the microscale to the nanoscale: hydroxyapatite crystals align longitudinally to form mineral nanofibrils. These nanofibrils, in turn, align longitudinally, aggregating into thicker fibrils, ultimately forming prisms and interprismatic substance. The prisms assemble into prismatic bands that exhibit different arrangements across the thickness of the enamel layer (Cui & Ge, 2007) (Fig. 3.38).

3.2.1 Millimeter Level

As paleoanthropologists well know, teeth represent the most durable part of the digestive system. Their size, shape, internal architecture, and resulting mechanical properties reveal a great deal about the diets of extinct species and the evolving dietary adaptations throughout their evolutionary history (Schwartz et al., 2020).

Mammalian teeth have evolved in response to their diverse diets, exhibiting a wide range of configurations. Some herbivorous mammals intentionally expose dentin to produce enamel with sharp crests for breaking down tough but soft foods. In contrast, the teeth of large primates, including humans, are bunodont, featuring rounded cusps with varying thicknesses of outer enamel.

Tooth shape is a crucial factor in determining resistance to damage accumulation. At the macroscopic level, the most important factors are tooth size and enamel thickness (Lee et al., 2010). Enamel thickness, in combination with its distribution on the molar crown, plays a significant role in the study of hominid evolution (gibbons, gorillas, chimpanzees, and humans). The three-dimensional distribution pattern has been recognized as having significant value in terms of functionality, development, and phylogenetic information (Kono & Suwa, 2008).

Enamel structure in mammals has evolved to optimize masticatory stresses, with modifications occurring not only at the nano and micrometer levels but also at the millimeter level. Enamel in the occlusal region adapted according to various dietary habits, with varying thicknesses to respond to demands and enable each dental element to perform its function as effectively and for as long as possible.

Chimpanzees, for instance, have thinner occlusal enamel compared to gorillas or orangutans, reflecting their primarily ripe fruit-based diet. In chimpanzees, the enamel thickness on the lateral portions of the crown is generally greater than that on the occlusal surface (Kono & Suwa, 2008). This may indicate their occasional need to consume leaves and tougher foods when ripe fruit is unavailable (Constantino et al., 2009).

In the molars of australopithecines, there was not only a proportional increase in tooth size but also an increase in enamel thickness. This thicker enamel was necessary for chewing hard and tough foods, whose fragmentation required hours of lateral chewing movements to break down their fibrous components (Lieberman, 2011) (Fig. 3.39).

In mammals, the single replacement of teeth, aimed at maintaining stable occlusal relationships, subjected the teeth to greater selective pressure to preserve their enamel, which became thicker compared to that of crocodiles or dinosaurs. The persistence of adult dentition (and masticatory stresses) imposed thicker enamel, ensuring the functional longevity of teeth throughout an individual's lifespan (Maas & Dumont, 1999).

Exceptions in enamel thickness among mammals can be observed in rodents and odontocetes, but both cases have functional justifications. Rodents have continuously growing incisors, so excessive wear is not a concern. For odontocetes, whose teeth are covered by a thin layer of enamel (Goldberg et al., 2014), the thinness of

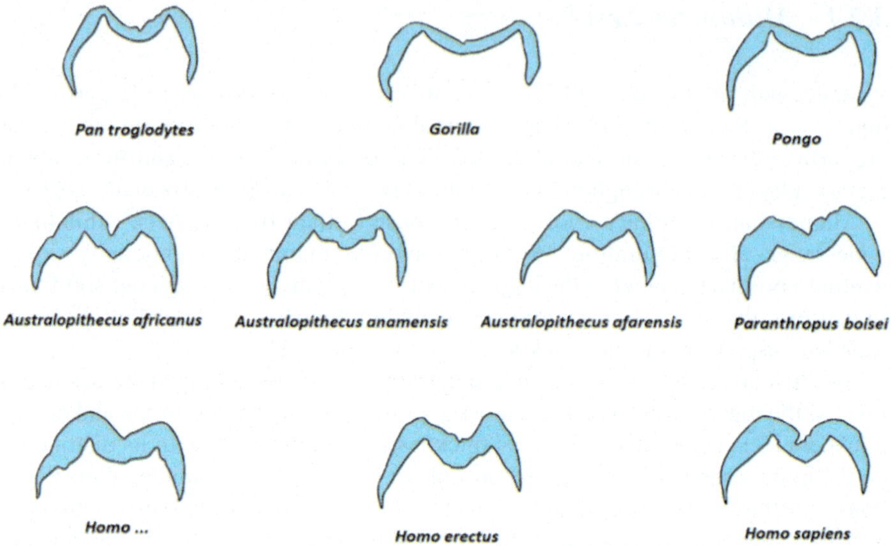

Fig. 3.39 Differentiated thicknesses in anthropoid apes, australopithecines, and humans. Note how the enamel thickness is thinner in anthropoid apes and how in *Paranthropus boisei* the enamel has a significantly greater thickness than in other hominids. Drawing freely adapted from: "Enamel thickness trends in Plio-Pleistocene hominin mandibular molars"

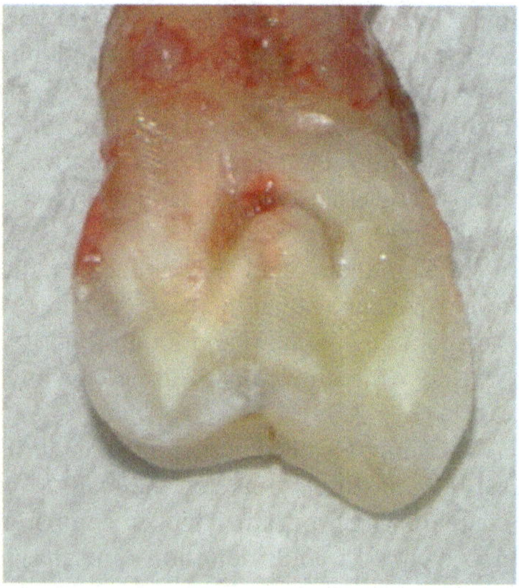

Fig. 3.40 Fresh section of a human tooth: the pattern of the occlusal surface is replicated in the enamel-dentine junction and in the roof of the pulp cavity. Note how the thickness of the enamel covering the working cusp is reduced due to the greater wear to which this occlusal area is subjected

the enamel is linked to reduced biomechanical requirements due to their feeding habits (Loch et al., 2014, 2015). Although enamel development does not simply mirror that of the underlying dentin, the outer surface of the tooth replicates and almost maps the subtle variations in dentin surface (Hakkinen et al., 2019). This ensures that enamel provides uniform support (Fig. 3.40).

In humans, the hardness and stiffness of enamel vary across different parts of the crown (Braly et al., 2007). For example, enamel near the outer surface has a high density (96% hydroxyapatite), which decreases to around 84% near the enamel-dentin junction (Maas & Dumont, 1999). In human molars, both the hardness and Young's modulus of enamel decrease by over 50% from the surface to the enamel-dentin junction. These variations likely originate from changes in local chemistry, microstructure, and prism orientation. Hydroxyapatite constituent concentrations are higher on the rigid occlusal surface and decrease as you move towards the softer enamel-dentin junction (Cuy et al., 2002; Jeng et al., 2011). This gradient allows for selective variation in both resistance and elasticity in different structural portions that require different biomechanical responses. The gradual transition of resilience between the rigid enamel layer and the more elastic dentin prevents stress concentrations in this area.

Teeth can be viewed as highly resistant biocomposite structures, capable of withstanding cumulative deformation and fracture. Their morphological features at both macroscopic and microscopic levels contribute to containing catastrophic damage spread. Nature's strategy is to contain potentially inevitable damage (Lawn et al., 2010).

In a rigid material, varying thicknesses tend to generate different elastic responses within it, causing stress. Similar to how a metal-ceramic fixed prosthesis must have a shape that allows for constant thicknesses of the rigid ceramic layer to avoid potential fractures, the dentin's morphology enables consistent enamel thickness, except in areas subjected to greater stress, where a thicker layer is required to resist wear more effectively.

3.2.2 Micrometer Level

The microstructure of enamel is the result of millions of years of evolution and has diversified among different species, evolving adaptively in response to functional demands and closely tied to the physiological function of dentition and food properties (Teaford et al., 2007).

Lower vertebrates' enamel possesses a simpler structure known as aprismatic enamel or parallel crystalline enamel. This represents the most primitive type of enamel, characterized by a single level of hierarchy represented by multiple crystallites. Enamel without prisms is mainly observed in reptile teeth, which typically serve the physiological role of tissue perforation and prey immobilization, without necessarily requiring opposing dental arches to come into contact (Line & Novaes, 2017) (Fig. 3.41).

The change in mastication-related dietary habits, accompanied by the resulting stress on dental elements, led to a modification of enamel microstructure, transitioning from the aprismatic enamel found in reptiles to the prismatic enamel seen in mammals. Prismatic enamel evolved as a consequence of differentiation in the cell that forms the individual enamel prism, the ameloblast (Yilmaz et al., 2015).

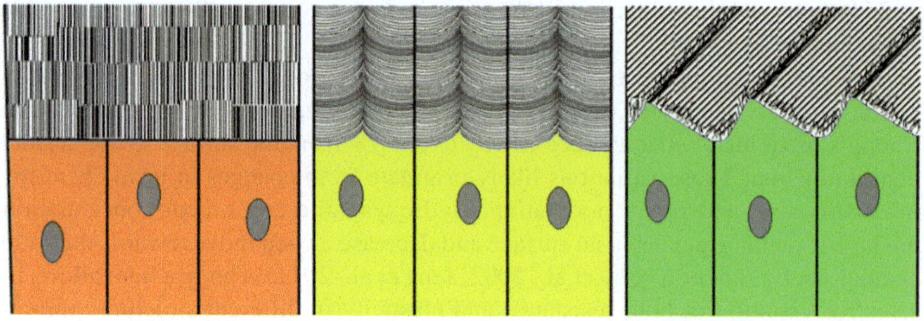

Fig. 3.41 Left: Aprismatic enamel; Center: Pseudoprismatic enamel; Right: Prismatic enamel

At the micrometer level, two distinct entities are identified: the prismatic unit and the interprismatic space. The enamel prisms are not sharply defined or isolated from the surrounding component but are interconnected with the crystallites (the fundamental building blocks at this hierarchical level) of both entities. These connections between prismatic crystallites and those of the sheaths also facilitate, at a lower dimensional level, the transition in resilience that mitigates dangerous force concentrations resulting from abrupt elasticity gradients.

The prisms are surrounded by a discontinuous interface, formed by contact points between mineral components (crystallites) alternating with empty spaces. This interprismatic matrix is often referred to as interprismatic enamel and differs from prismatic enamel, not only due to the presence of a protein component but also in the orientation of its crystallites and the bundles in which these crystallites are 'packaged.' This peripheral space around the prisms represents an average of approximately 57% (± 15) of the surface, and the contact between prisms and interprismatic space occurs either through crystallites that simply fill the prismatic boundary or through continuous crystallites that, by curving, connect interprismatic enamel to the prisms. This organization is of significant importance in understanding the mechanical properties of dental enamel (Fig. 3.42).

Each enamel prism follows a sinuous path that begins near the amelodentinal junction and extends toward the outer surface of the enamel. The course of enamel prisms, including any crossings or bends, reflects the trajectories followed by ameloblasts during amelogenesis (Lynch et al., 2011). The arrangement and orientation of prisms are complex and reflect the intricate processes that provide biomechanical resistance to masticatory stresses (Jia et al., 2014), allowing for a gradual transition from 'point-like' contacts (occlusal or interproximal) to progressively larger contact areas. The reciprocal rubbing of these surfaces facilitates the crushing of food particles (Zhou et al., 2013).

In many mammalian species, prismatic and interprismatic enamel can exhibit different organizations, forming the next level of hierarchy, known as the *Schmelzmuster* (Koenigswald & Clemens, 1992; Koenigswald & Sander, 2020). Within enamel, two predominant structures are found. In the inner part near the enamel-dentin junction, prisms cross each other, forming *Hunter-Schreger bands*,

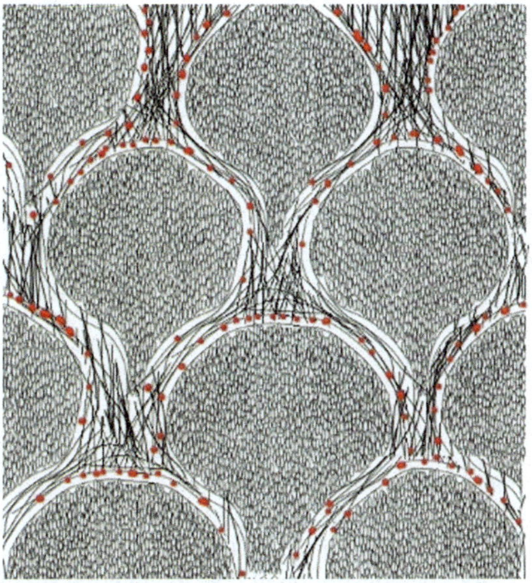

Fig. 3.42 Transversely sectioned prisms: note the directional homogeneity of the crystallite bundles within the prisms, orthogonal to the section plane, and the different course of the bundles in the interprismatic space. In red, the points of union between extraprismatic bundles and prisms are represented. Drawing freely adapted from: "The geometrical structure of interfaces in dental enamel", Koldehoff et al., 2020

undulations created by the complex and sinuous paths of ameloblasts during the deposition of the substance that will form the prisms. In the outer enamel regions, prisms become straight and parallel to each other, forming what is known as radial enamel. Some objections in the literature confirm the direction of prisms toward the outer enamel surface but question whether they are perpendicular to the occlusal surface of the tooth, except at the tips of cusps and the edges of interproximal crests (Fernandes & Chevitarese, 1991).

In any case, the course of prisms toward the outer occlusal surface allows both prisms and crystallites to terminate in a generally perpendicular manner to the surface, exposing their cross-sections towards it. This characteristic, as we will see, plays a crucial role in abrasive wear processes (Koldehoff et al., 2020). So, not only prisms but also their fundamental components, the crystallites, tend to arrange themselves perpendicular to the dental surface (Boyde et al., 1988; He & Swain, 2008). This arrangement, which is also present in aprismatic enamel, is essential in mammals (Xia et al., 2017) to create, through occlusal wear, crushing surfaces with sharp edges, necessary for enhancing chewing efficiency (Ungar, 2017).

Primitive enamel prisms or prism-like structures have been reported in therapsid reptiles from around 180 million years ago and other reptilian groups (Torii, 1998), but prismatic enamel is primarily a feature of mammals.

The interprismatic sheath exhibits a different orientation of crystallites and a higher organic component compared to the prismatic region, enabling significantly greater stress dissipation capacity. The elastic modulus and hardness of the prismatic region are significantly higher than those of the interprismatic sheath (Ang et al., 2012). The alternation between prisms and interprismatic substance inhibits the propagation of horizontal fractures (Crompton et al., 1994; Shimizu et al., 2005;

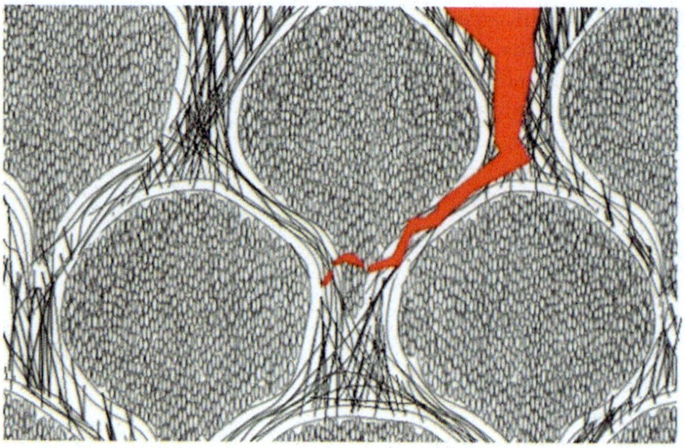

Fig. 3.43 Fracture line (in red) that tends to propagate with greater difficulty thanks to the diversified biomechanical response of the prismatic and interprismatic components

Popowics & Herring, 2006; Bajaj & Arola, 2009). The presence of the interprismatic area with its minute amounts of protein residues allows limited slipping between adjacent prisms, reducing stress and opposing the propagation of fracture lines (He & Swain, 2007) (Fig. 3.43).

As a demonstration of how the internal structure and the biological protein component influence the properties of mammalian dental enamel, it should be noted that dental enamel is much more flexible than an equivalent volume of geological hydroxyapatite (Xu et al., 1998; Maas & Dumont, 1999) and approximately three times stronger (White et al., 2001).

The formation of enamel prisms is linked to the development of a secretory process in the terminal portion of the ameloblast called the Tomes' process. This process, along with the acquisition of a wavy shift during amelogenesis, can be considered key innovations in mammalian evolution. The modification of the secretory surface of the ameloblast is a rare and perhaps unique example of evolution acting not at the level of tissue reorganization but deeper, at the cellular structure level (Line, 2005). The evolutionary aspect of the Tomes' process was achieved through rearrangements of cytoskeletal components in the distal secretory portion of ameloblasts (Lesot et al., 1982; Nishikawa & Kitamura, 1983).

In enhancing enamel's ability to resist fracture, as mentioned earlier, a biomechanically significant role is played by variations in the course of prisms caused by the Retzius' striae (or incremental lines) and *Hunter-Schreger bands (HSb)*.

The *Retzius' striae* form due to brief physiological pauses in enamel secretion, resulting from circadian variations in prism growth rate. This creates a wavy course in the enamel, and their emergence on the surface creates numerous small transverse ridges known as *perikymata* (Fig. 3.44). These ridges can wear away over time. Each ridge takes approximately 6–12 days to form, and their number provides an indication of the time it took to form the crown.

HSb are an optical phenomenon observed when the cut surface of enamel is viewed under polarized light microscopy. This phenomenon is due to the crossing of prism orientations, as they change direction and intersect with each other. Within each band, the prisms share the same orientation, which differs from the adjacent band.

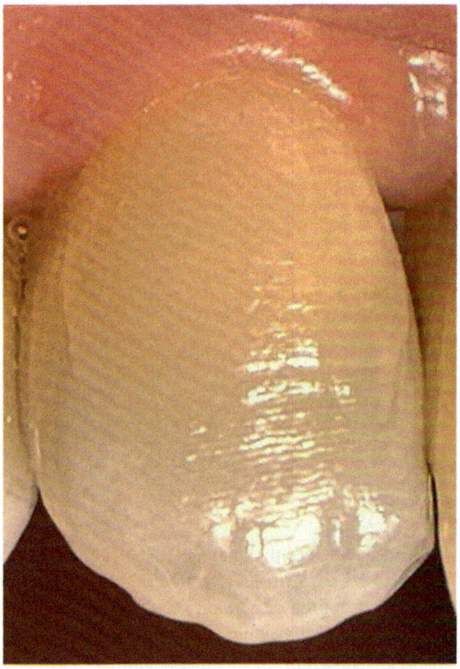

Fig. 3.44 Perikymata on the surface of the enamel of a conoid upper lateral

These variations in prism course enhance enamel's resistance to fractures because each undulation disrupts and dissipates the lines of force generated during chewing, preventing them from becoming axial and inhibiting the propagation of a crack throughout the enamel thickness. HSb has evolved to optimize the tooth's resistance to wear, abrasion, and fracture (Lynch et al., 2011). Similarly, in dentin, the wavy course of tubules (see Fig. 248) serves the same biomechanical need for the dissipation of potentially harmful forces.

The course of enamel prisms from the amelodentinal junction to the enamel surface varies, transitioning from a three-dimensional spiral pattern in the inner part to an approximately parallel and straight course in the outer part (Macho et al., 2003) (Fig. 3.45).

Different enamel structural patterns respond differently based on the course of prisms. The straight and radial arrangement of prisms in outer enamel contributes to greater rigidity and resistance to penetration, while the more interwoven arrangement of prisms in inner enamel allows for greater adaptability to occlusal loads (Jiang et al., 2003; Jia et al., 2014).

Damage that compromises enamel's lifespan not only initiates and propagates near the superficial occlusal contacts but also begins at the enamel-dentin junction. Therefore, the crossing of prisms near this junction provides internal resistance, inhibiting the extension of fractures, as observed. The crossed microstructure imparts toughness, stopping cracks that may occur at the enamel-dentin junction and transforming them into a kind of "fibrous" fracture, as often seen in wooden structures (Lucas et al., 2008).

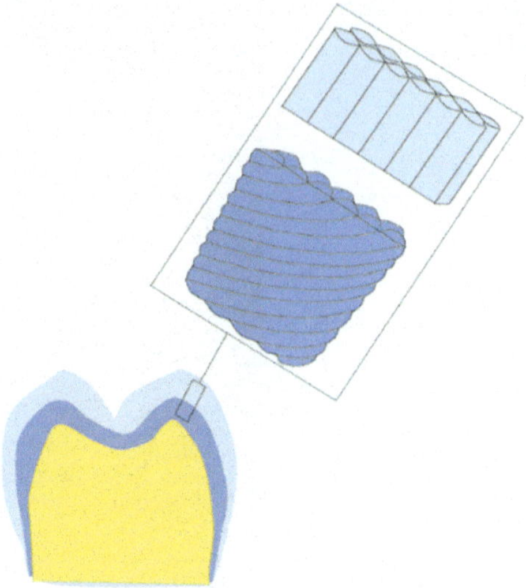

Fig. 3.45 Different course of the prisms in relation to the depth of the enamel layer. The outer enamel acts as a protective layer, thanks to its high resistance to contact deformation; the inner enamel has a greater capacity for plastic deformation with better resistance to breakage, correlated with more effective energy dissipation. Drawing freely adapted from "Finite element analysis of the cyclic indentation of bilayer enamel", Jia et al., 2014

3.2.3 Nanometric Level

The nanostructure of enamel represents the third key level, and perhaps the most important one, in understanding the response of enamel surface to masticatory loads (Xia et al., 2017). The nanoscale dimension of mature enamel is characterized by crystallites, defined as the fundamental components of enamel: mineral crystals that are approximately hexagonal, ribbon-like, measuring 50–70 nanometers in width, 20–25 nm in thickness, with a length well beyond 100 nm (Daculsi et al., 1984; Moradian-Oldak, 2012; Xia et al., 2017).

Each Tomes' process of every ameloblast produces millions of crystallites perpendicular to the cell's secretory surface (Anderson & Creanor, 2016). These crystallites are linked together by a thin protein layer (less than 2 nm) that "glues" the hydroxyapatite crystallites together, forming enamel rods arranged in parallel. Within these rods, the crystallites are oriented along the axis of the rods themselves (Zheng et al., 2011) (Fig. 3.46).

Each enamel rod has approximately the same diameter as the secretory process of the ameloblast, and it is believed to be the product of a single cell. The enamel rod is considered the fundamental organizational unit of mammalian enamel.

The crystallites are stacked and packed on top of each other and held together, as mentioned earlier, by a thin 2 nm layer of hydrated protein. This protein layer uses its elastic polymeric backbone to keep the crystallites bound together, forming enamel prisms.

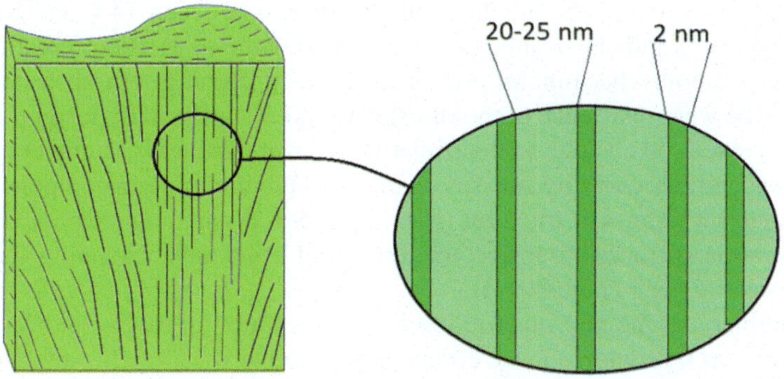

Fig. 3.46 Internal configuration of an enamel prism. Drawing freely adapted from "Contact induced deformation of enamel", He and Swain (2007)

The organization of the hierarchical structure of enamel seems to repeat itself at each dimensional level (Cui & Ge, 2007). In fact, both the strength and elastic modulus of enamel differ as we move to higher hierarchical levels, with both values decreasing by a factor of 2 as we progress to the next hierarchical level. When we move to a lower dimensional level, this significant variation is primarily attributed to the increased percentage of protein content per unit volume, leading to a progressive transformation from a brittle, rigid, and high-strength material to a less rigid and brittle material that can still withstand stress more effectively (Bechtle et al., 2012).

The protein matrix is composed of 90% amelogenin, along with ameloblastin, amelotin, and biglycans. Amelogenin monomers, measuring 4–6 nm, self-assemble to form nanospheres (Fincham et al., 1995), typically ranging in size from 12–18 nm but sometimes aggregating into larger nanospheres of 60–70 nm (Engelberth et al., 2018). At the genetic level, one of the main differences in mammalian amelogenin compared to lower vertebrates lies in significantly longer amino acid sequences in mammals. These sequences create a true "evolutionary hotspot" (Sedwick, 2009).

While mature enamel contains minimal organic material, during its formation, matrix proteins make up about one-third of its volume. In the case of immature enamel, amelogenin is the primary protein involved in its formation. It plays a crucial role in regulating the mineralization and structural organization of enamel by stabilizing and guiding the initial aggregation of the mineral component into linear chains. These prenucleation clusters subsequently fuse together, forming needle-like mineral particles that organize into bundles of crystallites, thus defining the characteristic nanoscale structural organization of forming enamel (Fincham et al., 1995; Moradian-Oldak & Goldberg, 2005; Fang et al., 2011; Bromley et al., 2011).

Pure hydroxyapatite crystallites behave as brittle ceramics, and it is the organic component that interconnects them that is key to understanding the nanoscale response of enamel surfaces to chewing forces and the inevitable wear caused by chewing. The organic component provides greater toughness and greater plastic deformation, allowing individual nanofibers to bend and the protein layer to compress (Xia et al., 2017).

The hardness and elastic modulus of sheaths are between 74% and 53% lower than the hardness and elastic modulus of hydroxyapatite rods. Therefore, the organic component, among the mineral portions, acts as a cushion that absorbs chewing impacts. The wear resistance of the sheaths, which is significantly lower than that of the interspersed rods, is due to their relatively lower hardness (Young et al., 1987), which concentrates chewing stresses on the rigid tubular mineral component. When subjected to such stress, it fractures (Fig. 3.47), thus dissipating stresses, preventing the generation of more extensive fractures, and efficiently activating the enamel wear process (Zheng et al., 2011).

In summary, at the micrometric level, the alternation of more rigid components like the prisms with more elastic components like the sheaths (this alternation also occurs at the nanometric level with interprismatic rods and crystallites enveloped in protein-rich enamel components) is one of the structural factors that tends to limit the propagation of fractures. Another crucial structural feature for crack resistance is the wavy pattern of enamel microstructures: the Retzius lines, Hunter-Schreger bands, and nodular enamel all act as damping components, like preloaded springs that act as crack stoppers. The same concept applies to dentin, where tubules exhibit a characteristic wavy pattern.

The inter-rod proteic portions have reduced mechanical characteristics and are more prone to deformation and wear, leaving the more mineralized and rigid portions more exposed to microfractures in their components, thus dissipating forces that could trigger destructive cracks. This nanometric mechanism plays a fundamental role in the overall behavior of enamel: the countless contacts experienced by its surface result in nanometric modifications that progressively accumulate, eventually becoming visible wear facets to the naked eye.

The response of the nanoscale structures in dental enamel to the forces of chewing can be broken down into three distinct modes: detachment, plastic deformation, and fragmentation. These aspects vary in intensity as the applied pressure increases, and each contributes to the nanoscale wear response of enamel (He & Swain, 2009; Xia et al., 2017).

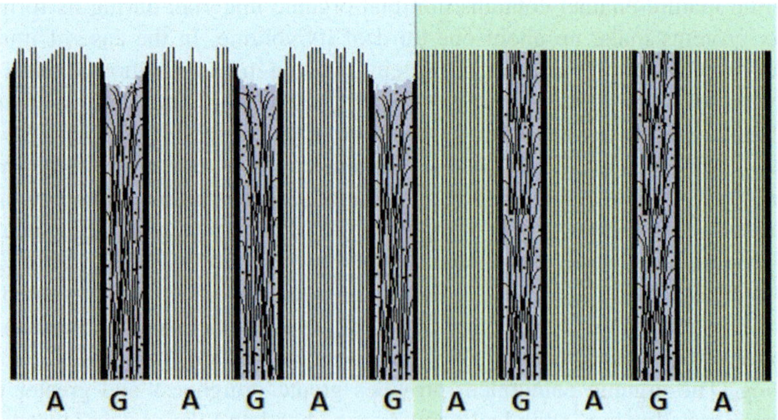

Fig. 3.47 Alternation of crystallite rods (A) and sheaths of union (G), richer in protein material; on the left, an area subjected to wear is represented, on the right, an area not yet worn

- *Detachment*: This occurs when the contact pressure exceeds the strength of the "protein adhesive" between the hydroxyapatite rods that make up the enamel's nanofibers. In this case, individual nanoparticles are "pulled away" and become debris.
- *Plastic Deformation*: Despite ceramics typically being very rigid materials that tend to fracture rather than undergo plastic deformation, enamel, considered a bio-ceramic, can allow individual nanofibers to flex. This is made possible by the cushioning effect of the hydrated protein layer within the enamel matrix. This elastic deformation capability enables the mineral component of enamel to maintain its size and shape under pressure.
- *Fragmentation*: This happens when the pressure exceeds the strength of the strong hydrogen bonds between the protein nanoparticles that compose the enamel's sheath, which have a diameter of around 20 nm.

The critical load required to trigger these responses depends not only on the applied force but also on the force's direction concerning the long axes of the enamel nanofiber rods (Fincham et al., 1995). It has also been observed that surface-level crystallites can decrease in size with increasing chewing forces (Xia et al., 2017).

There's a hypothesis that the mechanism of action of the protein component in the interfiber space in enamel acts similarly to the protein portion in bone tissue. In bone, protein filaments store and elastically dissipate most of the energy through breaking "sacrificial" bonds and stretching the molecular structures' folds. Weaker bonds break (and within certain limits, reform) before stronger bonds rupture. The energy required to break these weaker bonds and subsequently elongate the molecular structures increases the overall energy needed to fracture the material, thereby enhancing its toughness (Fantner et al., 2005).

With the formation of enamel prisms, in addition to its strength, enamel gained anisotropy, a characteristic of a material that responds differently depending on the direction of the force acting on it. Enamel's anisotropy allows it to modulate its wear, making it a fundamental feature for combining high fracture resistance with the ability to adapt morphologically to changing chewing demands. The ultrastructure of enamel forms the basis for its anisotropic nature, and in particular, the orientation of hydroxyapatite nanofibers influences enamel abrasion (Maas & Dumont, 1999) with distinct responses to stresses depending on the force's intensity and direction (Fig. 3.48).

Of course, enamel wear isn't solely influenced by its structure. It's also tied to a complex interplay between tooth shape, enamel thickness, food particle hardness, generated chewing force, and its direction in relation to the tooth surface (Young et al., 1987; Stern et al., 1989).

Enamel exhibits the highest fracture resistance against forces acting perpendicular to the microfibril axis (Zaytsev & Panfilov, 2015). However, the enamel surface, when exposed, doesn't possess high resistance to abrasion, which is more evident compared to other prismatic areas (Boyde, 1984). The orientations of wear facets indicate the movements of the jaw and the specific stresses exerted on food during chewing (Fig. 3.49), thus reflecting species-specific dietary habits (Hiiemae, 1978).

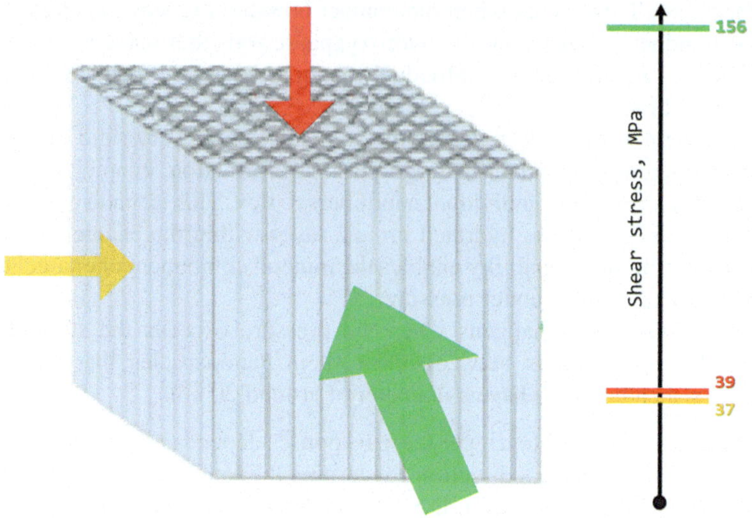

Fig. 3.48 Different response of the enamel ultrastructure depending on the direction of the force acting on it. Green arrow: force acting on a plane perpendicular to the mineral structures and sheaths; orange arrow: force acting on a plane perpendicular to the mineral structures but parallel to the sheaths; red arrow: force acting on a plane parallel to the mineral structures and sheaths; the graph highlights how a force that primarily stresses the heads of the prisms or mineral fibrils, or the protein sheaths, encounters less resistance given their greater fragility: this mechanism, associated with the radiality that enamel presents in the superficial layer, allows targeted wear of surfaces based on functional stresses. On the right, differentiated break points according to loads with different directions

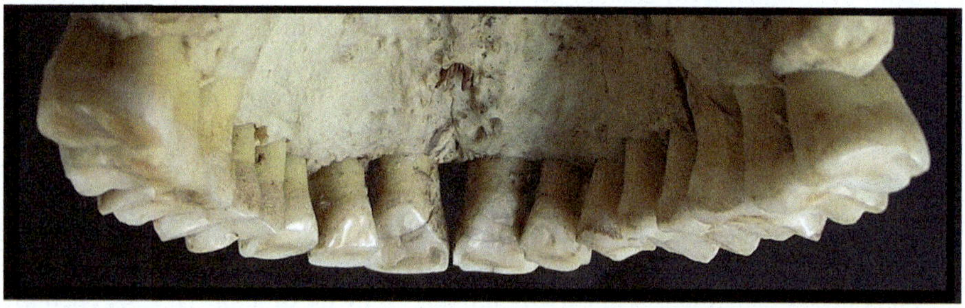

Fig. 3.49 Harmonious wear of occlusal surfaces. Skull 4235 from the Museum of Anthropology, University of Florence

3.3 Masticatory Cycle

To thoroughly understand dental wear, one must consider how occlusal surfaces interact with each other. The masticatory cycle emerged with mammalian chewing approximately 250 million years ago, involving the horizontal plane in mandibular dynamics, with movements varying depending on the animal's diet. Although the influence of food type varies among species, motion analyses and simultaneous electromyograms clearly show that food has a significant effect on the number of chewing cycles required to prepare each bolus for swallowing, the shape of successive cycles, and muscle activity (Gans et al., 1978).

In humans, molar wear development follows a pattern common to primates and other omnivorous mammals. Hominoids (the superfamily of primates including humans and anthropoid apes) share a common wear pattern due to significant similarities in crown morphology, occlusion, and chewing dynamics. In *Homo sapiens*, molar wear is distributed more evenly among hunter-gatherers, with a relatively low angle of wear planes in advanced wear. Among farmers, a narrower wear pattern and a tendency to develop oblique wear planes are observed. This is attributed to a reduction in dietary toughness and fibrousness associated with intensive cereal harvesting and food preparation using grinding tools (Smith, 1985).

The masticatory cycle has been observed in most mammals, including those with unusual feeding postures such as bats (Greet & De Vree, 1984) or sloths (Naples, 1985), and it is not species-specific as it can be found in many species, including herbivores (Hiiemae, 1978) and insectivores. By changing the angle of entry, the mandible can adapt to the different mechanical properties of food. However, the overall speed of masticatory movement in individual species is not highly flexible and is controlled by the autonomic nervous system (Hiiemae, 1984). The chewing movement, while having some variability, is always characterized by both vertical and transverse components (Fig. 3.50).

In various species, during the initial phase of food processing before the cyclic chewing movement begins, the food mass located between the teeth prevents the occlusal surfaces from coming into contact. The mandible initiates essentially exploratory movements, characterized by primarily vertical cycles, referred to as "puncture crushing" (Weijs, 1994). During this phase, there is no contact between opposing teeth. Therefore, once food is introduced into the oral cavity and before the actual chewing cycle begins, there is an initial treatment phase. During this phase, there is a predominance of the material to be ingested, which serves to assess the consistency of the food and set the minimum force required for crushing. This avoids excessive movements and contacts that could potentially damage the teeth.

Puncture crushing is a preparatory phase in which rhythmic "vertical" movements at the occlusal plane do not reach maximum muscle force. During this phase, the cusps' tips may incidentally come into contact with hard portions of the food, leading to both wear (Fig. 3.51) and microfractures (Fig. 3.52). Over time, these phenomena can result in blunting of the cusp tips, referred to as *tip-crushing*.

At the top of the cusps, the enamel takes on a distinctive configuration, a spiral arrangement of adjacent prisms that are still perpendicular to the cusp tip (Fernandes & Chevitarese, 1991), known as gnarled enamel or nodular enamel (Kodaka et al., 1996) (Fig. 3.53).

The groups of interwoven prisms with a spiral and irregular path on the cusp tips should be biomechanically interpreted as a configuration capable of providing greater resistance to the strong stresses applied to these areas. Nodular enamel is more resistant than normally structured enamel (Fig. 3.54).

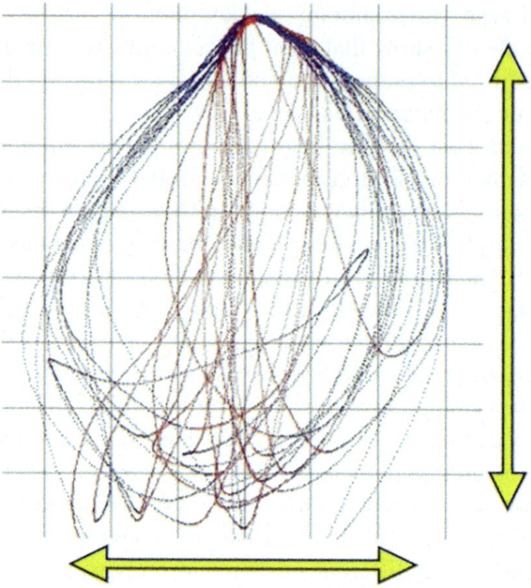

Fig. 3.50 Frontal plane: kinesiographic trace drawn from the movement of the interincisal point during chewing cycles, in humans: in each chewing cycle, there is a vertical component given by the opening and closing of the jaw and a horizontal component given by the initial lateral movement of the jaw and the subsequent medial translation in the closing phase. Red traces: opening movement; blue traces: closing movement. Note the symmetry between the right and left parts of the 'ball', significant of an alternate monolateral chewing

Fig. 3.51 Wear of the tips of the upper working cusps. Skull 3781 from the Museum of Anthropology, University of Florence

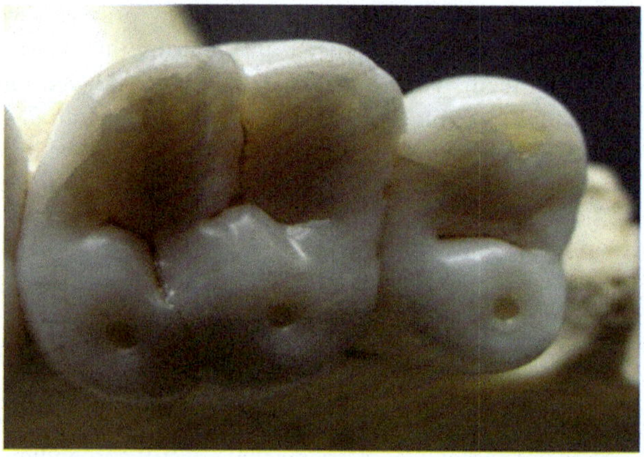

Nodular enamel, with its increased hardness and resistance, is functional in withstanding the higher stresses that concentrate on the cusp tips. Only after the exploratory and puncture crushing phase, which may be absent in the case of soft and/or thin foods (Hiiemae, 1984), do the molars begin the phase of reducing food into a bolus (Gerstner et al., 2011). Once the initial treatment is performed, the bolus

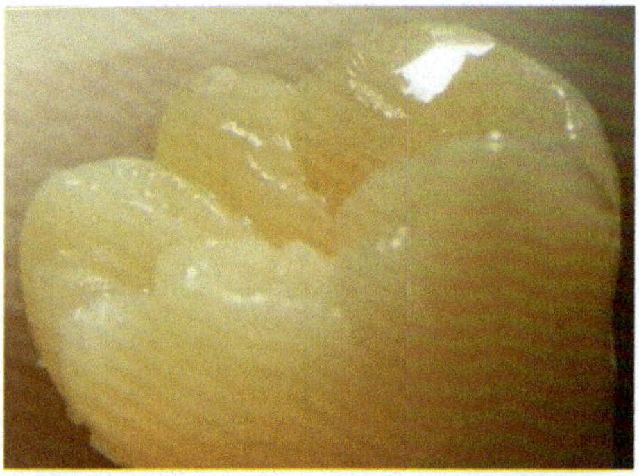

Fig. 3.52 Fractures of the tips of the working cusps in a lower molar. Note the shiny surface of a wear facet

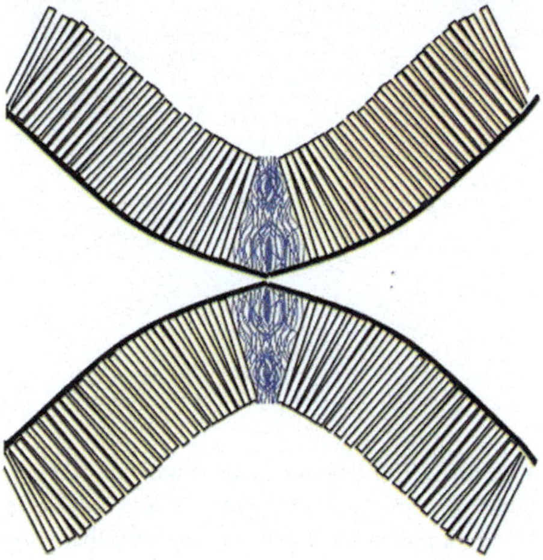

Fig. 3.53 Nodular enamel on the cuspal tip. This structure provides particular mechanical resistance, especially useful at points of intense stress

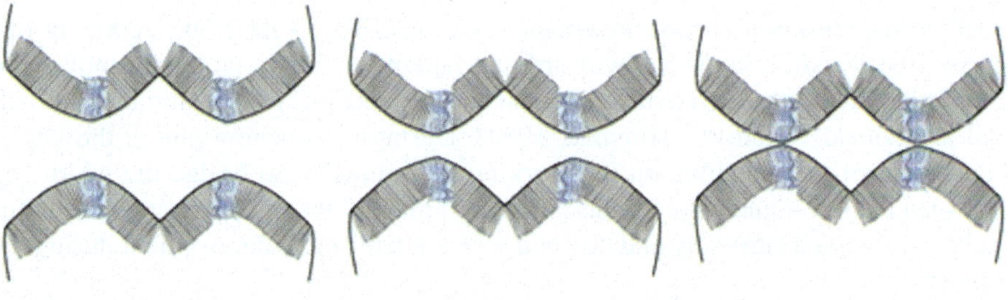

Fig. 3.54 Phase of puncture crushing in which the cuspal tips test the consistency of food

moves to the posterior teeth, initiating the actual chewing cycle, which has been divided differently by various authors. We will adopt the division of the chewing cycle into a sequence of three successive phases: the opening phase, the closing phase, and the power stroke (Fig. 3.55).

Opening phase: When food is introduced, the mouth opens, and the dental arches move apart to reach the minimum opening required for effective chewing of the

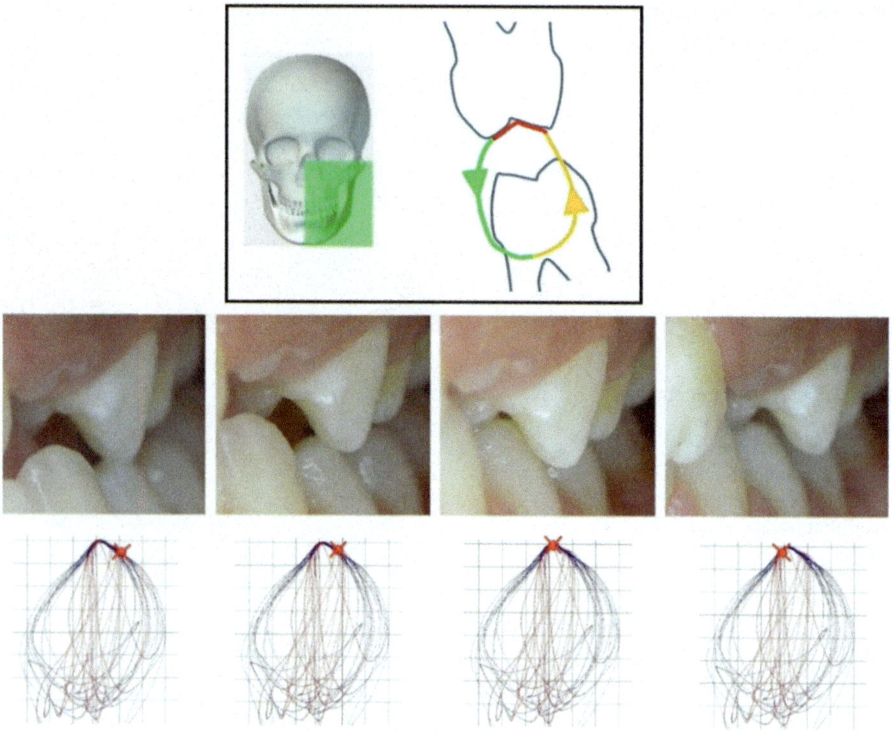

Fig. 3.55 Top: Phases of the chewing cycle: green, opening phase; yellow, closing phase; red, chewing power stroke. In the center: photographic sequence of the entry and exit phases of the chewing power stroke; at the bottom, the same sequence in a kinesiographic trace on the frontal plane

food bolus, with minimized movement variations (Miyawaki et al., 2000). In this phase, the mandible moves downward and laterally, with variations in movement amplitude due to the different food consistencies and species-specific and individual anatomical variations (Hiiemae, 1984). During the opening phase, the lateral movement of the mandible on the working side allows it to bridge the width gap between the lower and upper arches. This alignment of the molars enables the bolus to be compressed vertically and crushed with a lateral-medial movement during the power stroke.

Closing phase: The mandible rapidly moves upward and medially to bring the dental arches closer together for mechanical interaction with the food. This phase typically ends when there is contact between tooth-food-tooth, with the mandible deviated laterally towards the working side (Popowics & Herring, 2006).

Power stroke: The dental arches mechanically engage with each other through the interposed food, cutting and crushing the food until reaching the point of least verticality in the cycle. Initially, this point depends not only on the maximum dental interlocking but also on the volume of the interposed particles. During the power stroke, the mandible moves anteriorly and medially toward the midline, completing the cycle (Popowics & Herring, 2006). In this phase, the movement is partly guided by the inclined planes of the molar surfaces (Herring, 1993) and proprioception of the anterior guides (Fig. 3.56). It is characterized by strong contractions of the mandible's adductor muscles: the temporalis, masseters, and external pterygoids.

As the teeth are about to intercept food particles, the speed slows down because an error at this stage could result in trauma with potential damage to dental integrity. The closer you get to the point of maximum intercuspation, the less controllable and modifiable the movement becomes, with the guidance of the inclined planes of the occlusal table predominating over the vectors of the closing muscles. Precision in these movements is necessary to maximize grinding effectiveness while simultaneously avoiding traumatic contacts and nociceptive stimuli originating from the teeth, thus preventing damage to the structures of the stomatognathic system (Piancino & Kyrkanides, 2018).

During this phase, food is crushed, leading to contacts between teeth and ingested substances, as well as contacts between teeth themselves, resulting in wear facets on the cuspal slopes and basin surfaces (Bhullar et al., 2019). In human mastication, cycles tend to follow a typical pattern for our species, although they may vary depending on the anatomical characteristics of the individual and the consistency of the food being processed. The development of the power stroke on the working side also significantly involves the muscles on the balancing side, generating tensions within the bony structure of the entire mandibular body. The power phase is the most interesting one since the muscular forces of mastication are exerted on the occlusal surfaces with their full intensity (Fig. 3.57).

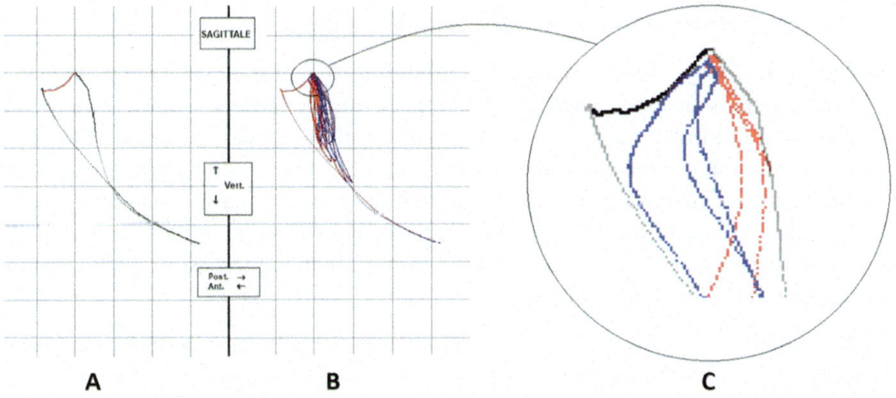

Fig. 3.56 A Posselt diagram (cf. also fig. 390 and the accompanying text) sagittal view. B Posselt diagram with traces of chewing cycles. C Enlargement of the first three chewing power strokes. Black line: incisal guide; blue line: entry trajectory of the cycle; red line: exit trajectory of the cycle. During the power stroke, the blue trajectory reaching maximum intercuspation follows the constraint of the anterior guidance

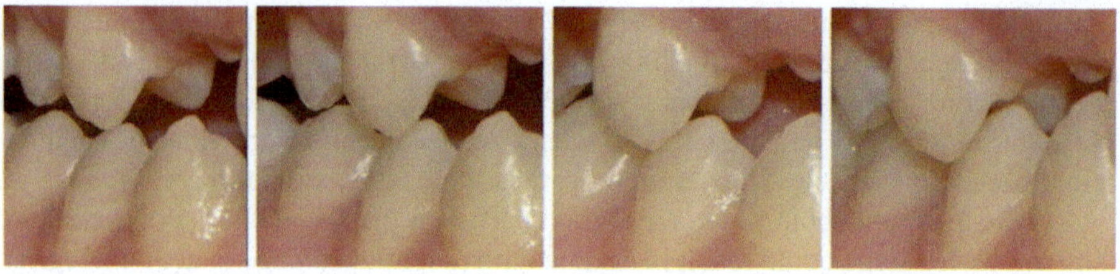

Fig. 3.57 Power stroke phase: a functional moment during which there is both maximum muscular power to achieve food disintegration and maximum proprioceptive control to prevent damage to dental structures

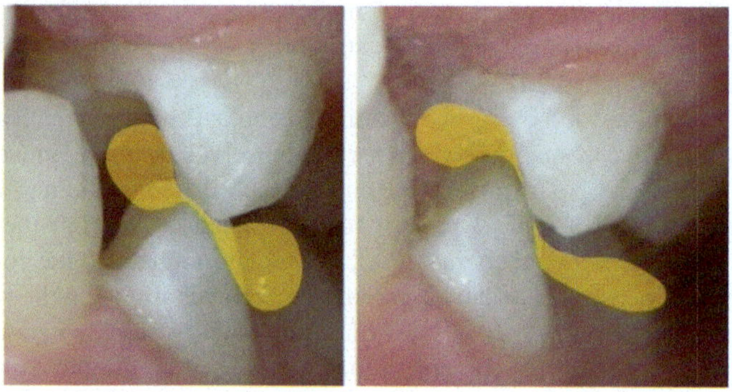

Fig. 3.58 Simultaneous cutting and crushing action during the entry phase of the power stroke: the cutting/guidance slopes wear down

The power stroke is further divided into two parts:

- ***Phase 1: Entry,*** during which the minimum distance between the arches is reached. This distance varies depending on the consistency of the interposed food and gradually decreases as the bolus is chewed until reaching maximum intercuspation (Fish & Mendel, 1982; Hiiemae, 1984). In this phase, the cutting surfaces slide against each other. Although there is also some crushing action in this phase, cutting and slicing dominate during the entry of the power stroke (Hylander, 1980; Wall et al., 2006) rather than compression and crushing (Figs. 3.58, 3.59, 3.60, 3.61).

Nevertheless, all of these abilities characterize the tribosphenic molar and functionally define it as the preferred tool for mammalian chewing.

During the entry phase of the power stroke, the cutting surface is subjected to tangential forces due to the contact between opposing teeth and the bolus (three-body wear) or between dental surfaces themselves (two-body wear). These forces tend to create interprismatic fracture lines, which are prevented by the strong resistance of the prisms to these stresses (He & Swain, 2008). In contrast, during the entry phase of the power stroke, the enamel on the crushing surfaces of the supporting cusps is exposed to perpendicular forces to the surface and parallel to the prism

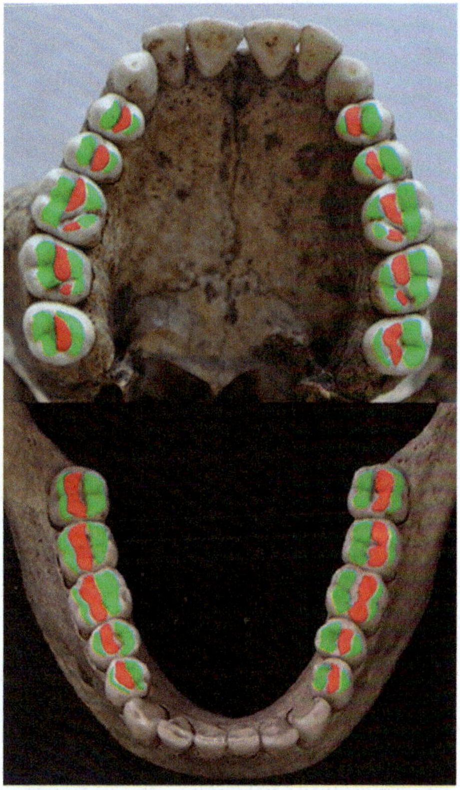

Fig. 3.59 In green, the cutting/guidance slopes; in red, the crushing slopes; specimen from the Museum of Anthropology of the University of Florence

axes due to the reciprocal approach of opposite cusp faces, resulting in the crushing and grinding of the bolus (Shimizu et al., 2005). The wear facets on the supporting surfaces are associated with the intense actions of food grinding, with the cusp slopes moving reciprocally toward opposite basins, like the jaws of a vise (Fig. 3.62).

- ***Phase 2—Exit phase***, in which the active slopes of the supporting cusps rub against each other and move apart to start a new chewing cycle. In humans, some primates, and ruminants, the masticatory muscles continue to work beyond the point of minimum verticality reached with the bolus compression in the entry phase (Gans et al., 1978). However, in other mammals and some other primates, the importance of this functional moment has been questioned due to a sharp decrease in muscular force once maximum intercuspation is passed (Hylander, 1980; Wall, 1997). In phase 1 of the power stroke, the majority of the closing muscular force is generated, and it has a much longer duration than phase 2 (Wall et al., 2006). In this exit phase of the power stroke, the enamel on the supporting slopes, which was previously subjected to forces with a direction parallel to the prism axes, is now exposed to cutting forces with rubbing in a direction perpendicular to the prism axes (Fig. 3.63).

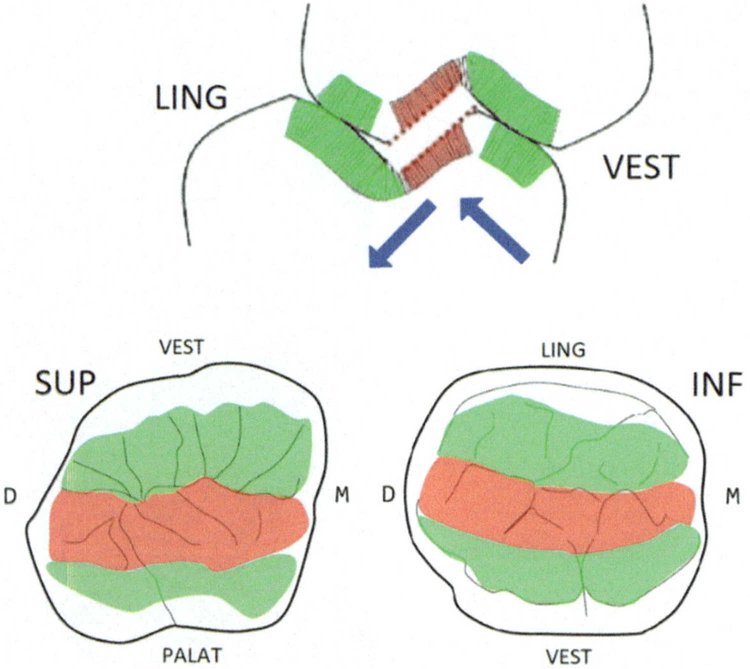

Fig. 3.60 Wear facets, related to tangential rubbing against the surfaces of occlusal slopes, can be distinguished according to the functional phases of the chewing cycle: phase 1 (entry), green facets; phase 2 (exit), red facets

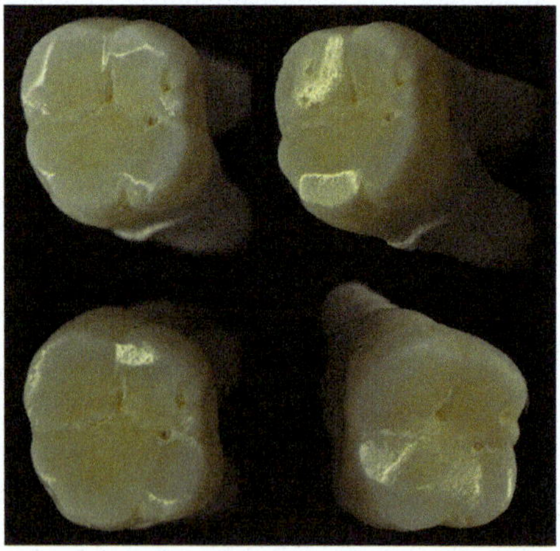

Fig. 3.61 Upper first molar (from a 70-year-old subject): wear has matured the functional cutting/guidance, working/crushing, and balancing slopes active in different phases of the chewing cycle. Note the evident impossibility of that theoretical tripod contact that, according to certain occlusal theories, a cusp should have with its antagonist fossa

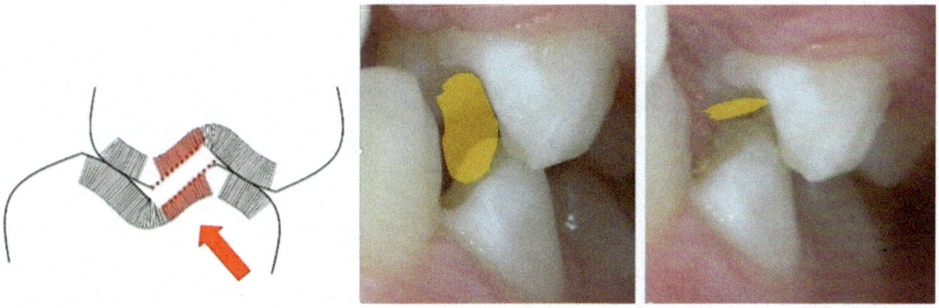

Fig. 3.62 Crushing action during the entry phase of the power stroke, in which the cutting/guidance slopes are subjected to compression

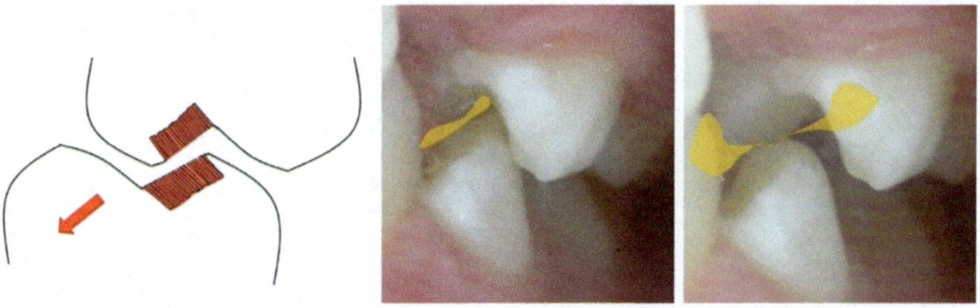

Fig. 3.63 Exit phase of the masticatory power stroke, during which food is crushed between the two working cusps

In humans, the active slopes of the working cusps (inner slope of the palatal cusps of the upper premolars and molars and the inner slope of the buccal cusps of the lower antagonists) are heavily stressed during the power stroke of chewing: in the entry phase, intense forces are exerted axially with respect to the prisms, while in the exit phase, forces act tangentially to the prism axes (Fig. 3.64).

Although cutting action is typically associated with phase 1 and compressive and grinding action with phase 2, both aspects actually occur in phase 1 of the entry. The different directions of occlusal forces, such as those parallel to cutting surfaces or those perpendicular to compressive surfaces, differently regulate functional wear in different movements of the chewing cycle (Hiiemae & Crompton, 1985). Even though there are no structural differences in enamel among different regions, differentiated wear of cusp inclines occurs during chewing based on the varying prism angles on these surfaces (Maas & Dumont, 1999).

On surfaces more affected by functional stresses—namely, the active slopes of working cusps—wear is more intense (Fig. 3.65), decreasing, in chronological order, from the first to the third molar, with the initial formation of a helical plane and later with the reversal of the Wilson curve (Osborn, 1982, 1993a, b).

Wear inevitably involves, and proportionally to their biomechanical engagement, the guiding surfaces of anterior teeth as well. Canines, the primary guiding teeth for

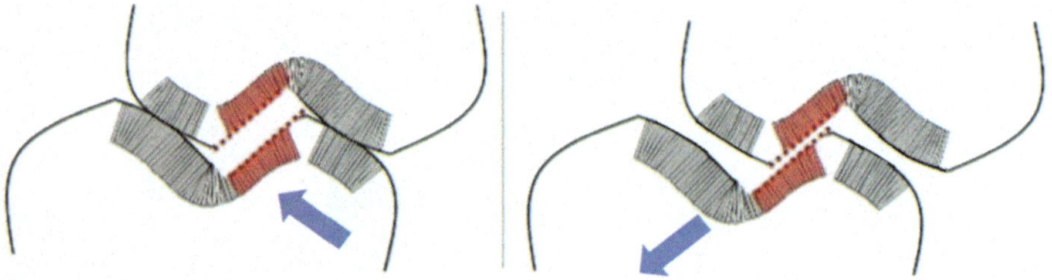

Fig. 3.64 In the entry and exit phases of the power stroke, the tips of the prisms on the active surfaces of the working cusps first experience axial force and then tangential force, accumulating greater stress compared to other occlusal surfaces

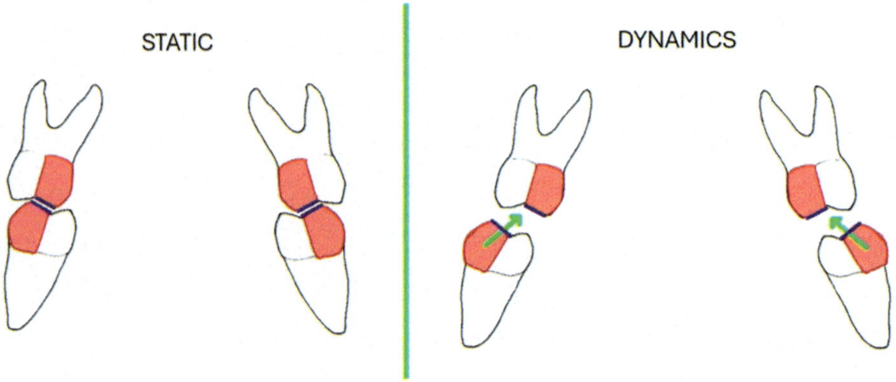

Fig. 3.65 Greater involvement of the inner slopes of the upper palatal cusps and lower vestibular cusps in static (support) and dynamic (working) functions

Fig. 3.66 Harmonic wear of occlusal surfaces. Skull 4235 from the Museum of Anthropology of the University of Florence

lateral movements, are no exception, and their functional inclines wear over time, leading to the involvement of posterior cutting surfaces in guiding function (Fig. 3.66) and thus achieving the progressive and physiological transition from pure canine guidance to group function (Figs. 3.67, 3.68).

Examining the effectiveness of dental engagement during the power stroke of chewing, which occurs when the bolus is being cut and crushed by the cutting and stamping cusps of each masticating tooth on the working side, it becomes intuitive to understand in which of the two illustrated situations (Fig. 3.69) the bolus will be treated more effectively.

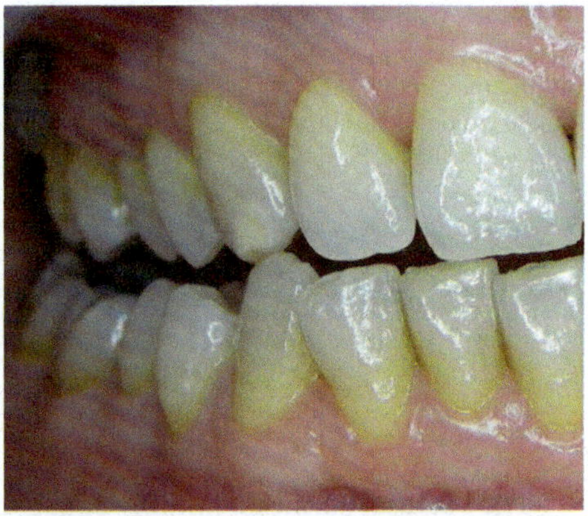

Fig. 3.67 Pure canine guidance of lateral movements. Note the functional abrasion on the cusp of the upper canine

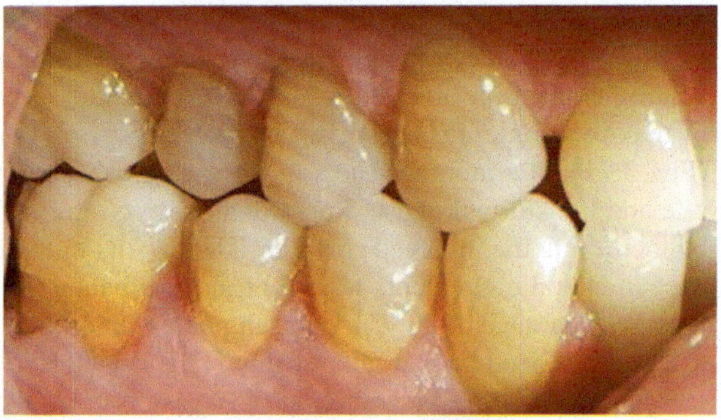

Fig. 3.68 Entry phase of the chewing cycle: group guidance

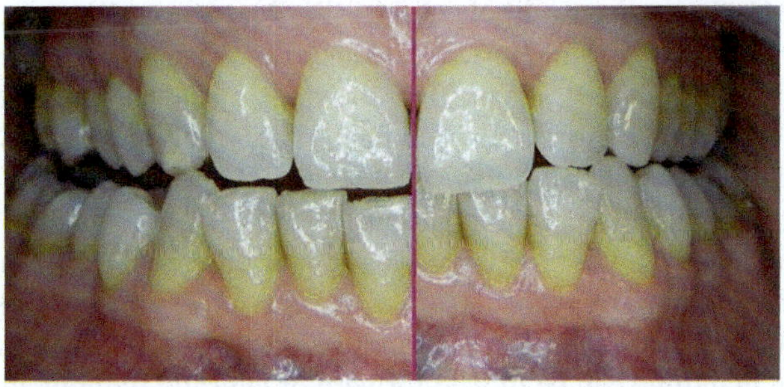

Fig. 3.69 Left hemi-arch: pure canine guidance. The chewing force begins with a single contact on a non-chewing tooth: the chewing elements will come into action only after having traveled part of the functional path inactive. Right hemi-arch: group function. The chewing force begins with the engagement of all the chewing teeth to maximize the masticatory effectiveness of the working hemi-arch; chewing will preferentially occur on this side, both for better food breakdown and for better distribution of stress on the dental elements

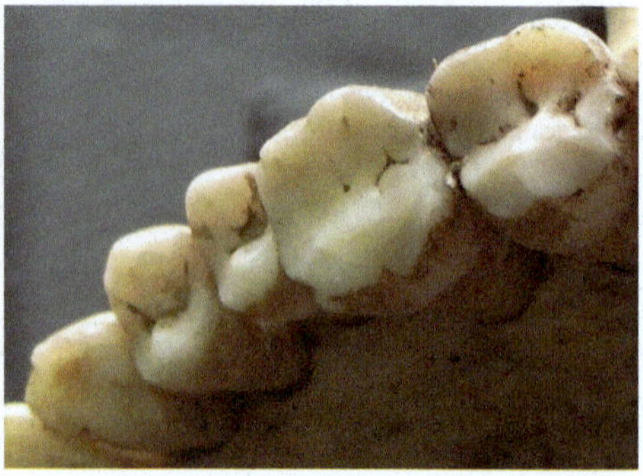

Fig. 3.70 Working and balancing wear facets. Skull 46 from the Anthropology Museum of the University of Florence

The occlusal surfaces, initially engaged in function, increase their extent through abrasion, simultaneously reducing the slope of their occlusal incline. This determines the functional co-optation of analogous surfaces of adjacent teeth (Fig. 3.70).

Mutually protected occlusion, where canines alone guide lateral movements preventing contact on other teeth, as well as incisors in protrusion and posterior teeth in maximum intercuspation, which prevent strong contacts on the anterior group (Fig. 3.71), in a mouth subject to physiological maturation represents only a juvenile and transitional phase. Functional wear progressively shapes the dental arches by increasing the involved surfaces. This results in the establishment of group guidance in lateral movements with contralateral balancing light brushing and the gradual disappearance of the Christensen phenomenon in protrusion, eventually leading to a head-to-head incisal relationship (Kaifu, 1996).

The Christensen phenomenon (CP) results from the separation of the posterior teeth during protrusive movement of the mandible and is closely related to the structure of the TMJ (Temporomandibular Joint) (Parilov, 1989). We can consider CP as the relationship between the anterior guidance represented by the incisal overbite and the posterior guidance represented by the slope of the glenoid fossa's roof in the mandibular fossa. This relationship is characteristic of humans and is not present in other primates. As we will see, it is part of a system that coordinates not only mandibular movement but also cranial position.

Overbite and the slope of the glenoid fossa's roof are both determinants of occlusion and tend to change over time, reducing in parallel and reciprocally. At birth, the articular eminence has not yet formed, and the first few months of life are the only time when the mandible can be moved forward or laterally without any lower movement (Katsavrias, 2002). This horizontal dynamic is functional for mandibular movements during breastfeeding.

The eruption of incisors in the dental arch (both deciduous and permanent) is characterized by the presence of an incisal overbite, which gradually decreases in past populations until, with advanced wear, it reaches edge-to-edge occlusion (Kaifu et al., 2003) (Fig. 3.72).

The lingual tipping of central incisors is a movement that, due to the action of the orbicularis muscle, causes them to lose their labial inclination. The anterior teeth of both maxillae are initially inclined labially, but with wear of the incisal edges and mandibular adaptation, they lose this inclination and gradually become almost

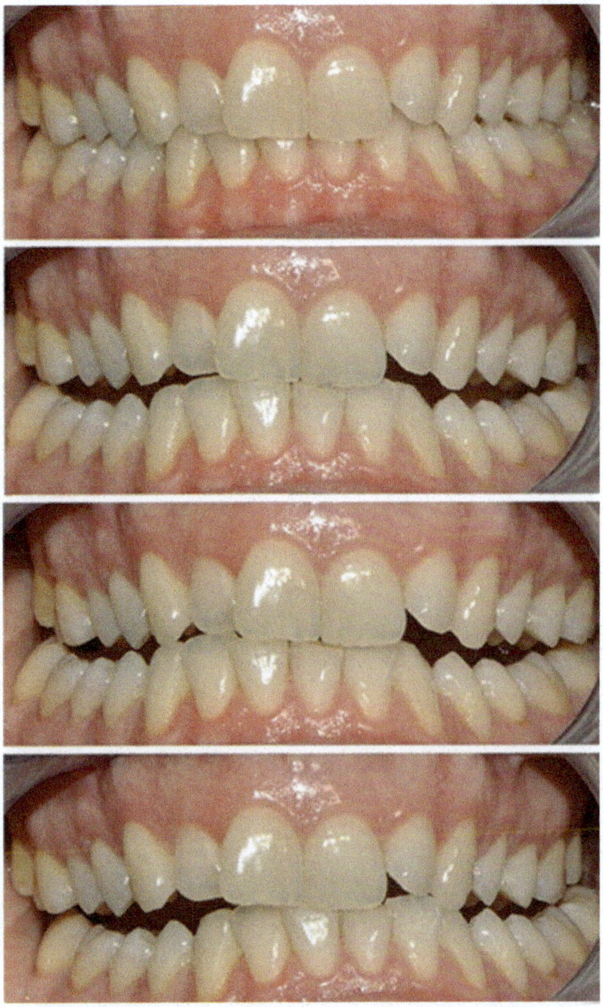

Fig. 3.71 Canines and centrals disengaging the other teeth in the posterior sectors. First photo from the top: maximum intercuspation; second photo from the top: protrusive; third photo from the top: right lateral movement; fourth photo from the top: left lateral movement

vertical. Lingual tipping has been observed in great apes, some fossil hominins like *Australopithecus robustus*, and even in *Homo neanderthalensis*. Worn and flat occlusal surfaces and anterior head-to-head occlusion are ubiquitous in the dentitions of prehistoric humans. With the progressive decrease in overbite, a position of matching worn edges will be reached, perhaps more appropriately considered as incisal surfaces (Kaifu et al., 2003) (Fig. 3.73).

These changes play a role in the closure of interproximal spaces, which occurred due to a substantial reduction in the mesiodistal dimensions of the crowns, resulting from friction and subsequent interproximal wear (Kaifu et al., 2003). In deciduous dentition, this phenomenon of occlusal advancement would be preparatory to achieving a first-class occlusion with the first permanent molars. It's important to note that the modern diet does not trigger these friction mechanisms that our species evolved with.

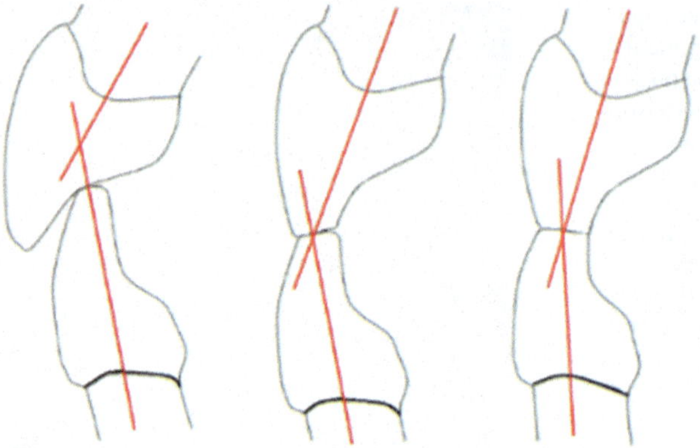

Fig. 3.72 Progressive reduction of the overbite until reaching an edge-to-edge relationship, associated with lingual tipping due to the progressive inclination of the anterior teeth of both jaws, until they become almost vertical. Drawing freely adapted from "Tooth wear and the 'design' of the human dentition: a perspective from evolutionary medicine," Kaifu et al., 2003

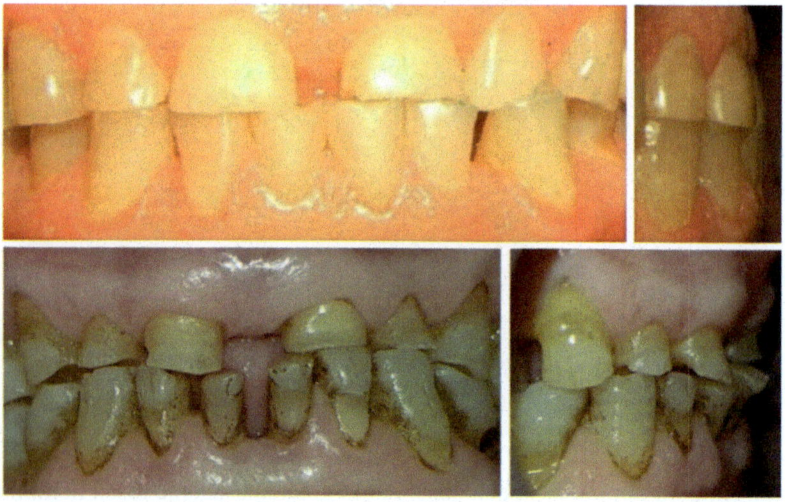

Fig. 3.73 Edge-to-edge relationship of the incisor group resulting from the wear of the incisal edges and the mesialization of the teeth

3.4 Helical Plane and Inversion of Compensation Curves

As we have seen, increased stress on the working cusps leads to greater wear compared to other occlusal areas, and this wear also varies based on the eruption chronology of individual teeth, meaning in relation to the working time of the dental element: the working cusp of the first molar reduces its height first and more than

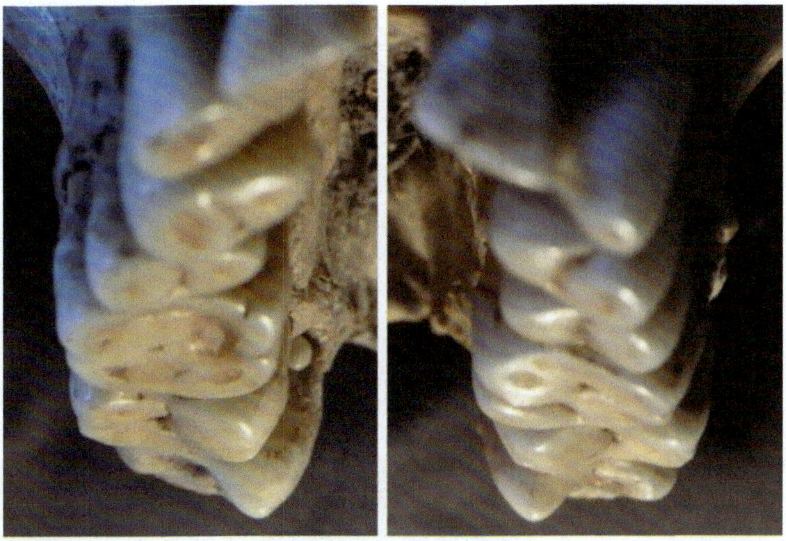

Fig. 3.74 Differentiated abrasion of the working cusps of the molars. Note (in this figure and the following ones) the progressive decrease in wear of the supporting/working cusp, from the first molar, to the second, to the third. Skull 3371 from the Anthropology Museum of the University of Florence

that of the second, and the same progression occurs with the third (Figs. 3.74, 3.75, 3.76, 3.77).

In dental prosthetic reconstruction, compensation curves are a significantly important factor. Between 1890 and 1930, many authors sought to identify physiological canons for these curves, proposing various geometric shapes to describe the ideal arrangement of teeth in humans. These included the circular curve of von Spee (1890), Bonwill's triangle (1899), Wilson's curve (1911), Monson's sphere (1920), Campbell's helix (1925), and Hanau's quintet (1926) (Osborn, 1993a, b).

The compensation curves consist of the *von Spee curve* in the sagittal plane and the *Wilson curve* in the frontal plane. The von Spee curve was described in 1890 by the anatomist Ferdinand Graf von Spee in his article "*Die Verschiebungsbahn des Unterkiefers am Schädel*" (The path of mandibular displacement on the skull). He highlighted that the occlusal shape of the teeth (and their position) changes in accordance with the movement of each other, meaning in relation to the dynamics of the jaws (Bimler, 2007). This article (Spee, 1890) argued that compensation curves, in light of the force vectors that develop during chewing, represent not only modifications of the occlusal plane to avoid dynamic interferences but also dental orientations that allow the axes and occlusal surfaces to align with masticatory movement (Dickerson et al., 2001).

The Wilson curve was described by George H. Wilson in 1911 as a concave curve in the mandibular arch and a convex curve in the maxillary arch, created in the

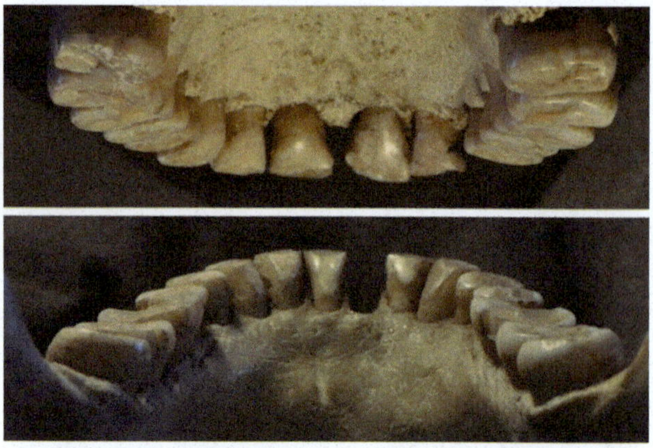

Fig. 3.75 Differentiated abrasion of the working cusps of the molars. Skull 710 from the Anthropology Museum of the University of Florence

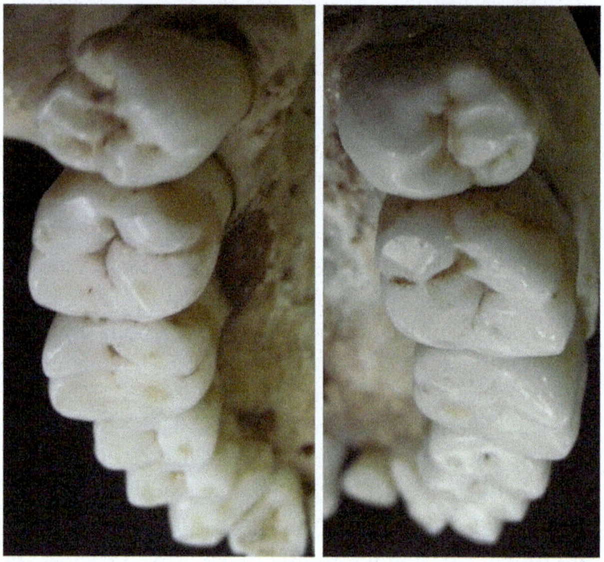

Fig. 3.76 Differentiated abrasion of the working cusps of the molars. Note how the wear increases from the first to the third molar. Skull 3104 from the Anthropology Museum of the University of Florence

frontal plane by the inclination of posterior teeth to avoid possible balancing interferences (Barrera et al., 2013) (Fig. 3.78).

The von Spee and Wilson curves represent the harmonization of teeth and their occlusal surfaces with species-specific and individual masticatory movements. They are usually associated with hominids, although curves in the sagittal plane are present in nearly all carnivores and herbivores (Wall, 1999), and the correlation between the orientation of molar teeth and the superficial masseter has been observed in a wide range of primates and other mammals (Osborn, 1982). In the frontal plane, many mammals with grinding dentitions have an often inverse occlusal curvature (Osborn, 1993a, b). Also, in the sagittal plane, the von Spee curve is present in almost all mammals, both carnivores and herbivores, in a functional relationship between the orientation of molars and the superficial masseter (Osborn,

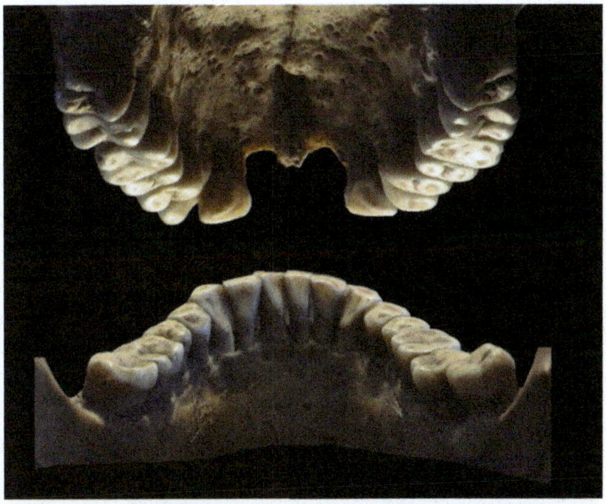

Fig. 3.77 Figure 211. Differentiated abrasion of the working cusps of the molar group. Skull 3798 from the Anthropology Museum of the University of Florence

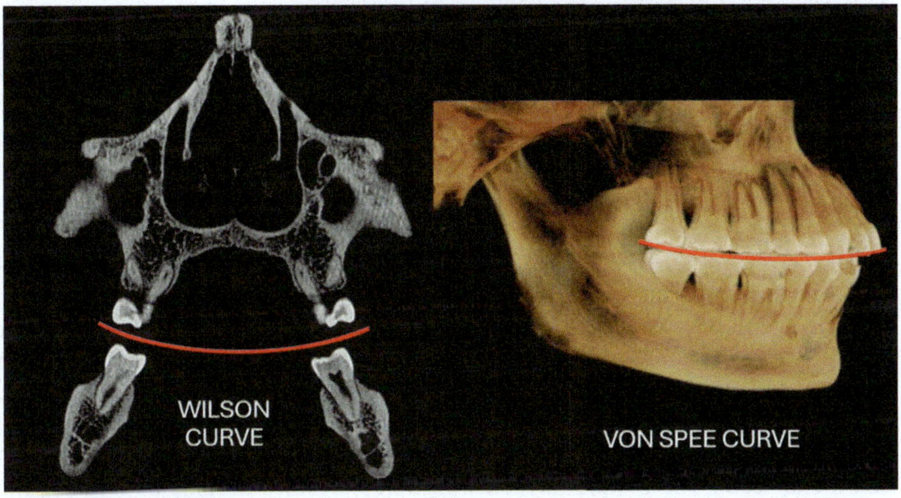

Fig. 3.78 Compensating curves. CT scans of specimens from the Anthropology Museum of the University of Florence

1982). Since this curve takes on the same shape in most mammals (Fig. 3.79), it likely serves the same function (Marshall et al., 2008)—even though some of them, such as bats (Bimler, 2007) or sloths, exhibit a downward-inverted curve (a curious coincidence with the unique habitual posture of both these animals).

The course and modeling of occlusal planes are part of a functional scheme composed of various anatomical components (occlusion determinants), which influence and coordinate with each other. The occlusal surface of teeth, their inclination, and their position are regulated by masticatory trajectories, and vice versa.

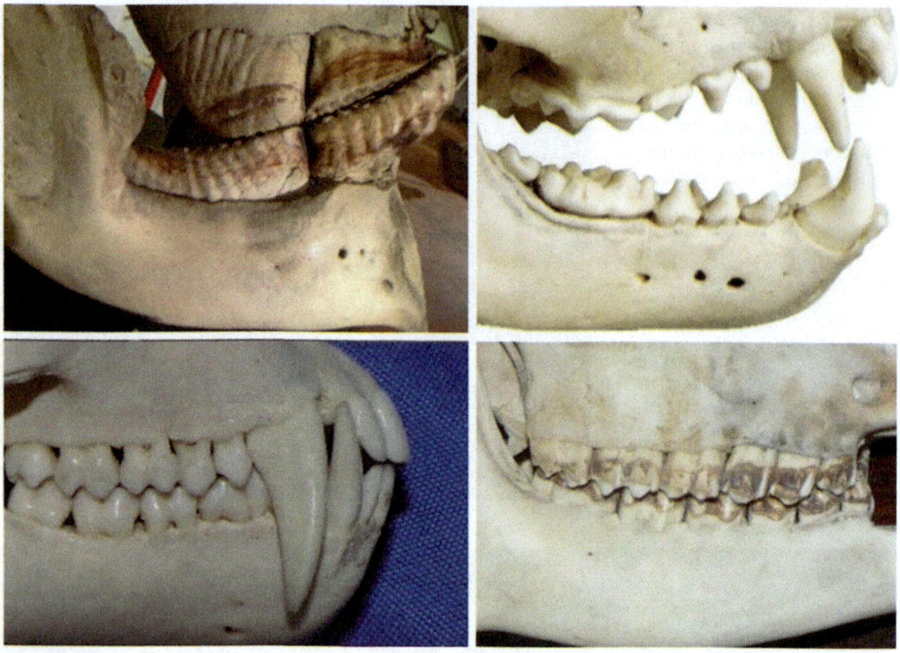

Fig. 3.79 Curvilinear course of the occlusal plane in sagittal perspective. From left to right and top to bottom: elephant, badger, cercopithecus, antelope. Collection of the Zoology Museum of the University of Florence

The shortening of the face along the evolutionary path from *Australopithecus afarensis* to *Homo sapiens* resulted in an increased axial inclination of the molars. This change did not apply to *Australopithecus afarensis*, where the first premolar lost its cutting form to become a masticatory tooth more similar to its function in humans. With *Australopithecus africanus*, the size of the molars increased further, and in the robust australopithecines, the masticatory apparatus became even more specialized, with an almost disproportionate increase in premolar (molarization) and molar teeth. The molars in robust australopithecines became as large as the thumbnail on our thumb (see Fig. 41); in humans, the third molar is roughly the size of the pinky nail), while there was a reduction in anterior teeth, including the canines. Such an increase in chewing surface area corresponded to the development of a well-developed muscle mass, which later led to the formation of bony ridges on the skull and wide zygomatic arches to accommodate the temporal muscle. With *Homo habilis*, there was a change in diet, and a process of reducing the volume of teeth began.

The progressive inversion of the Wilson curve is physiologically evident in a dentition that has been subjected to tough and consistent foods (Smith, 1986). This aspect of wear deserves attention because every dental element in the arch undergoes abrasion related to its spatial position, the functional movements it experiences, and the duration of time it spends in the mouth. However, the inclination of

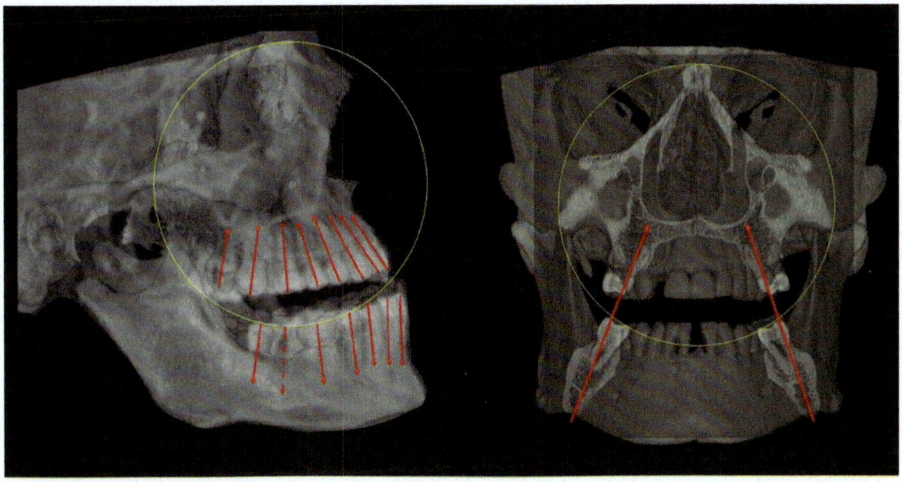

Fig. 3.80 Occlusal surfaces, dental axes, periodontal ligament, and alveolar bone compose a system that harmonizes with the dynamics of mastication, allowing the function to be performed by dissipating the intense forces developed by the powerful masticatory muscles. Cone beam CT scan performed on a skull from the Anthropology Museum of the University of Florence

the teeth is one of the primary mechanisms guiding the distribution of wear during the chewing phases (Oxilia et al., 2018).

Just as a circle, even when subjected to different directions, retains one of its specific properties: minimal resistance to sliding (Bimler, 2007), the dental axes in the three spatial planes structure the occlusal plane into curved surfaces, closely resembling spherical caps. They arrange individual dental elements in a way that optimally dissipates friction and chewing forces (see Fig. 3.80).

Each tooth within its alveolar process is a biomechanically passive element capable of adapting its position in the arch until it reaches equilibrium of the forces acting upon it (Cárabe Fernández, 2014). Since the direction of bite force is rarely perpendicular to the occlusal surface, during masticatory function or even in maximum intercuspation, vectors of force with different directions simultaneously arise. Multidirectional forces on an irregular surface can be accurately reduced to a single force vector (see Figs. 3.81, 3.82). Therefore, even numerous muscle forces can be simplified to a single resultant force (Osborn, 1993a, b).

The lingual inclination of posterior teeth concerning the dominant component of the force exerted by the muscles on them demonstrates that axial alignment is parallel to the upward and inward traction force of the medial pterygoid muscle (Silvestri et al., 2008) (see Fig. 3.83).

So, what is the purpose of these so-called compensation curves, von Spee's and Wilson's? These characteristic orientations of dental axes serve to accommodate the condylar trajectories along the lateral-lateral development of the articular eminence. In humans, the development of the articular eminence in this direction allows and supports the antero-medial movement of the balancing condyle during chewing

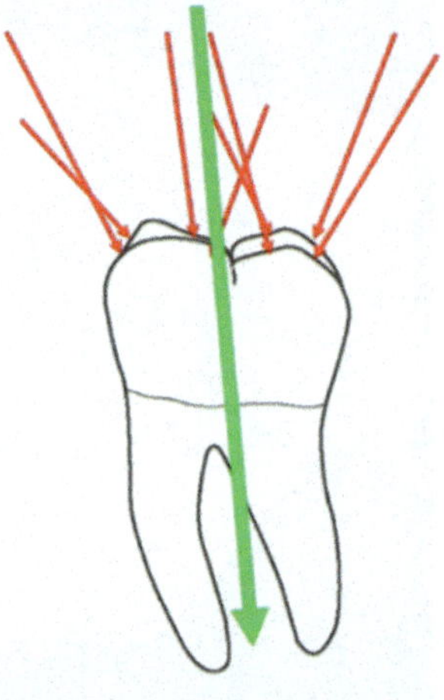

Fig. 3.81 The multidirectionality of the forces acting on a tooth will express in real time a resultant that will progressively determine the position of the tooth itself and the orientation of its axis

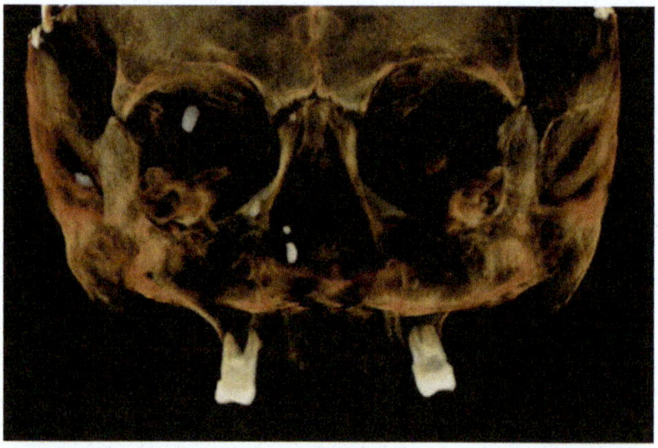

Fig. 3.82 Image of a skull on the frontal plane: note the inclination axis of the molars. Cone beam CT scan of a specimen from the Anthropology Museum of the University of Florence

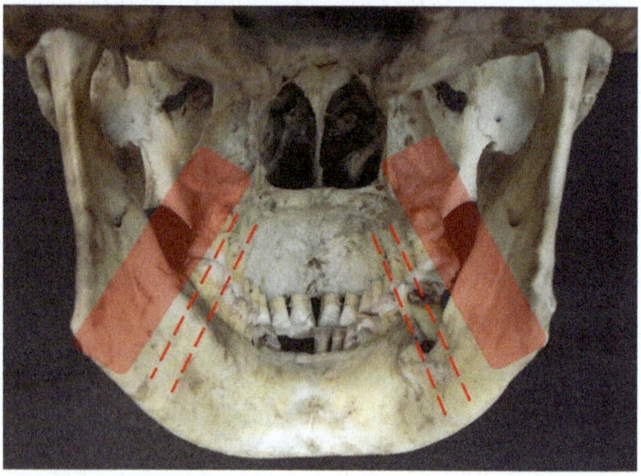

Fig. 3.83 Image of a skull on the frontal plane: parallelism between the inclination of the molars and the axis of the internal pterygoid muscles. Specimen from the Anthropology Museum of the University of Florence

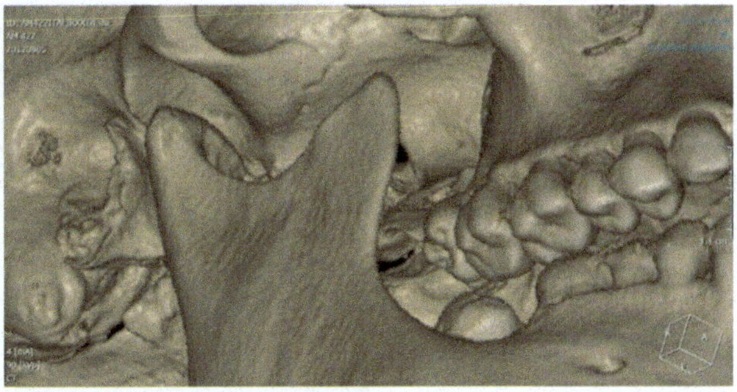

Fig. 3.84 Note how, despite the poor morphological congruence of the TMJ (temporomandibular joint), the functional surfaces of the condyle and fossa exhibit two characteristic facets, almost mirror images, medial and lateral, in relation to the occlusal surfaces. Cone beam CT scan performed on a specimen from the Anthropology Museum of the University of Florence

(Wall, 1999; Barrera et al., 2013). Compensation curves reduce the possibility of traumatic dental interferences, ensuring that the teeth on the balancing side do not impact each other during the power phase, hindering the action of the working teeth (Wall, 1999). At the same time, they support the balancing joint (see Fig. 3.84).

Indeed, compensation curves reduce the separation of the arches during the functional movements of incision, tearing, and chewing, facilitating posterior dental balance to protect the TMJ from higher functional loads (Hylander, 1979a, b, c). They also coordinate occlusal surfaces with masticatory function, harmonizing the inclination of the glenoid roof of the TMJ, the vectors of the masticatory muscles, the cusp height, the overbite, and the inclination of the occlusal plane. The Wilson curve determines the arrangement of dental axes in response to the functional vectors of the superficial masseters and internal pterygoids (see Fig. 3.85) (Hylander, 1979a, b, c; Cárabe Fernández, 2014; Silvestri et al., 2008) and adapts to them through wear until inversion occurs.

Compensation curves create a curved occlusal plane that allows only a few teeth to occlude at a time, concentrating and optimizing occlusal force (Popowics & Herring, 2006). The von Spee curve, being more related to mandibular antero-posterior movements, is more stable than the Wilson curve (Karani et al., 2018) because its shape is less influenced by masticatory cycles. Although very few studies have been conducted to determine the long-term stability of the von Spee curve, it can physiologically change with age (Kumar & Tamizharasi, 2012), decreasing due to a reduction in posterior disclusion relative to a decrease in overbite (Ahmed et al., 2011). In the same individual, compensation curves can differ from one side to the other because, even though anatomical symmetry is often assumed, it is not always the case: asymmetry in masticatory function over time can lead to structural adaptation, resulting in asymmetry in both form and function (see Figs. 3.86, 3.87).

The von Spee curve can be pathologically altered due to tooth rotation, extrusion, or more general tooth movement, and planned serial extractions in orthodontic

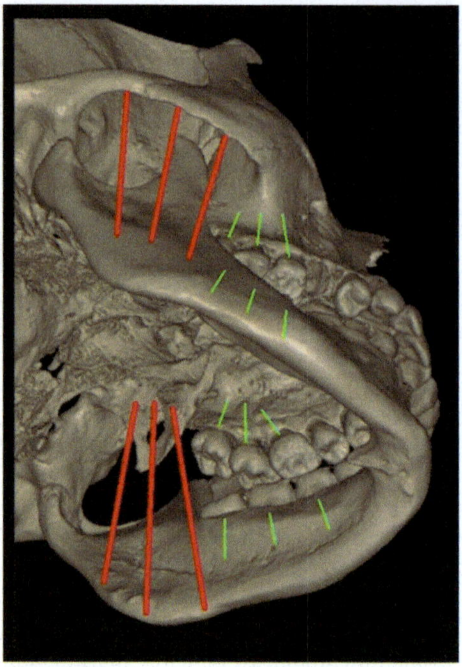

Fig. 3.85 Correspondence between the force vectors of the masseter and internal pterygoid muscle fibers (red lines) and the dental axes of the upper and lower molars (green lines). Cone beam CT scan performed on a specimen from the Anthropology Museum of the University of Florence

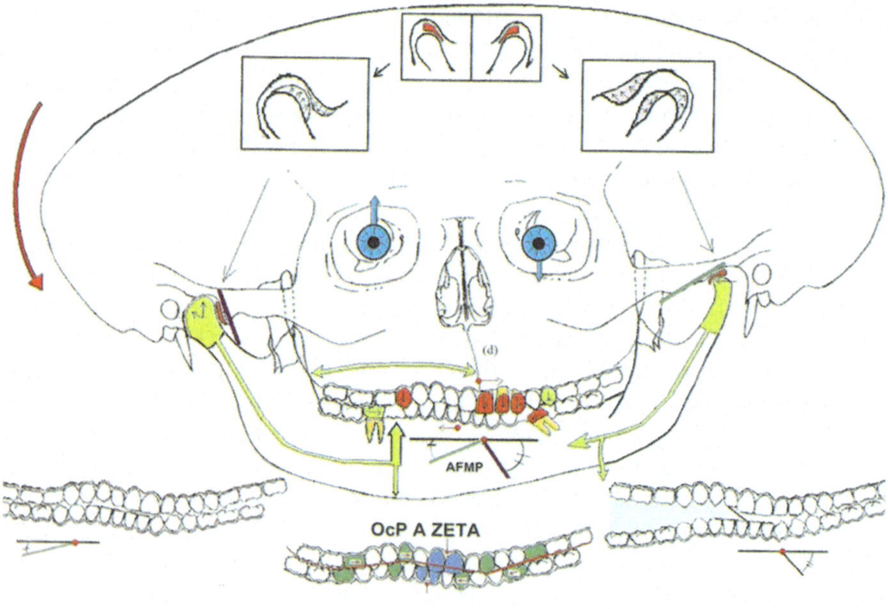

Fig. 3.86 Adaptation of maxillofacial structures to asymmetric chewing guided by a different entry phase of the chewing force. Drawing adapted from "Evolution of the functional dynamic relationship between occlusion and TMJ," S. Dettori, A. Confaloni, 2001

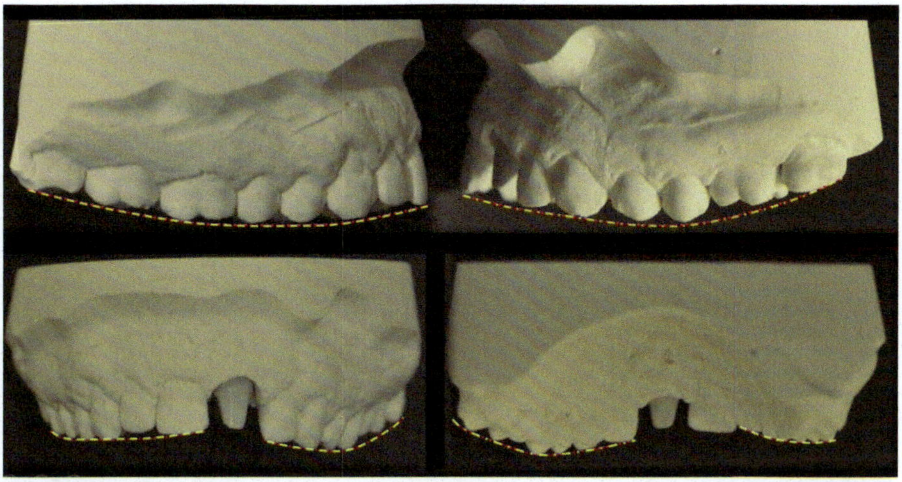

Fig. 3.87 Working model for a crown on the upper left incisor. Asymmetric development of the two hemimaxillae and the curves of von Spee and Wilson

treatments can lead to steeper occlusal curves (Feldman et al., 2015; Mintenko et al., 2020).

If a correct von Spee curve is not reestablished during the complete mouth rehabilitation, there can be precontacts in maximum intercuspation and interferences during mandibular movements, which can jeopardize the health of the masticatory system (Osborn, 1987; Mohan et al., 2011). Abnormal muscle activities, especially of the masseter and temporal muscles, may also develop as a result, leading to pathological wear, periodontal issues, restoration fractures, and temporomandibular joint dysfunction (Williamson & Lundquist, 1983) (see Fig. 3.88).

Managing the von Spee curve is a critical factor in the stability of complete dentures (Farella et al., 2002) and also plays a significant role in the medium to long-term success of implant-supported prostheses (Karani et al., 2018).

The von Spee curve differs between brachyfacial and dolichofacial individuals, and the bite force is directly related to the depth of the von Spee curve: brachyfacial individuals with a slight antero-posterior occlusal curvature exhibit higher forces (Sakaguchi et al., 2012) (see Fig. 3.89).

Individuals with flatter von Spee and Wilson curves show better food grinding and bolus mixing capabilities (Fueki et al., 2013). Occlusal plane curves are harmonized with stomatognathic function, and often, patients with dysfunctional temporomandibular joint disorders (clicks and joint blocks) exhibit excessive occlusal curvatures, significantly greater than those of healthy subjects (Ito et al., 1997; Fidler et al., 1995). The von Spee curve appears deeper in Class II Division 2 malocclusion (Ahmed et al., 2011) (see Fig. 3.90).

The occlusal surfaces of properly structured arches subject to physiological functional movements tend to wear following precise three-dimensional patterns: the crushing surfaces of the supporting cusps of molar teeth experience greater wear

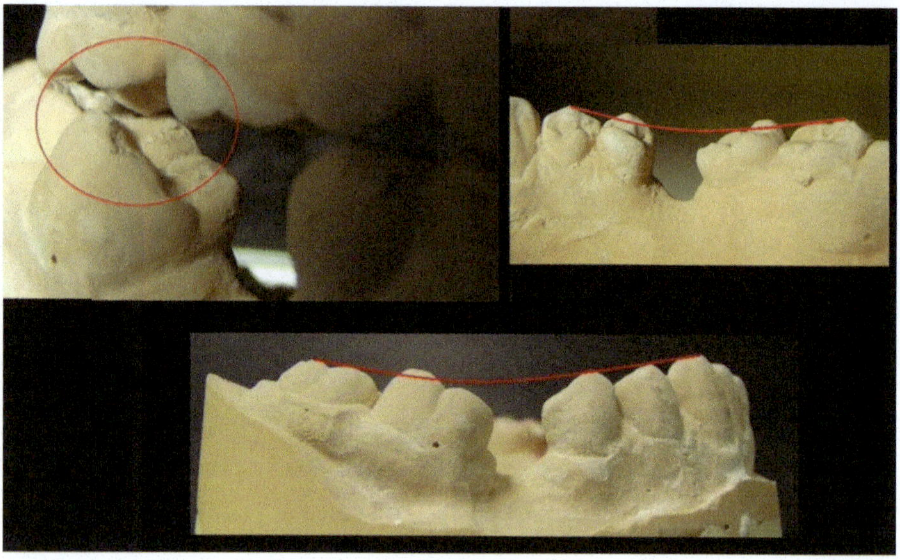

Fig. 3.88 The loss of the first lower molar has induced mesial inclination of the second molar and distal inclination of the second premolar. Note how the inclination of the second molar causes an elevation of the distal cusp, leading to premature contacts in static and interferences in dynamic

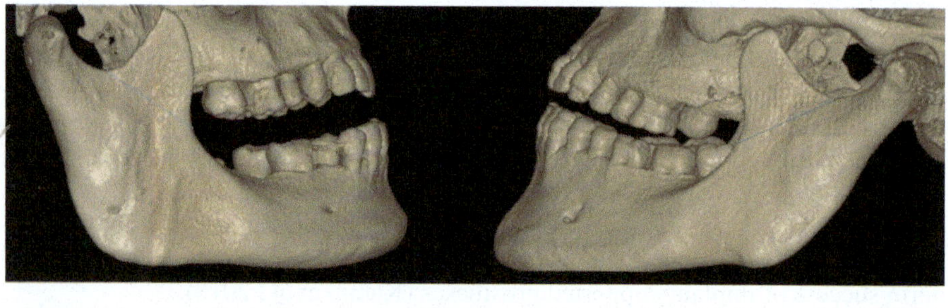

Fig. 3.89 Different courses of the curves of Spee in a brachyfacial individual and a dolichofacial individual. Cone beam CT scan performed on skulls from the Anthropology Museum of the University of Florence

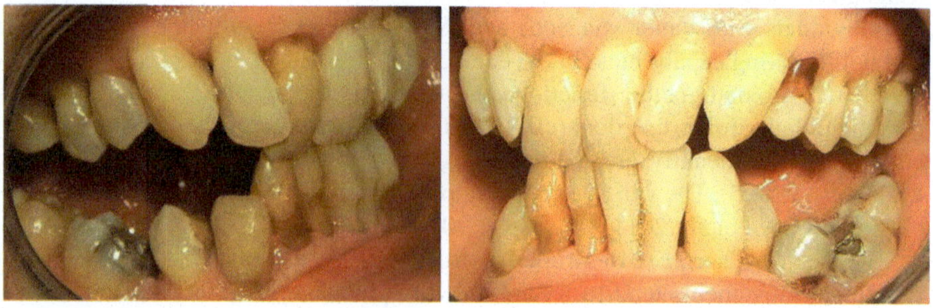

Fig. 3.90 Significant Class II malocclusion with pronounced deep bite and marked curve of Spee

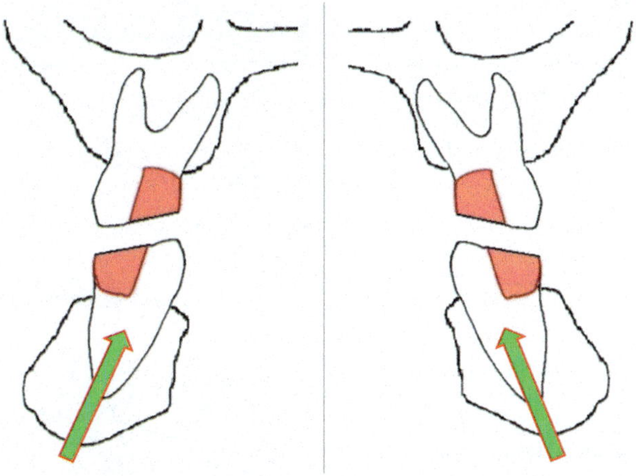

Fig. 3.91 Wear of the supporting cusps

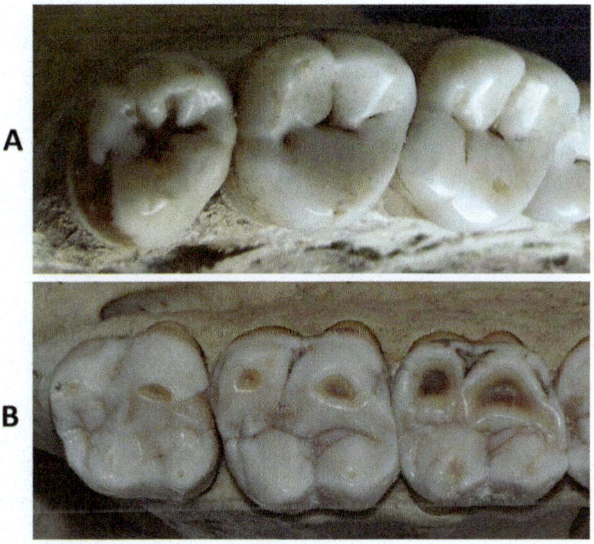

Fig. 3.92 Wear progressively decreases from the first to the third molar, in relation to their time of work in the mouth. A: *Homo sapiens* (Skull 3049 from the Anthropology Museum of the University of Florence); B: gorilla (*Gorilla gorilla*) (Specimen 2808, collection of the Zoology Museum of the University of Florence)

compared to other occlusal surfaces, precisely due to the different intensity and direction of chewing forces (see Fig. 3.91).

The differentiation of abrasion on different teeth is also related, as mentioned at the beginning of this section, to their eruption chronology (see Figs. 3.92, 3.93).

With increasing age and dental wear, the inclination of mandibular occlusal surfaces increased toward the vestibule (toward the palate for upper teeth), mainly manifesting on the first molars—the first to erupt—and then involving the second and finally the third molars (Richards & Brown, 1986) (see Fig. 3.94). In populations with a diet of very tough foods, this pattern tended to appear even in deciduous dentition (see Fig. 3.95).

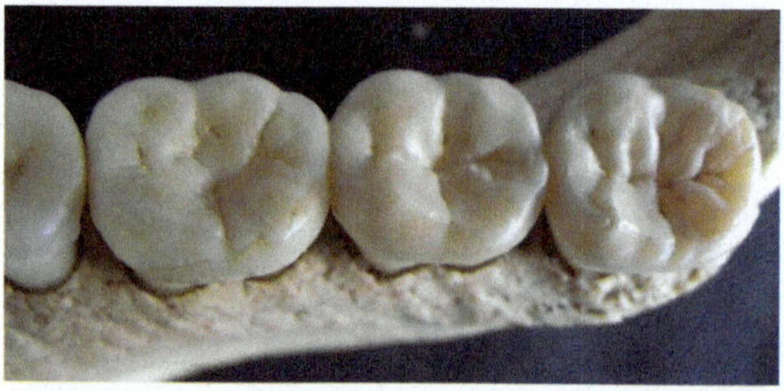

Fig. 3.93 Progressive decrease in wear from the first to the third molar. Skull 5586 from the Anthropology Museum of the University of Florence

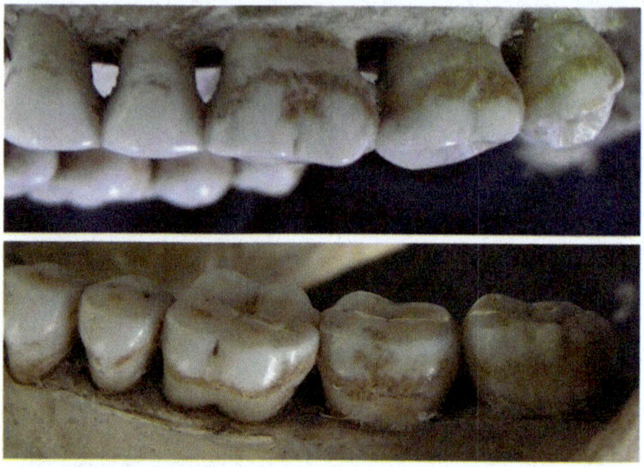

Fig. 3.94 Progressive decrease in wear from the first to the third molar. Skull 3049 from the Anthropology Museum of the University of Florence

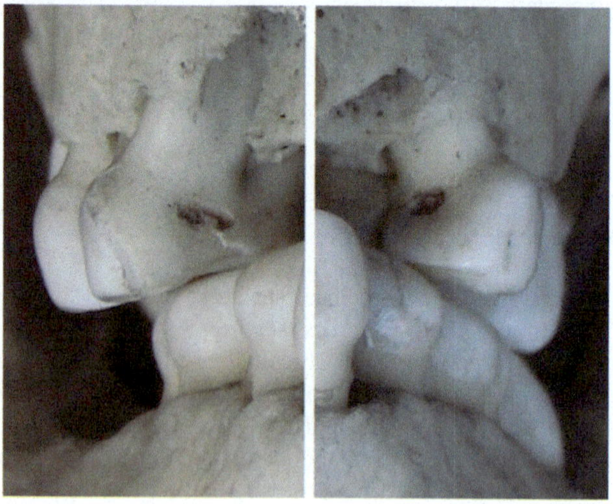

Fig. 3.95 Wear of deciduous teeth. Skull 6153 from the Anthropology Museum of the University of Florence

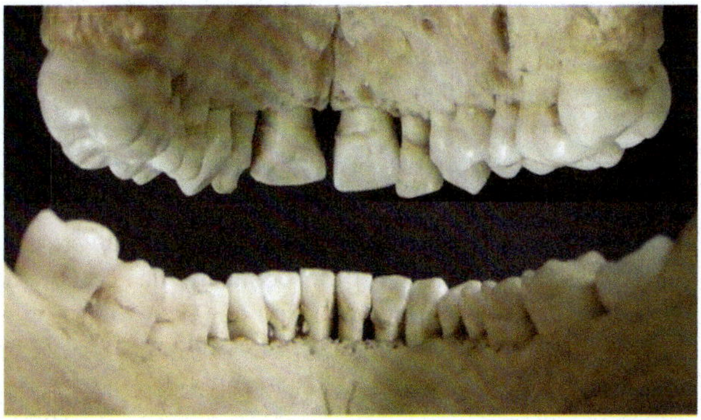

Fig. 3.96 Unworn occlusal surfaces with a helical pattern. Skull 372 from the Anthropology Museum of the University of Florence

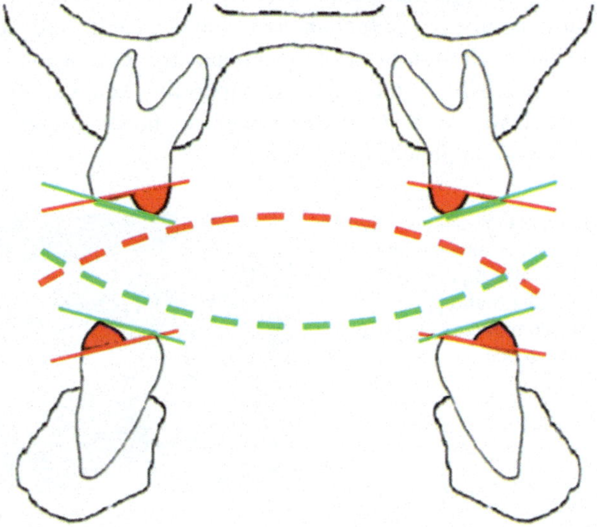

Fig. 3.97 Section on the frontal plane of the helical planes. Green: helical plane of an unworn dentition; red: inverse helical plane after abrasion

So, the differential action of abrasive wear shapes the occlusal plane and its curves, modifying them and giving them a characteristic pattern referred to as 'helical.' Even in an unworn dentition, the arrangement of dental cusps and incisal edges in three-dimensional space conforms the occlusal surfaces to a curvature with a helical pattern. These surfaces exhibit an arrangement that adapts to a sphere with a diameter of 8 inches (20.32 cm) and a center at the level of the glabella, as described by George S. Monson in 1920 (see Fig. 3.96).

In this discussion, we will refer to the occlusal plane that occurs in an unworn dentition due to the normal arrangement of dental elements along the curves of von Spee and Wilson as the "helical plane." To distinguish the inversion of the occlusal surface alignment related to wear, we will refer to it as the "Inverse Helical Plane" (IHP) (see Fig. 3.97).

The IHP was first described in 1923 by Campbell in Australian Aborigines (Campbell, 1923) and is a condition of advanced wear on posterior teeth where the

Wilson curve is reversed on the first molars, flattened on the second molars, and present in its classical form on the third molar. This specific wear is caused by increased functional stress on the supporting cusps of molars and has been described in various populations (Native Americans, Egyptians, Siamese, Neolithic and recent Caucasians, Inuit, African Bushmen) and found in many recent and fossilized human dentitions (Osborn, 1982).

The helical pattern of an intact dentition changes over time, progressively reversing its slope, starting with the occlusal surfaces of the working cusps of the first molars. Since they erupt first, they begin to wear down before the others. This will lead to a reversal of the inclination of the tooth's occlusal plane, inducing a buccal slope for the lower teeth (see Fig. 3.98) and a palatal slope for the upper teeth, developing complementary opposing surfaces (Smith, 1972)

In the same dentition, the eruption times, varying between the first, second, and third molars, play a fundamental role in wear differences (see Fig. 3.99), which lead to the maturation of the helical plane due to abrasion. This continuous wear gradually intensifies, and when the supporting cusps of the third molars have become sufficiently worn down throughout the molar dentition, the reversal of the Wilson curve is completed (Osborn, 1982).

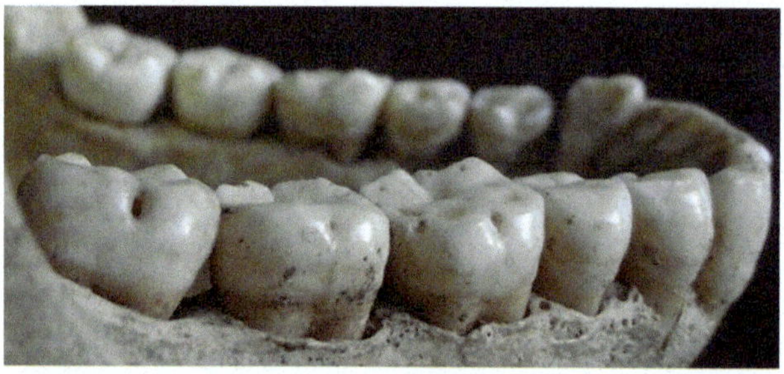

Fig. 3.98 Initial inversion of the helical plane on the first molar. Skull 5602 from the Anthropology Museum of the University of Florence

Fig. 3.99 Progression of the helical plane abrasion on the second and third molars. Skull 2985 from the Anthropology Museum of the University of Florence

The distribution of wear within a population can be variable, and a group with a highly abrasive diet will tend to develop the reversal of the Wilson curve more rapidly than another. In past populations, the diet was much more abrasive than today's, resulting in pronounced wear even at a very early age. In the modern industrialized world, even elderly individuals show much milder dental wear compared to past societies (Benazzi, Kullmer, et al., 2013).

So, in populations with a less abrasive diet, the intermediate phase of the helical plane will tend to have a longer duration and will be more prevalent in an adult population. If the diet's abrasiveness decreases, the formation of a helical plane and the reversal of the Wilson curve are rarely observed within a population. These are the current conditions in populations with a 'Westernized' diet (Osborn, 1982).

In *Homo sapiens*, the first molar erupts and immediately starts to wear down at least 12 years before the third molar. This was sufficient time in pre-industrial populations to quickly achieve a reversal of the Wilson curve. At the eruption of the third molar, the second molar already appears flattened, and the first molar is significantly worn, resulting in the formation of the PEI. With abrasive chewing due to a diet consisting of tough and abrasive foods, even the third molars quickly experience a reversal of the curve, leading to a short duration of the helical plane and the rapid development of a reversed Wilson curve in most people over the age of 20 (Osborn, 1982).

Based on the decreasing wear gradient from the first to the third molar, various chronological classifications have been created to estimate the age at the time of death of individuals belonging to past populations. However, great care must be taken when considering the relationship between tooth age and wear because the latter varies significantly among different ethnic groups depending on dietary habits and food preparation, especially after the first 'food revolution' that occurred during the transition from hunter-gatherer to agricultural diets (Deter, 2009). The assessment of a helical plane cannot be used as a taxonomic indicator because numerous variables contribute to its existence (Figs. 3.100, 3.101).

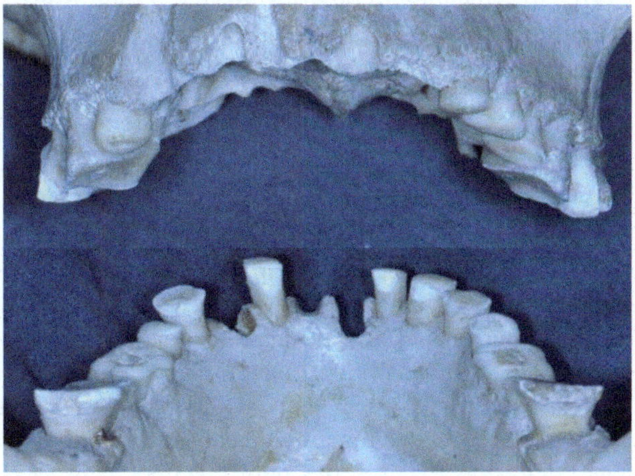

Fig. 3.100 Inversion of the Wilson curve. Skull 3525 from the Anthropology Museum of the University of Florence

Fig. 3.101 Inversion of the Wilson curve. Skull 3525 from the Anthropology Museum of the University of Florence

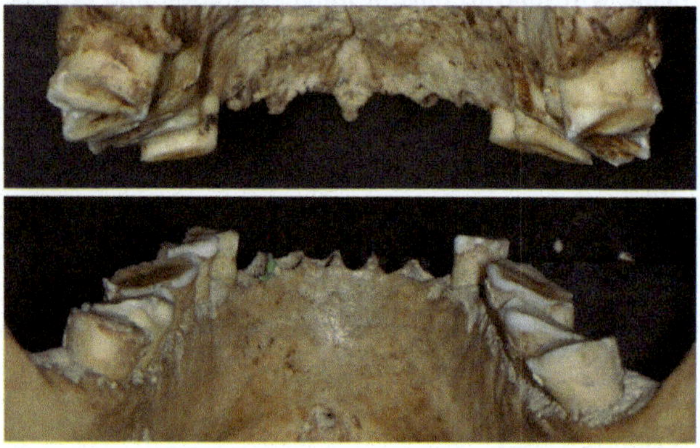

Fig. 3.102 Destruction of the dental structure caused by intense abrasive wear. Skull 3215 from the Anthropology Museum of the University of Florence

So, wear shapes occlusal surfaces, adapting them and making them congruent with function. However, this reciprocal balance between form and function cannot be maintained indefinitely. Just as it is wrong to think that evolution is constantly progressing towards an ideal improvement, it is equally wrong to believe that masticatory function leads to an indefinite and progressive morphological refinement in dental elements.

Although wear that occurs during normal chewing stimulates biological responses from supporting tissues that compensate for tissue loss, such as the reparative formation of reactionary dentin and the obliteration of dentinal tubules (Mrör, 2001), the adaptability of structures has a limit. Beyond this limit, wear that functionally sculpts the form also leads to its collapse, driving both form and function towards deterioration (Fig. 3.102).

Wear affects dentin, which is less mineralized and resistant than enamel. This leads to accelerated wear and the disappearance of morphological characteristics of the occlusal surface, eventually resulting in surface flattening and exposure of the pulp chamber (Fig. 3.103).

Once occlusal wear exposes the pulp chamber in the oral cavity, bacteria infect the pulp, leading to necrosis (Fig. 3.104).

After colonizing the endodontic structures, the bacterial contamination spreads to the supporting tissues, causing infection. Bacterial contamination beyond the root apex leads to irreversible damage to the dental element's periodontium. This damage can extend to adjacent elements, propagating an infection that, in a natural state, can evolve into a severe pathological condition that endangers an individual's life. Once abrasion exposes the pulp chamber (Fig. 3.105), the endodontic system

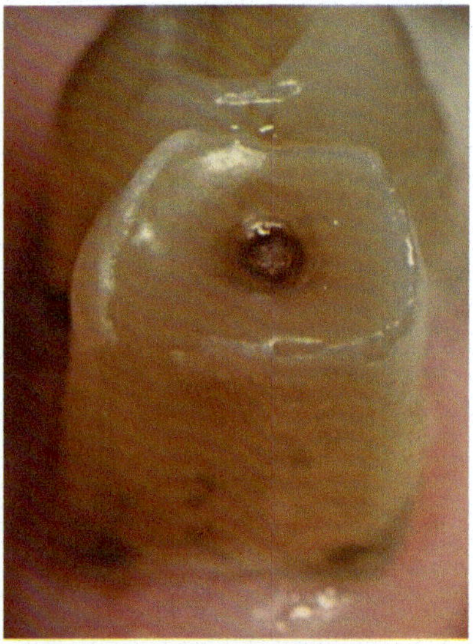

Fig. 3.103 Destruction of the crown of a lower central incisor with exposure of the pulp chamber and subsequent bacterial colonization of the endodontium

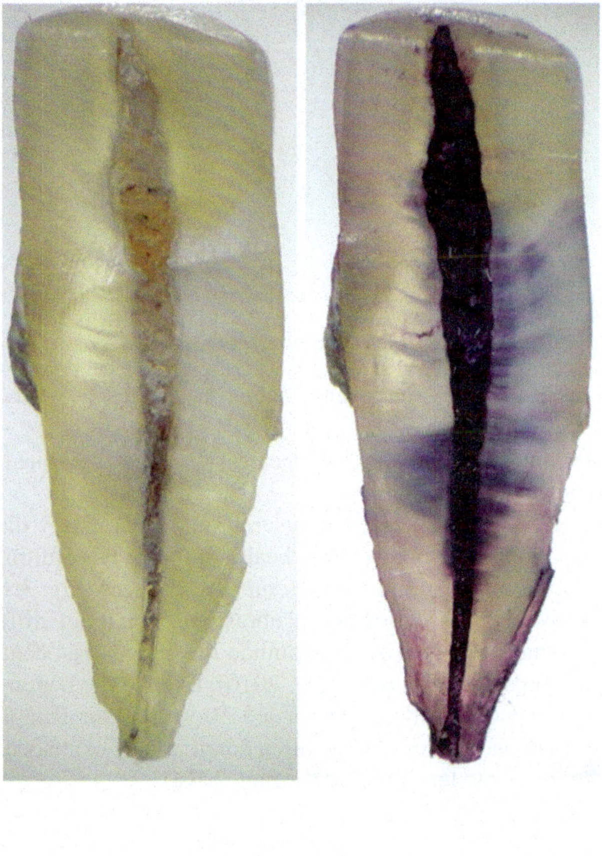

Fig. 3.104 Lower central incisor in which wear has caused exposure of the pulp chamber; the section highlights the macroscopic structure of the endodontium. On the left, the tooth sectioned; on the right, the same section with the endodontic content stained with bacterial plaque disclosing agent

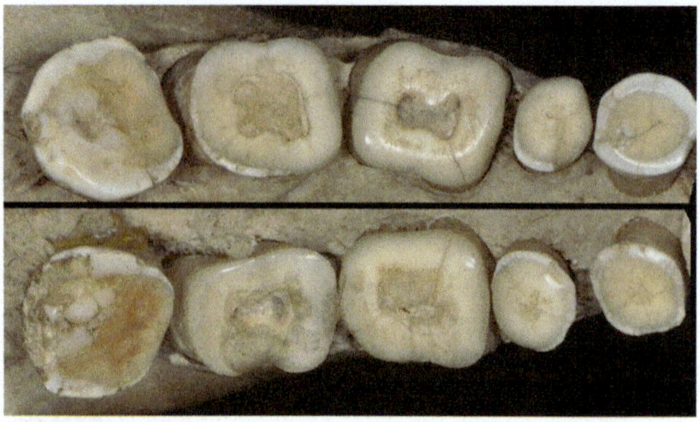

Fig. 3.105 Exposure of the pulp chamber in teeth destroyed by abrasion. Skull 3126 from the Anthropology Museum of the University of Florence

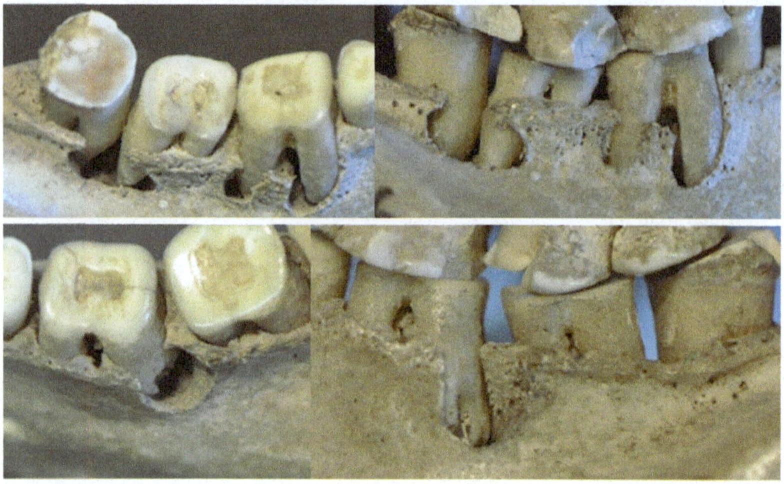

Fig. 3.106 Bacterial infection surpasses the apical foramen, spreading from the endodontium to the periodontium and alveolar bone, causing severe and irreversible damage to the structure and function of mastication. Skull 3126 from the Anthropology Museum of the University of Florence

becomes the pathway for bacteria to reach periapical tissues, infecting them and causing destruction.

As wear progresses, it severely and irreversibly damages not only masticatory function but also the overall health of the body, resulting in a situation that seriously compromises an individual's survival chances (Fig. 3.106).

When considering tooth abrasion throughout different eras, it's important to remember that the average human lifespan in past centuries was much shorter than it is today (Benoît & Granat, 2010). Even at the beginning of the twentieth century, life expectancy barely exceeded 40 years. Our Paleolithic bodies, which evolved slowly and progressively, often adapt poorly or inadequately to the habits and lifestyles of modern life (Lieberman, 2014).

Aging alters bones, muscles, the nervous system, organs, and teeth; the body loses its agility and swiftness. This aging process is fully evident only in the human species. Animals die earlier because as they age, the reduction in their physical abilities puts them at a competitive disadvantage in the struggle for survival. For humans, medicine, a healthy lifestyle, diet, and the environment allow for an almost unnatural aging process, placing individuals (and society) in a position to confront individual health issues and social challenges that in the past literally did not have time to manifest.

Teeth play a significant role in determining the lifespan of mammals. When teeth become unusable, the animal is condemned to death by starvation or other conditions related to the difficulty of feeding (Flower, 1925; Popowics & Herring, 2006; Ungar 2008). Therefore, in a natural state, we can think of enamel as a factor of programmed obsolescence. Once the quantity of life necessary to wear down the enamel is exhausted, the animal has a lower chance of survival and must make way for a more suitable individual.

Aesthetics and function are closely related to the individual's age. The work of the dentist should not vainly pursue the dental aesthetics of one's twenties but should first interpret the anatomical and functional structure of the patient in relation to their age. Since form and function are two sides of the same coin (Liem, 2001), emphasizing the importance of aesthetics now would seem like a mere rhetorical exercise. It is not about burying Caesar or singing his praises, advocating the primacy of one over the other. In reconstructive work, the absence of either one represents, regardless, a failure.

The conceptual dichotomy between form and function is ancient, dating back at least to Aristotle, and it is futile and petty to debate the primacy of one over the other today. Morphology clearly shows us that organisms are networks of interaction and integration, not linear chains of cause and effect. Our task is to integrate them by expanding their boundaries, not to separate them (Schwenk, 2000).

However, we must be aware that the increasingly dominant aesthetic model tends to impose solutions passively accepted, if not desired, by a significant portion of the population. It is perplexing to see mature gentlemen and equally mature ladies flaunting anachronistic youthful dentition in advertising images, as if they have gracefully weathered the last forty years—perhaps with a portrait that ages in the attic in their place! The pursuit of aesthetic results must always be supported by an understanding of their function, and vice versa. A functional and balanced reconstruction of the occlusal surface is necessary for the assessment and long-term success of orthodontic, conservative, implant, and prosthetic treatments (Benazzi et al., 2011).

The lack of maturation of occlusal surfaces is indicative of masticatory hypofunction induced by a diet that has radically changed over the last century. The introduction of foods different from those with which our Paleolithic bodies—as we have seen—evolved slowly and progressively (Lieberman, 2014) has resulted in a reduction in growth stimuli for the bony bases. These underdeveloped bases are the primary cause of a range of gnathological, orthodontic, periodontal, and muscular problems.

Individuals with strong masticatory muscles exhibit greater facial morphological homogeneity, in contrast to those with weaker masticatory muscles who display significant interindividual variability. Reduced mastication accentuates phenotypic variability (Kiliaridis, 1995), which represents an attempt to adapt the stomatognathic apparatus to a biomechanically deficient environment. A system that has evolved to develop and mature in response to intense demands cannot remain unaffected by a sudden decrease in the activity that has characterized its evolutionary path.

References

Ahmed, I., Nazir, R., & Ahsan, T. (2011). Influence of malocclusion on the depth of curve of Spee. *JPMA. The Journal of the Pakistan Medical Association, 61*(11), 1056–1059.

Anderson, P., & Creanor, S. (2016). Enamel. *Essential Clinical Oral Biology, 23*.

Ang, S. F., et al. (2012). Comparison of mechanical behaviors of enamel rod and interrod regions in enamel. *Journal of Materials Research, 27*(2), 448–456.

Armfield, B. A., et al. (2013). Development and evolution of the unique cetacean dentition. *PeerJ, 1*, e24.

Bajaj, D., & Arola, D. D. (2009). On the R-curve behavior of human tooth enamel. *Biomaterials, 30*(23–24), 4037–4046.

Barrera, J. M., et al. (2013). Wilson maxillary curve analyzed by cbct. A study on normocclusion and malocclusion individuals. *Medicina Oral, Patologia Oral y Cirugia Bucal, 18*(3), e547.

Bechtle, S., Özcoban, H., Lilleodden, E. T., Huber, N., Schreyer, A., Swain, M. V., & Schneider, G. A. (2012). Hierarchical flexural strength of enamel: Transition from brittle to damage-tolerant behaviour. *Journal of the Royal Society Interface, 9*(71), 1265–1274.

Begg, P. R. (1954). Stone age man's dentition: With reference to anatomically correct occlusion, the etiology of malocclusion, and a technique for its treatment. *American Journal of Orthodontics and Dentofacial Orthopedics, 40*(4), 298–312.

Benazzi, S., et al. (2011). Using occlusal wear information and finite element analysis to investigate stress distributions in human molars. *Journal of Anatomy, 219*(3), 259–272.

Benazzi, S., et al. (2012). Brief communication: Comparing loading scenarios in lower first molar supporting bone structure using 3D finite element analysis. *American Journal of Physical Anthropology, 147*(1), 128–134.

Benazzi, S., Kullmer, O., Schulz, D., Gruppioni, G., & Weber, G. W. (2013). Individual tooth macrowear pattern guides the reconstruction of Sts 52 (*Australopithecus africanus*) dental arches. *American Journal of Physical Anthropology, 150*, 324–329.

Benazzi, S., Nguyen, H. N., Kullmer, O., & Hublin, J. J. (2013). Unravelling the functional biomechanics of dental features and tooth wear. *PLoS One, 8*(7), e69990.

Benazzi, S., Nguyen, H. N., Schulz, D., Grosse, I. R., Gruppioni, G., Hublin, J. J., & Kullmer, O. (2013). The evolutionary paradox of tooth wear: Simply destruction or inevitable adaptation? *PLoS One, 8*(4), e62263.

Benoît, R., & Granat, J. (2010). The canine in the dento-alveolar system. *Journal of Dentofacial Anomalies and Orthodontics, 13*(1), 11–39.

Berkovitz, B., & Shellis, R. P. (2018). *The teeth of mammalian vertebrates* (1st ed.). Academic Press. eBook ISBN: 9780128028193, Hardcover ISBN: 9780128028186, Published Date: 17th August 2018.

Berry, D. C., & Poole, D. F. G. (1976). Attrition: Possible mechanisms of compensation. *Journal of Oral Rehabilitation, 3*(3), 201–206.

Bhullar, B.-A. S., et al. (2019). Rolling of the jaw is essential for mammalian chewing and tribosphenic molar function. *Nature, 566*(7745), 528.

Bimler, B. (2007). Die "Spee'sche Kurve" bei einigen Säugetieren.

Boyde, A. (1984). Dependence of rate of physical erosion on orientation and density in mineralised tissues. *Anatomy and Embryology, 170*(1), 57–62.

Boyde, A., et al. (1988). Basis of the structure and development of mammalian enamel as seen by scanning electron microscopy. *Scanning Microscopy, 2*(3), 1479–1490.

Brace, C. L. (1977). Occlusion to the anthropological eye. *The Biology of Occlusal Development, 7*, 179–209.

Braly, A., et al. (2007). The effect of prism orientation on the indentation testing of human molar enamel. *Archives of Oral Biology, 52*(9), 856–860.

Bromley, K. M., Kiss, A. S., Lokappa, S. B., Lakshminarayanan, R., Fan, D., Ndao, M., Evans, J. S., & Moradian-Oldak, J. (2011). Dissecting amelogenin protein nanospheres: Characterization of metastable oligomers. *Journal of Biological Chemistry, 286*(40), 34643–34653.

Brown, W. A. B., & Chapman, N. G. (1991). The dentition of red deer (*Cervus elaphus*): A scoring scheme to assess age from wear of the permanent molariform teeth. *Journal of Zoology, 224*(4), 519–536.

Campbell, T. D. (1923). *Dentition and palate of the Australian aboriginal from observations on the skull: A study in physical anthropology and dental pathology*. Diss.

Cárabe Fernández, L. (2014). Análisis tridimensional de la inclinación bucolingual de primeros y segundos molares superiores en los distintos patrones faciales.

Carlson, S. J. (1990). Vertebrate dental structures. In *Skeletal biomineralization: Patterns, processes and evolutionary trends* (pp. 531–556). Wiley.

Chang, Y.-H., et al. (2009). Mechanical interactions of cuspal-coverage designs and cement thickness in a cusp-replacing ceramic premolar restoration: A finite element study. *Medical & Biological Engineering & Computing, 47*(4), 367.

Chun, K. J., Choi, H. H., & Lee, J. Y. (2014). Comparison of mechanical property and role between enamel and dentin in the human teeth. *Journal of Dental Biomechanics, 5*.

Constantino, P. J., et al. (2009). The influence of fallback foods on great ape tooth enamel. *American Journal of Physical Anthropology: The Official Publication of the American Association of Physical Anthropologists, 140*(4), 653–660.

Cooper, J. S., & Poole, D. F. G. (1973). The dentition and dental tissues of the agamid lizard, *Uromastyx*. *Journal of Zoology, 169*(1), 85–100.

Corruccini, R. S. (1984). An epidemiologic transition in dental occlusion in world populations. *American Journal of Orthodontics, 86*(5), 419–426.

Corruccini, R. S., & Beecher, R. M. (1982). Occlusal variation related to soft diet in a nonhuman primate. *Science, 218*(4567), 74–76.

Corruccini, R. S., et al. (1983). Epidemiological survey of occlusion in North India. *British Journal of Orthodontics, 10*(1), 44–47.

Crompton, A. W., Wood, C. B., & Stern, D. N. (1994). Differential wear of enamel: A mechanism for maintaining sharp cutting edges. In V. L. Bels, M. Chardon, & P. Vandewalle (Eds.), *Biomechanics of feeding in vertebrates* (pp. 321–346). Springer.

Cui, F.-Z., & Ge, J. (2007). New Observations of the Hierarchical Structure of Human Enamel, from Nanoscale to Microscale. *Journal of Tissue Engineering and Regenerative Medicine, 1*(3), 185–191.

Cuy, J. L., et al. (2002). Nanoindentation mapping of the mechanical properties of human molar tooth enamel. *Archives of Oral Biology, 47*(4), 281–291.

Daculsi, G., Menanteau, J., Kerebel, L. M., & Mitre, D. (1984). Length and shape of enamel crystals. *Calcified Tissue International, 36*(1), 550–555.

Deter, C. A. (2009). Gradients of occlusal wear in hunter-gatherers and agriculturalists. *American Journal of Physical Anthropology: The Official Publication of the American Association of Physical Anthropologists, 138*(3), 247–254.

Dettori, S., Confaloni, A., & Pazzola, T. (2001). *RNO: evoluzione del rapporto dinamico funzionale tra occlusione ed ATM*. Mercurio editore.

Dickerson, W. G., Chan, C. A., & Carlson, J. (2001). The human stomatognathic system: A scientific approach to occlusion. *Dentistry Today, 20*(2), 100.

Dickson, G. C. (1970). The natural history of malocclusion. *The Dental Practitioner and Dental Record, 20*(6), 216–232.

D'Incau, E., Couture, C., & Maureille, B. (2012). Human tooth wear in the past and the present: Tribological mechanisms, scoring systems, dental and skeletal compensations. *Archives of Oral Biology, 57*(3), 214–229.

D'Incau, E., & Saulue, P. (2012). Understanding dental wear. *Journal of Dentofacial Anomalies and Orthodontics, 15*(1), 104.

Dumont, E. R. (1995). Enamel thickness and dietary adaptation among extant primates and chiropterans. *Journal of Mammology, 76*, 1127–1136.

Engelberth, S. A., Bacino, M. S., Sandhu, S., Li, W., Bonde, J., & Habelitz, S. (2018). Progression of Self-Assembly of Amelogenin Protein Supramolecular Structures in Simulated Enamel Fluid. *Biomacromolecules., 19*(10), 3917–3924. https://doi.org/10.1021/acs.biomac.8b00808

Evans, A. R., & Fortelius, M. (2008). Three-dimensional reconstruction of tooth relationships during carnivoran chewing. *Palaeontologia Electronica, 11*(2), 1–11.

Fang, P.-A., et al. (2011). Hierarchical self-assembly of amelogenin and the regulation of biomineralization at the nanoscale. *Proceedings of the National Academy of Sciences, 108*(34), 14097–14102.

Fantner, G. E., et al. (2005). Sacrificial bonds and hidden length dissipate energy as mineralized fibrils separate during bone fracture. *Nature Materials, 4*(8), 612–616.

Farella, M., Michelotti, A., van Eijden, T. M., & Martina, R. (2002). The curve of spee and craniofacial morphology: A multiple regression analysis. *European Journal of Oral Sciences, 110*, 277–281.

Feldman, E., et al. (2015). Mandibular changes secondary to serial extractions compared with late premolar extractions and controls. *American Journal of Orthodontics and Dentofacial Orthopedics, 148*(4), 633–640.

Fernandes, C. P., & Chevitarese, O. (1991). The orientation and direction of rods in dental enamel. *The Journal of Prosthetic Dentistry, 65*(6), 793–800.

Fidler, B. C., et al. (1995). Long-term stability of Angle Class II, division 1 malocclusions with successful occlusal results at end of active treatment. *American Journal of Orthodontics and Dentofacial Orthopedics, 107*(3), 276–285.

Fincham, A. G., et al. (1995). Evidence for amelogenin "nanospheres" as functional components of secretory-stage enamel matrix. *Journal of Structural Biology, 115*(1), 50–59.

Fish, D., & Mendel, F. (1982). Mandibular movement patterns relative to food types in common tree shrews (*Tupaia glis*). *American Journal of Physical Anthropology, 58*, 255–269.

Flower, S. S. (1925). Contributions to our knowledge of the duration of life in vertebrate animals. Reptiles.—III. *Proceedings of the Zoological Society of London, 95*(3) Blackwell.

Fortelius, M. (1985). Ungulate cheek teeth: Developmental, functional, and evolutionary interrelations. *Acta Zoologica Fennica, 180*, 1–76.

Fueki, K., et al. (2013). Association between occlusal curvature and masticatory movements with different test foods in human young adults with permanent dentitions. *Archives of Oral Biology, 58*(6), 674–680.

Gans, C., de Vree, F., & Gorniak, G. C. (1978). Analysis of mammalian masticatory mechanisms: Progress and problems. *Anatomia, histologia, embryologia, 7*(3), 226–244.

Gaspard, M. (1979). *L'appareil manducateur et la manducation. Première partie: anatomie descriptive, ontogénèse et phylogénèse de la mandibule humaine.* Julien Prélatédit.

Gerstner, G., Madhavan, S., & Crane, E. (2011). Mammalian oral rhythms and motor control. In V. Klika (Ed.), *Biomechanics in applications*. InTech.

Goldberg, M., et al. (2014). Comparative studies between mice molars and incisors are required to draw an overview of enamel structural complexity. *Frontiers in Physiology, 5*, 359.

Gould, S. J. (2002). *I have landed*. Harvard University Press.

Greet, D. G., & De Vree, F. (1984). Movements of the mandibles and tongue during mastication and swallowing in *Pteropus giganteus* (Megachiroptera): A cineradiographical study. *Journal of Morphology, 179*(1), 95–114.

Gross, M. D. (2008). Occlusion in implant dentistry. A review of the literature of prosthetic determinants and current concepts. *Australian Dental Journal, 53*(Suppl 1), S60–S68.

Hakkinen, T. J., et al. (2019). Modeling enamel matrix secretion in mammalian teeth. *biorxiv*. https://doi.org/10.1101/525162

Harris, C. K., Broussard, J. P., & Keska, J. K. (2002). Determination of wear in a tribosystem. In *Proceedings of the ASEE Gulf-South Western annual conference*.

He, L. H., & Swain, M. V. (2007). Contact induced deformation of enamel. *Applied Physics Letters, 90*(17), 171916.

He, L. H., & Swain, M. V. (2008). Understanding the mechanical behaviour of human enamel from its structural and compositional characteristics. *Journal of the Mechanical Behavior of Biomedical Materials, 1*(1), 18–29.

He, L. H., & Swain, M. V. (2009). Nanoindentation creep behavior of human enamel. *Journal of Biomedical Materials Research Part A: An Official Journal of The Society for Biomaterials, The Japanese Society for Biomaterials, and The Australian Society for Biomaterials and the Korean Society for Biomaterials, 91*(2), 352–359.

Herring, S. W. (1993). Functional morphology of mammalian mastication. *American Zoologist, 33*(3), 289–299.

Herzberg, F., & Holic, R. (1943). An anthropologic study of face height. *American Journal of Orthodontics and Oral Surgery, 29*(2), 90–100.

Hiiemae, K. M. (1978). Mammalian mastication: A review of the activity of the jaw muscles and the movements they produce in chewing. In P. M. Butler & K. A. Joysey (Eds.), *Development, function and evolution of teeth* (pp. 359–398). Academic Press.

Hiiemae, K. (1984). Functional aspects of primate jaw morphology. In D. J. Chivers, B. A. Wood, & A. Bilsborough (Eds.), *Food acquisition and processing in primates* (pp. 257–281). Springer.

Hiiemae, K. M., & Crompton, A. W. (1985). Mastication, food transport and swallowing. In M. Hildebrand, D. Bramble, K. Liem, & D. Wake (Eds.), *Functional vertebrate morphology* (pp. 262–290). Belknap Press.

Hogervorst, T., Bouma, H. W., & De Vos, J. (2009). Evolution of the hip and pelvis. *Acta orthopaedica, 80*(suppl 336), 1–39.

Hunt, E. E., Jr. (1961). Malocclusion and civilization. *American Journal of Orthodontics, 47*(6), 406–422.

Hylander, W. L. (1979a). The functional significance of primate mandibular form. *Journal of Morphology, 160*(2), 223–239.

Hylander, W. L. (1979b). Mandibular function in *Galago crassicaudatus* and *Macaca fascicularis*: An in vivo approach to stress analysis of the mandible. *Journal of Morphology, 159*(2), 253–296.

Hylander, W. L. (1979c). An experimental analysis of temporomandibular joint reaction force in macaques. *American Journal of Physical Anthropology, 51*(3), 433–456.

Hylander, W. L. (1980). Functional anatomy. In B. G. Sarnat & D. M. Laskin (Eds.), *The temporomandibular joint: A biological basis for clinical practice* (3rd ed., pp. 85–113). Thomas.

Ishiyama, M. (1987). Enamel structure in odontocete whales. *Scanning Microscopy, 1*(3), 1071–1079.

Ito, H., Okimoto, K., Mizumori, T., Terada, Y., & Maruyama, T. (1997). A clinical study of the relationship between occlusal curvature and craniomandibular disorders. *International Journal of Prosthodontics, 10*, 78–82.

Jeng, Y.-R., et al. (2011). Human enamel rod presents anisotropic nanotribological properties. *Journal of the Mechanical Behavior of Biomedical Materials, 4*(4), 515–522.

Jia, Y., et al. (2014). Finite element analysis of the cyclic indentation of bilayer enamel. *Journal of Physics D: Applied Physics, 47*(17), 175401.

Jiang, Y., Spears, I. R., & Macho, G. A. (2003). An investigation into fractured surfaces of enamel of modern human teeth: A combined SEM and computer visualisation study. *Archives of Oral Biology, 48*(6), 449–457.

Kaidonis, J. A. (2008). Tooth wear: The view of the anthropologist. *Clinical Oral Investigations, 12*(1), 21–26.

Kaifu, Y. (1996). Edge-to-edge bite and tooth wear. *Bulletin of the National Science Museum Series D (Tokyo), 22*, 45–54.

Kaifu, Y. (1999). Changes in the pattern wear from prehistoric to recent periods in Japan. *American Journal of Physical Anthropology, 109*(4), 485–499.

Kaifu, Y., Kasai, K., Townsend, G. C., & Richards, L. C. (2003). Tooth wear and the "design" of the human dentition: A perspective from evolutionary medicine. *American Journal of Physical Anthropology: The Official Publication of the American Association of Physical Anthropologists, 122*(S37), 47–61.

Karani, J., Idrisi, A., Mistry, S., Badwaik, P., Pai, A., & Yadav, S. (2018). Comparative evaluation of the depth of curve of Spee between individuals with normal dentition and individuals with occlusal wear using conventional and digital software analysis techniques: An in vivo study. *Journal of Indian Prosthodontic Society, 18*, 61–67.

Katona, T. R. (2009). An engineering analysis of dental occlusion principles. *American Journal of Orthodontics and Dentofacial Orthopedics, 135*(6), 696–6e1.

Katsavrias, E. G. (2002). Changes in articular eminence inclination during the craniofacial growth period. *The Angle Orthodontist, 72*(3), 258–264.

Kaul, S. S., & Corruccini, R. S. (1984). The epidemiological transition in dental occlusion in a North Indian population. In J. R. Lukacs (Ed.), *The people of South Asia* (pp. 201–216). Springer.

Kay, R. F. (1981). The nut-crackers: A new theory of the adaptation of the Ramapithecine. *American Journal of Physical Anthropology, 55*, 141–151.

Kemp, A. (2002). Hyaline tissue of thermally unaltered conodont elements and the enamel of vertebrates. *Alcheringa, 26*(1), 23–36.

Kemp, T. S. (1982). *Mammal-like reptiles and the origin of mammals*. Academic Press.

Kiliaridis, S. (1995). Masticatory muscle influence on craniofacial growth. *Acta Odontologica Scandinavica, 53*(3), 196–202.

Kim, S. K., et al. (2001). A study of the effects of chewing patterns on occlusal wear. *Journal of Oral Rehabilitation, 28*(11), 1048–1055.

Kim, Y., Oh, T. J., Misch, C. E., & Wang, H. L. (2005). Occlusal considerations in implant therapy: Clinical guidelines with biomechanical rationale. *Clinical Oral Implants Research, 16*, 26–35.

Kodaka, T., Abe, M., & Higashi, S. (1996). Alternative appearance of spiral prisms in the gnarled enamel of human teeth. *The Journal of Showa University Dental Society, 16*(1), 1–3.

Koenigswald, W. V. (2020). Evolutionary trends in the differentiation of mammalian enamel ultrastructure. In *Tooth enamel microstructure: Proceedings of the enamel microstructure workshop, University of Bonn, Andernach, Rhine, 24–28 July 1994*. CRC Press.

Koenigswald, W. V., & Clemens, W. A. (1992). Levels of complexity in the microstructure of mammalian enamel and their application in studies of systematics. *Scanning Microscopy, 6*(1), 195–217.

Koenigswald, W. V., & Sander, P. M. (2020). Schmelzmuster differentiation in leading and trailing edges, a specific biomechanical adaptation in rodents. In *Tooth enamel microstructure: Proceedings of the enamel microstructure workshop, University of Bonn, Andernach, Rhine, 24–28 July 1994*. CRC Press.

Koldehoff, J., Swain, M. V., & Schneider, G. A. (2020). The geometrical structure of interfaces in dental enamel: A FIB-STEM investigation. *Acta Biomaterialia, 104*, 17–27.

Kono, R. T., & Suwa, G. (2008). Enamel distribution patterns of extant human and hominoid molars: Occlusal versus lateral enamel thickness. *Bulletin of the National Museum of Nature and Science, Series D, 34*, 1–9.

Kumar, K. P. S., & Tamizharasi, S. (2012). Significance of curve of Spee: An orthodontic review. *Journal of Pharmacy & Bioallied Sciences, 4*(Suppl 2), S323.

Larsen, C. S. (2002). Post-Pleistocene human evolution: Bioarchaeology of the agricultural transition. In P. S. Ungar & M. F. Teaford (Eds.), *Human diet: Its origin and evolution* (pp. 19–36). Bergin and Garvey.

Lawn, B. R., Lee, J. J.-W., & Chai, H. (2010). Teeth: Among nature's most durable biocomposites. *Annual Review of Materials Research, 40*, 55–75.

Lee, J. J.-W., et al. (2010). Properties of tooth enamel in great apes. *Acta Biomaterialia, 6*(12), 4560–4565.

Lesot, H., et al. (1982). Immunofluorescent localization of vimentin, prekeratin and actin during odontoblast and ameloblast differentiation. *Differentiation, 21*(1–3), 133–137.

Lieberman, D. (2011). *The evolution of the human head.* Harvard University Press.

Lieberman, D. (2014). *The story of the human body: Evolution, health, and disease.* Vintage.

Liem, K. F., et al. (2001). *Functional anatomy of the vertebrates: An evolutionary perspective.* No. 596 FUN.

Line, S. R. P., & Novaes, P. D. (2017). The development and evolution of mammalian enamel: Structural and functional aspects. *Journal of Morphological Sciences, 22*(2).

Loch, C., Kieser, J. A., & Fordyce, R. E. (2015). Enamel ultrastructure in fossil cetaceans (Cetacea: Archaeoceti and Odontoceti). *PLoS One, 10*(1), e0116557.

Loch, C., et al. (2014). Elemental and chemical characterization of dolphin enamel and dentine using X-ray and Raman microanalyzes (Cetacea: Delphinoidea and Inioidea). *Journal of Structural Biology, 185*(1), 58–68.

Lombardi, A. V. (1982). The adaptive value of dental crowding: A consideration of the biologic basis of malocclusion. *American Journal of Orthodontics, 81*(1), 38–42.

Lucas, P. W. (2004). *Dental functional morphology: How teeth work.* Cambridge University Press.

Lucas, P., Constantino, P., Wood, B., & Lawn, B. (2008). Dental enamel as a dietary indicator in mammals. *BioEssays, 30*(4), 374–385.

Lucas, P. W., et al. (2013). Mechanisms and causes of wear in tooth enamel: Implications for hominin diets. *Journal of the Royal Society Interface, 10*(80), 20120923.

Lynch, C. D., et al. (2011). Hunter–Schreger band patterns and their implications for clinical dentistry. *Journal of Oral Rehabilitation, 38*(5), 359–365.

Maas, M. C., & Dumont, E. R. (1999). Built to last: The structure, function, and evolution of primate dental enamel. *Evolutionary Anthropology: Issues, News, and Reviews: Issues, News, and Reviews, 8*(4), 133–152.

Macho, G. A., Jiang, Y., & Spears, I. R. (2003). Enamel microstructure—A truly three-dimensional structure. *Journal of Human Evolution, 45*(1), 81–90.

Manzi, G. (2013). *Il grande racconto dell'evoluzione umana.* Società Editrice il Mulino.

Marshall, S. D., et al. (2008). Development of the curve of Spee. *American Journal of Orthodontics and Dentofacial Orthopedics, 134*(3), 344–352.

Mew, J. R. C. (2004). The postural basis of malocclusion: A philosophical overview. *American Journal of Orthodontics and Dentofacial Orthopedics, 126*(6), 729–738.

Mintenko, R., Kennedy, D. B., Aleksejuniene, J., Hannam, A. G., & Yen, E. H. (2020). Mandibular dental changes following serial and late extraction of mandibular second premolars. *The Angle Orthodontist, 90*(2), 187–193.

Misra, A., Sharma, R., Gulati, S., Joshi, S. R., Sharma, V., Ghafoorunissa, Ibrahim, A., Joshi, S., Laxmaiah, A., Kurpad, A., Raj, R. K., Mohan, V., Chandalia, H., Krishnaswamy, K., Boindala, S., Gopalan, S., Bhattiprolu, S. K., Modi, S., Vikram, N. K., … Bhardwaj, S. (2011). Consensus dietary guidelines for healthy living and prevention of obesity, the metabolic syndrome, diabetes, and related disorders in Asian Indians. *Diabetes Technology & Therapeutics, 13*(6), 683–694.

Miyawaki, S., et al. (2000). Effect of food size on the movement of the mandibular first molars and condyles during deliberate unilateral mastication in humans. *Journal of Dental Research, 79*(7), 1525–1531.

Mohan, M., D'Souza, M., Kamath, G., & Parolia, A. (2011). Comparative evaluation of the curve of spee in two age groups and its relation to posterior teeth disclusion. *Indian Journal of Dental Research, 22*, 179.

Molnar, S. (1971). Human tooth wear, tooth function and cultural variability. *American Journal of Physical Anthropology, 34*(2), 175–189.

Molnar, S., & Gantt, D. G. (1977). Functional implications of primate enamel thickness. *American Journal of Physical Anthropology, 46*(3), 447–454.

Molnar, S., McKee, J. K., & Molnar, I. (1983). Measurements of tooth wear among Australian Aborigines: I. Serial loss of the enamel crown. *American Journal of Physical Anthropology, 61*(1), 51–65.

Moore, W. J., Lavelle, C. L. B., & Spence, T. F. (1968). Changes in the size and shape of the human mandible in Britain. *British Dental Journal, 125*, 163–169.

Moradian-Oldak, J. (2012). Protein-mediated enamel mineralization. *Frontiers in bioscience (Landmark Edition), 17*, 1996–2023. https://doi.org/10.2741/4034

Moradian-Oldak, J., & Goldberg, M. (2005). Amelogenin supra-molecular assembly in vitro compared with the architecture of the forming enamel matrix. *Cells Tissues Organs, 181*(3–4), 202–218.

Moss, M. (2013). *Salt, sugar, fat: How the food giants hooked us*. Random House.

Mrör, I. A. (2001). Pulp-dentin biology in restorative dentistry. Part 5: Clinical management and tissue changes associated with wear and trauma. *Quintessence International, 32*(10).

Naples, V. L. (1985). Form and function of the masticatory musculature in the tree sloths, *Bradypus* and *Choloepus*. *Journal of Morphology, 183*(1), 25–50.

Nishikawa, S., & Kitamura, H. (1983). Actin filaments in the ameloblast of the rat incisor. *The Anatomical Record, 207*(2), 245–252.

Osborn, J. W. (1982). Helicoidal plane of dental occlusion. *American Journal of Physical Anthropology, 57*(3), 273–281.

Osborn, J. W. (1987). Relationship between the mandibular condyle and the occlusal plane during hominid evolution: Some of its effects on jaw mechanics. *American Journal of Physical Anthropology, 73*, 193–207.

Osborn, J. W. (1993a). A model to describe how ligaments may control symmetrical jaw opening movements in man. *Journal of Oral Rehabilitation, 20*(6), 585–604.

Osborn, J. W. (1993b). Orientation of the masseter muscle and the curve of Spee in relation to crushing forces on the molar teeth of primates. *American Journal of Physical Anthropology, 92*(1), 99–106.

Oxilia, G., Bortolini, E., Martini, S., Papini, A., Boggioni, M., Buti, L., Figus, C., Sorrentino, R., Townsend, G., Kaidonis, J., Fiorenza, L., Cristiani, E., Kullmer, O., Moggi-Cecchi, J., & Benazzi, S. (2018). The physiological linkage between molar inclination and dental macrowear pattern. *American Journal of Physical Anthropology, 166*(4), 941–951.

Parfitt, G. J. (1960). Measurement of the physiological mobility of individual teeth in an axial direction. *Journal of Dental Research, 39*(3), 608–618.

Parilov, V. V. (1989). The relationship of Christensen's phenomena to the structure of the temporomandibular joints. *Stomatologiia, 68*(2), 49–50.

Pereira, C. B., & Evans, H. (1975). Occlusion and attrition of the primitive Yanomami Indians of Brazil. *Dental Clinics of North America, 19*(3), 485–498.

Piancino, M. G., & Kyrkanides, S. (2018). *La funzione masticatoria nei morsi incrociati monolaterali*. ed. edi ermes. ISBN 9788870516319.

Pileicikiene, G., & Surna, A. (2004). The human masticatory system from a biomechanical perspective: A review. *Stomatologija, 6*(3), 81–84.

Popowics, T. E., & Herring, S. W. (2006). Teeth, jaws and muscles in mammalian mastication. In V. L. Bels (Ed.), *Feeding in domestic vertebrates: From structure to behaviour* (pp. 61–83). CABI.

Price, W. A. (1945). *Nutrition and physical degeneration: A comparison of primitive and modern diets and their effects*.

Price, W. A. (1934). Relation of nutrition to dental caries among Eskimos and Indians in Alaska and Northern Canada. *Journal of Dental Research, 14*, 227–229.

Price, W. A. (1936). Field studies among some African tribes on the relation of their nutrition to the incidence of dented caries and dental arch deformities. *Journal of the American Dental Association, 23*, 876–890.

Proença, H. H. F. A., et al. (2014). A 3D computerized tomography study of changes in craniofacial morphology of Portuguese skulls from the eighteenth century to the present. *International Journal of Stomatology & Occlusion Medicine, 7*(2), 33–45.

Proffit, W. R., Fields, H. W., Jr., & Sarver, D. M. (2006). *Contemporary orthodontics*. Elsevier Health Sciences.

Richards, L. C., & Brown, T. (1986). Development of the helicoidal plane. *Human Evolution, 1*(5), 385–398.

Riga, A., Boggioni, M., Papini, A., Buzi, C., Profico, A., Di Vincenzo, F., Marchi, D., Moggi-Cecchi, J., & Manzi, G. (2020). In situ observations on the dentition and oral cavity of the Neanderthal skeleton from Altamura (Italy). *PLoS One, 15*(12), e0241713.

Sahni, A. (1981). Enamel ultrastructure of fossil mammalia: Eocene Archaeoceti from Kutch. *Journal of the Palaeontological Society of India, 25*, 33–37.

Saito, P., Hitoshi, K., Nobuo, I., Yasunari, F., Yohei, T., & Tarou, M. (2013). Development of hybrid three-dimensional jaw motion diagnostic system for occlusion therapy support (Analysis of contact condition of upper and lower teeth using individual hybrid jaw bone model). *Journal of the Japan Society of Mechanical Engineers C, 79*(807), 4121–4130.

Sakaguchi, K., et al. (2012). Relationship between occlusal curvatures and bite force in humans. *Orthodontic Waves, 71*(3), 79–84.

Saratti, C. M., Rocca, G. T., Vaucher, P., Awai, L., Papini, A., Zuber, S., Di Bella, E., Dietschi, D., & Krejci, I. (2021). Functional assessment of the stomatognathic system. Part 1: The role of static elements of analysis. *Quintessence International (Berlin, Germany: 1985), 52*(10), 920–932.

Sarnäs, K. V. (1957). Growth changes in skulls of ancient man in North America: An X-ray cephalometric investigation of some cranial and facial changes during growth in the Indian Knoll skeletons. *Acta Odontologica Scandinavica, 15*(3), 213–271.

Schwartz, G. T., McGrosky, A., & Strait, D. S. (2020). Fracture mechanics, enamel thickness and the evolution of molar form in hominins. *Biology Letters, 16*(1), 20190671.

Schwenk, K. (Ed.). (2000). *Feeding: Form, function and evolution in tetrapod vertebrates*. Elsevier.

Sedwick, C. (2009). Amelogenin: A protein to smile about. *PLoS Biology, 7*(12), e1000263.

Shimizu, D., Macho, G. A., & Spears, I. R. (2005). Effect of prism orientation and loading direction on contact stresses in prismatic enamel of primates: Implications for interpreting wear patterns. *American Journal of Physical Anthropology: The Official Publication of the American Association of Physical Anthropologists, 126*(4), 427–434.

Sicher, H. (1953). The biology of attrition. *Oral Surgery, Oral Medicine, Oral Pathology, 6*(3), 406–412.

Silvertown, J. (2018). *A cena con Darwin: cibo, bevande ed evoluzione*. Bollati Boringhieri.

Silvestri, A., Mazzoli, V., & Ferraris, L. (2008). La curva di Wilson. Concetti gnatologici e implicazioni terapeutiche. *Mondo Ortodontico, 2*, 101.

Slavicek, R. (2002). *Organo Masticatorio*. Gamma medizinischwissenschafttliche. fortbildungs-ag.

Smith, B. H. (1986). Development and evolution of the helicoidal plane of dental occlusion. *American Journal of Physical Anthropology, 69*(1), 21–35.

Smith, P. (1972). Diet and attrition in the Natufians. *American Journal of Physical Anthropology, 37*(2), 233–238.

Smith, R. J. (1985). Functions of condylar translation in human mandibular movement. *American Journal of Orthodontics, 88*(3), 191–202.

Spee, F. G. (1890). Die verschiebungsbahn des unterkiefers am schadel. *Archives fur Anatomie und Physiologie, 16*, 285–294.

Stern, D., Crompton, A. W., & Skobe, Z. (1989). Enamel ultrastructure and masticatory function in molars of the American opossum, *Didelphis virginiana*. *Zoological Journal of the Linnean Society, 95*(4), 311–334.

Teaford, M. F., Smith, M. M., & Ferguson, M. W. J. (Eds.). (2007). *Development, function and evolution of teeth*. Cambridge University Press.

Teaford, M. F., & Walker, A. (1983). Dental microwear in adult and still-born guinea pigs (*Cavia porcellus*). *Archives of Oral Biology, 28*(11), 1077–1081.

Tiwari, B., et al. (2014). Occlusal concepts in full mouth rehabilitation: An overview. *The Journal of Indian Prosthodontic Society, 14*(4), 344–351.

Torii, S. (1998). Origin of enamel prisms and Hunter-Schreger bands in reptilian enamel. *Connective Tissue Research, 38*(1–4), 45–51.

Ungar, P. S. (2008). Material science: Strong teeth, strong seeds. *Nature, 452*, 703–705.

Ungar, P. S. (2010). *Mammal teeth: Origin, evolution, and diversity*. JHU Press.

Ungar, P. S. (2015). Mammalian dental function and wear: A review. *Biosurface and Biotribology, 1*(1), 25–41.

Ungar, P. S. (2017). *Evolution's bite: A story of teeth, diet, and human origins*. Princeton University Press.

Varrela, J. (1990). Effects of attritive diet on craniofacial morphology: A cephalometric analysis of a Finnish skull sample. *The European Journal of Orthodontics, 12*(2), 219–223.

Wall, C. E. (1997). The expanded mandibular condyle of the Megaladapidae. *American Journal of Physical Anthropology: The Official Publication of the American Association of Physical Anthropologists, 103*(2), 263–276.

Wall, C. E. (1999). A model of temporomandibular joint function in anthropoid primates based on condylar movements during mastication. *American Journal of Physical Anthropology, 109*(1), 67–88.

Wall, C. E., et al. (2006). Phase II jaw movements and masseter muscle activity during chewing in *Papio anubis*. *American Journal of Physical Anthropology: The Official Publication of the American Association of Physical Anthropologists, 129*(2), 215–224.

Waltimo, A., & Könönen, M. (1995). Maximal bite force and its association with signs and symptoms of craniomandibular disorders in young Finnish non-patients. *Acta Odontologica Scandinavica, 53*(4), 254–258.

Weijs, W. A. (1994). Evolutionary approach of masticatory motor patterns in mammals. In V. L. Bels, M. Chardon, & P. Vandewalle (Eds.), *Biomechanics of feeding in vertebrates* (pp. 281–320). Springer.

Werth, A. J. (2000). Feeding in marine mammals. In K. Schwenk (Ed.), *Feeding: Form, function and evolution in tetrapod vertebrates* (pp. 475–514). Elsevier.

White, S. N., et al. (2001). Biological organization of hydroxyapatite crystallites into a fibrous continuum toughens and controls anisotropy in human enamel. *Journal of Dental Research, 80*(1), 321–326.

Will, M. J., Franzblau, E. B., & Kelley, A. E. (2003). Nucleus accumbens μ-opioids regulate intake of a high-fat diet via activation of a distributed brain network. *Journal of Neuroscience, 23*(7), 2882–2888.

Williamson, E. H., & Lundquist, D. O. (1983). Anterior guidance: Its effect on electromyographic activity of the temporal and masseter muscles. *Journal of Prosthetic Dentistry, 49*, 816–823.

Woda, A., Gourdon, A. M., & Faraj, M. (1987). Occlusal contacts and tooth wear. *Journal of Prosthetic Dentistry, 57*(1), 85–93.

Wolpoff, M. H. (1971). Interstitial wear. *American Journal of Physical Anthropology, 34*(2), 205–227.

Xia, J., et al. (2017). Enamel crystallite strength and wear: Nanoscale responses of teeth to chewing loads. *Journal of the Royal Society Interface, 14*(135), 20170456.

Xu, H. H., et al. (1998). Indentation damage and mechanical properties of human enamel and dentin. *Journal of Dental Research, 77*(3), 472–480.

Yilmaz, E. D., Schneider, G. A., & Swain, M. V. (2015). Influence of structural hierarchy on the fracture behaviour of tooth enamel. *Philosophical Transactions of the Royal Society A: Mathematical, Physical and Engineering Sciences, 373*(2038), 20140130.

Young, W. G., McGowan, M., & Daley, T. J. (1987). Tooth enamel structure in the koala, *Phascolarctos cinereus*: Some functional interpretations. *Scanning Microscopy, 1*(4), 1925–1934.

Zaytsev, D., & Panfilov, P. (2015). Anisotropy of the mechanical properties of human dental enamel. *Materials Letters, 159*, 428–431.

Zhang, M., & Kelley, A. E. (2000). Enhanced intake of high-fat food following striatal mu-opioid stimulation: Microinjection mapping and fos expression. *Neuroscience, 99*(2), 267–277.

Zheng, S. Y., Zheng, J., Gao, S. S., Yu, B. J., Yu, H. Y., Qian, L. M., & Zhou, Z. R. (2011). Investigation on the microtribological behaviour of human tooth enamel by nanoscratch. *Wear, 271*(9–10), 2290–2296.

Zhou, Z.-R., et al. (2013). *Dental biotribology*. Springer.

Zum Gahr, K. H. (1987). Classification of wear processes. In *Microstructure and wear of materials* (1st ed., pp. 80–131). North Holland.

Chapter 4
The Stomatognathic System in Human Evolution: Bipedalism and Encephalization

Abstract This chapter explores the crucial role of bipedalism and encephalization in human evolution, focusing on the profound anatomical and functional transformations affecting the stomatognathic system. Through a comparative analysis between non-human primates and *Homo sapiens*, the text highlights how upright posture and bipedal locomotion required a reorganization of the spinal column, pelvis, and masticatory apparatus. The verticalization of the spine and cranio-pelvic alignment demanded significant adaptations in cervical musculature and vertebral morphology, with important consequences for cranial biomechanics and balance.

In parallel, encephalization led to an expansion of the neurocranium and a retraction of the splanchnocranium, resulting in a reshaping of the dental arches and mandible. The chapter also examines the impact of these changes on occlusal dynamics and overall body posture. Special attention is given to the integration between the central nervous system, the musculoskeletal structure, and masticatory function, outlining a complex evolutionary framework in which bipedalism and encephalization mutually influence each other in a decisive manner.

With feet firmly planted on the clouds—Ennio Flaiano

Even today, it's still quite common to read or hear about the "theory of evolution," with the term "theory" preceding the concept of evolution. This might have been acceptable during the Victorian era, but today, it sounds decidedly inappropriate, if not anachronistic.

Evolution is documented like any other scientific phenomenon, well-founded, much like our knowledge of the Earth's revolution around the Sun. In this sense, we can define evolution as a "fact"—science doesn't deal in certainties—and thus, "fact" should be understood as a proposition with so many confirmations as to render irrational the position of anyone denying its provisional agreement (Gould, 2002).

The evolution of humans presents two unique developmental characteristics, with bipedalism being the first chronologically. Bipedalism is not to be understood as temporary ambulation on the hind limbs, which is also common among anthropoid apes, but rather as a definitive departure from being quadrupeds, associated with an upright posture. Overlapping with this enormous evolutionary change in hominids was a gradual encephalization, marked by the growth of a massive brain

A. Papini, *On the Origin of Human Masticatory Function*,
https://doi.org/10.1007/978-3-031-98664-2_4

Fig. 4.1 Change in posture and cranial and bodily balance between the silhouettes of an anthropoid ape and a human. Red: neurocranium; light blue: splanchnocranium. In green, the support bases for quadruped and biped (Diagram freely adapted from a drawing by Yoel Rak)

volume and its bony container. These two characteristics have defined our evolutionary path, resulting in numerous mutual adaptations, even concerning the stomatognathic system.

The portion of the skull not intended to contain the brain has also adapted to these postural changes, contributing to a new balance of a head that's becoming increasingly heavy, perched on a more delicate and vertical spine, stabilized by a less powerful neck musculature (Fig. 4.1).

The stomatognathic system in *Homo sapiens* has unique characteristics because human bipedalism in an upright posture is unique, as is our extreme and globular encephalization.

4.1 Bipedalism

The story of humans acquiring bipedalism, associated with an upright posture, is a tale of a series of modifications involving the support and locomotion systems. Bipedalism is an incredibly rare form of locomotion not only among mammals but also among current and extinct primates (Manzi, 2013). We are the only living mammalian species to move this way, and such an unusual behavior has resulted in a series of highly specialized anatomical features. Bipedalism and an upright posture are closely interconnected but not entirely interdependent. In fact, an upright posture can exist without bipedalism. The ability to hold the trunk somewhat upright is common in many primate species, such as the anthropoid apes, which have a semi-erect posture, known as "clinograde," with a forward curve.

The evolution of the upright posture was progressively adopted by our ancestors after the split between the human evolutionary line and that of chimpanzees. Only humans have developed the ability to maintain an upright stance while simultaneously adopting bipedalism as the obligatory form of locomotion. In humans, the spine and pelvis had to adapt to a new relationship with gravity through an enlargement and verticalization of the pelvis. This adaptation also resulted in the appearance of distinctive sagittal curves in the spine, accompanied by profound changes in the muscles supporting the vertebral column. Although all members of the *Homo* genus have been exclusively bipedal, this is by no means a uniquely human characteristic.

Throughout time, there have been numerous species with a wide variety of types of bipedalism, including archosaurs that used their tails as counterbalances, all modern birds, and even some contemporary anthropoid apes. However, most bipedal species, except for birds, only occasionally adopt this awkward and taxing position for short periods and distances. For example, chimpanzees use their arms as counterweights because they lack spinal curves, leaving them in an anterior imbalanced position.

In *Homo sapiens*, there are many morphological traits that are unique within the order of primates, but the upright posture and bipedal gait are among the first that appeared in our lineage; we can hypothesize that hominid bipedalism established itself when we were still forest-dwelling anthropoid apes, around 5–6 million years ago, and only 3–4 million years ago this type of locomotion, once acquired, became useful for new adaptations in the open savannah (Manzi, 2013). The earliest evidence is represented by four species (*Ardipithecus kadabba, Ardipithecus ramidus, Sahelanthropus tchadensis, Orrorin tugenensis*) whose limited fossil remains show direct or indirect anatomical evidence of upright posture and bipedal gait, which could also have been the result of parallel evolution to that of humans.

In other words, an upright posture and bipedal gait may have independently evolved multiple times in different lineages at the base of the human evolutionary tree, all within a closed forest environment. This suggests that these two characteristics evolved without any connection to life in the open savannah, meaning they do not represent, as was believed until a few years ago, a single adaptation in response to hominids occupying these open territories, an occupation that occurred a few million years later. Therefore, an upright posture and bipedal gait serve as examples of pre-adaptation, of an 'exaptation,' a structure that evolved in a certain context and with a certain function, and was later 'reused' for different purposes in a modified environment (Moggi-Cecchi, 2017).

The early bipedal hominins did not stand upright on two legs simply to free their hands; rather, they likely did so to procure food more efficiently and reduce the energy expenditure of walking. Our adaptation to bipedal walking enables us to cover greater distances more efficiently and with less muscular involvement in the back, pelvis, buttocks, and thighs compared to anthropoid apes (Lieberman et al.,

2002). Human walking reduces the energy cost of quadrupedal and bipedal walking in chimpanzees by approximately 75% (Sockol et al., 2007).

The earliest representatives of the human evolutionary lineage, such as *Ardipithecus ramidus*, even while still living in forested environments, began to develop bipedal locomotion methods. This adaptation was facilitated by the extreme diversification of locomotion among primates. Chimpanzees and gorillas, for example, adopted *knuckle-walking*, a characteristic quadrupedal method in which the forelimbs keep the fingers partially flexed, allowing the body's weight to be borne on the knuckles. Upright posture and bipedal gait were the initial distinguishing features of the human species. *Ardipithecus ramidus, Ardipithecus kadabba, and tugenensis* were among the first forms to exhibit these postural characteristics. It's possible that they may not be in a direct evolutionary line with us but rather represent divergent branches of the evolutionary tree.

The early ancestors of the human genus developed primarily in the African continent, predominantly within the Great Rift Valley. This geographical area stretches from present-day Ethiopia to the southernmost tip of the continent. It is characterized by mountainous terrain, valleys, large lakes, and vast stretches of grassland interspersed with trees, scrubland, and remaining patches of gallery forests along riverbanks. The unique geological features of this region, combined with environmental changes that intensified around 6 million years ago, made this enormous tectonic rift distinct from other areas along the same latitudinal band. To the west lay, and still lie, dense equatorial forests, while in the southeastern region, progressive aridification replaced remnants of forest and scrubland with savannah.

As mentioned earlier, bipedalism initially developed when our ancestors were still anthropoid apes, partly due to a climatic shift that reduced forested areas in favor of more open landscapes. The gradual fragmentation of the once-uniform forest cover into smaller, isolated patches surrounded by tall grass forced early hominins to further develop this unique form of locomotion.

The australopithecines, a group of primates with clear ties to the human lineage, were characterized by what is termed "optional" bipedalism. This highlights that, in addition to walking on their hind limbs, they retained the ability to climb trees (a trait inherited from more ancient forms). This versatility in locomotion allowed them to access food resources such as fruits and seeds and seek shelter in trees during the night. This diverse locomotion repertoire likely evolved in response to gradual climatic changes responsible for the progressive reduction of forest cover.

Among the australopithecines, the oldest is *Australopithecus anamensis*, from which it is plausible that *Australopithecus afarensis*, likely a key branch on the hominin phylogenetic tree (Manzi, 2013), originated. The most famous specimen of this species is the partial skeleton of a female individual, better known by the nickname "Lucy." Speaking of the imagery of the evolutionary tree, in one of his reflections, Darwin wrote that it was not the representation that satisfied him the most. Geometrically, evolutionary processes appear more akin to a coral, which has a less symmetrical structure and seems to expand irregularly. However, it actually develops its branches more vigorously where there are greater chances of survival.

The emergence of strong selective factors, such as more open environments with reduced tree cover, led hominin communities to exhibit highly differentiated responses, including adaptations to various diets. On one hand, there were the so-called “robust” australopithecines, characterized by a pronounced specialization of their masticatory apparatus. On the other hand, the lineage of our early ancestors saw the appearance of the first species attributed to the genus *Homo*, characterized by a trend toward a reduced masticatory apparatus and an increase in brain size, along with enhanced manipulative abilities.

The “robust” australopithecines of the genus Paranthropus were characterized by the reduction in the size of anterior teeth (incisors and canines), contrasted with enlarged premolars and molars, associated with powerful masticatory muscles and marked development of bony insertions for these muscles. This is evidenced by the existence of a sagittal bony crest on the top of the skull. The oldest species in the hominin lineage include *Paranthropus aethiopicus*, along with *Paranthropus boisei* and *Paranthropus robustus*.

The other lineage comprises the species conventionally attributed to the genus *Homo*: *Homo habilis* and *Homo rudolfensis*. In these species, a marked increase in absolute cranial capacity can be observed, along with a trend toward reduced tooth size. The increase in average brain size compared to older species (*Australopithecus africanus* and *Australopithecus afarensis*) is one of the reasons these fossils have been assigned to the genus *Homo*. However, it should be noted that the limited evidence of postcranial remains for *Homo habilis* suggests limb proportions and structure not dissimilar to the australopithecines. This, among other factors, has led some authors to include *Homo habilis* and *Homo rudolfensis* within the genus *Australopithecus*.

While the classification of *Homo habilis* and *Homo rudolfensis* within the genus *Homo* is not universally accepted, there is no doubt that the species *Homo erectus* and related taxa (such as *Homo ergaster* and the fossils from Dmanisi) belong to the genus Homo. The specific name *Homo erectus* refers to a large body of fossils found in Africa and Asia, showing significant morphological variation. In fact, some authors describe the African forms as belonging to *Homo ergaster,* reserving the name H. erectus for the Asian forms. Together, they are also referred to as *Homo erectus sensu lato* and can reasonably be considered the result of geographic variation within a species that, shortly after its origin, dispersed from Africa to Asia over a relatively short period. In *Homo erectus*, both brain size and body size are larger than in *Homo habilis,* and the body’s overall structure suggests that its mode of locomotion was exclusively bipedal (obligate bipedalism), in contrast to what is known for *Homo habilis* today.

Current evidence suggests that Asia was primarily inhabited by *H. erectus*, with the exception of a species endemic to the island of Flores known as *Homo floresiensis*. In Europe, the presence of the species *Homo antecessor* has been documented in the Gran Dolina cave site in Spain. This species is now considered related to the common ancestor of Neanderthals and modern humans. The rich fossil evidence from Europe is generally grouped under the specific name *Homo heidelbergensis*, characterized by primarily cranial morphological features that can also be found in later Neanderthals. The combined evidence from European and African fossils forms the basis for various interpretations. Some researchers consider all of them as part of *Homo heidelbergensis*, attributing the existing differences to geographic variation. Others recognize two distinct species: *Homo heidelbergensis* in Europe and *Homo rhodesiensis* in Africa. However, in both interpretative models, it is acknowledged that the European form gave rise to *Homo neanderthalensis* (Neanderthals), while the African form led to *Homo sapiens (*modern humans).

In any case, regardless of the taxonomic interpretation, the diversity of human species is undeniable, making the old model of a single continuous human lineage leading to *Homo sapiens* clearly unsustainable. The emerging picture of human evolution is one of a complex process, far from a linear model in which older species evolved into others in a single lineage, from *Australopithecus* to *Homo sapiens*. The diversity of hominids, at any given time, has always been high, and most of these species are not directly related to the origin of modern humans but rather represent evolutionary "dead ends." Even the recent phases of evolution within the genus *Homo* are emerging as characterized by an incredibly complex evolutionary pattern, with repeated events of dispersal, hybridization, competition, and extinction of different human species. What is certain (at least up to the present day) is that by the end of the Pleistocene, *Homo sapi*ens was the only human species left on Earth (Moggi-Cecchi, 2017).

This gradual transition from a forest environment to a savannah environment led the early representatives of our evolutionary lineage (e.g., *Ardipithecus ramidus*) to adopt a form of facultative bipedalism, which was probably later embraced by the early hominids (the australopithecines and their precursors) as a stable yet alternative mode of locomotion. It allowed them to move more securely across the increasingly open savannah spaces that separated the remnants of the forest. This form of non-obligate bipedalism was an additional adaptation alongside arboreal locomotion. Only with the genus *Homo* did bipedalism become the sole mode of locomotion available to us, with its various applications, including walking, running, and jumping (Manzi, 2013).

The dietary adaptation of various species of australopithecines to an environment increasingly composed of grasslands with sparse shrubs and plants, where they could find roots, dried seeds, and fibrous fruits, required increased masticatory effort to consume these foods. Significant modifications in dentition also occurred to accommodate a diet of tougher and more fibrous foods (Di Vincenzo & Manzi, 2013). These adaptations included the thickening of enamel and a reduction in the volume of incisors and canines, while the volume of posterior teeth (premolars and molars) expanded considerably, especially in the genus Paranthropus.

The stability and ergonomics of human bipedalism, which allow for carrying loads or running, are facilitated by a stable horizontal line of sight, essential for proper stabilization of the semicircular canals and maintaining balance. The curved shape of the human spine enables efficient weight distribution (Manzi, 2013) and compensates for the necessary rotations and counter-rotations of the pelvis and shoulders during walking. The vertebral curves, acting like pre-compressed springs, tend to reduce the oscillations that would affect the visual and vestibular system during walking. Furthermore, in an upright position, the hips and knees form a straight line with a long and vertical femoral bone, a hollow trochlea to stabilize the kneecap during walking, and an enlarged proximal end of the tibia. The non-prehensile foot serves to push the body, with its arches helping to absorb shocks during walking and provide additional propulsive force.

The pelvis, often considered the "first pelvic vertebra," plays an essential role in the human musculoskeletal system by supporting the trunk and the spine. However, it must serve conflicting purposes: it needs to be small, rigid, and solid to provide effective support for the spine, but it must also allow for the passage of the baby during childbirth. To reconcile these conflicting demands, the human pelvis has evolved to be less vertical and wider, with broad iliac wings that accommodate stronger gluteal muscles capable of providing effective leverage. This adaptation helps stabilize the pelvis laterally during walking and maintain an upright position (Le Huec et al., 2011) (Fig. 4.2).

While numerous variations and specializations of teeth can exemplify how selection has modified anatomical structures that, although originating from a similar configuration, serve different functions, the spine is one of the clearest examples of a structure constrained by functional factors that limit morphological variations. This limitation prevents drastic changes in the overall structure that could result in significant alterations of its original function.

The modularity of a structure promotes the independence of individual segments belonging to different developmental regions. These modules are differently integrated and functionally linked to each other, and they can also evolve relatively independently. This modularity has allowed for diverse evolutionary trajectories (Arlegi et al., 2018).

Adaptations in both the cranial and post-cranial skeleton to bipedalism in an upright posture have been numerous. However, it is worth noting the modifications that occurred in the vertebral column, which, as the main support axis of the skeletal structure, significantly increased its flexibility. This increase was achieved through an increase in the number of lumbar vertebrae, the creation of curves that accommodate the walking motion, and structural changes in vertebral bodies and processes, which adapted in response to the new muscular dynamics.

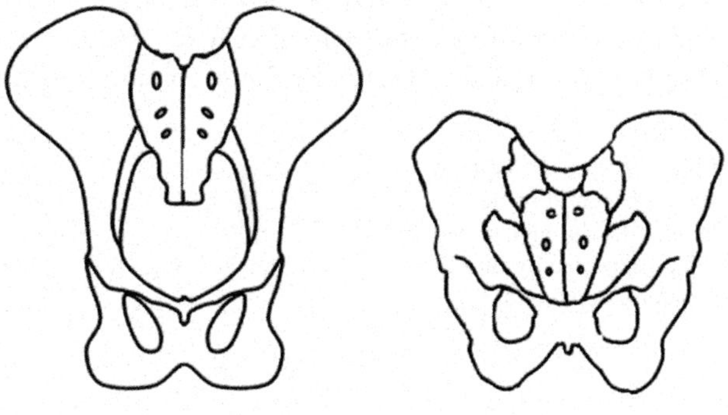

Fig. 4.2 Different shape of the pelvis in chimpanzee (left) and *Homo sapiens* (right). The pelvises are depicted proportionally to equal height between *Homo* and chimpanzee

The development of terrestrial life depended on the evolution of limbs that could support the weight of animals in a higher-gravity environment while enabling movement (Hogervorst et al., 2009). As vertebrates transitioned from water to land, supporting the weight of the body and skull became a significant challenge. The vertebral column had to resist not just the compression exerted by the surrounding liquid environment but also the gravitational forces acting on it (Walker & Liem, 1994). In amphibians and early reptiles, limbs gradually became larger and stronger but were still flexible and laterally oriented, much like the fins of fish. However, this arrangement was energetically costly for terrestrial locomotion. The lateral positioning of the limbs relative to the body required continuous muscular effort to prevent the trunk from collapsing to the ground (Fig. 4.3).

The vertebral column is a distinctive feature of vertebrates. It facilitates locomotion through lateral flexion in many fish, amphibians, and reptiles, as well as through sagittal flexion (flexion and extension) in many mammals (Williams et al., 2016). To move efficiently on land, locomotion must minimize unnecessary body movements. The entire body mass can be thought of as concentrated at a point where gravity acts, and quadrupeds must project their center of mass within the polygon defined by their limbs on the ground to maintain balance.

In amphibians and early reptiliomorphs, limbs served more for propelling the body than supporting it. They lifted the body off the ground and then projected it forward. The vertebral column had to (and still has to) withstand both vertical flexion due to gravity and horizontal movements used for locomotion. The lateral flexibility of the vertebral column creates an undulating movement of the trunk, which represents a dynamic vestige of the propulsion used by their fish ancestors (Fig. 4.4).

During these movements, where the limbs alternately bear weight, the force of gravity tends to rotate the trunk in opposite directions (Walker & Liem, 1994) (Fig. 4.5). This type of locomotion is not only energetically inefficient but also hinders respiration. Lateral flexion of the trunk compresses one lung alternately, diverting stale air to the other lung and reducing respiratory efficiency.

The transition from water to land occurred gradually, with terrestrial locomotion initially complementing swimming rather than entirely replacing it. The vertebral column gradually became more robust, serving as a support beam to transfer the body's weight to the limbs. It was no longer necessary for the head and trunk to function as a single unit, as air represented a much less dense medium for movement than water. As a result, in early terrestrial vertebrates, the head became more independent and free to move relative to the trunk, thanks to the specialization of the cranial portion of the vertebral column. In amphibians, only one cervical vertebra,

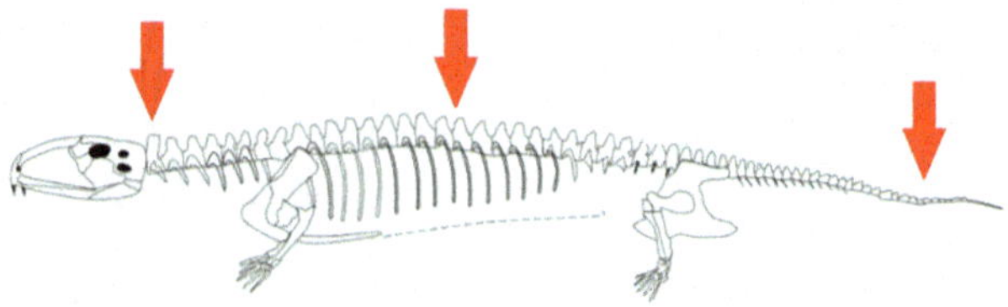

Fig. 4.3 In amphibians and early reptiles, the vertebral column also had to be supported by muscles capable of counteracting gravity

Fig. 4.4 Undulatory movement of the vertebral column (colored lines) on the horizontal plane, characteristic first of fish and then of amphibians and reptiles. Note the horizontal oscillation of the body's center of gravity (colored circles) in amphibians/reptiles during locomotion

Fig. 4.5 Torsion of the trunk during the locomotion of a pre-mammalian terrestrial vertebrate: the alternation of the support side creates a torsion of the vertebral column on the coronal plane

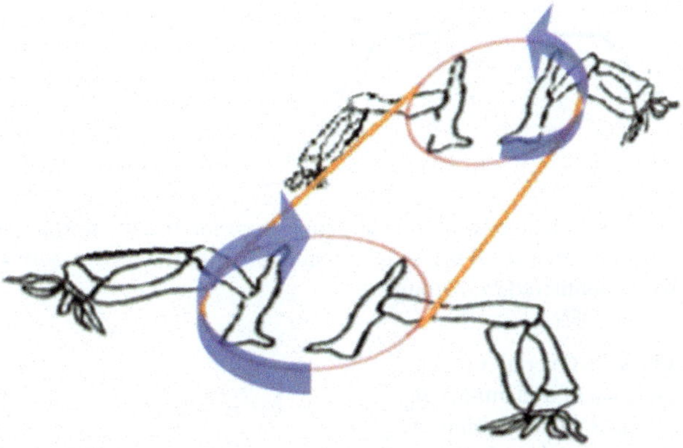

the atlas, differentiated in the cranial region of the vertebral column. This allowed the head limited movement relative to the trunk. The first reptiles evolved from amphibians around 350 million years ago. As they became more mobile and expanded into terrestrial environments, their vertebral columns became increasingly robust. The modular differentiation of the vertebral column into regions became more pronounced, and in the cervical region, the number of vertebrae increased, enabling the head to move in multiple directions. The cranio-vertebral articulation further developed, essentially representing the modification of an intervertebral joint. Subsequently, the second cervical vertebra, the axis (epistropheus), also underwent modification. The first two cervical vertebrae formed a universal joint, giving the head an exceptional range of movement.

The specialization of the vertebral column reached its pinnacle in mammals. While other cervical vertebrae remained less morphologically distinctive (Walker & Liem, 1994). Mammals evolved from reptiles approximately 250 million years ago. In this class of vertebrates, the axial skeleton became both sturdy and flexible. It effectively distributed the body's weight to the limbs and allowed for significant head movement. The limbs shifted beneath the body, aligning perpendicular to the vertebral column to enhance stability during quadrupedal locomotion (Fig. 4.6).

In quadrupedal mammals, the vertebral column's axis remained parallel to the ground, upheld against gravity by the powerful paravertebral muscles (Fig. 4.7).

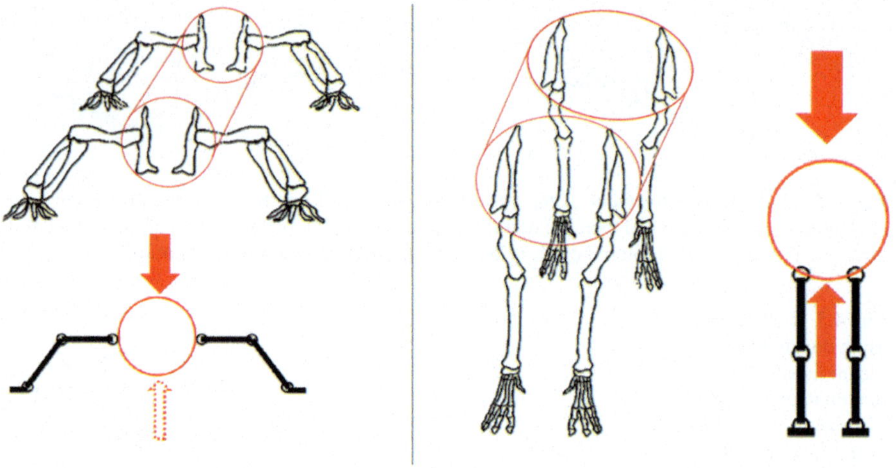

Fig. 4.6 Change in limb emergence from the trunk: from a lateral position in reptiles (left) to a position underneath the body in mammals (right). Gravity, acting on the trunk, is much more effectively countered in mammals

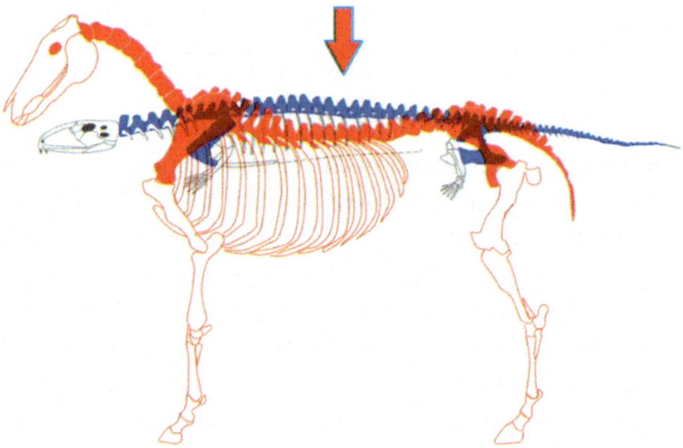

Fig. 4.7 Overlay of the skeleton of a mammal (in red) and an amphibian (in blue): even though the arrangement of limbs in mammals passes underneath the body, the vertebral column remains perpendicular to the gravitational line and depends on powerful paravertebral musculature to support proper posture

This maintained a pronograde posture, with the body's primary axis parallel to the ground. However, the evolution of limb placement and improved static support offered more efficient dynamic movement. For instance, during rapid movement, the vertebral column can flex in the sagittal plane, ensuring that the center of gravity remains along the body's axis as it rises and falls (Fig. 4.8). These adaptations also necessitated significant changes in the musculature, with increased power to support the vertebral column as a whole, especially the cervical region.

With few exceptions, all mammals possess seven cervical vertebrae, regardless of neck length. For instance, giraffes have notably elongated necks, but they still have only seven cervical vertebrae, except for the atlas (C1) and axis (C2). In contrast, cetaceans, such as whales and dolphins, have reduced and sometimes fused cervical vertebrae, adapting to their return to an aquatic environment. Mammals also

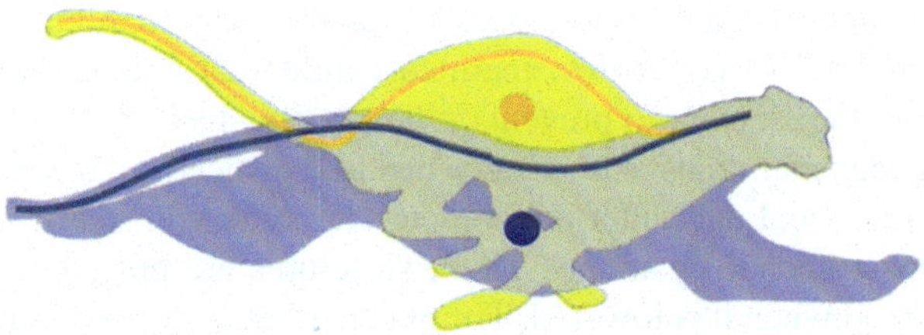

Fig. 4.8 Sprint of a cheetah with movement of the vertebral column in the vertical plane. Note the vertical oscillation of the vertebral column (colored lines) and the body's center of gravity (colored circles) during locomotion (Drawing freely adapted from "Functional anatomy of the vertebrates: an evolutionary perspective," Walker & Liem, 1994)

developed the odontoid process, a protrusion of the axis (C2) that inserts into the vertebral canal of the atlas (C1). It acts as a pivot for rotational movements of the head. In the remaining part of the vertebral column, the vertebrae underwent morphological differentiation, creating thoracic and lumbar regions (Walker & Liem, 1994).

In quadrupeds, when at rest, the active movements allowed to the atlanto-occipital joint are primarily extension (flexion occurs passively assisted by gravity). Consequently, the posterior cervical musculature in these animals is notably strong, even for merely supporting the head.

In humans, the entire vertebral column becomes more vertical, with less pronounced curvatures compared to quadrupeds. The head is positioned directly above the cervical spine, almost passively supporting its weight. These factors have led to a significant reduction in the musculature of the nape and neck in humans (Goel et al., 2011).

In mammals, the head is elevated by the cervical vertebrae, forming an arch supported by the dorsal cervical muscles, ligaments, and the spinous processes of the vertebrae. These spinous processes act as lever arms to transmit muscular forces, opposing gravity, and supporting the head. The spinous processes of the anterior thoracic and cervical vertebrae are nearly perpendicular or slightly inclined backward concerning the line of action of the muscles that insert onto them. Their inclination and length are related to the strength of the musculature attaching to them. An increase in their length corresponds to an increase in the lever arm and mechanical advantage.

Among modern mammals, structural variations in the vertebral column closely correspond to different postural behaviors. In quadrupeds, the cervical region is subjected to significant gravitational forces. The head is part of an equilibrium system known as the "cervico-cephalic balance," perfectly adapted to different types of locomotion and species' habits. This system adjusts the center of gravity to accommodate various movements and positions maintained by the animal (Walker & Liem, 1994).

The vertebrae are connected by sturdy longitudinal ligaments that run along the dorsal surfaces and the tips of the spinous processes. The epaxial muscles link pairs of vertebrae (interspinal muscles) or extend over several vertebral segments, creating a solid yet flexible complex. The vertebral column, while not constituting a static support system, can be likened, in a way, to a bridge. The muscles and ligaments that support the neck and back of quadrupedal mammals can be biomechanically compared to guy wires, inclined cables that extend from the top of a support, holding the element to prevent collapse (Fig. 4.9).

In quadrupedal mammals with heavy heads that need to lower their heads to graze, we find cervical and anterior thoracic vertebrae with particularly long processes and robust nuchal ligaments. These structures are put under tension by the weight of the head when it is lowered. The elastic energy stored is released when the head is raised, contributing to muscular action. The greatest stress on these muscle-ligament structures primarily occurs along the sagittal plane. Similar to a wooden beam that resists better under a unidirectional load when placed in a cutting position, the spinous and transverse processes take on a blade-like shape, increasing their width on this plane and, consequently, their strength (Walker & Liem, 1994).

Associated with a cervical and thoracic vertebral structure adapted to support the weight of the head and neck, a powerful muscular and ligamentous system developed, allowing for easy execution of extensive flexion-extension, lateral inclination, and circumduction movements. In mammals, extensive neck rotation primarily occurs in the relationship between the atlas and axis vertebrae. This relationship remains consistent in primates and, as we will see later, in humans.

Of course, given the pronograde posture, in quadrupeds, the vertebral column does not serve as a support for the head as it does in humans. Instead, it behaves like an extending beam that extends beyond the front limbs. The cervical ligament is crucial in passively assisting the dorsal neck muscles in supporting the weight of the head and neck. The cervical structure of quadruped mammals can be represented as a "bowstring" that is tensed and relaxed, following the movements of the head (Walker & Liem, 1994). The load of the head induces a permanent flexion moment on the neck, which is countered by the powerful nuchal muscles and ligaments. The lower part of the neck is biomechanically connected to the front limb through a pattern that results in a tripartite functional modularity of the neck (Fig. 4.10).

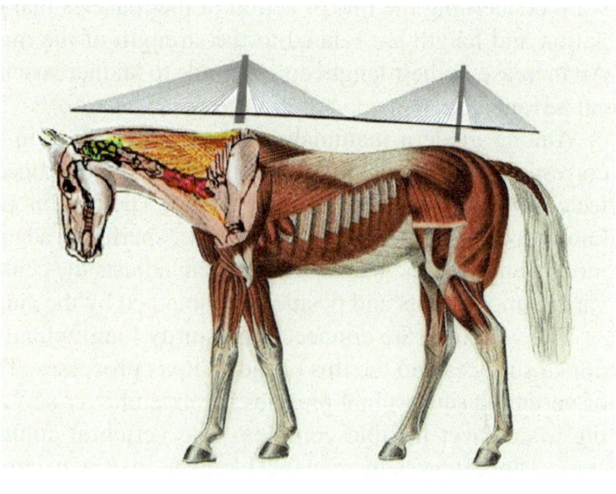

Fig. 4.9 Cervical bone, muscular, and ligamentous structure of a horse; in orange, the bundles of cervical ligament. Above, the structure of a cable-stayed bridge, with the deck suspended, supported by a series of stays (cables) fixed to support pylons

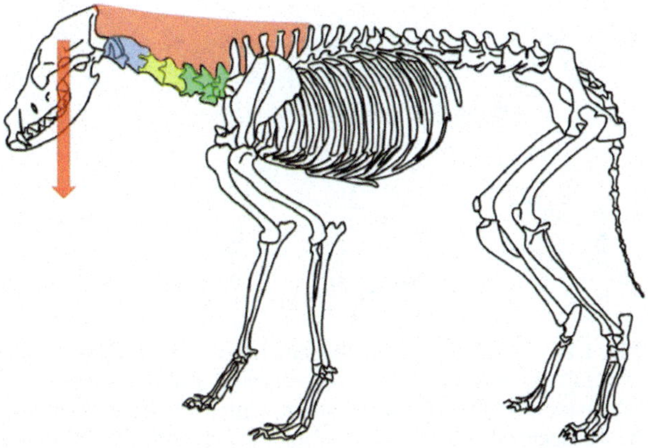

Fig. 4.10 Weight of the head impacting the cervical vertebral column and its musculotendinous apparatus

However, interspecific adaptations of cervical vertebrae can occur in mammals. For example, the variation in the length of the cervical spine as a whole is one of the main sources of interspecific disparity in neck structure. Nevertheless, the number and proportions of cervical vertebrae remain consistent. The way mammals achieve a very long or very short neck is fundamentally different from that of other amniotes, in which neck disparity is achieved by varying the number of vertebrae (Arnold, 2020).

The primates that separated from other mammals approximately 65 million years ago were quadrupeds with tails, an extension of the vertebral column. This tail has been retained by modern non-anthropoid monkeys, except for the anthropoid ones. The general architecture of their bodies was and is quite similar to that of canids, characterized by a pronograde posture, which is typical of non-anthropoid monkeys.

The powerful cervical muscle-ligament structure in pronograde follows a constant pattern because the biomechanical challenge of supporting the head remains constant. In the quadruped position, gravity acts on the head in a hanging and horizontally oriented manner, resulting in greater activation and a collective response of the nuchal muscles. Electromyographic studies on macaque nuchal muscles have observed negligible muscle activity during upright resting positions with a vertically oriented neck. However, significantly higher muscle activation is evident in the quadruped position in non-anthropoid primates. Despite the extreme morphological variability among species, the cranio-vertebral connection pattern remains unchanged in non-anthropoid primates, with seven cervical vertebrae specialized in rotation and flexion-extension movements in the atlas and axis vertebrae. The posterior placement of the foramen magnum also remains consistent in these primates (Fig. 4.11).

The remarkably low level of interspecific variation in the number of cervical vertebrae in mammals has puzzled biologists for over a century and a half. In contrast, birds, reptiles, and amphibians exhibit considerable variation in this regard, and the number of vertebrae varies in other regions of the vertebral column in mammals as well. The cervical region of the vertebral column appears to be an

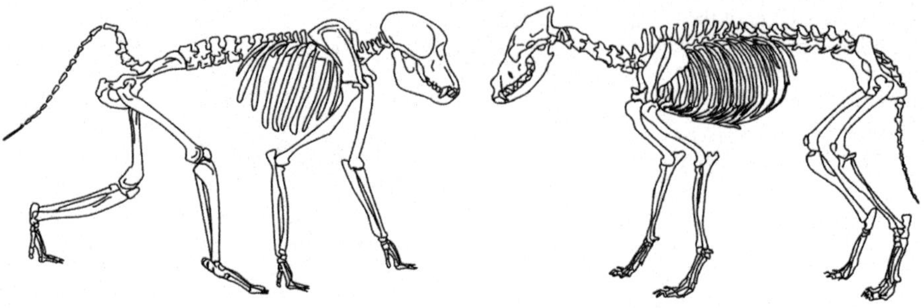

Fig. 4.11 Left: Skeleton of a baboon (*Papio anubis*). Right: Skeleton of a wolf (*Canis lupus*). The skeletal structure in non-anthropomorphic primates remains essentially unchanged compared to the quadrupedal mammalian pattern, with a vertebral column parallel to the ground and the head held suspended in anterior extension by powerful cervical musculature

evolutionary constraint that resists variation. The number of cervical vertebrae is determined during the early stages of organogenesis and is strongly conserved in mammals because any variation in the number of cervical vertebrae likely has serious repercussions for an individual's survival (Galis et al., 2006).

Anthropoid primates, which separated from other primates approximately 20 million years ago, underwent changes in their body posture. They transitioned from a pronograde posture, where the vertebral column was essentially parallel to the ground, to a clinograde posture. While the vertebral column tended to become more vertical in the clinograde posture, it did not consistently achieve a fully upright posture. The hind limbs remained flexed, while the anterior limbs, particularly elongated in chimpanzees and gorillas, led to their characteristic *knuckle-walking* gait, where they walk on the knuckles of their hands (Fig. 4.12).

Despite the clinograde posture, the skeletal arrangement of the cervical spine does not differ significantly from that of quadrupeds (Fig. 4.13). Even the neck muscles remain robust and developed, as they continue to counteract gravity by supporting the head, which remains hanging in relation to the vertebral column.

In bipedalism with an upright posture, the cranio-cervical arrangement undergoes a radical change, and the neck muscles play a more regulatory role than supporting the skull on the vertebral column (Fig. 4.14).

The vertebral column in gorillas, chimpanzees, bonobos, and orangutans is particularly rigid. In contrast, in humans, the lower back is more mobile, adapted to the lordosis in a fully upright posture, both of which are characteristics necessary for bipedalism. Compared to humans, anthropoid apes have a ribcage closer to the pelvis and a shorter lumbar spine. The number of lumbar vertebrae in existing African apes varies from three to four, in contrast to humans who have five lumbar vertebrae. Humans have elongated their lumbar spine during the transition to bipedalism (Fig. 4.15).

Since both chimpanzees (*Pan troglodytes*) and gorillas (*Gorilla gorilla)* have the same vertebral distributions in the thoracic, lumbar, and sacral regions, it has been hypothesized that the last common ancestor (LCA) between apes and humans also

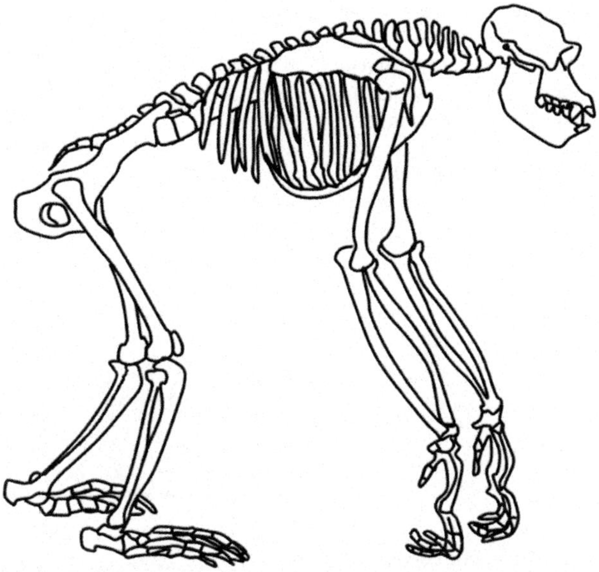

Fig. 4.12 Skeleton of a chimpanzee (*Pan troglodytes*) in a knuckle-walking posture. Note the characteristic quadrupedal locomotion with support of the forelimbs on partially flexed fingers

Fig. 4.13 Chimpanzee (*Pan troglodytes*) in knuckle-walking (Drawing by Marco Boggioni)

had a similarly short lower vertebral column. Recent studies suggest that the last common ancestor (LCA) between apes and humans had a long lumbar vertebral column, with at least six lumbar vertebrae. The short lumbar columns observed in all three African apes are believed to have evolved independently and only after their separation from humans. These apes likely inherited their short lumbar

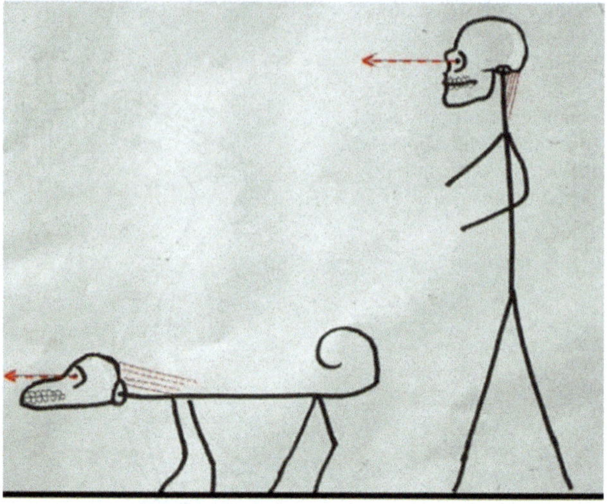

Fig. 4.14 In bipedalism in the upright position, cervical musculature plays more of a role in regulating cranial posture than in anti-gravity support (Drawing by Yoel Rak)

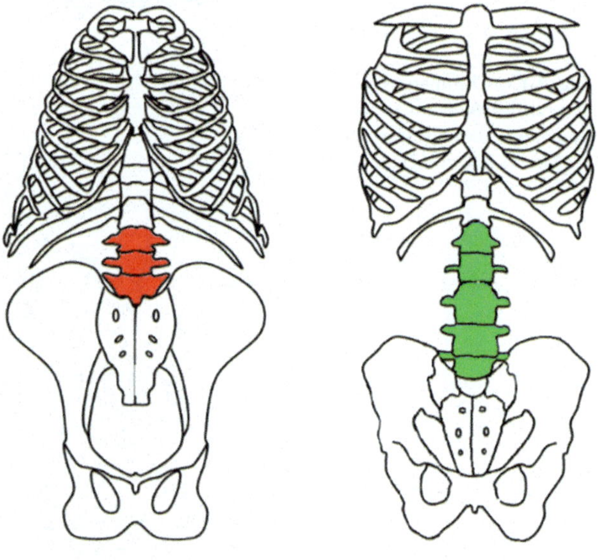

Fig. 4.15 Left: Pelvis, vertebral column, and thorax of a chimpanzee; right: in *Homo sapiens*. In humans, the number of lumbar vertebrae increases, making the vertebral column more elastic, flexible, and better suited to bipedal walking in an upright posture

columns from an ancestor with only three or four lumbar vertebrae. Early australopithecines, as well as the first representatives of the *Homo* genus and modern humans, already had elongated lumbar sections consisting of five vertebrae (McCollum et al., 2010).

In anthropoid apes, the typically rigid and immobile lower back prevents any lumbar lordosis, which is the main reason for their less graceful (at least in the human eye) bipedal gait with bent hips and knees. The rigid "compaction" between the abdomen and chest provides biomechanical advantages for both climbing and swinging through trees. It also reduces the risk of spinal injuries in case of falls.

However, it sets the characteristic bipedal gait of species like chimpanzees, bonobos, and gorillas. In these species, bipedal walking is characterized by flexed hip and knee joints, primarily due to the reduced length and mobility of the lumbar spine, rather than functional limitations imposed by the hip joint itself. Nevertheless, non-human primates exhibit various and distinctive locomotion patterns that influence their postcranial morphologies (Fig. 4.16).

Australopithecines had an axial skeleton that comprised a mosaic of features, including primitive traits common to apes and characteristics shared with humans. Their versatile locomotor capabilities were useful for moving easily in diverse environments, both in trees and open fields, enabling a form of bipedalism, albeit different from that of modern humans. These australopithecines did not have adaptations for long-distance walking and running, indicating a "compromise bipedalism." This likely represented an early stage in the long process that established the functional link between the pelvis and the vertebral curvatures characteristic of modern humans.

The human spine reflects the constraints imposed by its upright posture in relation to gravity. The morphology and biomechanics of the human spine, along with its relationships with the pelvis and skull, are directly related to bipedalism (Fig. 4.17). The pelvis widened and straightened, spinal sagittal curves appeared, and paraspinal muscles underwent significant reorganization. The formation of these vertebral curvatures is influenced not only by upright posture and bipedalism but also by the size and shape of the pelvis and thorax, encephalization, and even the

Fig. 4.16 Knuckle-walking posture of a Chimpanzee (*Pan troglodytes*) (Drawing by Marco Boggioni)

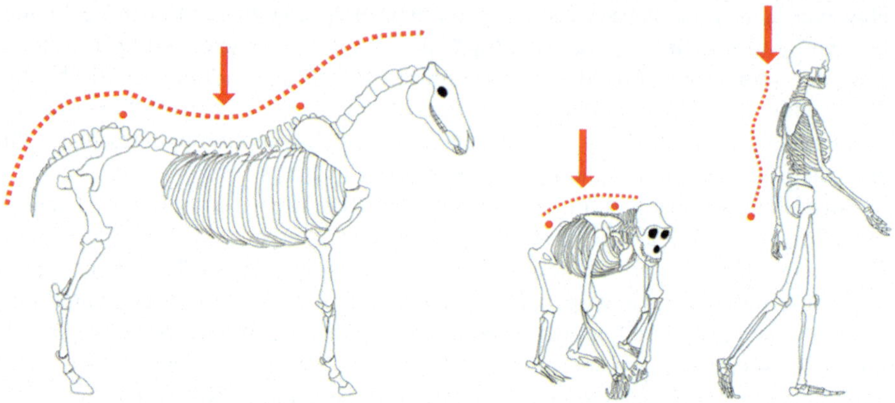

Fig. 4.17 Unlike other vertebrates, the human vertebral column experiences gravitational force parallel and not perpendicular to the spinal axis; this leads to compression of vertebral and intervertebral elements. In H. sapiens, a large and heavy globular skull is positioned at the top of the column, balancing thanks to the interaction of muscular chains acting on different planes

masticatory apparatus. This complex interplay of biomechanical interactions is referred to as the "cranio-spino-pelvic balance" (Been et al., 2017).

In the evolutionary history of primates, *Homo sapiens* only appears in the last 3% of the total time. The last common ancestor with chimpanzees (and bonobos) lived around 250,000 generations ago, approximately 6 million years ago. Bipedalism and encephalization are unique characteristics of human evolution. While many species may exhibit bipedal walking to some extent, humans are the only primates to have made it their exclusive form of locomotion. In humans, upright posture and bipedal gait required the alignment of the spine with the head (Fig. 4.18), placed on the gravitational line in relation to the pelvis and lower limbs, with constant control of the center of gravity within a small base of support.

The human spine became vertical but not straight, with curves that, besides being functional for thoracic and abdominal anatomy, allowed for better weight distribution. The kyphoses and lordoses can be interpreted as pre-compressed springs that stretch and return to their original position during movement, playing a role in cranial stabilization and effective dissipation of forces acting on the spine.

The various segments of the vertebral column exhibit sagittal plane curvatures, either with anterior convexity (lordosis) or posterior convexity (kyphosis). These spinal curves increase resistance to vertical loads by up to 17 times, directing deformations in predetermined directions that can be immediately controlled by muscle contraction (Izzo et al., 2013). The sagittal curves of the vertebral column represent an evolutionary response to the requirements of upright posture and bipedal gait associated with minimizing energy expenditure. These curves also develop during ontogenetic growth. The dorsal kyphosis is the only sagittal spinal curve present at birth, while the cervical and lumbar lordoses develop with the raising of the head and the assumption of an upright posture and bipedal walking.

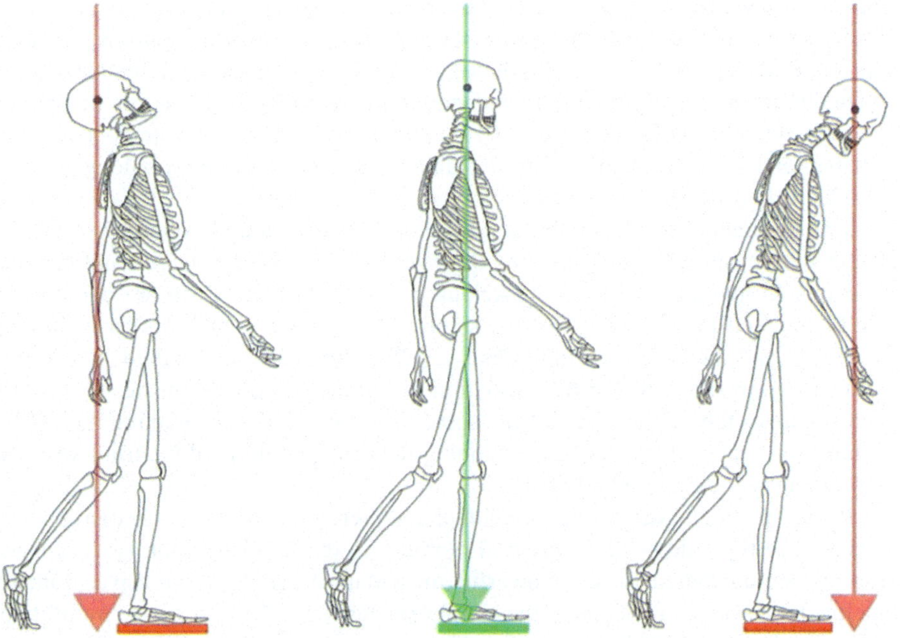

Fig. 4.18 Proper cranio-cervical alignment is essential to maintain the center of cranial mass within the small plantar support area

While *Homo habilis* had a different upper body balance compared to modern humans, *Homo erectus* already exhibited a different gait pattern, with body proportions and thoracic shape entirely human-like, making them "obligate bipeds" with a posture and mode of locomotion nearly identical to ours. During this phase of hominid evolution, which involved the transition from occasional to permanent bipedalism, the functional connection between the vertebral column and the pelvis that characterizes humans was established. Humans have adapted so well to an upright posture that it requires only 7% more energy compared to a supine or lateral resting position. In contrast, for example, dogs in their natural quadrupedal position have significantly higher energy expenditure compared to their resting posture, presumably due to the naturally flexed position of their hind limbs (Abitbol, 1988).

In humans, muscles under the control of the nervous system provide active stabilization of the vertebral column, ensuring its balance both in an upright posture and during the alternating eccentric movement of bipedal walking and running. When a human walks in an upright position, the adaptation of the lumbar lordosis aligns the body's center of mass with the ground contact area, allowing for highly energy-efficient upright walking (Hogervorst et al., 2009).

Compared to the last common ancestor with large African apes, the functional connection between the pelvis and the vertebral column likely became established with *H. erectus*, facilitating the development of vertebral characteristics that would have facilitated the formation of lordosis in the curvatures of the spine. Naturally, the upper part of the spine was also involved in this process of integration and adaptation of cervical and cranial morphology because natural selection tends to maximize efficiency in maintaining the balance of the entire body, including the crucial balance of the head (Tardieu et al., 2017). Humans specialized in a single locomotor pattern, and as a result, their entire cranial and postcranial morphology became "exclusively dedicated" to bipedalism in an upright posture.

Among mammals, the lumbosacral region of the vertebral column is typically the most relevant when investigating the correlation between structure and postural position. In humans, the musculoskeletal demands of bipedalism are particularly well expressed in this portion of the spine. The increase in the number of lumbar vertebrae and the height of individual vertebral bodies have resulted in a longer lumbar region, allowing for the formation of lordosis. Lordosis acts as a transition and stabilizes the thoracic part of the spine over the pelvic part, making it easier to control the center of gravity by positioning the center of mass of the trunk over the hips, thereby facilitating body balance.

It also prevents subjecting individual intervertebral joints to excessive shear forces. The flip side of this increased mobility in the lumbar region is its reduced robustness, attributed to less powerful musculature and less extensive, dorsally inclined transverse processes of the vertebrae (Fig. 4.19). The increased mobility of the spine has come at the cost of its robustness, and conditions like scoliosis and spondylolisthesis are virtually absent in large apes (Plomp et al., 2015).

The morphology of the vertebral column is directly correlated with and has an influence on both body posture and locomotion. With bipedalism, the area of balance has dramatically reduced compared to quadrupeds. During locomotion, to efficiently maintain an upright posture, humans must shift the entire weight of the head and trunk onto the hip joint of the supporting leg, while the other leg is swung forward to balance the center of gravity shifting out of the support area (Fig. 4.20) (Marchi, 2016). To assist with the balance of bipedalism in an upright posture, the pelvis widened, and the attachment of the gluteal muscles shifted laterally, with a function that evolved from propulsion to stabilizing bipedal walking, especially during phases when only one foot is in contact with the ground. Additionally, the sacrum bone widened, and a large hip joint formed. Feet, which were once

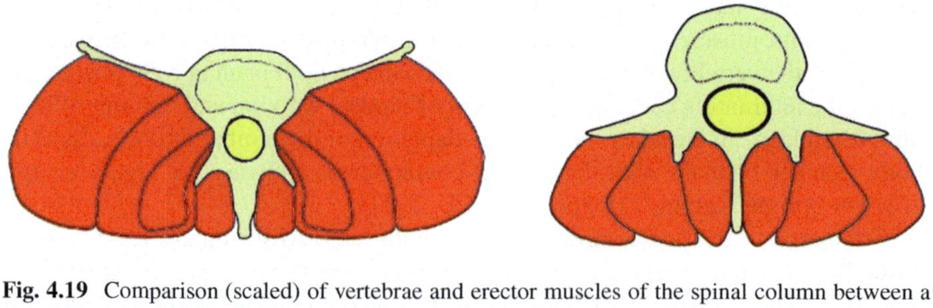

Fig. 4.19 Comparison (scaled) of vertebrae and erector muscles of the spinal column between a cercopithecoid (left) and a human (right): the reduced extension of the transverse process in humans has significantly reduced the muscles inserting onto it

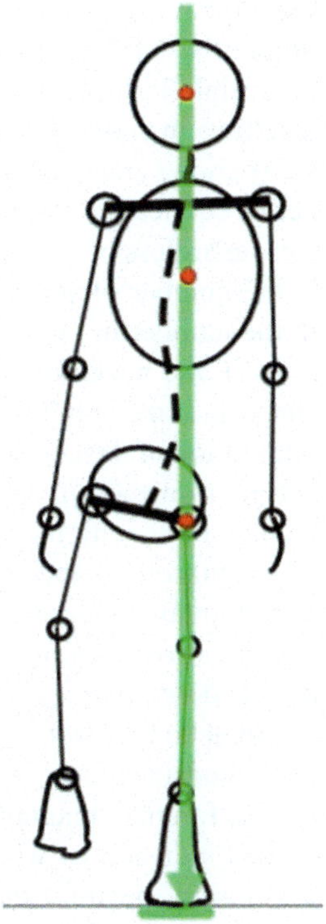

Fig. 4.20 The flexibility of the lumbar spine stabilizes bipedal walking, helping to keep the centers of cranial and thoracic masses as close as possible to the gravitational line that impacts the area of the supporting foot

prehensile, became formidable structures for dynamic support, and the femur and tibia brought the knee closer to the body's center of gravity (Senter, 2010).

These characteristics not only facilitate trunk support but also efficiently transfer weight to the feet through the legs, both in a static position and while walking or running. Our legs are much longer than our arms, we have knee joints that can fully extend and lock, and specialized feet with a robust ankle region and an arched structure designed to dissipate the impacts that this type of locomotion entails, turning them into formidable dynamic support structures as mentioned earlier.

In the condition of obligate bipedalism, many functions useful for climbing were lost, such as long arms, opposable big toes, or curved finger bones, or special adaptations in the shoulder joint. This loss of characteristics specific to adaptation for arboreal life is as important as the acquisition of specialized adaptations for bipedalism because it indicates strong directional selection for this particular type of terrestrial locomotion exclusively (Aiello & Dean, 1990; Harcourt-Smith, 2010).

The lumbar region has good flexion-extension but less rotation than the thoracic region. The thoracic region experiences mechanical resistance from the ribcage, which significantly limits its mobility, even though the thoracic vertebrae, despite the presence of ribs, retain good mobility on the horizontal plane. In the thoracic region, the rotational movement, unlike the lumbar region, has a wide range (40°), thanks to a different orientation of the interapophyseal articular facets compared to the lumbar region, where rotation is hindered by the internal and posterior orientation of these facets. However, as the years go by, costal cartilages ossify, and chondrocostal elasticity decreases, transforming the ribcage into an almost rigid block, resulting in reduced movements in this region. Regarding the cervical spine, its study shows an advantage over the lumbar spine because, as we have seen, the number of cervical vertebrae (seven) is remarkably constant in almost all mammal species regardless of the length of the neck, whereas the lumbar spine shows differences in the number of vertebrae even within the same genus. This evolutionary stability in the number of cervical vertebrae among mammals means that anatomical variations in the neck among different species depend only on the modification of the shape and size of the vertebrae, but not their number, a factor that itself promotes the integration of the cervical spine with the underlying portions (Arlegi et al., 2018). A variation in the number of vertebrae in this region would likely require changes incompatible with the viability of surrounding tissues and anatomical structures.

In anthropoid apes, the muscular structure of the cervical region remains similar and maintains the same powerful structure as that of quadrupeds, as the clinograde posture also has to support the "suspended" skull against gravity. The foramen magnum and its support fulcrums, the occipital condyles, remain in the posterior area of a cranial box far from any static equilibrium, which forces the posterior musculature into a constant "arm wrestling" with gravity.

The foramen magnum, along with the occipital condyles, provides a direct osteological indicator of where and how the cervical spine articulates with the skull. In the upright posture, the human head is positioned above the trunk with an axis essentially parallel to the long axis of the vertebral column: a distinctive position that reflects the erect posture of the trunk (orthograde). In humans, the foramen magnum shifts below the cranial box to a position much closer to the gravitational vertical of the center of mass of the skull, which in *Homo sapiens* is located near the sella turcica, allowing the head to rest above the upright vertebral column with minimal muscular effort.

The cervical spine connects the skull to the post-cranial skeleton, and one of its main functions is to balance and stabilize the head while maintaining a stable field of vision. This functional relationship is structured as a system of first-class levers, in which the head is balanced almost in equilibrium on its fulcrum, formed by the atlanto-occipital joint. The weight of the portion of the skull anterior to the condylar support is countered by the forces of the nuchal muscles. This morphology reduces the distance between the center of mass of the head and its fulcrum, favoring efficient postural control of the head-balancing muscles and allowing a reduction in their cross-sectional areas, but with a concurrent decrease in vertebral robustness (Been et al., 2019). The cranial portion anterior to the occipital condyles is significantly correlated with the variation in cranio-caudal dimensions of the cervical vertebrae: as the anterior part of the skull elongates, the cervical vertebrae also lengthen cranio-caudally (Fig. 4.21).

Primates with more pronounced heads and necks exhibit a greater mechanical advantage from an increase in the length of the spinous process in their cervical vertebrae. This elongation allows for the insertion of powerful muscles required to

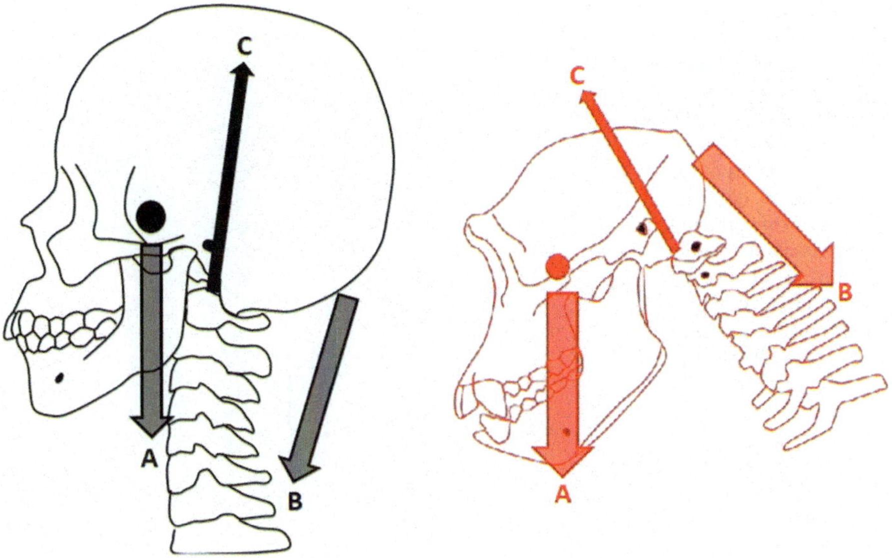

Fig. 4.21 Forces balancing the head. Colored circles: center of mass. (*A*) Force of gravity. (*B*) Force of nuchal muscles. (*C*) Resulting force

support and stabilize the spine, which would otherwise be highly unstable even under very light loads. Among primates, there is a correlation between cervical morphology and postural behaviors. Large African apes with protruding heads and necks, such as gorillas and chimpanzees, display cervical vertebrae with longer and more horizontally oriented spinous processes. In contrast, in modern humans, these processes appear notably shorter and oriented more towards the tail. This difference is attributed to the reduced nuchal muscle mass resulting from the human head being not protruding forward but balanced on a small support area at the base (see Fig. 4.22).

Within the *Homo* genus, the nuchal ligament appears, which is absent in chimpanzees and probably also in australopithecines. The nuchal ligament functions like an elastic band that crosses the midline of the neck, connecting the occipital nuchal line to the spinous processes of the cervical vertebrae. It stabilizes and supports the head during walking, serves as an attachment site for nuchal muscles, limits and controls flexion, acts as a shock absorber for loads, and also functions as a significant proprioceptive structure for the head, playing a functional role in bipedal locomotion.

The absence of the nuchal ligament in large apes is functionally linked to their distinct posture, even though this ligament is commonly found in many non-primate mammals that have adapted to running (Been et al., 2019). In humans, the nuchal ligament plays a crucial role in cervical lordosis by reducing the muscular effort needed to maintain head balance. A well-developed cervical lordosis, compared to other primates, is associated with greater freedom of neck movement during flexion and extension.

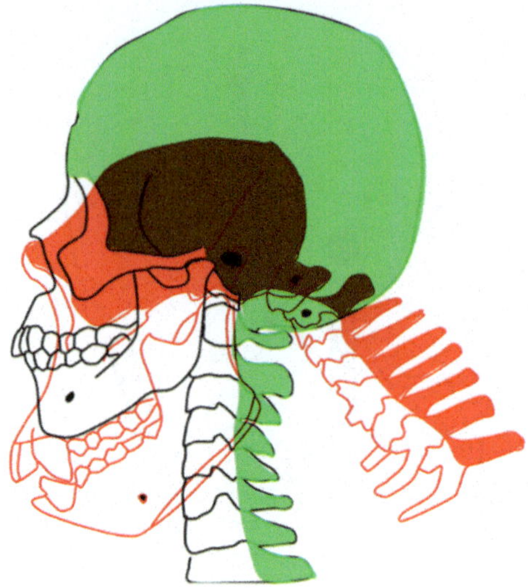

Fig. 4.22 Overlay of profiles of chimpanzee and *Homo sapiens*, obtained by aligning the auditory meatuses of the two skulls. Note, in addition to the different volumes of the neurocranium, how the diversity in the orientation of the vertebral columns and the extension of the spinous processes is closely related to different posture and muscle mass

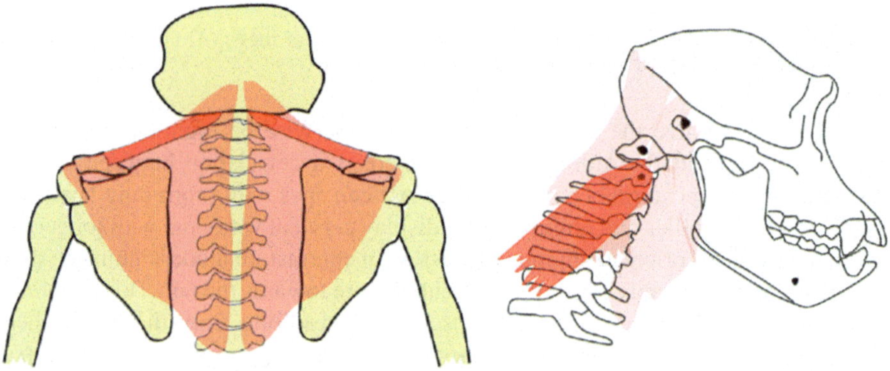

Fig. 4.23 Powerful dorsal musculature of the chimpanzee. Light red: trapezius muscles; dark red: atlanto-clavicular muscles, not present in humans

In contrast to large apes, the human spine is highly mobile (Lovejoy, 2005). The cervical structure of anthropoid primates differs from that of humans not only in terms of bone morphology but also in musculature. In large apes, neck muscles are more robust, powerful, and bulky than in humans. Additionally, many mammals, including all anthropoids, possess a muscle that humans have lost: the atlanto-clavicular muscle, connecting C1 to the clavicle. This muscle elevates the clavicle and contributes to increased cervical rigidity (see Fig. 4.23). It remains unclear when this muscle was lost in the human lineage, but this loss is likely associated with the shift from quadrupedal to bipedal locomotion (Aiello & Dean, 1990).

The early representatives of the *Homo* genus had relatively larger vertebral bodies compared to their body sizes. Along with the australopithecines, they exhibited more horizontally oriented spinous processes not only compared to modern humans but also to chimpanzees. This complex evolutionary scenario is closely tied to the biomechanical relationship between the spinous processes and the movements of the head and upper limb (Arlegi et al., 2017).

As just outlined, there are intimate and profound anatomical and functional correlations between the skull and the vertebral column. In humans, cranial morphology has been shaped in response to an upright posture and obligatory bipedalism. These changes have not only affected the neurocranium but also the splanchnocranium, along with the structure of the stomatognathic apparatus.

Our stomatognathic apparatus is phylogenetically ‘recent,’ and in human evolution, bipedalism and encephalization are much more interconnected with dental occlusion than previously believed. The new structural adjustments involving the spine, mandible, and cranium, placed atop the vertebral column, have produced a cascade of somewhat unpredictable effects (Raia et al., 2018). These effects will be the subject of the next paragraphs.

4.2 Encephalization

The modifications of the teeth and the maxillofacial region that occurred during the transition to upright bipedal posture, necessary to balance the skull on a nearly vertical vertebral axis, intersected with those brought about by increasing encephalization, which in *Homo sapiens* (and Neanderthals) became extraordinary compared to ancestral species.

The scientifically indisputable presence of evolutionary sequences in fossil records stands as one of the primary lines of evidence for evolution (Senter, 2010). One need only consider the close correlation between teeth and the stomatognathic apparatus with the postural system to dispel any debate. Those who argue against these relationships (even denying the possibility of a dysfunctional interaction between dental occlusion and the temporomandibular joint) advocate their positions solely based on statistical evaluation methods concerning the number and frequency of citations in the literature on this topic. Such an approach overlooks (or remains unaware of) the profound connections, not only in terms of evolutionary evidence but also in anatomy (both systematic and comparative), histology, embryology, and functionality. These connections highlight how the masticatory apparatus is an integral part of the postural system.

Around two to two and a half million years ago, the ‘early *Homo*’ emerged from the australopithecines, retaining a somewhat similar skeleton while undergoing significant changes in cranial morphology, facial structure, and dentition. There was an increase in cranial size, reaching a volume of approximately 650 cubic centimeters. This was accompanied by a new arrangement of the cranial vault and facial bones, with less elongated dental arches, more slender mandibles, and teeth devoid of the extreme premolar and molar expansion that characterized the australopithecines, especially the robust ones (Manzi, 2013).

In vertebrates, brain size increase is associated with both 'passive' growth linked to body size increase, resulting in minor cerebral rearrangements, and 'active' growth related to diet. A diet rich in energy can enable the development of a larger brain with enhanced processing capacity (Aboitiz, 1994).

The skull is divided into the neurocranium, which houses and protects the brain and sensory organs like the eyes, ears, and olfactory system, and the splanchnocranium, which supports the digestive and respiratory systems. The evolution of the skull serves as a prime example of how form changes with function. In early agnathans, ancient jawless vertebrates, the splanchnocranium initially played a role in branchial respiration to maintain a constant water flow and later became an integral part of the feeding processes (Walker & Liem, 1994) (see Fig. 4.24).

When comparing brain sizes across different mammal species, it's usually observed that brain mass tends to adapt to body size, growing more slowly than other body parts. Larger species tend to have relatively smaller brains. However, it's undeniable that an increase in brain mass, despite being energetically costly, results in significant differential reproductive success among individuals (Harvey & Pagel, 1992). This trend changes in anthropoid primates like chimpanzees, who, compared to mammals of similar size, possess brain volumes roughly twice as large. In modern humans, this dimension increases sixfold (Hogervorst et al., 2009).

The acquisition of a larger brain by our ancestors had far-reaching consequences. When we observe modern humans, we can see that, unlike all other mammals, their skulls are unusually large. This is associated with significant characteristics such as upright posture coupled with the ability to walk on two legs. The combination of a large brain (and the large skull that houses it) with bipedal locomotion makes our species unique in the animal kingdom. At the same time, it has posed numerous challenges in terms of bodily organization during the evolutionary adaptations required (Marchi, 2016).

The man has many more muscles in the head and neck than any other primate (Powell et al., 2018), but in this order of mammals he presents relatively less powerful musculature, which however supports the heavier skull. A greater number of muscles, less powerful musculature, and a heavier head seem to be three factors apparently in contrast with each other but actually highlight how, in *Homo* in general and in *Homo sapiens* in particular, the development of the muscles of the head

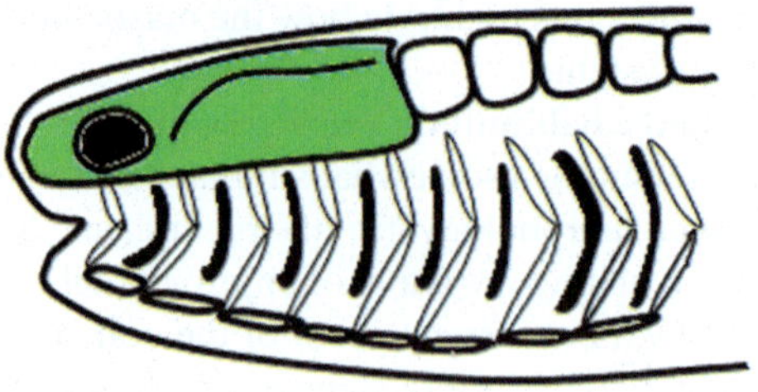

Fig. 4.24 Head of agnatha, an ancient vertebrate lacking jaws. In agnatha, the cranial box (green) interacted with the branchial system; the conventional hypothesis is that the maxillaries evolved from the first branchial arch

and neck has evolved towards versatile mobility rather than rigid stability. The increase in the number of muscles increases the complexity of the vectorial system, creates greater dynamic redundancy, which has allowed, despite a loss of system stability, the increase in adaptability of this vital functional district.

In this refined balancing system of a spherical and heavy skull, supported and balanced on a verticalized spinal column and supported by a highly mobile vertebral musculature, the muscles that move the jaw, directly or indirectly connected to the skull, actively participate in its statics and dynamics. Our cranio-cervical district constitutes a system that distinguishes itself from that of other primates, characterized by a less mobile spinal column, posturally inclined forward, with a single primary kyphotic curve and more powerful musculature (Williams & Russo, 2015). In humans, this vertical spinal column, with multiple curves, highly mobile and adaptable, balances through the pelvis with the lower limbs, creating an unstable bipedal support: all this has made necessary the adaptation of numerous anatomical components, both cranial and post-cranial.

However, while there are numerous variations in post-cranial anatomical features, more than one in three modifications occur at the maxillofacial level: these include changes in mandible shape, the development of the internal symphyseal area around the genial tubercles, sagittal inclination of the symphysis, presence/absence of the chin (particularly developed in *H. sapiens*), orientation of pre- and post-canine roots, incisor sizes, diastema between the lower canine and the first lower premolar, lateral wear facets on the lower canine and the first lower premolar, and size of the first lower premolar (Senter, 2010). There are also changes in the less elongated and narrower parabolic dental arches and thicker enamel.

The palate widens, dental arches become more parabolic, canine crown volume reduces, and canine diastemas are lost. Zygomatic arches are resized in relation to a reduction in masseter and temporal muscles (see Fig. 4.25).

The progressive encephalization of the *Homo* genus, coupled with a shift in the relationship between the cranial vault and a less protruding facial skeleton (Manzi, 2013), led to further adaptation of the great occipital foramen at the base of the neurocranium. This adaptation had already occurred with the evolution of bipedalism in an upright posture. In the *Homo* genus, the occipital foramen moves forward while the protrusion of the facial skeleton decreases (see Fig. 4.26).

This different cranial configuration required a profound resizing of the cranio-cervical bone and muscle structure, finely modulated by new balances between the posterior and anterior neck musculature. With the forward shift of the foramen magnum, occipital condyles came closer to the center of gravity of the head, reducing the workload on the posterior muscles needed to keep the head upright. However, it also necessitated a synergistic collaboration with muscle bundles in the anterior region of the skull and neck (see Fig. 4.27).

The process of brain expansion over the last 2 million years has been remarkable in both its trajectory and speed. Our cranial capacity has tripled compared to our bipedal ancestors and other anthropoid apes (the split between the human evolutionary line and that of the chimpanzee occurred between 5 and 7 million years ago), increasing from around half a liter to the current 1.5 L (Manzi, 2017). Brain growth

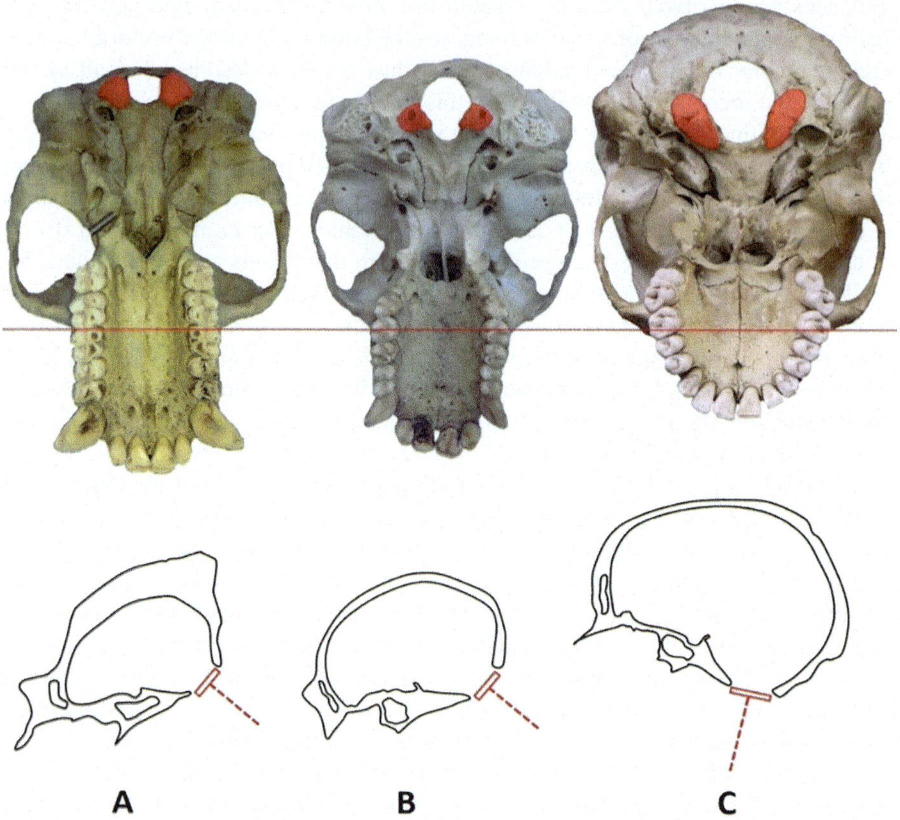

Fig. 4.25 (**a**) Gorilla (*Gorilla gorilla*) and (**b**) Chimpanzee (*Pan troglodytes*), Zoology Museum of the University of Florence; (**c**) *Homo sapiens*, Anthropology Museum of the University of Florence. In the upper part of the photo, the solid red line marks the boundary between the anterior origin of the zygomatic arch and the anterior portion of the maxillaries, highlighting the different protrusion of the facial mass in various species: note the increase in the surface area of the occipital condyles to support greater cranial mobility, freed from powerful muscle-articular constraints; in the lower part of the figure, the cranial profiles of the aforementioned species: note the forward and downward shift of the foramen magnum and the verticalization of the vertebral column

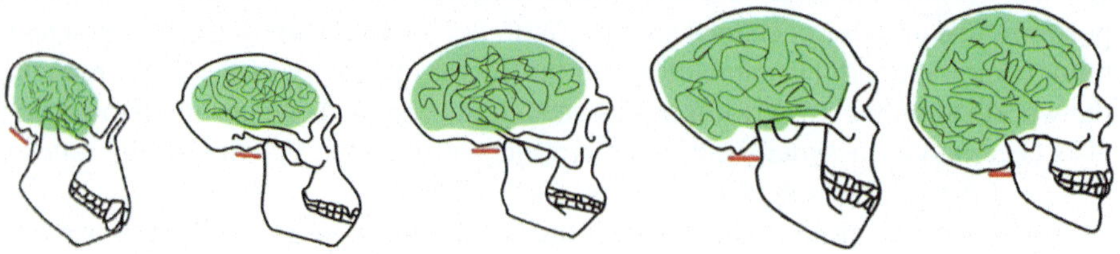

Fig. 4.26 Progressive encephalization and modification of spatial relationships between neurocranium, facial skeleton, and foramen magnum: the latter moves increasingly closer to the cranial balance center. From left to right: chimpanzee; *Australopithecus; Homo erectus; Homo neanderthalensis; Homo sapiens*

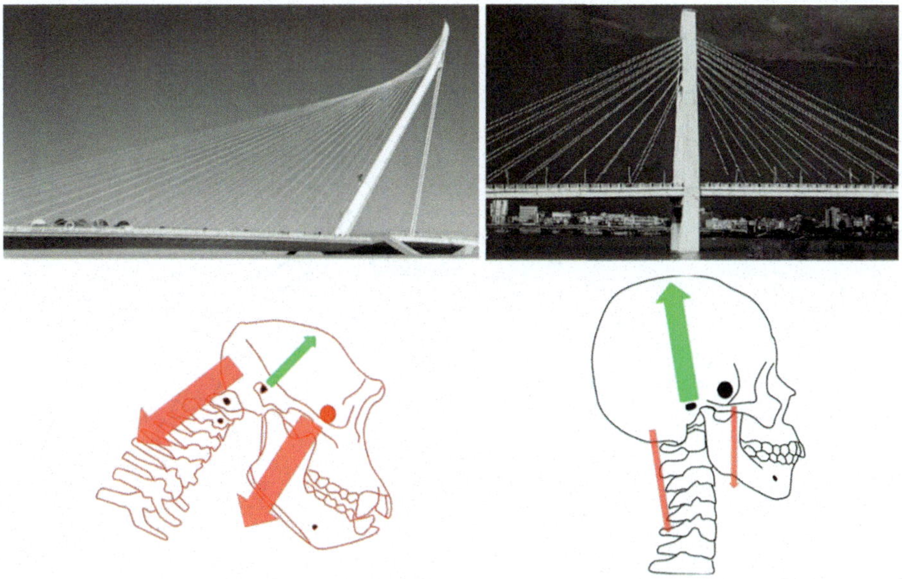

Fig. 4.27 Modification, in *Homo sapiens* compared to a quadrupedal chimpanzee, of the vectors of cervical musculature, masticatory musculature, and spine support vector

was also supported by dietary changes driven by environmental pressures. Around 3 million years ago, in the Rift Valley region, further aridification of the climate expanded the savanna, creating an abundance of plant resources. This led to a rapid increase in herbivore populations and, consequently, their predators. Primates had new opportunities to feed on the remains of carnivore kills.

This environmental backdrop presented the *Australopithecus* genus with an evolutionary crossroads, resulting in the adoption of two distinct strategies. On one hand, it led to the genus *Paranthropus*, which developed substantial cranial structural modifications: sagittal crests to increase the attachment area for temporal muscles, larger and more powerful masticatory muscles and jawbones, and a significant increase in the volume of posterior teeth. The *Paranthropus* lineage specialized in a particular diet, especially during seasons when typical frugivore food resources were scarce. They were forced to subsist on seeds, roots, and tubers—foods that required extensive and vigorous mastication. The other evolutionary branch led to the early representatives of the *Homo* genus, known as the *early Homo,* whose skeleton retained many of the archaic characteristics of the australopithecines and *Paranthropus*. *Homo habilis*, for instance, can be considered an australopithecine with a more voluminous brain (see Fig. 4.28).

In the dietary adaptation of australopithecines that followed the evolutionary path of early *Homo*, there was a progressive increase in meat consumption. This was made possible by scavenging carcasses left on the ground by large carnivorous animals. Feeding on carrion provided access to highly energy-rich foods like fat, muscles, brain, liver, and bone marrow.

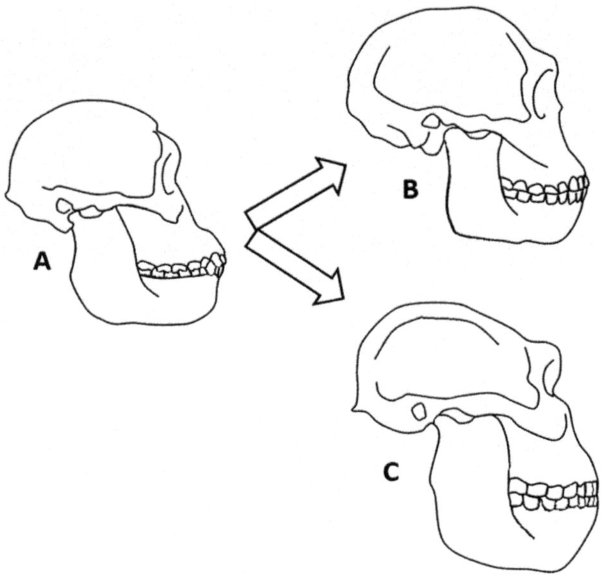

Fig. 4.28 (**a**) *Australopithecus africanus;* (**b**) *Homo habilis*; (**c**) *Paranthropus boisei*

Alongside the dietary shift, a genetic mutation occurred that allowed for brain expansion. The MYH16 gene, responsible for the formation of heavy chains of myosin in the masticatory muscles, was silenced. These muscles were so hypertrophic and powerful that they compressed the cranial vault during growth and development (Di Vincenzo & Manzi, 2013). The process of encephalization, which began around 2.4 million years ago, preceded the appearance of modern human cranial dimensions (Stedman et al., 2004). The transition from *Australopithecus* to *Homo* was initially marked by cranial changes such as larger brain volumes (approximately 650 cm^3), more slender mandibles and facial bones, a dental structure characterized by a reduction in the volume of premolars and molars, and dental arches that transitioned from a V-shape to a parabolic shape. From this evolutionary point onward, neurocranial growth continued until it reached extreme endocranial volumes in *Homo neanderthalensis* (1410 cm^3) and *Homo sapiens* (1350 cm^3) (see Fig. 4.29).

One of the key features that sets us apart from other primates, as well as from our direct ancestors and many of our closest extinct relatives (excluding Neanderthals), is a highly developed brain, three times larger than that of a chimpanzee or an australopithecine. In *Homo sapiens,* this mutation completely revolutionized the architecture of the brain, resulting in a substantially spherical morphology (Bruner et al., 2003). Our cranial structure is uniquely characterized by two structural autapomorphies (an autapomorphic character is a new and exclusive trait not shared with the closest species or common ancestral progenitors): a retraction of the facial skeleton and a globular neurocranium.

Around 200,000 years ago, our species, *Homo sapiens*, suddenly appeared in the Ethiopian region. These creatures had heads not only as large as Neanderthals but also rounder. Thanks to a genetic change that affected the developmental

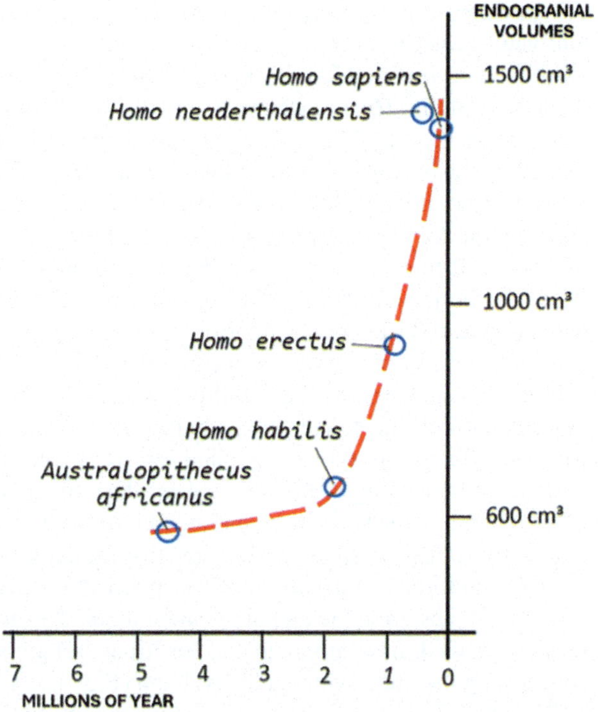

Fig. 4.29 Development of encephalic volumes in hominids

mechanisms maintaining a elongated cranial shape, *Homo sapiens* saw the structural constraints on the skull loosen. A cranial reconfiguration model guided by brain growth, oriented not along a sagittal axis but with a surprising upward expansion, took hold.

This marked cranial development posed a challenge during childbirth, due to the passage of the newborn's head through a birth canal adapted to bipedalism and an upright posture. Unlike all other mammals, in humans, the skull is already large at birth. The challenge posed by the growing mismatch between the voluminous cranial structure of the newborn and the altered birth canal was "resolved" through a kind of early birth from a cerebral perspective. It's as if the newborn, still cerebrally immature, continues its normal growth after birth. For about a year, the rate of cranial growth in a human newborn maintains the same progression seen during fetal life. Humans and other primates have the same ratio between cranial and body dimensions at birth. However, while in great apes and most mammals, the brain's growth rate experiences a decline after birth, in humans, it continues at the same embryonic pace until one year of age. This significant brain growth means that adult humans have a brain mass at least three times larger than they did at birth. It's only at the completion of the first year of life that human children reach a developmental state comparable to that of any other mammal at birth.

Before *Homo sapiens*, brain development followed a so-called archaic orientation, characterized by marked development in the antero-posterior direction. With the appearance of the early *H. sapiens* (Galway-Witham & Stringer, 2018), the cranial vault began to change, especially in the parietal region, giving the neurocranium a more rounded shape. This globular shape distinguishes the anatomically modern skull in *Homo sapiens* from that of various archaic groups of *Homo* (Lieberman et al., 2002). It's a unique shape that emerged through a transition guided by the expansion of the parietal area of the skull (Bruner et al., 2003).

In archaic humans, the human skull was massive with thick walls. The frontal bone featured a robust supraorbital torus (a bony ridge above the eye sockets), and the forehead retreated backward. The maxillofacial area was considerably large, wide, and projected forward, located anteriorly to the cranial vault. Over time, the facial skeleton shifted below the frontal region, meaning the cranial vault expanded anteriorly to accommodate the increased cranial volume, altering its relationship with the facial skeleton. The mandible, while still robust, retained a retrusive chin. The dental arch, no longer elongated as in australopithecines, became wide and parabolic. Teeth reduced in size, and incisors took on a characteristic shovel-like shape with thickening of the inner edges. The molars reversed the gradual increase in size from the first to the third molar, which was typical of australopithecines, with an exactly opposite sequence where the first molar became larger than the other two.

In *Homo sapiens*, the pattern of encephalization changed, with growth taking on a verticality absent in the archaic *Homo* model. This new shape of the cranial vault further changed the anatomical and functional relationships with both the vertebral column and the underlying facial skeleton (Fig. 4.30).

The cranial vault in archaic humans, when viewed from above, was characterized by maximum width located very low compared to the vault, a marked post-orbital constriction, a low and elongated vault (platycephaly), and a volume developed along an antero-posterior axis (Manzi, 2013) (Fig. 4.31).

The growth of cerebral volumes encapsulated within the membranous meningeal system generates constant pressure, which tends to separate the ossification centers of the vault (primarily those of the parietal bones, closed anteriorly and posteriorly by the frontal and occipital squamae), promoting the deposition of new bone along the suture margins. Simultaneously, the cells on the inner surface of the bone continuously remodel the shape and curvature of the skull. The final result of all these actions is that the endocranial cavity is shaped by the morphology and volume of the brain masses it contains.

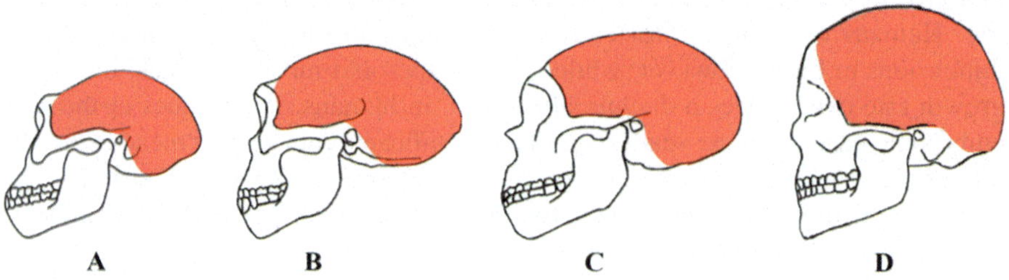

Fig. 4.30 (**a**) *Homo erectus.* (**b**) *Homo heidelbergensis.* (**c**) *Homo neanderthalensis.* (**d**) *Homo sapiens*

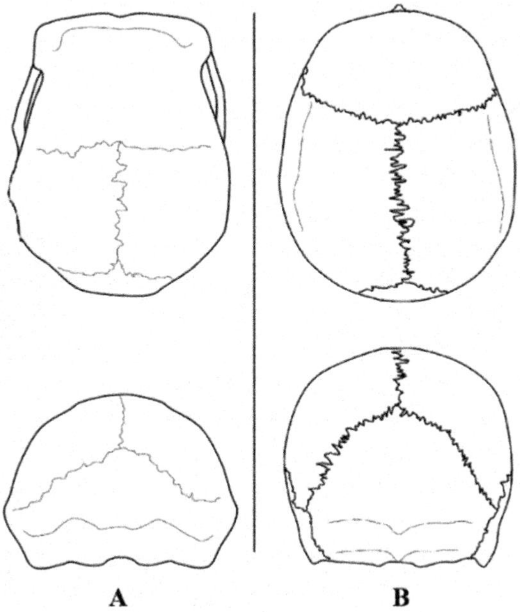

Fig. 4.31 Characteristic modifications of encephalization in H. sapiens: the neurocranium assumes a globular shape, further changing its relationships with the facial skeleton. (**a**) *Early Homo* cranial shape. (**b**) *Homo sapiens* cranial shape

Unlike the anthropoid apes, in which the face is large and the neural cranial capacity is relatively small, in *H. sapiens*, it's the opposite: our face is relatively small and located in the front of the cranial vault, while the neural cranial capacity has expanded in all directions, assuming a globular appearance in which the parietal and occipital bones together form a sagittal arch (Manzi, 2007) (Fig. 4.32).

In addition to the brain, the neurocranium houses the insertions of the masticatory muscles, which physically and functionally connect it to the mandible. In *Homo sapiens*, diet influences the variability among different parts of the masticatory apparatus and the overall cranial shape (Veneziano et al., 2018). The face and teeth underwent changes in shape, position, and functional relationships not only due to dietary shifts but also driven by postural adaptations induced by obligatory bipedalism in an upright posture and encephalization (Fig. 4.33).

The following figures highlight how not only the profiles of the neurocranium but also those of the facial complex have progressively changed from *Australopithecus afarensis* to *Homo erectus*—which can be considered the first representative of the human genus—to the most recent non-extinct human species, *Homo sapiens*. Typically, in this type of educational representation, the emphasis is placed on increasing encephalization, but equally significant is the progressive reduction of the facial complex and the stomatognathic apparatus. This retraction, stemming from significant changes in diet, also originates from the progressive encephalic expansion and adaptation to the challenging balance of the upright posture (Figs. 4.34, 4.35, 4.36).

In humans, the new encephalic volumes contributed to further modifications in the cranial-cervical relationship (Jenkins, 1969), which gained significant flexibility

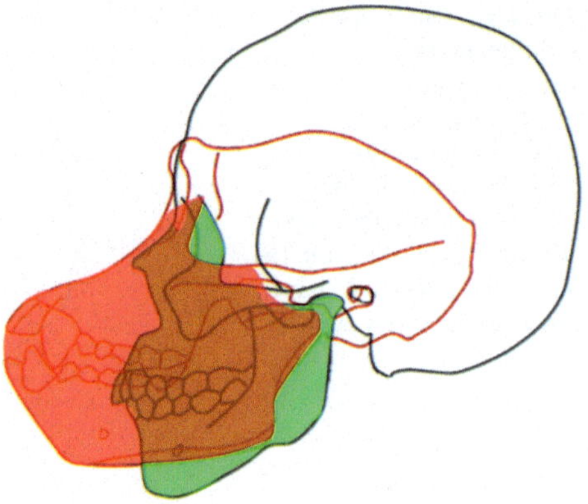

Fig. 4.32 The close correlation between neurocranium and splanchnocranium has meant that the volume and shape of the former have led to profound adaptations in the masticatory apparatus. In red: profile of chimpanzee neurocranium and facial skeleton; for *Homo sapiens*: in black, neurocranium profile, in green, facial skeleton

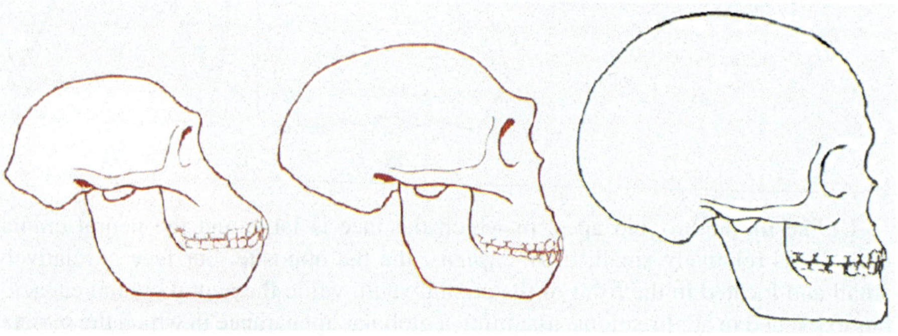

Fig. 4.33 Process of encephalization. From left: *Australopithecus afarensis*; *Homo erectus; Homo sapiens* (Drawing by Marco Boggioni)

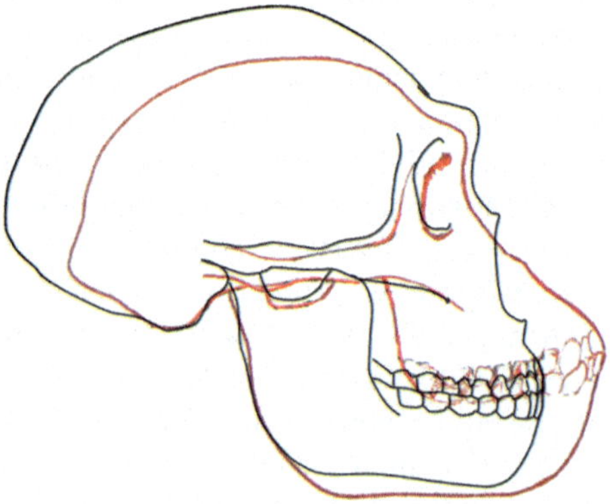

Fig. 4.34 Overlay of cranial profiles: *Australopithecus afarensis* (red) and *Homo erectus* (black). Note the increase in cranial vault size while maintaining an elongated antero-posterior shape and reduction of the facial skeleton (Drawing by Marco Boggioni)

compared to apes. As this spheroid species, weighing 5/7 kg, expanded, it found support on the occipital condyles. They feature small, curved, and inclined articular surfaces (Fig. 4.37), which interact with the concave surfaces of the atlas without

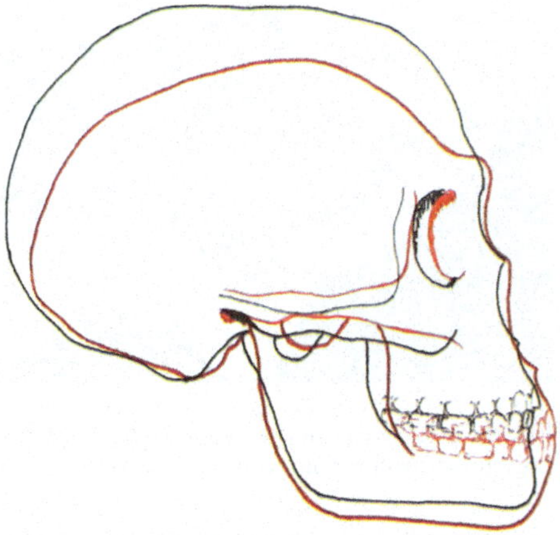

Fig. 4.35 Overlay of cranial profiles: *Homo erectus* (red) and *Homo sapiens* (black). Note the increase in cranial vault size, which loses the elongated antero-posterior aspect of the archaic skull for a globular shape, with further recession of the facial skeleton (Drawing by Marco Boggioni)

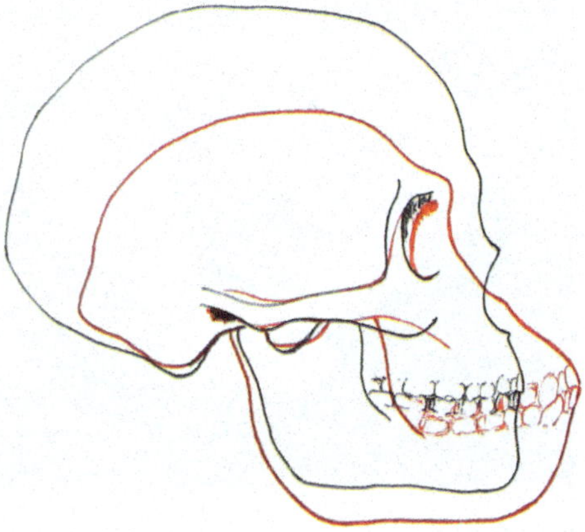

Fig. 4.36 Overlay of cranial profiles: *Australopithecus afarensis* (red) and *Homo sapiens* (black). Note the extreme encephalization and the pronounced recession of the facial skeleton (Drawing by Marco Boggioni)

intervertebral discs (similar to the joints between C1 and C2). The lack of intervertebral discs makes these surfaces nearly mirror-like, allowing for extremely precise and refined movements.

The occipital condyles support our cranial box, acting as a cradle, beneath which, but slightly more forward, is positioned our relatively small facial skeleton (Manzi, 2006).

The rotation of the cranium on the spine mainly occurs between C1 and C2, with the odontoid acting as the pivot point of the movement. It's between the atlas and the epistropheus that the rotary movement takes place, where the atlas, virtually almost solidary with the cranium, rotates around the epistropheus' tooth like a wheel rotates around its axis (Fig. 4.38).

The epistropheus' tooth doesn't only serve as a pivot point around which much of the head rotation occurs, but it's also crucial for accommodating flexion-extension movements. This is thanks to its articulation with the anterior arch of the atlas (Figs. 4.39 and 4.40) and its slight but significant posterior inclination.

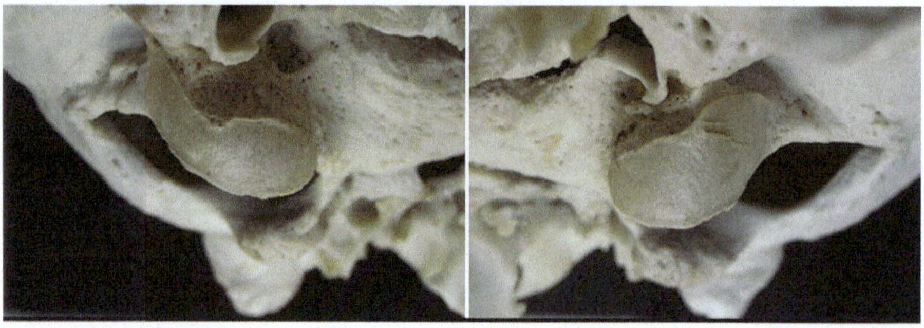

Fig. 4.37 Convex and inclined articular surfaces of the occipital condyles. Specimen from the Museum of Anthropology at the University of Florence

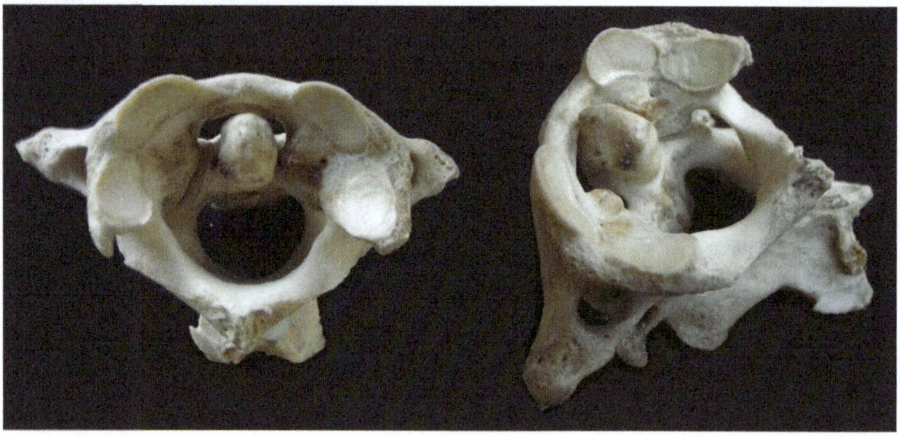

Fig. 4.38 Articular surfaces of C1 supporting the skull to accommodate primarily flexion-extension movements. Cranial rotation occurs mainly between C1 and C2. Specimen from the Museum of Anthropology at the University of Florence

When the occiput rotates on the atlas, it participates in a coordinated movement that also involves the rotation of the atlas on the epistropheus (Pang & Li, 2004), keeping in mind that vertebral rotation on the horizontal plane is always associated with movements on other planes. The schematic division of movement on individual planes is merely an educational tool since no movement is ever purely isolated. The articular surfaces are neither flat nor horizontal, and the path followed by the lateral masses of the atlas is associated with a vertical displacement of 2–3 mm, resulting in an inclination on the frontal plane (Fig. 4.41).

As previously noted, compared to anthropoid apes, in *Homo sapiens*, the head exhibits high mobility associated with a great flexibility of the neck. However, at the same time, it experiences reduced static stability due to a relative weakening of the neck muscles. The center of mass of the cranium, though close, does not lie on the gravitational line of the fulcrum represented by the occipital supports but is located

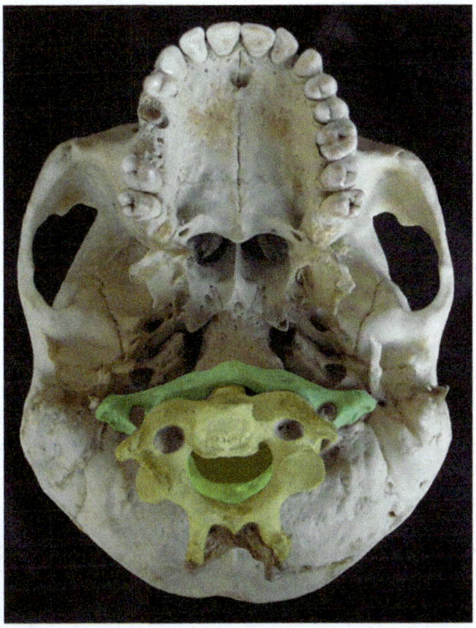

Fig. 4.39 Relationships between the skull, mandible, atlas (green), and axis (yellow). Specimen from the Museum of Anthropology at the University of Florence

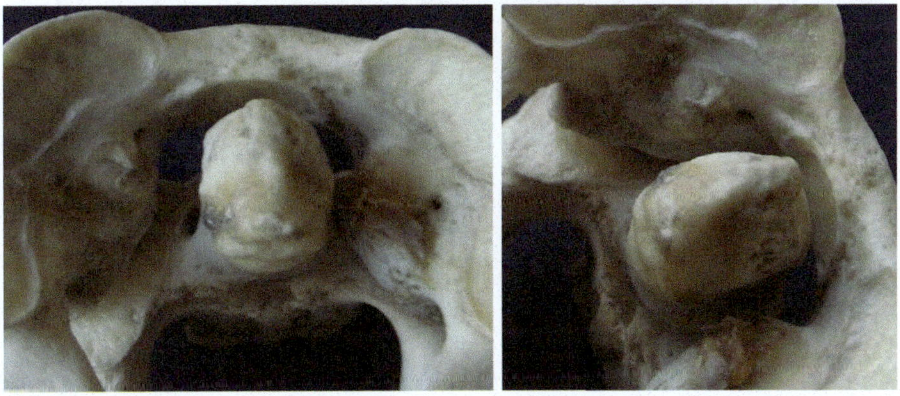

Fig. 4.40 The anterior surface of the odontoid process articulates with the anterior surface of the atlas, supporting flexion-extension movements of the head. Specimen from the Museum of Anthropology at the University of Florence

anterior to them at the level of the sella turcica. This is precisely why the head is naturally tilted forward: as anyone who has experienced a momentary lapse of sleep can attest, without the active support of the nuchal musculature, the head naturally flexes forward (Fig. 4.42).

The morphology of the splanchnocranium and the post-cranial structure reorganized themselves to accommodate the development of a very large brain as if it were now devoid of constraints dictated by bone growth. With the erect posture and anterior translation of the occipital foramen, there were multiple adaptations in the

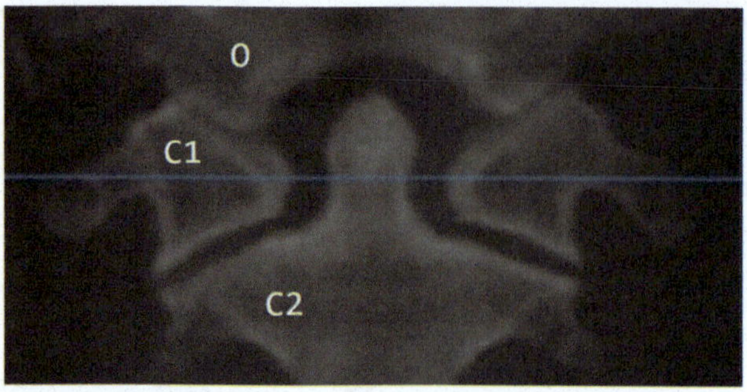

Fig. 4.41 CT scan in the coronal (frontal) plane of the occipital condyles (O), C1, and C2. Blue line: horizontal plane. The articular joints between the occiput and C1, and between C1 and C2, are inclined and not completely flat, so simple rotational movement in the horizontal plane triggers changes in the verticality and lateral inclination of the vertebrae

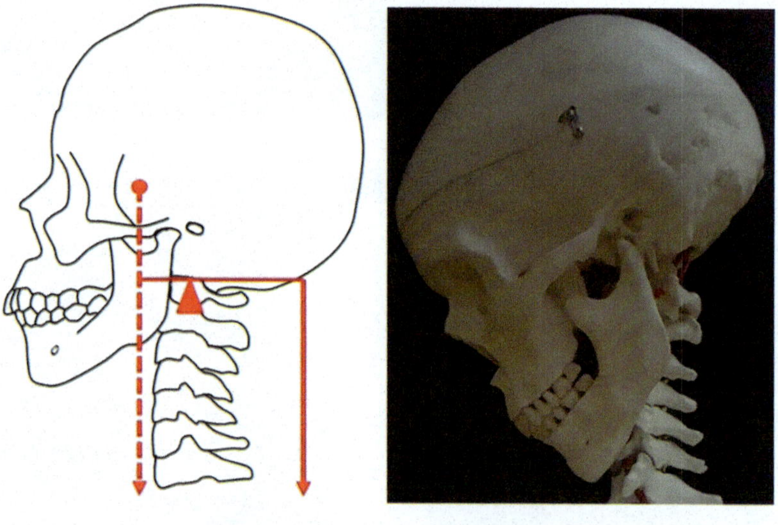

Fig. 4.42 Despite the foramen magnum in H. sapiens being below the cranial vault, support on the occipital condyles is always posterior to the cranial center of gravity

sub-occipital area: the extent and verticality of the nuchal plane were reduced, and compared to earlier hominins, the nuchal musculature dramatically reduced its insertion surface (Di Vincenzo & Manzi, 2013) (Fig. 4.43).

In the *Homo* genus, compared to australopithecines and even more so to chimpanzees, there is a combination of a reduced facial length with increased cranial projection posterior to the foramen magnum (Bramble & Lieberman, 2004). In *H. sapiens*, the 'globular' development of the posterior neurocranium and the vertical spinal column positioned beneath the cranium coincided with a reduction in the posterior muscular component, allowing for a more refined regulation of balance, aided by synergistic collaboration of the anterior musculature (Fig. 4.44).

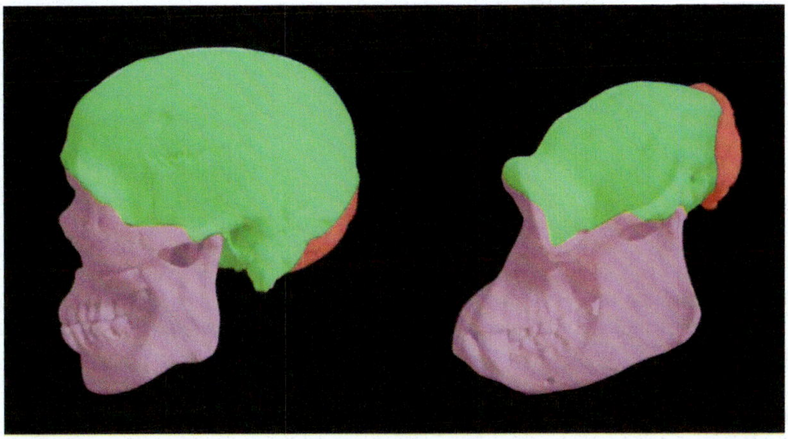

Fig. 4.43 Left: *Homo sapiens*; right: chimpanzee (*Pan troglodytes*). Reduction of the facial skeleton (violet), globular expansion of the neurocranium (green), reduction of the nuchal plane, insertion area of sub-occipital muscles (red)

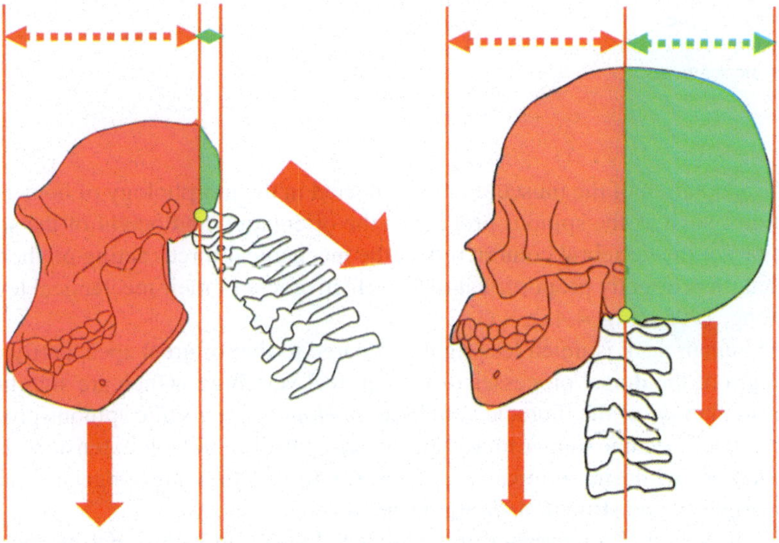

Fig. 4.44 Typical encephalization in H. sapiens, along with a reduction of the facial skeleton (which is located below the cranial vault) and the anterior-inferior shift of the foramen magnum, has radically changed muscular balances, completely altering the relationships between anterior and posterior musculature in the cranio-cervical region. The encephalic and facial masses in *Homo sapiens* (on the right) redistribute, balancing more effectively on the fulcrum of the occipital condyles (yellow circle); the powerful cervical musculature in anthropoid apes reduces in *H. sapiens* (red arrows); in red, the anterior cranial mass anterior to the support fulcrum, in green, the posterior cranial mass posterior to the support fulcrum: note the pronounced imbalance in the ape-like skull

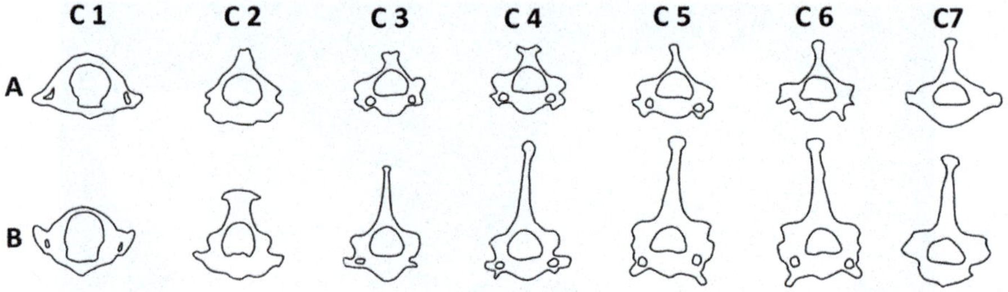

Fig. 4.45 Comparison between human cervical vertebrae (**a**) and those of the chimpanzee (**b**). Note the different lengths of the spinous processes, insertion surfaces for sub-occipital and neck musculature

Fig. 4.46 Left: Cervical spine in *Homo sapiens*; right: Cervical spine of chimpanzee (*Pan troglodytes*). The extension and orientation of cervical spinous processes reflect the strength of the musculature exerted on these bone segments

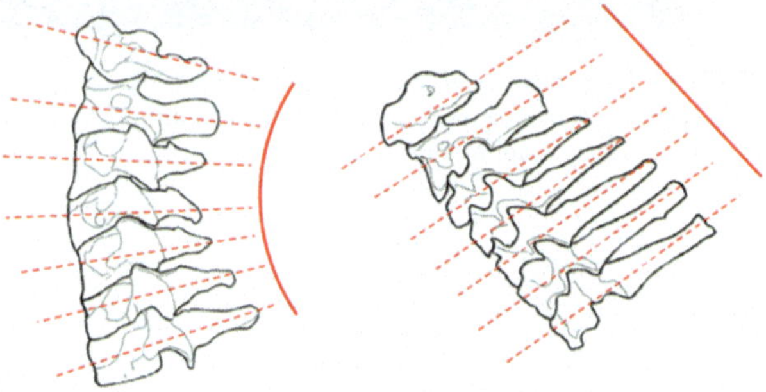

The adaptation of the musculature is reflected in the morphology of bony muscle insertions, such as the spinous processes: the orientation and height of the spinous processes of the vertebral column reflect the mechanical forces acting on them, and their length is directly proportional to the effectiveness of their mechanical leverage (Kardong, 1995) (Fig. 4.45).

The human neck is much less rigid compared to that of great apes, partly due to the length of the dorsal processes of the cervical vertebrae, which are significantly longer than those of the thoracic vertebrae: in *Homo sapiens*, the spinous processes of C1 and C2 became longer to achieve better control in flexion-extension and cranial rotation, while in chimpanzees, from C3 to C7, they are longer to maintain greater rigidity and structural strength (Fig. 4.46).

The shape and characteristics of the nuchal plane provide information about posture and the relationship between the head and the vertebral column, reflecting the size and orientation of the neck muscles (Wolpoff, 1971). The nuchal plane is delimited posteriorly by the *Inion* (the most posterior point of the occipital bone protuberance and the point of insertion of the nuchal ligament and the trapezius muscle) and anteriorly by the *Opisthion* (the posterior margin of the occipital foramen); the nuchal plane is more extensive and prominent in chimpanzees and gorillas, where significant bony ridges, known as crests, provide attachment points for powerful neck muscles (Fig. 4.47).

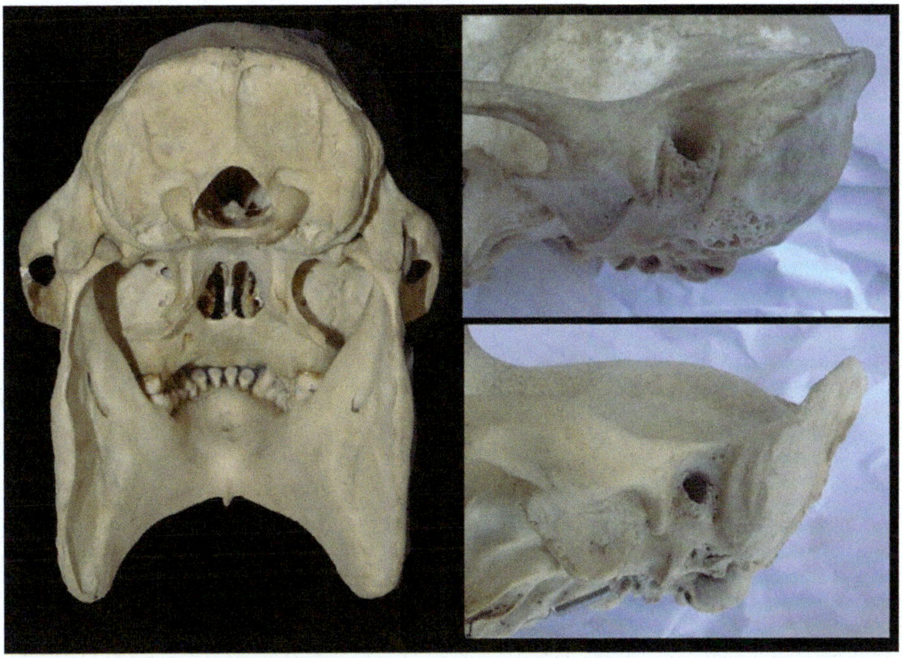

Fig. 4.47 Nuchal plane. Left: Howler monkey (*Alouatta seniculus*), specimen 4668: note the impressions left on the nuchal plane by powerful cervical musculature. Top right: Chimpanzee (*Pan troglodytes*); bottom right: Gorilla (*Gorilla gorilla*): the bony surface and its ridges indicate the strength of nuchal muscles. Specimens from the Museum of Zoology at the University of Florence

In these primates the developed nuchal plane provides attachment for powerful neck muscles that hold the skull in extension, putting it in traction posteriorly. This robust musculature, strongly anchored on an extensive nuchal plane, finds an equally effective biomechanical support in the long lever arms provided by the significant spinous processes of the cervical vertebrae, which counterbalance the weight of the forward-protruding frontal mass (Manzi, 2013) (Fig. 4.48).

In these primates In the human skull, especially the globular one of *H. sapiens*, the nuchal plane reduces in relation to the decreased muscular connections between the shoulders, neck, and head (Fig. 4.49).

Beyond the reduced bulk and therefore greater space within which to express movement, the reduction in the length of the spinous processes changes the muscle biomechanics by varying the distance of muscle attachment from the joint fulcrum. The farther a muscle attachment moves from the joint fulcrum, the less extensive and explosive (immediate and massive contraction) the movement that contraction can impart to the bone segment, and the more intense the force applied. Conversely, the closer the muscle attachment is to the fulcrum, the greater and more explosive (with the same contraction) the movement that muscle contraction can impart to the bone segment, but the weaker its force will be. The reduction of cranial levers in the

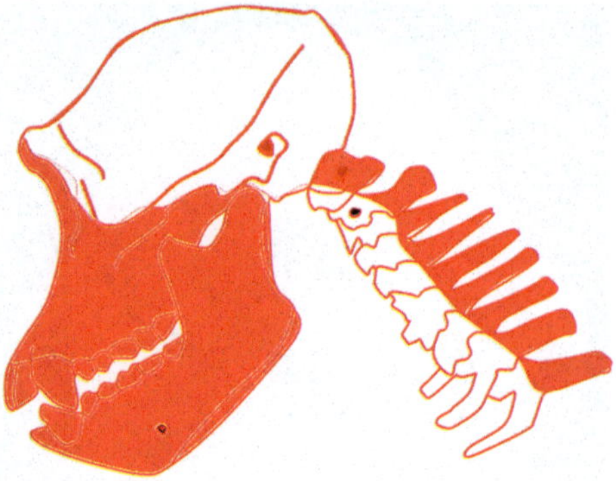

Fig. 4.48 The size of cervical spinous processes is directly proportional to the effort required to counterbalance the weight of the head

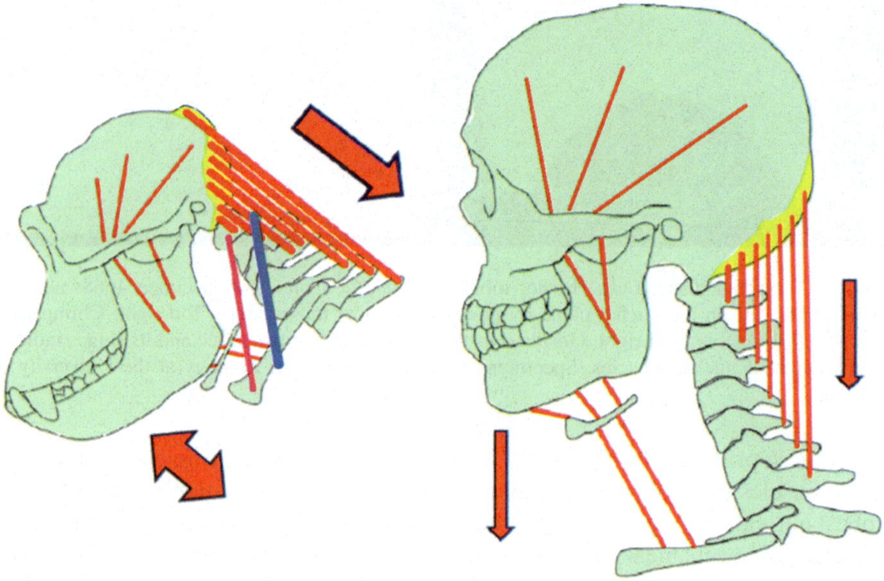

Fig. 4.49 Consequently to the reduction of cervical musculature in the genus *Homo*, the length of spinous processes decreased, establishing a different balance between the anterior and posterior muscle chains. Red lines: direction of muscle bundles; blue line: the atlanto-clavicular muscle present in chimpanzees but absent in humans; yellow area: nuchal insertion area of sub-occipital musculature

posterior processes decreases the power and control of the neck muscles but enhances their mobility (Fig. 4.50).

The masticatory muscles, along with the supra- and infra-hyoid muscles, took on the role of antagonists to the neck muscles to maintain the balance of the skull on the sagittal plane (Fig. 4.51).

On this plane, the position of the skull is influenced by the reciprocal relationships of the mouth-opening and closing muscles, together with the group of

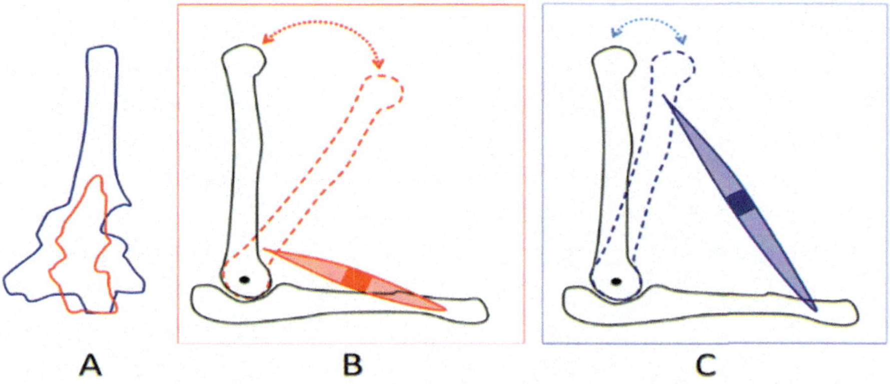

Fig. 4.50 (**a**) Lateral perspective of overlaid silhouettes of C5; red *Homo sapiens*, light blue chimpanzee (*Pan troglodytes*): note how in the same vertebral body, the chimpanzee has proportionally much longer spinous processes. (**b**) Muscle insertion near the joint fulcrum; under equal muscle contraction (colored band), there is greater range of motion and less force compared to **c**

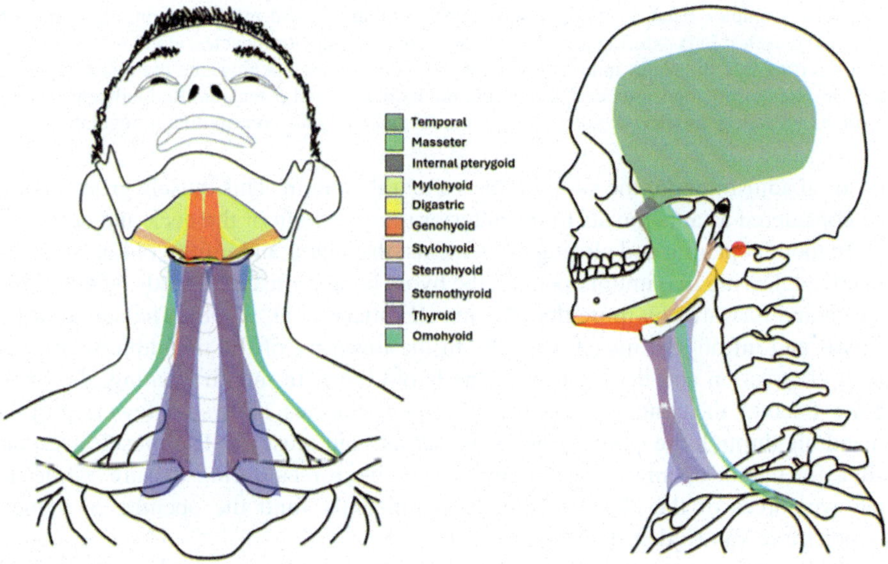

Fig. 4.51 Anterior muscle chain of the neck and head. Muscles involved in opening and closing the mouth actively participate in balancing the skull on the spinal column

infrahyoid muscles. They play a role in the flexion of the head, regulated by the sternocleidomastoid muscles, the long neck muscles, the small and large anterior rectus muscles, and the sub-occipital muscles (Fig. 4.52). These sub-occipital muscles, which are short, powerful, and sensitive, attach superiorly to the occiput and inferiorly to the first two cervical vertebrae. They are responsible for the fine

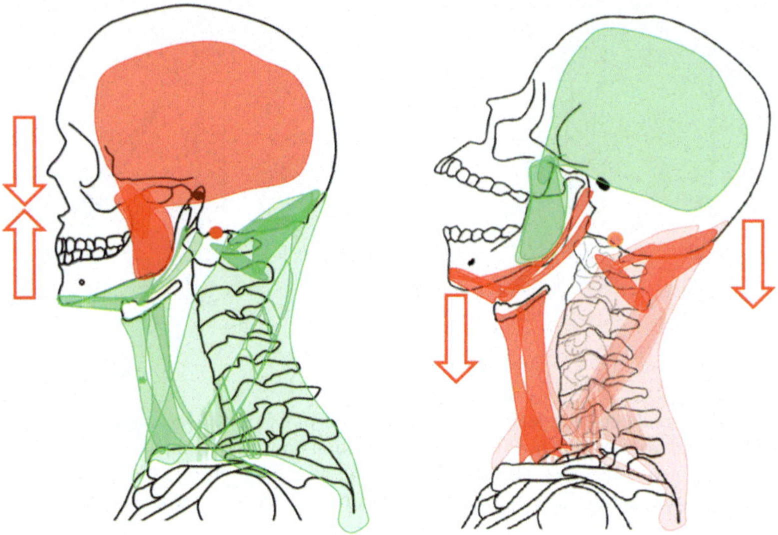

Fig. 4.52 Complex balance between masticatory, suprahyoid and infrahyoid muscles, and sub-occipital muscles: Left—during jaw closure, the main jaw adductor muscles, along with jaw elevation inserting into the skull, induce flexion; Right—during jaw opening, the main jaw depressor muscles are assisted by sub-occipital musculature inducing skull extension. Red indicates agonist muscles activating movement; green indicates antagonist muscles contributing to regulation

postural adjustment of the skull on the vertebral column. This muscle group should be considered as antagonists to the anterior muscle chain of the neck and head.

In mouth opening and closing movements, the supra- and infrahyoid muscles are involved. During maximum opening, the hyoid bone is raised upwards by more than 3 mm and forwards by more than 1.5 mm (Hannam et al., 2008). The action of the supra- and infrahyoid muscles results in the lowering of the mandible relative to their contraction and the fixation of the hyoid bone. In mouth opening, the hyoid bone, fixed by the contraction of the infrahyoid muscles, allows the lowering of the mandible through the contraction of the suprahyoid muscles. The muscles responsible for mouth opening exert negligible passive forces during closure, while the passive forces of the closing muscles significantly limit the opening movement (Koolstra & Van Eijden, 1997).

In humans, the complex interactions among the various components of neck muscles allow for dynamic redundancy, enabling a high degree of independence between trunk and head musculature. Even during activities like walking or running, independent counter-rotations of the chest and arms stabilize the skull (Bramble & Lieberman, 2004).

In humans, in contrast to chimpanzees and probably australopithecines, neck mobility compensates for trunk movements, facilitating head stabilization (Fig. 4.53).

In primates, bipedal walking is hindered by strong muscle connections that tend to prevent independent counter-rotations of the chest and arms, minimizing axial rotation of the head (Bramble & Lieberman, 2004).

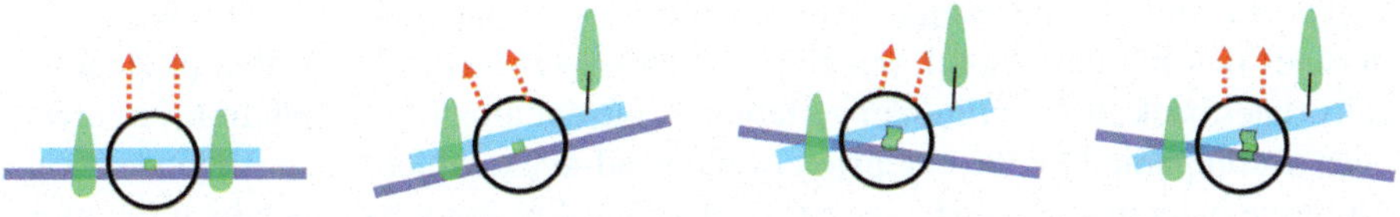

Fig. 4.53 Schematic representation of gait and adaptations of the pelvis (light blue line), shoulders (blue line), and head (large circle) during walking (dashed red arrows: direction of gaze): four images, from left to right. Starting position. Right step with homolateral rotation of the pelvis: if the thoracic section of the spine does not counterrotate the shoulders, they would remain parallel to the pelvis causing significant imbalance. Counter-rotation to balance the shoulders: if the cervical spine does not counterrotate the head, gaze stabilization would be ineffective; like balancing scales, the contralateral upper limb to the step moves forward and the ipsilateral limb moves backward. Counter-rotation of the head relative to shoulder rotation, to maintain stable gaze and vestibular system

Both the head and the trunk can assist the nervous system in stabilizing spatial and orientation references (Dunbar et al., 2004) necessary for locomotion. Stability and adaptation rely on visual, vestibular, and somatosensory input, which provide continuously updated information about body orientation in space and how to move through complex natural environments while maintaining balance. The visual and vestibular systems play a predominant role in locomotion: vision is crucial for adapting gait patterns to uneven terrain and avoiding obstacles, while the vestibular system responds to rotational and translational head accelerations, providing information about head orientation in space, contributing to posture, and guiding neck and trunk movements during locomotion.

During walking, quadrupeds stabilize the three cervical segments in the three planes of space, while humans perform wide head rotations around their transverse axis (pitch) and vertical axis (yaw). In humans, during locomotion, the head compensates for step oscillations with compensatory horizontal rotations of up to twenty degrees (≤20°), synchronized with each step. When we visually scan our surroundings, our head often rotates much more than 20°, but when we run or jump, larger head rotations (>20°) are avoided. During walking, it's usually unnecessary to keep one's gaze fixed on a specific target because wild animals visually inspect their environment to locate food, predators, or other dangers (Dunbar et al., 2008). The head or trunk must be rotationally stabilized to provide the brain with a frame of reference for the spatial orientation of the entire body, and it has been hypothesized that bipedalism in humans has developed forces that increase the tendency for individual vertebrae to rotate (Janssen et al., 2010). Measurements of axial spine rotation have been taken during walking: after the step's extension, the thoracic vertebra can rotate up to 5°, while the fifth lumbar vertebra can reach 6° in the opposite direction, with some movement in the frontal plane as well (Gregersen & Lucas, 1967; Lumsden II & Morris, 1968). Although with different dynamics in various segments of the spine, horizontal rotation is never pure and is always associated with a change in verticality, as intervertebral articulation surfaces are generally neither flat nor horizontal.

The functional connection between the sub-cranial posterior muscles and the muscles that regulate mouth opening and closing is a correlation that phylogenetically dates back to deep time: in vertebrates, the opening of the mandible has always been accompanied by the extension of the head-neck, which has been present since primitive paleoniscoids fish, ancestors of all current bony fish, in which the mouth was opened by the epiaxial muscles that lifted the head (Fig. 4.54).

This association between raising the skull and lowering the lower jaw due to the activation of epiaxial muscles has also been maintained in sharks, evolved teleost fish, and amphibians during vertebrate evolution. Even in the transition from aquatic to terrestrial environments, cranial extension remained in the rapid mouth-opening phase (Walker & Liem, 1994). The correlation between cervical muscles and the movements of mouth opening and closing, both to stabilize the neck and head joints against excessive stresses during predation and to make biting more effective and powerful, is ancient. For example, in dinosaurs, this feeding style involved a powerful flexion of the skull (movement of the upper teeth towards the mandible) to sink the upper teeth into the prey's flesh, piercing it with the 'cranial dentition'. The upper teeth of the maxilla were wielded like daggers by the neck muscles (Samman, 2006). The functional interaction between mouth movements and the muscles in the neck region has also been retained in mammals. In rats, functional connections between the masticatory system and the head motor system were demonstrated through trigeminal muscle afferents to neck muscles (Zeredo et al., 2003). In rabbit chewing, head movements and neck muscle activities showed synchronized rhythm with mandibular movement (Igarashi et al., 2000).

This functional relationship between the temporomandibular system and the cranio-cervical system is also maintained in humans. The movement of mouth opening (MdA) is not solely due to the mandible but is the result of the

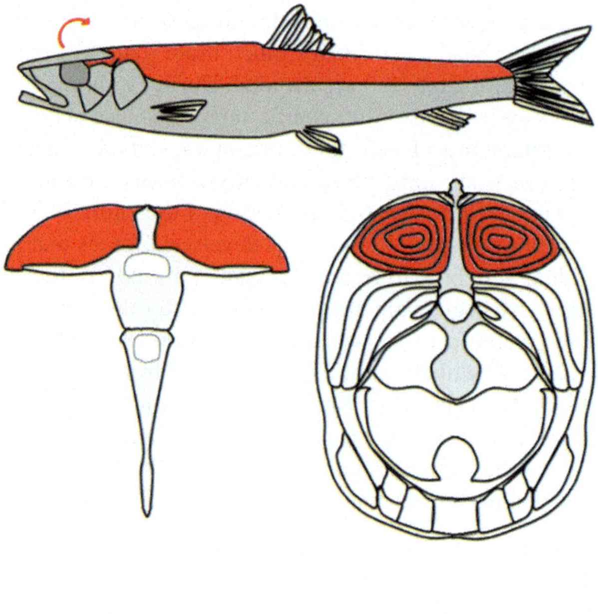

Fig. 4.54 Epaxial muscles in fish: their bilateral contraction extends the head contributing to mouth opening

simultaneous activation of these two regions, with simultaneous movements of both the TMJ and the atlanto-occipital joint and cervical joints (Eriksson et al., 2000; Kohno et al. 2001, 2001). The lowering of the mandible is accompanied by cranial extension involving the cervical spine, which follows this movement (Eriksson et al., 1998). When we think about chewing movements, we often overly simplify by imagining the skull as if it were fixed in space, with the mandible solely responsible for the entire range of motion. In reality, in humans, the skull moves on the fulcrum of the occipital condyles, not only during maximum mouth opening (Eriksson et al., 2000) (Fig. 4.55).

Furthermore, if we attempt to extend the head to its maximum while keeping the dental arches in contact until reaching the limit of the movement, this limit can only be surpassed by opening the mouth. When the mandibular adductors relax, the anterior muscle chain stretches, allowing the skull to achieve additional extension (Fig. 4.56).

The experimental induction of various cranio-cervical postures affects the values of maximum mouth opening, confirming the biomechanical relationship between the cranio-cervical region and the dynamics of the temporomandibular joint (Chen, 1998).

During mouth opening (MdA), the initiation of head acceleration lags behind that of the mandible by approximately 8 milliseconds, while the activity of the sternocleidomastoid muscles precedes that of the head by about 18 milliseconds. These results suggest that head movement is not initiated by peripheral reflexes but by pre-programmed central commands, and this may be relevant to the muscular discomfort experienced in the neck and shoulder regions of patients with stomatognathic disorders (Torisu et al., 2001). The simultaneous movements of the mandible and the head during fetal yawning suggest that these motor programs are innate

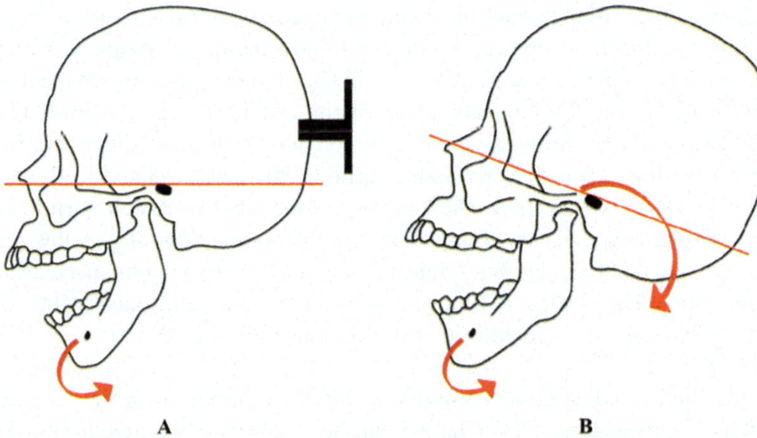

Fig. 4.55 (**a**) Hypothetical and non-physiological mouth opening with the skull fixed at an immobile point in space; the degree of opening is solely due to mandible lowering. (**b**) Real and physiological mouth opening, with head extension contributing to the degree of opening

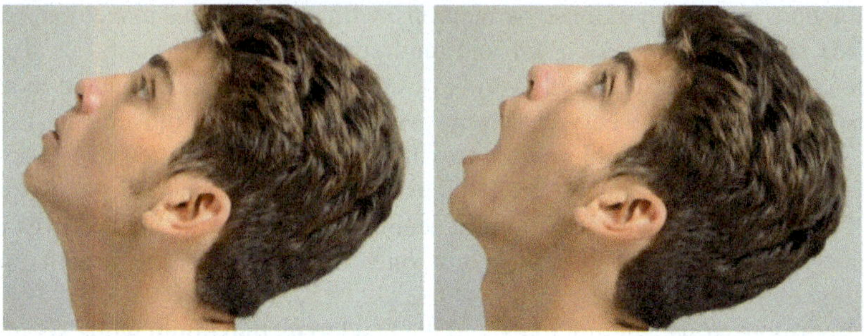

Fig. 4.56 Sagittal plane: Relationships between mandibular and cervical musculature. Left: Maximum head extension with mouth closed; Right: Maximum head extension with mouth open. Note the greater head extension once the muscles closing the mouth are relaxed

(Zafar, 2000), with a high degree of spatiotemporal coherence between mandibular movements and head-neck movements (Schindler et al., 2010; Zafar, 2000).

Even in the closing movement, neck muscles are involved. For instance, during sudden strong clenching efforts, there is bilateral co-activation of the sternocleidomastoid muscles, with intensity progressively proportional to the activation of the masseter muscles (Clark et al., 1993). In humans, during the simple movement of mouth opening and closing—a primitive dynamic present in pre-mammalian vertebrates—the skull follows the mandibular movements. This also occurs during unilateral chewing with simultaneous movements of the head and neck. In the horizontal plane, which pertains to mandibular movements and is a hallmark of mammalian mastication, the synergy between mandibular and cervical musculature can be demonstrated with a head rotation test. This test can be performed by asking a subject (free from significant muscle tension) to execute the maximum rotation movement. When the rotation reaches its limit, it can be exceeded by performing the maximum lateral displacement of the mandible on the same side as the head rotation (Fig. 4.57).

In humans, during chewing, small head movements accompany mandibular movements. The upper incisal point moves in the opposite direction to the mandible through a periodic and rhythmic head movement, synchronized in coordination with mandibular movement. Simultaneous extension and flexion movements of the head and neck have been recorded, with an amplitude primarily related to the size and consistency of the chewed food. The behavior of the head-neck unit during chewing is modulated in response to changes in the proprioception of the jaw muscles (Eriksson et al., 2000; Häggman-Henrikson et al., 2006; Häggman-Henrikson & Eriksson, 2004). The vertical component of head movement remains within 10% of mandibular movement, in rhythmic coordination with mandibular motion (Kohno et al. 2001).

Chewing induces head extension due to the co-contraction of the sternocleidomastoid and trapezius muscles. Changes in the position of the mandible can affect the resistance of neck muscles, and variations in the contraction of masticatory muscles influence the posture of the entire body. Changes in mandibular position are capable of modulating postural stability during an upright posture. To maintain

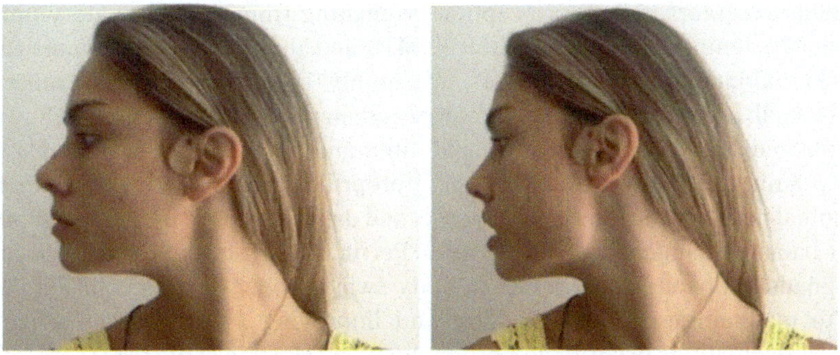

Fig. 4.57 Synergy between mandibular and cervical musculature in the horizontal plane: If maximum head rotation is combined with mandibular movement towards the side of rotation, there will be an increase in the head rotation limit

postural stability, the body's balance system requires information about the position and movement of the body in space. Motor activity involving the oral cavity exerts a strong influence on the overall motor function of the body (Miyahara, 1991).

Dental occlusion is the primary determinant of mandibular statics and dynamics. When dental occlusal surfaces come into contact with each other, they create the supports, guides, and constraints through which the mandibular musculature can express its range of motion in relation to the upper arches and, therefore, the skull. There is a sensory-motor relationship between the maxillary and cervical regions, and occlusal changes affect the position of the skull and its relationships with the spine, leading to a redistribution of stresses on the cervical spine. The number and position of dental contacts (the position of the mandible, as mentioned earlier) can also influence the resistance of neck muscles (Alghadir et al., 2015). A centered mandibular position relative to the upper maxilla results in greater symmetry in the contraction patterns of the sternocleidomastoid muscles (Zafar et al., 2019).

It is a common experience that during intense physical exertion, individuals clench their jaws to increase muscular effort. So much so that the saying "gritting one's teeth" has entered common language when faced with a challenging task. Voluntary clenching of the dental arches enhances the muscular strength of the entire body. Athletes often clench their teeth during strength and power exercises, generating facial and neck muscle tension to gain an ergogenic advantage (Ebben et al., 2008). Voluntary clenching contributes to improving bodily stability and is commonly observed in conjunction with voluntary movements that require significant efforts, such as weightlifting (Takada et al., 2000).

The proprioceptors in the neck can play a significant role in maintaining the posture of the head and body, both in static and dynamic situations. This control occurs through the integration of sensory information from various sources, including vestibular, visual, proprioceptive, and cutaneous receptors, combined with motor commands to provide the necessary feedback for balance maintenance.

When standing upright, the human body constantly undergoes small and continuous oscillations to maintain balance, thanks to continuous afferents from vestibular receptors and proprioception originating from the masticatory muscular system, the temporomandibular joint (TMJ), and the periodontal ligament receptors. Periodontal mechanoreceptors are highly sensitive (receptors innervating healthy teeth are sensitive to passive forces greater than 0.66 newtons). They play a key role in discriminatory sensory capabilities and the control of mandibular statics and dynamics (Grieznis et al., 2010). This proprioceptive function is ensured by the trigeminal nerve, the largest cranial nerve, and disturbances in its afferent inputs can affect balance maintenance (Gangloff & Perrin, 2002). Information from the stomatognathic apparatus can influence body sway. Proper dental occlusion contributes to maintaining postural balance and improving stability when unexpected postural fluctuations occur in an upright position (Hosoda et al., 2007).

Modifications in the vertical dimension of occlusion can influence the distribution of stress in the cervical spine, although the considerable variability in individual morphological characteristics prevents any statistical assessment (Sforza et al., 2006). The lateral inclination of the occlusal plane, causing an imbalance between the masticatory muscles on both sides, also affects the cervical spine and plays a compensatory role in posture control (Shimazaki et al., 2003).

During a masticatory simulation, biomechanical effects on the upper vertebrae have been quantified. Stress was concentrated on the odontoid process facet of the atlas joint, with an intensity on the vertebral joint about one-thousandth of the force exerted on the occlusal surfaces (Motoyoshi et al., 2000).

Movements of the mandible always correspond to cranial movements, which, in turn, lead to adaptations of the cervical spine. The relationship between the occlusal surfaces of the lower arch (mandible) and the upper arch (skull), apart from determining the statics and dynamics of the temporomandibular joint (Klineberg & Murray, 1999), also influences the statics and dynamics of the skull and its interactions with the vertebral column. These functional relationships are not only anatomical or ontogenetic but, as we have seen, have their evolutionary origins in deep time. Such evidence should encourage us to consider occlusion not merely as a relationship between dental arches but as a more complex relationship involving the mandible, skull, and cervical region.

With obligatory bipedalism in an upright posture and the progressive encephalization of the *Homo* genus, associated with the distinctive globular shape of the neurocranium in *H. sapiens*, our dental occlusion takes on an important proprioceptive role for the position of the mandible and the skull in space. This complex of joints (occlusal joint, temporomandibular joint, and cranio atlantoidea joint) can be thought of as a multi-articular system that, with its complementarity, regulates our posture, helping us maintain our precarious balance. This balance results from our obligatory and disadvantageous bipedal support, upright posture, and a heavy spherical skull perched on a verticalized vertebral column supported by relatively weak musculature.

4.3 Occlusion Determinants

As we have seen, during human evolution—from archaic human forms to the present day—bipedalism, encephalization, and diet have redefined the anatomical and functional relationships among various cranial structures to optimize and balance specific functions. Within the anatomical district of the masticatory apparatus, some of these anatomical features have been grouped into a set called 'occlusion determinants.' Each of them can be thought of as a gnathometric entity capable of influencing a collective functional pattern. These various entities coordinate with each other to regulate masticatory function (Dickerson et al., 2001). The study of occlusion determinants has led to the statistical identification of their average value, from which each individual deviates to a greater or lesser extent. Indeed, these entities vary individually and characterize function both due to their genetic basis and individual morphological adaptation during life. Here are the occlusion determinants:

Condylar guidance (inclination of the roof of the mandibular fossa).

- Incisal guidance (overbite).
- Intercondylar distance.
- Bennett movement.
- Cuspal height.
- Inclination of the occlusal plane.
- Depth of compensatory curves.

Within this set of measurable entities in degrees or millimeters, condylar guidance, intercondylar distance, and Bennett movement are referred to as fixed determinants because their existence is not strictly linked to the presence of teeth (Molina et al., 1980).

The anatomical and functional components that regulate mandibular dynamics were described and grouped together in 1926 by Rudolph L. Hanau in an attempt to create tools that reproduce mouth movements to achieve balanced occlusion for the stability of complete removable dentures during chewing (Hanau, 1926). In subjects with complete edentulism who are rehabilitated with complete removable dentures (but the same applies to implant supported complete dentures), non-fixed determinants disappear, and the only fixed variable characterizing occlusion that we need to record is the slopes of the glenoid fossa roof in the temporomandibular joint (TMJ). The plural form is appropriate because in a subject who has had a fixed monolateral chewing function for part of their life, asymmetric roof slopes of the mandibular fossae will have developed in response to this asymmetric function.

In prosthetic reconstructions, it will be necessary to harmonize occlusal determinants with the one that interacts most in its correct functioning (the slopes of the glenoid roofs, i.e., condylar guidance), to avoid destabilizing interferences for the removable prosthesis during mandibular movements (Figs. 4.58 and 4.59). In the case of implant prostheses, such interferences can harmfully load the implants, their osseous support, or musculoskeletal structures.

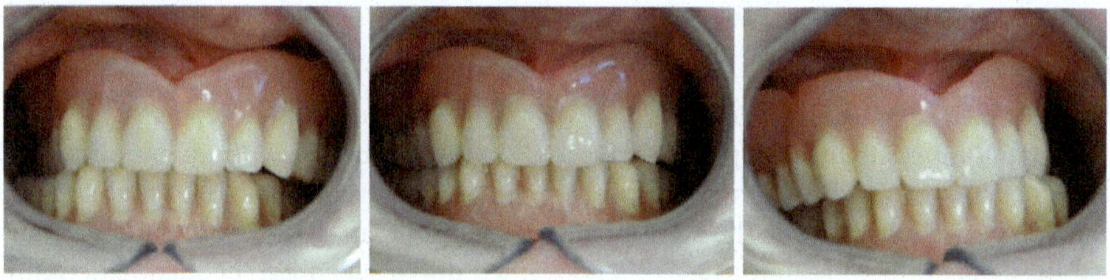

Fig. 4.58 Total prosthesis in an edentulous patient with incorrect occlusal determinants setting: During lateral movements, the lower prosthesis destabilizes, displaces, and drags the upper prosthesis along with it

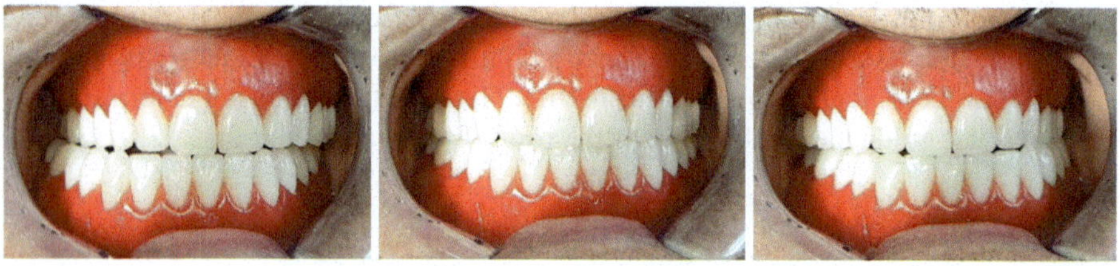

Fig. 4.59 Wax setup trial of a total prosthesis in an edentulous patient with correct occlusal determinants setting: During lateral movements, the upper and lower prostheses stabilize each other, both with a working-side group function and a balancing-side glide. This prevents twisting forces or destabilizing moments: as seen in the wear chapter, in the teeth of human populations in the past, this occlusal pattern was quickly established; bilateral balanced occlusion, now mistakenly considered "total prosthesis occlusion," is the result of a physiological maturation of occlusal contacts, capable of stabilizing all components of the masticatory system

The determinants that characterize occlusion have been implicated in the adaptations induced by bipedalism and encephalization. Due to the interdependence between the jaw, skull, and the vertebral column, these parameters should be referred to as 'determinants of the mandibulo-cranio-cervical relationship.' These factors should thus be considered as 'proprioceptive reciprocities' that, in addition to regulating chewing, assist the system in the challenging task of balancing an oversized and heavy globular skull on an underlying vertical vertebral column.

The aforementioned cranial modifications, discussed in the following paragraphs, can be summarized as follows:

- *Reduction of the postglenoid process*: This occurs as the distance between the tympanic cavity and the mandibular fossa decreases.
- *Presence of Civinini's canal* (Cdc; or Huguier's canal): This results from the development and persistence of communication between the tympanic cavity and the mandibular fossa, traversed by two ligaments connecting the mandible and the posterior part of the temporomandibular joint (TMJ) with the hammer bone.
- *Increased inclination of the roof of the mandibular fossa*: As the inclination of the glenoid roof intensifies, condylar dynamics become functionally related to the overlap of the upper incisor over the lower incisor (overbite).

- *Formation of the articular tubercle*: A slight bony protrusion formed, serving as the attachment point for the temporomandibular ligament.
- *Development of the temporomandibular ligament* (TML): This ligamentous component of the temporomandibular joint structured into two distinct bundles, crucial for the new joint dynamics.
- *Modification of dental volumes and disappearance of the C/P3 honing complex*: The canine tooth reduced in size, becoming a pivotal element for guiding horizontal movements.
- *Shortening of the mandible*: This occurred with the retrusion of the facial mass.
- *Expansion of the upper maxilla*: Along with the facial mass retrusion, there was an increase in the transverse diameter of the palate.
- *Alteration of the occlusal plane*: Achieved by adapting the occlusal plane to the changes in cranial structure.
- *Formation of the overbite/incisor overlap*: With the overlap of the upper incisors over the lower incisors, the overbite became a new proprioceptive area. Thanks to its extreme sensitivity, this dental group could regulate anteroposterior (incisal guidance) and lateral (canine guidance) movements, functionally interacting with the posterior joint guidance.

4.4 Reduction of the Postglenoid Process

In humans, the Postglenoid Process (PP) is a conical bony prominence formed by a lateral thickening of the posterior articular lip of the mandibular fossa. It is located directly in front of the external acoustic meatus and contributes to forming the posterior wall of the temporomandibular joint (TMJ). The PP steadily grows and reaches its final size around the age of 13. It is present in 79% of TMJs (Katsavrias, 2002). The PP separates the mandibular fossa from the external acoustic meatus, limiting posterior movements of the condyle (Sicher, 1951) (Fig. 4.60).

In most marsupial and placental mammals, the middle ear is enclosed by a bony structure known as the tympanic bulla or auditory bulla, which forms the floor of the middle ear and protects the space of the tympanic cavity (Tucker, 2017) (Fig. 4.61).

Between the tympanic bulla and the mandibular fossa, there is a sizable postglenoid process that separates these two structures, safeguarding the organs of hearing and balance from the lethal danger of a forceful condylar retrusion, which could occur during intense chewing forces (Fig. 4.62), or even more so during predation (Fig. 4.63).

In addition to the progressive proximity of the middle ear to the temporomandibular joint (TMJ), resulting from the disappearance of the tympanic bulla (which occurred in the anthropoids) and the significant reduction of the postglenoid process, modern humans are distinguished from other primates by the direct correlation between the sagittal flexion of the cranial base, the orientation of the petrous pyramid, and the enlargement of the brain (encephalization) with a shortened cranial base (Spoor, 1997). In primates, the Postglenoid Process (PP), despite its high

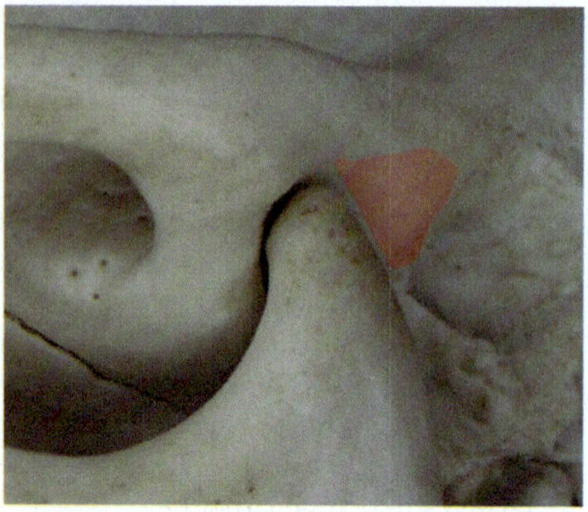

Fig. 4.60 Postglenoid process that delimits the mandibular fossa posteriorly and externally. Specimen from the Museum of Anthropology at the University of Florence

Fig. 4.61 Skull of a waterbuck or ellipsen waterbuck (*Kobus ellipsiprymnus*). The auditory bulla is highlighted in green. Specimen from the Museum of Zoology at the University of Florence

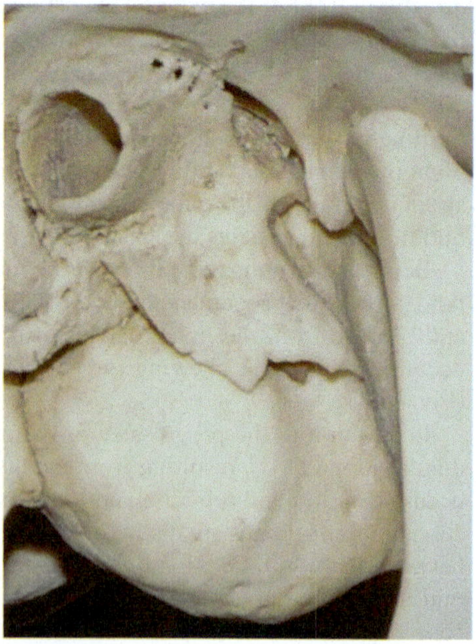

Fig. 4.62 Postglenoid process and auditory bulla of a waterbuck or ellipsen waterbuck (*Kobus ellipsiprymnus*). Specimen from the Museum of Zoology at the University of Florence

Fig. 4.63 Lion (*Panthera leo*): a developed postglenoid process intervenes between the middle ear and mandibular fossa to protect the structures of the middle and inner ear, essential for the animal's survival. Specimen from the Museum of Zoology at the University of Florence

morphological variability, comes into functional contact with the mandibular condyle during chewing (Wall, 1997) (Figs. 4.64 and 4.65).

In monkeys, the mandibular fossa is considered low, slightly inclined, and with poorly formed articular eminences (Fig. 4.66). In *Australopithecus africanus,* on the other hand, the glenoid fossa is posteriorly delimited by a well-developed postglenoid process, maintaining a significant distance from the external auditory meatus (Kieser, 1997).

Unlike humans, in anthropoid apes, when their teeth occlude, the condyle rests against the posterior face of the articular eminence (Aiello & Dean, 1990) (Fig. 4.67).

In anthropoid apes, the Postglenoid Process (PP) is connected to the anterior wall of the tympanic cavity, and the postglenoid area is considerably more voluminous and thicker compared to that of humans (Ashton & Zuckerman, 1954; Kimbel et al., 1986) (Fig. 4.68).

A significant feature of the temporomandibular joint (TMJ) in monkeys is the medial aspect of the joint, where the articular disc intervenes between the condyle head and the entoglenoid process. The surface of the condylar head closely follows the medial contour of the fossa (Fig. 4.69).

The evolutionary modifications of the mandibular fossa in the *Homo* genus, specifically in *H. sapiens*, involve not only the depth of the fossa, the reduction of the postglenoid process, or the development of the articular tubercle but also affect the inner wall of the fossa. In anthropoid apes, this wall is medially bounded by a bony structure called the entoglenoid process (Fig. 4.70).

In many anthropoid apes, there is limited lateral or medial movement of the condyle, whereas in modern humans, the mandible is not as tightly constrained from a dynamic perspective. It's as if the entoglenoid process has reduced its significance to allow for greater freedom of movement of the condyles (Fig. 4.71).

In anthropoid apes, during lateral movements of the mandible, the condyle interacts with the entoglenoid process, which forms the medial wall of the fossa. This

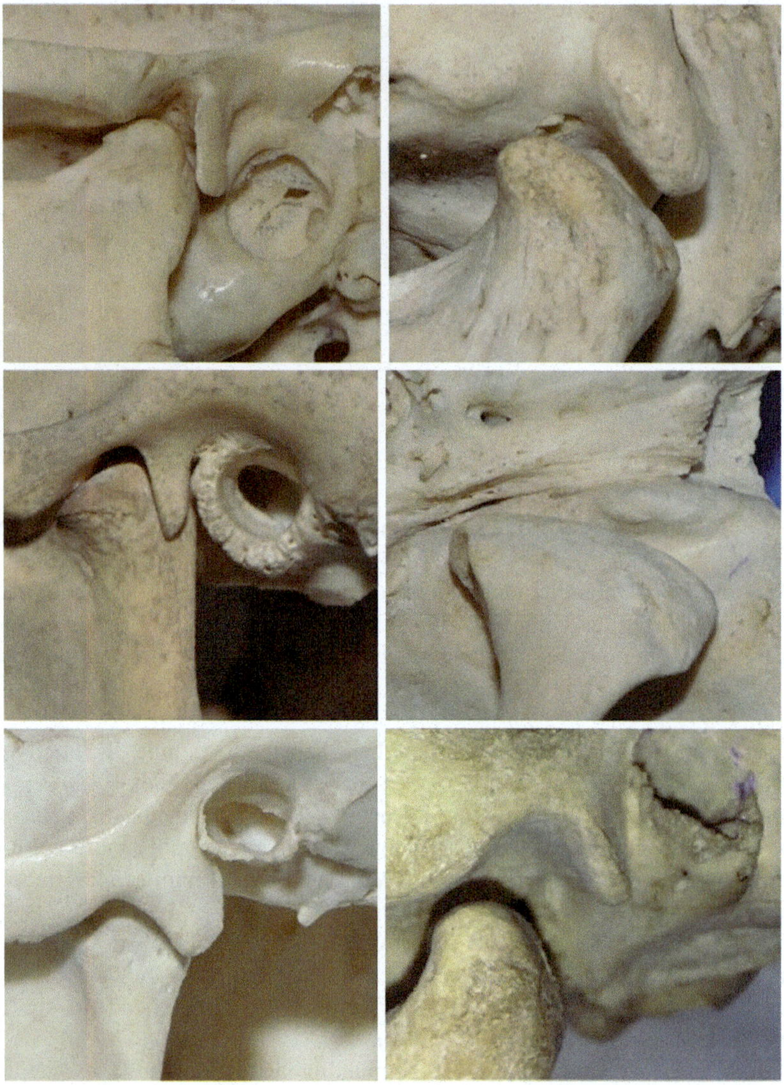

Fig. 4.64 Morphological variability of postglenoid processes. Clockwise from top left: *Cebus apella, Macaca sylvanus, Chlorocebus pygerythrus, Macaca mulatta, Pongo pygmaeus, Pan troglodytes*. Specimens from the Museum of Zoology at the University of Florence

might explain why the temporomandibular ligament in these primates is reduced or absent. The function of resisting lateral stresses on the working condyle is likely performed by the contralateral entoglenoid process (Fig. 4.72).

The presence of an extensive and robust external ligament in the joint capsule of the TMJ in these primates suggests that its primary function is not so much to resist lateral movements but rather to maintain close contact between the condylar head, disc, and articular eminence during the opening and closing movements of the

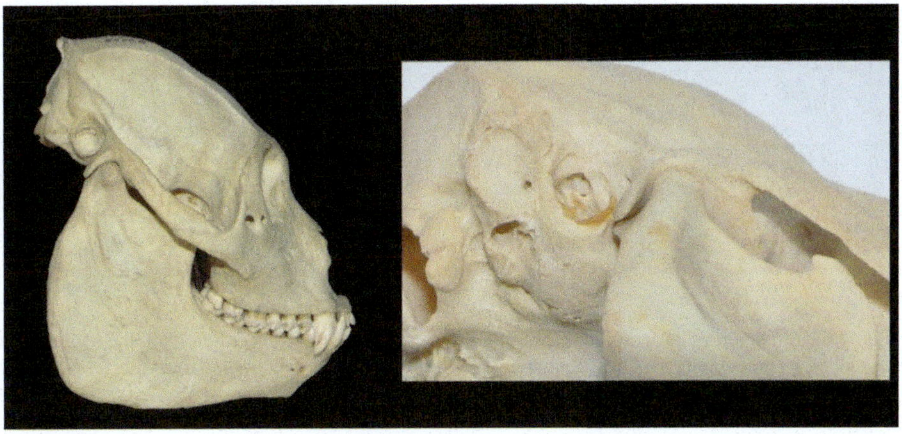

Fig. 4.65 Relationship between the condyle and postglenoid process in a red howler monkey (*Alouatta seniculus*). Specimen 4668 from the Museum of Zoology at the University of Florence

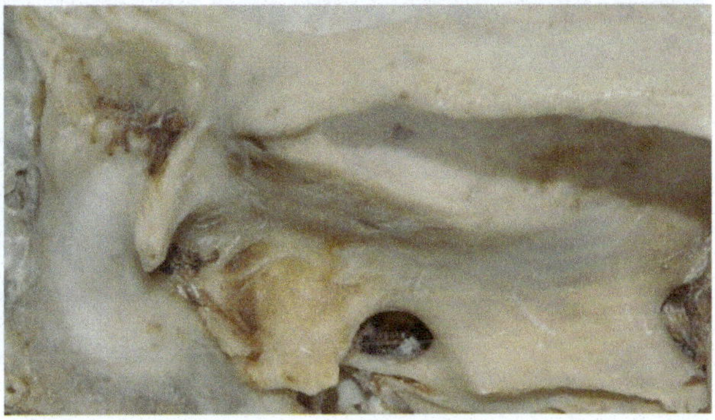

Fig. 4.66 TMJ (Temporomandibular Joint) of a gibbon (*Bunopithecus hoolock*). Note the slight inclination of the mandibular fossa roof. Specimen from the Museum of Zoology at the University of Florence

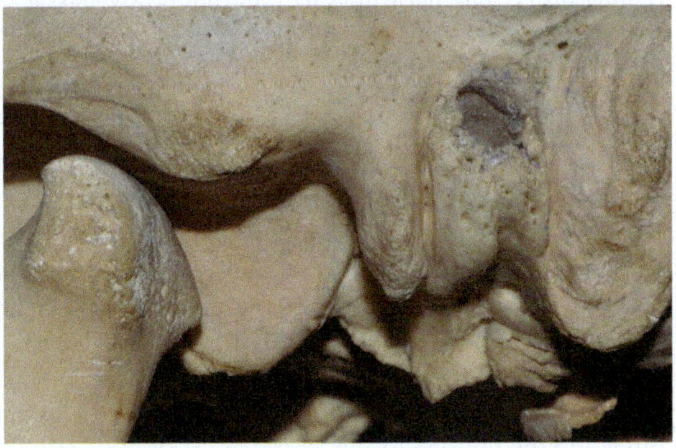

Fig. 4.67 Condyle, mandibular fossa, postglenoid process, and auditory meatus of a gorilla (*Gorilla gorilla*). Specimen from the Museum of Zoology at the University of Florence

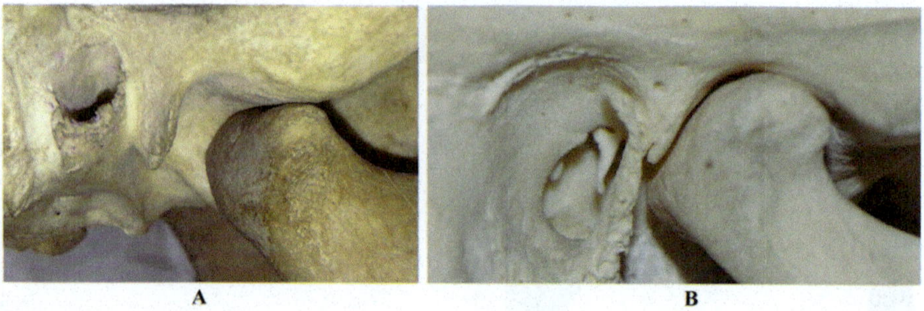

Fig. 4.68 Postglenoid process. (**a**) Chimpanzee (*Pan troglodytes*), specimen from the Museum of Zoology at the University of Florence; (**b**) *Homo sapiens*, specimen from the Museum of Anthropology at the University of Florence. The petrotympanic fissure delimits the part of the mandibular fossa pertaining to the squamous portion and separates it from the tympanic plate forming the posterior portion of the fossa

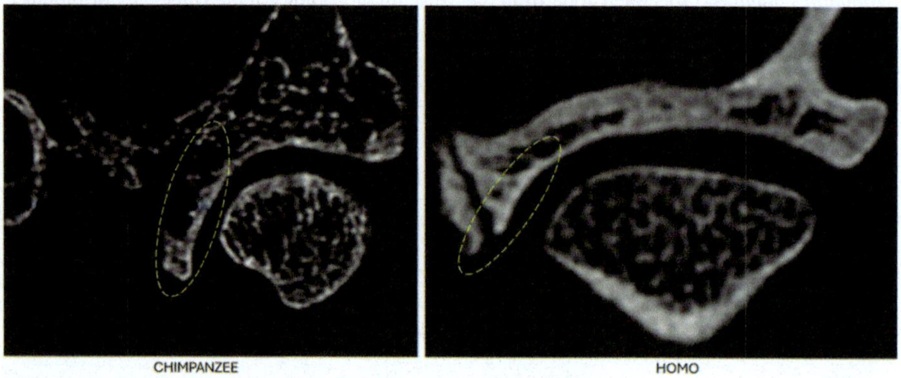

Fig. 4.69 Cone beam CT scan: coronal sections at condylar level of the TMJ. Left: Chimpanzee (*P. troglodytes*), specimen from the Museum of Zoology at the University of Florence; right: Human (*H. sapiens*), specimen from the Museum of Anthropology at the University of Florence. Note the different development of the entoglenoid process (highlighted in yellow), whose bony wall, especially in anthropoid apes, limits any medial displacement of the condyle

mandible (Aiello & Dean, 1990). In humans, this task is accomplished by the oblique portion of the temporomandibular ligament.

A relatively large entoglenoid process not only prevents excessive medial-lateral sliding of the condyle but can also assist in guiding its anteroposterior movement, favoring the sagittal sliding of the mandible (Wall, 1999) (Fig. 4.73).

The anatomical relationships between the external acoustic meatus, tympanic cavity, and TMJ changed with the expansion of the overlying neurocranium resulting from progressive encephalization. As these structures moved closer to each other and the volume of the Postglenoid Process (PP) reduced, the TMJ and middle ear achieved such close anatomical proximity that they were separated only by a

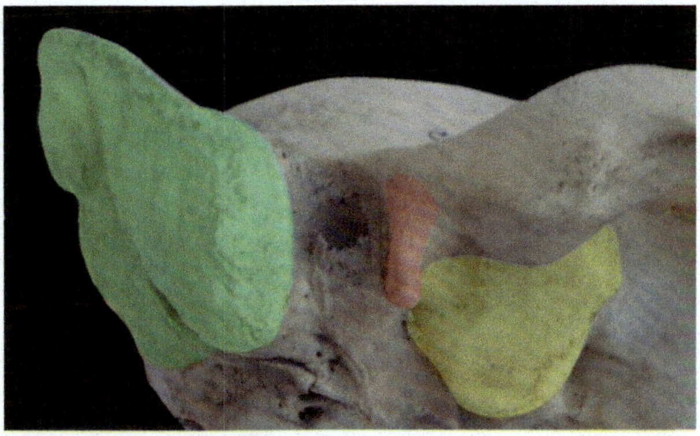

Fig. 4.70 Chimpanzee (*Pan troglodytes*). Lateral-inferior view of the mandibular fossa: an important entoglenoid process (yellow) delimits the fossa medially; also note the developed postglenoid process (red) and the prominent nuchal plate (green), the insertion area of powerful neck muscles. Specimen from the Museum of Zoology at the University of Florence

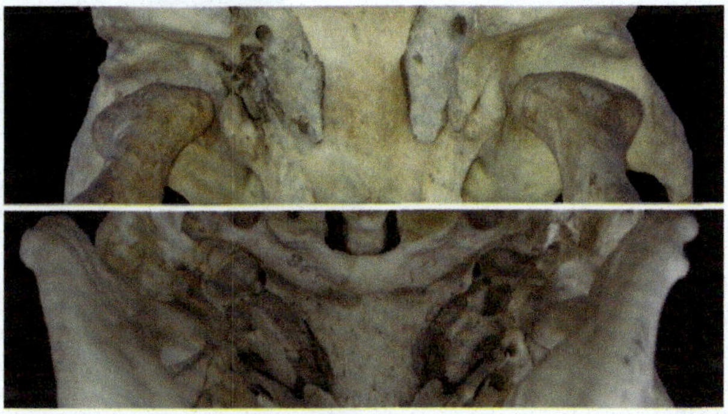

Fig. 4.71 Top: Entoglenoid process in chimpanzee (*Pan troglodytes*), specimen from the Museum of Zoology at the University of Florence; Bottom: Same perspective in *Homo sapiens*, specimen from the Museum of Anthropology at the University of Florence. Note how in chimpanzees, the mandibular fossa is "boxed in," posteriorly by the postglenoid process and medially by the entoglenoid process, while in humans these bony walls are reduced to almost disappearing

thin bony layer. From a gnathological perspective, this area became a potential critical zone (Figs. 4.74 and 4.75).

In humans, the Postglenoid Process (PP), which contributes to forming the posterior-lateral border of the mandibular fossa, consists of a lower projection of the squamous portion of the anterior temporal bone (Durband, 2008) (Fig. 4.76).

As previously described, in humans, there is no significant bony thickness in the postglenoid region to protect the delicate structures of the middle and inner ear (Fig. 4.77), which are crucial for survival. Protection against this potential fragility is achieved for the inner ear through thicker cortical bone layers that shield its structures. For the middle ear, it is ensured by the horizontal band of the temporomandibular ligament, which prevents posterior movements of the condyle from its

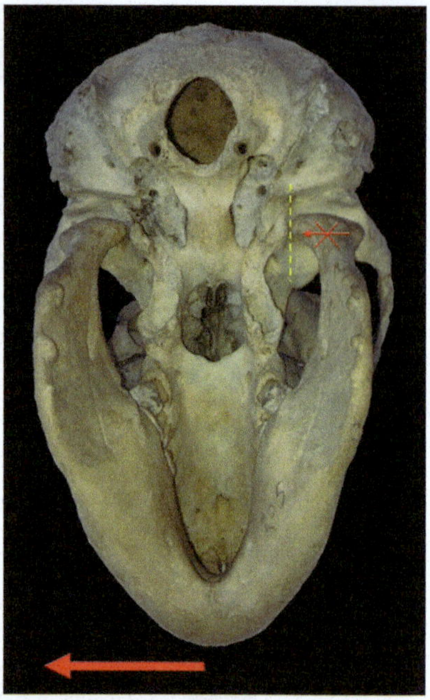

Fig. 4.72 Chimpanzee (*Pan troglodytes*). The entoglenoid processes (yellow dashed line) contain the medial movements (red arrow) of the condyle and the entire mandible. Specimen from the Museum of Zoology at the University of Florence

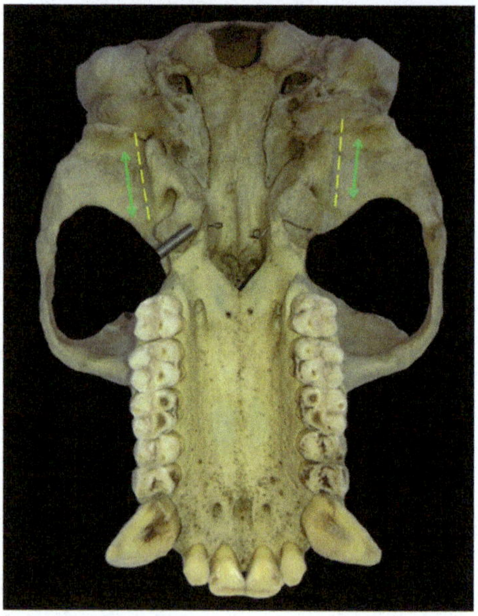

Fig. 4.73 Skull of a gorilla (*Gorilla gorilla*). A developed entoglenoid process facilitates protrusive movements, which in anthropoid apes are necessary for biting and peeling fruits that constitute a good percentage of their diet. Specimen from the Museum of Zoology at the University of Florence

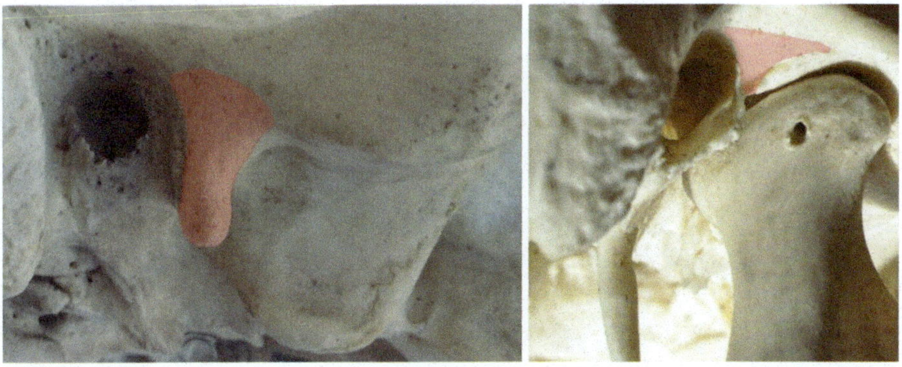

Fig. 4.74 Highlighted in red: Postglenoid processes. Left: Gorilla (*Gorilla gorilla*), specimen from the Museum of Zoology at the University of Florence; Right: *Homo sapiens*, specimen from the Museum of Anthropology at the University of Florence

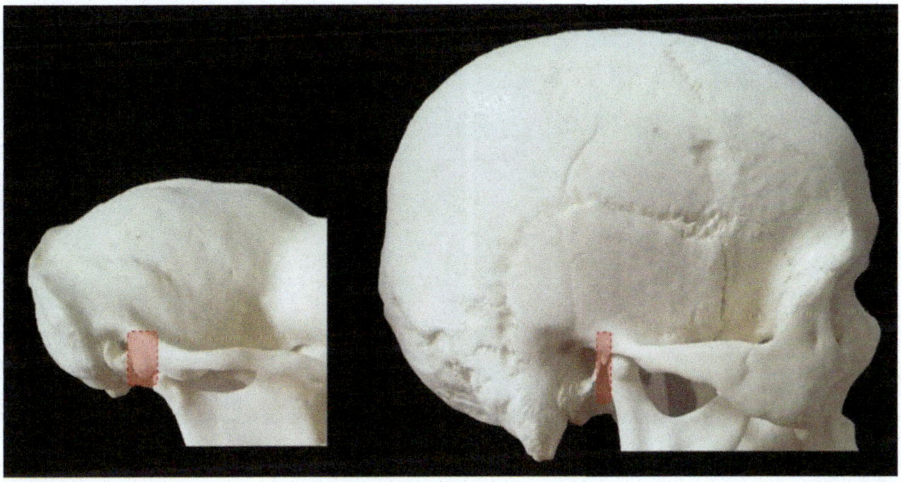

Fig. 4.75 Left: Chimpanzee skull (*Pan troglodytes*); Right: *Homo sapiens* skull. Encephalization modified the relationships between the mandibular fossa and the middle ear in humans: these two structures became markedly contiguous

homeostatic neutral position within the fossa toward the posterior portion. The structure of the horizontal portion of the temporomandibular ligament is designed to prevent uncontrolled retrusive movements of the condyle from creating compressive forces in joint areas ill-suited to bear them, considering the nature of soft tissues and the thickness of bone walls.

In the *Homo* genus, compared to anthropoid apes, there is a reduction, almost to the point of disappearance, of the bony constraints represented by the postglenoid and entoglenoid processes in the temporomandibular joints (TMJ). Meanwhile, the depth of the mandibular fossa increases, the slope of the glenoid roof becomes more

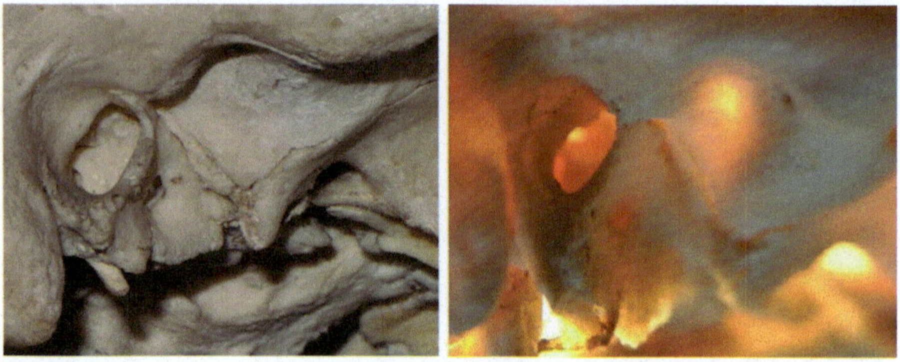

Fig. 4.76 Mandibular fossa: petrotympanic fissure dividing the tympanic plate (forming the posterior wall of the fossa) from the squamous portion, whose inferior projection forms the postglenoid process. On the right, the specimen photographed in transillumination to highlight the bone thicknesses. Skull 1364 from the Museum of Anthropology at the University of Florence

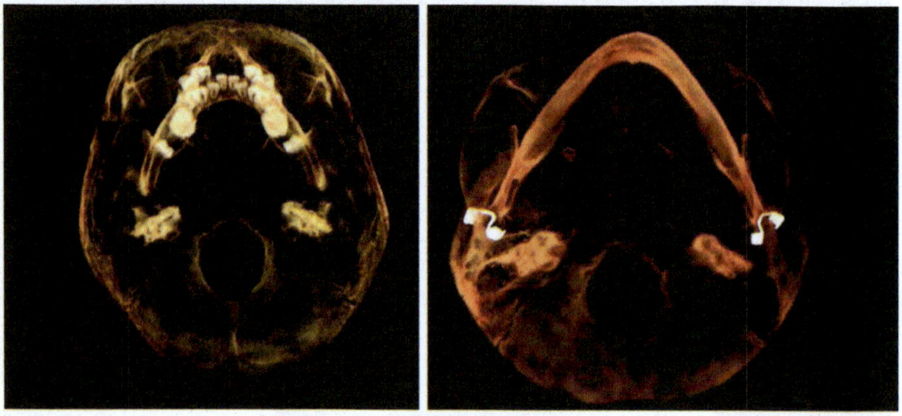

Fig. 4.77 Cranial tomographies of a child (top) and an elderly individual (bottom). Note the significant bone density protecting the structures of the middle and inner ear: this particular compactness develops early and persists throughout life. Cone beam CT scans of specimens from the Museum of Anthropology at the University of Florence

pronounced, and a bony elevation that defines the anterior region of the mandibular fossa, known as the articular tubercle, develops. The increased slope of the glenoid roof and the presence of the articular tubercle are related to cranial modifications induced by an upright posture. They compensate for the loss of rigidity in the glenoid roof due to the postural change, a rigidity that is typically maintained in flexed cranial posture seen in anthropoid apes.

The reduction of the postglenoid process can be understood in light of the profound structural rearrangements associated with extreme encephalization in *H. sapiens*. Similarly, the reduction of the entoglenoid process can be interpreted as a consequence of the need for greater joint mobility allowed by the absence of a significant bony constraint. In our lateral movements, the balancing condyle experiences medial displacement facilitated by a much-reduced entoglenoid process compared to that of anthropoid apes.

4.5 Presence of the Civinini Canal

The Civinini canal (Cdc), commonly known as Huguier's canal, connects the posterior medial part of the mandibular fossa to the tympanic cavity. In the sagittal plane, it has an hourglass shape, with an average minimum diameter of about 0.5 mm (Buffoli et al., 2014) (Fig. 4.78).

The canal was first described in 1828 by Filippo Civinini, an anatomist from Pistoia (IT), who, building on the studies of Andrea Comparetti from 1789, published his discovery in Pistoia in 1830 in a pamphlet titled "Linee anatomiche, di Filippo Civinini pistoiese (presso i Fratelli Bracali in Pistoia)." Four years later, in 1834, the French surgeon Pierre-Charles Huguier described the canal again, and both the attribution and the name of the anatomical structure were unjustly credited to him (Mudry, 2015). In this discussion, we will refer to this structure using the name of the person who first described it, namely the "Civinini canal."

In modern amniotes (birds, reptiles, and mammals), the middle ear includes an air-filled space known as the tympanic cavity through which sound travels via bony segments. The formation of a tympanic ear was a key innovation to resolve impedance mismatch problems in sound transmission that occurred during the transition from an aquatic to a terrestrial environment (Tucker, 2017).

Mammals are unique in having a chain of three small bones (stapes, incus, and malleus) within their auditory system. These bones transmit vibrations from the tympanic membrane to the sensory structures in the cochlea. In contrast, most non-mammalian tetrapods have a single columnar bony structure called the columella auris that connects the eardrum to the oval window. In mammals, the bone

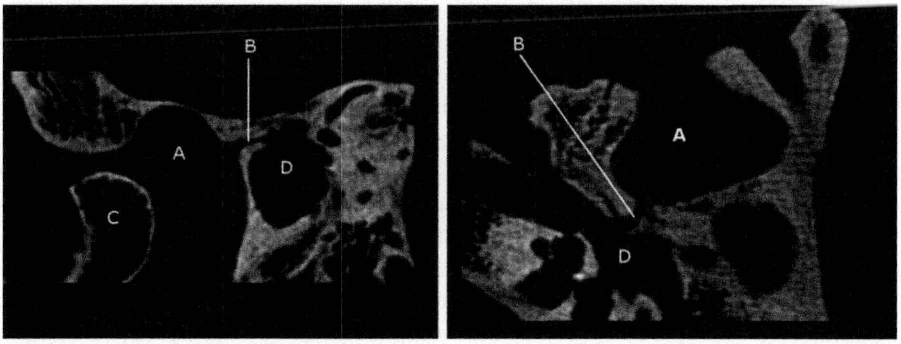

Fig. 4.78 Cone beam CT scan of the mandibular fossa: left sagittal section, right horizontal section. (*A*) Mandibular fossa; (*B*) Civinini canal; (*C*) Condyle; (*D*) Tympanic cavity. Specimen from the Museum of Anthropology at the University of Florence

surrounding the inner ear structures, once formed in utero, does not undergo postnatal growth or changes (Fricano, 2018). Unlike other bones, the ossicles are fully formed and reach their definitive size at birth. The shape and size of these bony structures are evidently under strong genetic control.

During development, the malleus, incus, and stapes ossify relatively early, fixing their dimensions at an early stage, unlike the rest of the skull and mandible. As we discussed previously, the sizes of these ossicles remain stable during the growth of other cranial structures. It has been suggested that this specific allometry plays a role in the posterior shift of this area during development, which is necessary for isolating the ear from the temporomandibular joint (Anthwal et al., 2013).

From an evolutionary perspective, the formation of the complex of auditory ossicles (a distinctive feature of mammals) cannot be separated from the development of the temporomandibular joint. In most mammals, the middle ear is enclosed in the *tympanic bulla* (Tucker, 2017), located on the ventral and posterior surface of the cranial base. The bulla, by enlarging the volumes of the middle ear, is believed to be a possible adaptation to maximize hearing at low frequencies (Fricano, 2018).

In mammals, the presence of both the tympanic bulla and a well-developed postglenoid process characterizes the anatomical relationship between the mandibular fossa and the ossicular chain, resulting in a greater distance between the temporomandibular joint (TMJ) and the malleus. An anatomical study conducted in dogs highlighted a ligamentous connection between the mandible and the middle ear, originating from remnants of Meckel's cartilage. However, it's important to note that the Spix's spine and the sphenoid spine do not exist in dogs, so interspecific comparisons of this ligament should only be made from an embryonic development perspective. In neonatal dogs, a ligamentous connection was demonstrated between the medial portion of the mandible and the extensive rostral process of the malleus. However, this process regresses during the animal's development, and in none of the adult specimens was a continuous ligamentous connection observed between the mandible/TMJ and the malleus. In fact, the presence of a true ligament between the mandible and the malleus has only been identified in newborn puppies in which the tympanic bulla has not yet developed. As the bulla develops and ossifies, the ligament becomes trapped between the bulla and the adjacent tympanic bone (the cranial bone of mammals, often fused with the temporal bone, which encloses part of the middle ear and supports the tympanic membrane). This results in a thin connective tissue sheet with most of the ligament fibers adhering to the bony surface. The opening movement and lateral displacement of the mandible have never caused any movement of the malleus. Compared to humans, the position of the mandibular joint in dogs is more anterior and further away from the ossicular chain, which does not allow for the persistence of ligamentous connections as observed in humans (Stevens-Sparks & Strain, 2014).

From an evolutionary perspective, the tympanic ring, on which the tympanic membrane is stretched, likely derives from the angular bone of the reptilian mandible. In primitive forms, it is represented by an incomplete ring, which subsequently forms and deepens to become a true external acoustic meatus, the external ear canal. This can be observed through the evolutionary stages from prosimians to non-anthropoid monkeys and then to anthropoid monkeys (Padoa, 1991).

In primates, the tympanic bulla remains only in prosimians, the most ancient representatives of this order, resembling the first primate species that appeared millions of years ago (Fig. 4.79).

This structure is retained in many primates but is lost in anthropoid apes and in the evolutionary lineage of the *Homo* genus. In the tympanic region, the loss of the bulla takes on the form of an external tubular meatus, the ear canal. The disappearance of the bulla in anthropoid apes and in the human evolutionary lineage allowed for the maintenance of ligamentous connections between the mandibular fossa and the tympanic cavity even in adulthood. As we have seen previously, in humans, the mandibular fossa is in close proximity to the posterior tympanic cavity, separated by a thin bony layer. The relationships between the middle ear and the mandibular fossa involve evolutionary, embryological, histological, anatomical, and functional aspects.

In an unpublished study conducted by the author on primate skulls from the collection of the Zoological Museum of La Specola in Florence, the skulls of nine anthropoid apes and 15 non-anthropoid monkeys were scanned to assess the presence of anatomical connections between the mandibular fossa and the tympanic cavity. In all anthropoid apes, a structure anatomically similar to the Civinini canal in humans was observed (Fig. 4.80), while in all non-anthropoid monkeys, there was an absence of a radiologically detectable connecting canal.

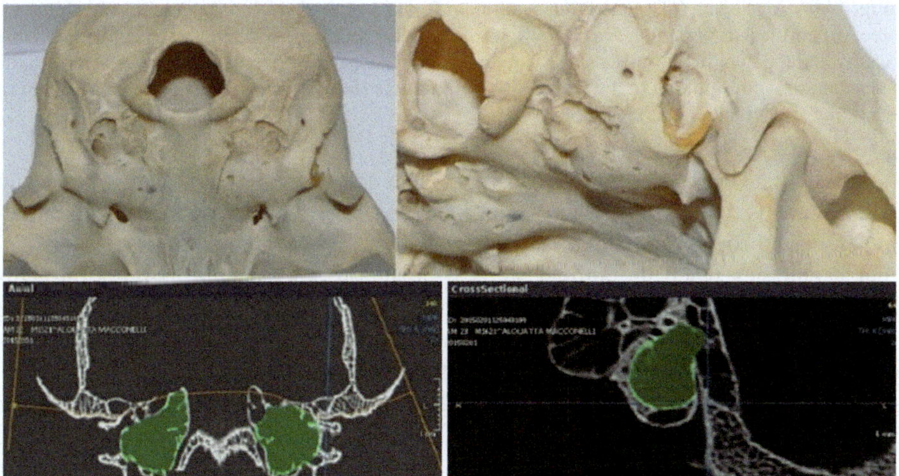

Fig. 4.79 Auditory bullae of Guianan howler monkey (*Alouatta macconnelli*): in the lower images (cone beam CT), note in the coronal section on the left and sagittal section on the right, the cavity of the tympanic bulla (in green). Specimen from the Museum of Zoology at the University of Florence

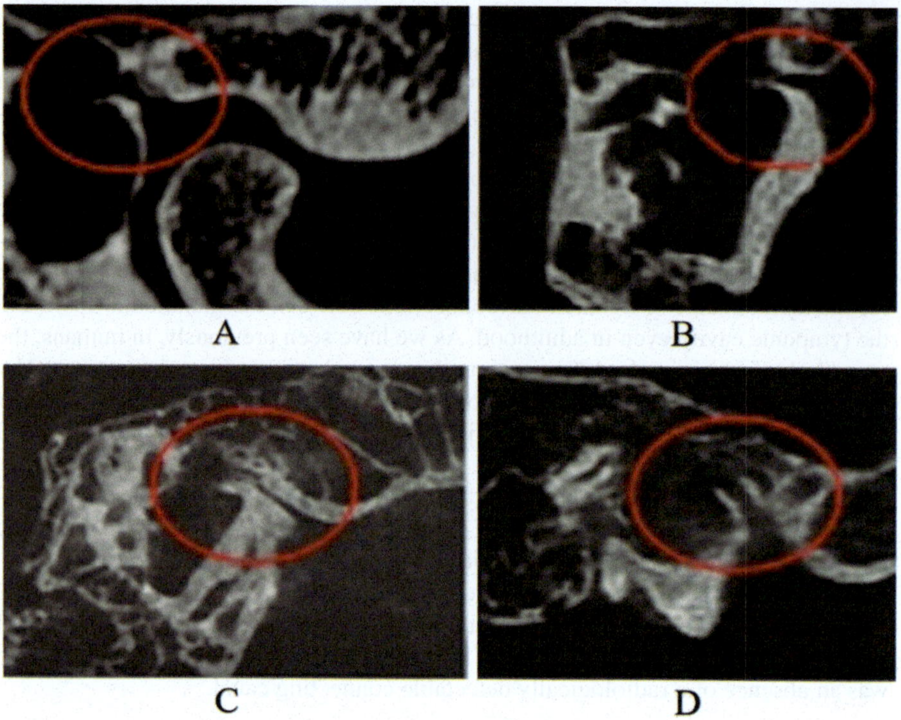

Fig. 4.80 Cone beam CT scan of the mandibular fossa in sagittal section: (**a**) human, (**b**) chimpanzee, (**c**) gorilla, (**d**) orangutan; in all sections of the anthropoid apes, a canal between the tympanic cavity and the mandibular fossa similar to that of humans can be observed. (**a**) Specimen from the Museum of Anthropology at the University of Florence; (**b**, **c**, **d**) specimens from the Museum of Zoology at the University of Florence

These data seem to suggest that the persistence of the Civinini canal (Cdc) and the ligaments that traverse it into adulthood may be related to the facultative bipedalism of anthropoid apes. In quadrupedal primates, this canal appears to be absent, but it emerges when the animal starts to abandon quadrupedal locomotion. This makes the hypothesis plausible that these ligaments somehow aid in the challenging process of bipedalism, conveying information about the mandibulo-cranial relationship to the central nervous system through the ossicular system.

As mentioned, anthropoid apes lose the bulla, and in humans, the postglenoid process undergoes progressive reduction. The well-developed tympanic bulla and postglenoid process can be considered anatomical factors that hinder the formation of a true ligament between the mandible and the malleus (Fig. 4.81).

In humans, the mechanical proximity of the mandible is one of the factors influencing the ossification process of the tympanic bone after birth. In the fetus at around 8 weeks, the tympanic ring begins to form from a poorly defined condensation of mesenchyme that becomes apparent in the lower region of the developing temporal bone's squama. Even though it is still in an early cartilaginous development stage, the bony mandible is already present in the region of Meckel's cartilage

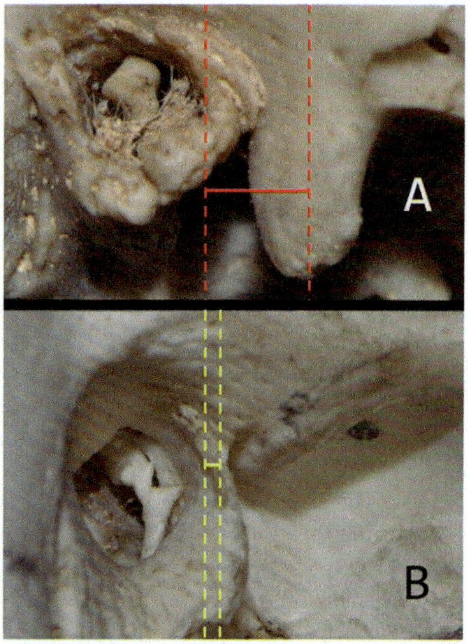

Fig. 4.81 Lateral view of the external auditory meatus, showing the presence of the malleus. (**a**) Guereza (*Colobus guereza*, non-anthropomorphic monkey), Museum of Zoology at the University of Florence; (**b**) *Homo sapiens*, Museum of Anthropology at the University of Florence. Note the reduction of the distance (solid horizontal line) between the posterior limit of the mandibular fossa and the anterior limit of the external auditory meatus (both indicated by dashed lines)

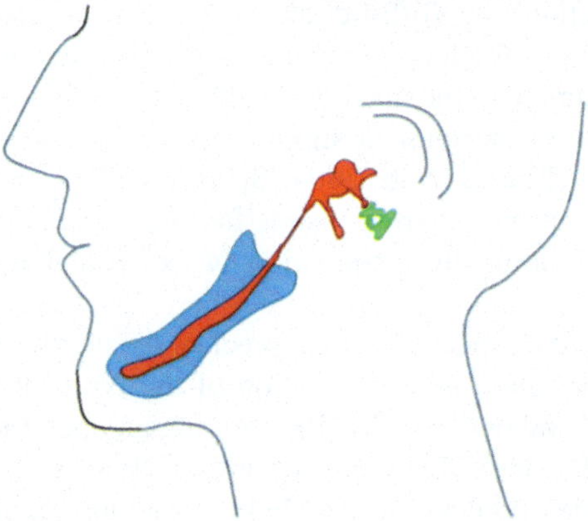

Fig. 4.82 Development of the incus and malleus from Meckel's cartilage. The connection between the malleus and Meckel's cartilage temporarily serves as the jaw joint in fetal development, allowing movements necessary for the correct formation of the temporomandibular joint. Red: Meckel's cartilage, malleus, incus; blue: forming mandibular bone; green: stapes

(Anson et al., 1955). The profound connections between these structures are established during their embryological development (Fig. 4.82) and persist into adulthood, with the petrotympanic fissure never fully closing (Buffoli et al., 2014; Rodriguez Vazquez et al., 1998).

During the 7th week of embryonic development, the primary mandibular joint forms between the malleus and incus in the dorsal and terminal segment of Meckel's cartilage. This joint briefly functions as the mandibular joint, allowing for mouth opening and closing movements. Mandibular movements at the incudo-malleolar joint begin during the 7th and 8th weeks of development (Humphrey, 1968). Following muscle contractions, primarily those of the external pterygoid muscle (PEs), the joint spaces of the temporomandibular joint (TMJ) are formed through a mechanism of mesenchymal tissue cavitation. This joint begins to play a predominant role in oral movements compared to the incudo-malleolar joint. From the cavitation phase onwards and during the maturation phase, you can observe the insertion of PEs muscle bundles on the condyle and the anterior two-thirds of the articular disc (Mérida Velasco et al., 1993).

Between the 9th and 11th weeks, the cavitation phase of the TMJ occurs, which corresponds to the initial formation of the lower joint cavity and the onset of condylar chondrogenesis. Ossification of the tympanic bone begins at the 9th week of gestation, with four ossification centers developing around the tympanic membrane. They fuse into a U-shaped bony structure that connects the tympanic bone with the petrous part of the temporal bone, forming the anterior wall of the tympanic cavity. The 11th week marks the start of the organization of the upper joint cavity, and during this embryonic development phase, swallowing begins (Mérida Velasco et al., 1999).

The muscle bundles of the lower portion of the PEs muscle insert directly onto the mandibular condylar fossa (the small depression located on the anterior part of the condylar neck fossa) without the interposition of tendon bundles to increase the precision of their action (De Toffol et al., 2004). During the prenatal period, PEs muscle activity is one of the factors regulating the ossification of the condylar process (Alves, 2008), and both the maxillofacial bones and the tympanic bone primarily grow in response to mechanical stimuli created by chewing, swallowing, and breathing after birth (Mao & Nah, 2004; Sperber, 1975). In humans, the external pterygoid is the primary muscle responsible for managing horizontal movements of the mandible, representing the acquisition of horizontal motion by mandibular dynamics.

Another fundamental aspect to fully appreciate the presence of the Civinini canal in humans is the formation and ossification of the tympanic bone. It begins in a multicentric manner and develops by first forming an open ring, which then gradually assumes a tubular shape that closes progressively (Fig. 4.83).

The developmental process of this bony structure seems to recapitulate, in embryonic development, the stages of evolutionary development, as described by the suggestive phrase coined by Ernst Heinrich Haeckel, "Ontogeny recapitulates phylogeny." Haeckel observed that in embryonic preparations of animals, even those very different from each other, the initial stages of development were very similar if not identical, as if the evolutionary path from fish to amphibians, reptiles, birds, and mammals had been imprinted in the early stages of embryonic development. However, this assertion, although realistic, is imprecise and lacks a solid scientific basis because ontogeny follows very different paths compared to a recapitulation of phylogeny. There is nothing predetermined for the developmental mechanisms of individual species, and this theory could imply the concept of a

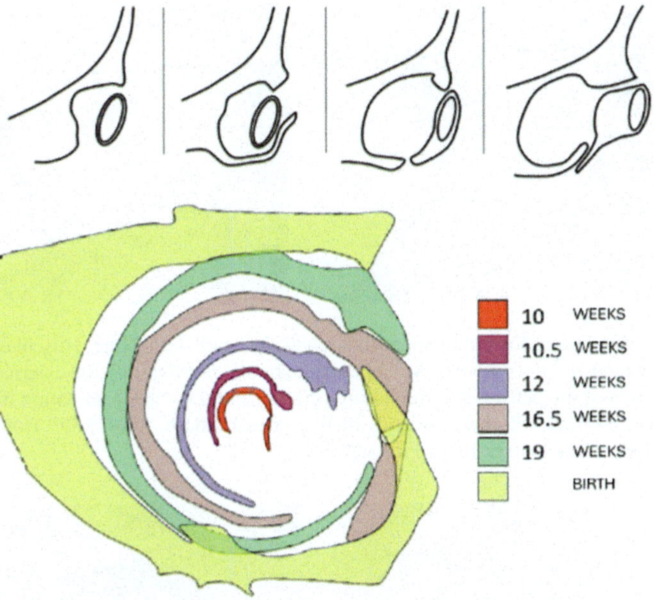

Fig. 4.83 Tympanic bone. Top: Phylogenetic development in primates (from the most ancient primates to anthropoids). Bottom: Ontogenetic development in humans (sagittal section)

positivistic evolution, a progressive and predestined ascent towards improvement, confusing evolution with progress (Manzi, 2013).

The ossification of the tympanic bone, as previously discussed, occurs in a circular manner and upwards through four ossification centers, which, by fusing the tympanic bone with the petrous part of the temporal bone, form the outer wall of the tympanic cavity that separates it from the mandibular fossa. The external auditory canal is formed by the fusion of these various ossification points.

In many children, a medial tympanic foramen persists (a communication present in the wall that separates the mandibular fossa from the middle ear), which gradually closes completely around the age of five (Hashimoto et al., 2011) (Fig. 4.84).

Due to an anomaly in the fusion of the different ossification centers of the tympanic ring, the separation between the mandibular fossa and the tympanic cavity may be incomplete, resulting in the persistence of a residual opening into adulthood, generally of elliptical shape, which is called *Huschke's foramen* (Fig. 4.85). The presence of Huschke's foramen has been found in more than one-tenth of the population, so it is essential to consider the possibility of its presence when performing TMJ (temporomandibular joint) or ear-related surgical procedures to avoid unexpected injuries. Dentists should consider Huschke's foramen as one of the possible diagnoses when dealing with patients presenting otological complications related to TMJ issues (Pekala et al., 2018).

In *Homo sapiens*, the bony lamina that forms the posterior wall of the mandibular fossa and separates it from the tympanic cavity is very thin (Figs. 4.86 and 4.87).

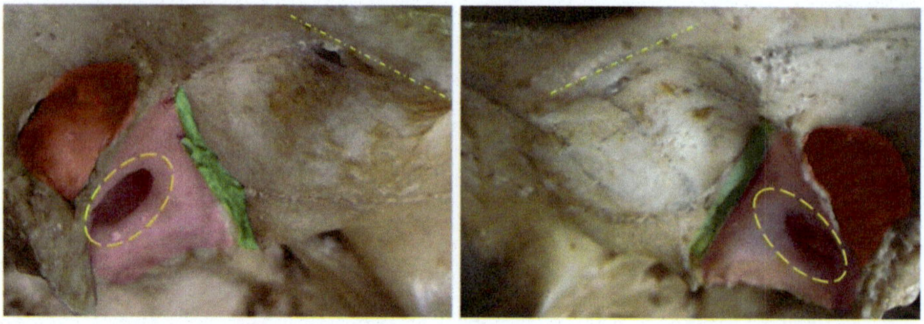

Fig. 4.84 Mandibular fossa of a approximately 4-year-old child. Note: Huschke's foramen (highlighted in yellow) on the posterior wall of the mandibular fossa (pink); the external auditory meatus (red); the area of the petro-tympanic fissure (green); and the slight inclination of the glenoid roof (yellow dashed line), all characteristics of this growth phase. Skull 2857 from the Museum of Anthropology at the University of Florence

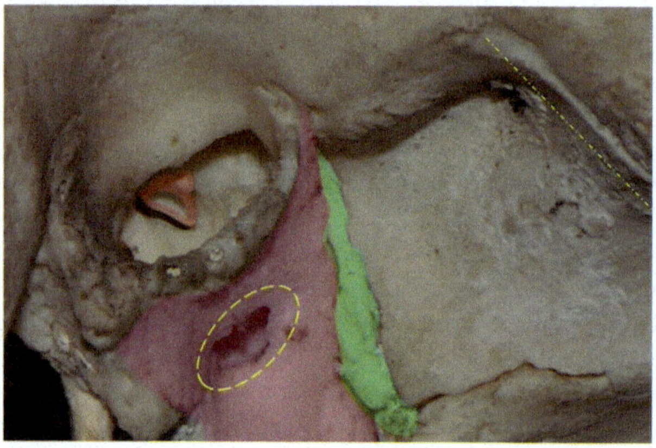

Fig. 4.85 Adult individual: lateral-inferior view of the mandibular fossa and external auditory meatus. Note: Huschke's foramen (highlighted in yellow) on the posterior wall of the mandibular fossa (pink); the stapes (red), visible through the external auditory meatus and still resting on the oval window on the medial wall of the tympanic cavity; the area of the petro-tympanic fissure (green), dividing the mandibular fossa into anterior and posterior portions; and the accentuated inclination (characteristic of adulthood) of the glenoid roof (yellow dashed line). Skull 1390 from the Museum of Anthropology at the University of Florence

As mentioned earlier, the posterior part of the mandibular fossa has a delicate bony structure that is ill-suited to resist the posterior movements of the condyle. The fragility of this anatomical volume can be considered one of the cranial adaptations related to the evolutionary processes of bipedalism and encephalization.

In contemporary humans, the delicateness of this area is challenged by the increasingly common underdevelopment of the maxillaries, often associated with an overbite. This anterior dental wall, represented by the upper incisors, forces the mandible to adopt trajectories during chewing that lead the condyle, during

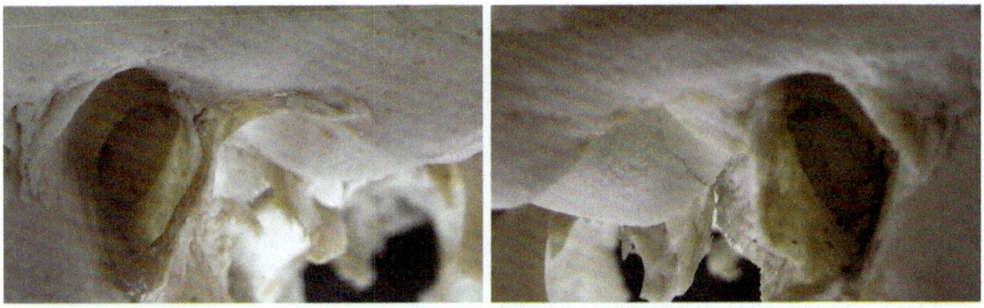

Fig. 4.86 Posterior wall of the mandibular fossa/anterior wall of the tympanic cavity: note the extreme thinness of this bony wall. Skull 327, collection of the Museum of Anthropology at the University of Florence

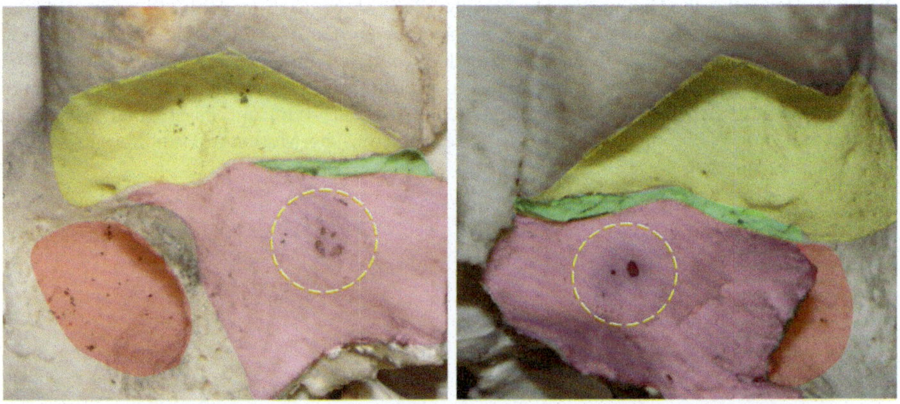

Fig. 4.87 Bottom view of the mandibular fossa: left image shows the right mandibular fossa (skull 1785); right image shows the left mandibular fossa (skull 4156). Note: the translucent area (highlighted in yellow) of the posterior wall (pink) with small rounded gaps; the anterior articular portion (yellow) with two characteristic functional facets (dashed lines); the external auditory meatus (red); the area of the petro-tympanic fissure (green). Museum of Anthropology at the University of Florence

maximum chewing effort, into posterior areas that are non-functional and therefore pathological in the mandibular fossa.

Under physiological conditions, the only communication that persists in adulthood between the mandibular fossa and the tympanic cavity is found in the medial portion of the petrotympanic fissure and is represented by Civinini's canal (Buffoli et al., 2014) (Fig. 4.88).

In the human body, within the Civinini canal, two distinct fibrous structures are clearly present: the disco-malleolar ligament (DML), which connects the tissues behind the eardrum to the hammer bone; and the anterior malleolar ligament (AML), which originates from the sphenomandibular ligament and links the hammer bone to the Spix's spine located on the inner side of the jawbone. Both of these ligaments attach to the front process of the hammer bone, which extends from the neck of this

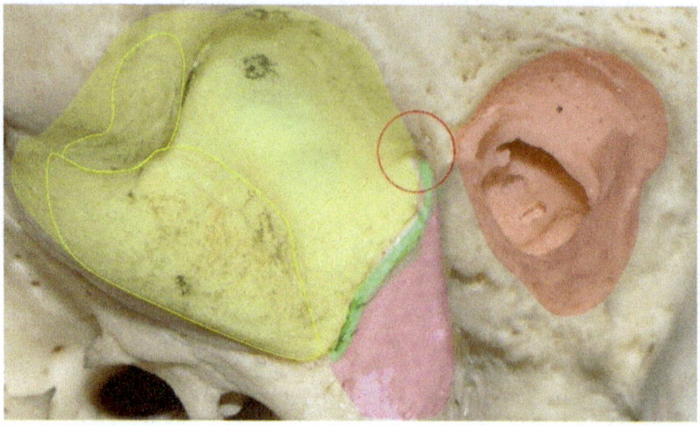

Fig. 4.88 Lateral-inferior perspective of the mandibular fossa and external auditory meatus (red): highlighted in red is the Civinini canal where the fibers of the disco-malleolar ligament and anterior malleolar ligament converge; note, in the anterior portion of the mandibular fossa (yellow) responsible for articulating with the corresponding condylar surfaces via the interposition of the meniscus, the two characteristic mirror-image functional facets (highlighted in yellow) present on the condylar articular surface; in green, the area of the petro-tympanic fissure, and in pink, the posterior portion of the mandibular fossa. Specimen from the Museum of Anthropology at the University of Florence

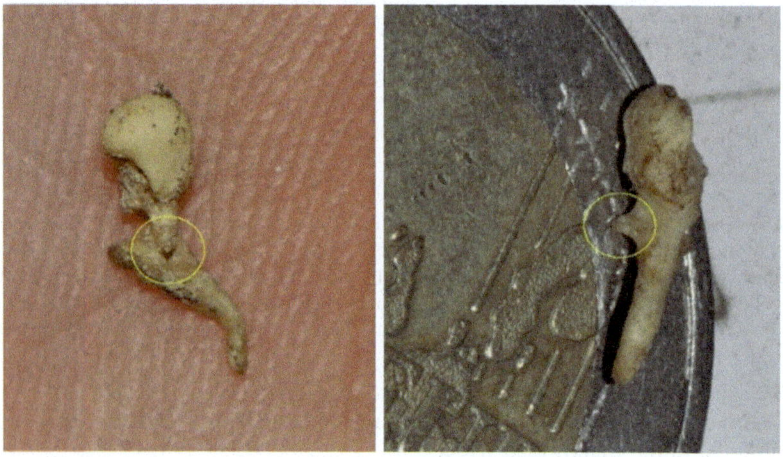

Fig. 4.89 Human malleus: highlighted in yellow is the anterior process. Specimen from the Museum of Anthropology at the University of Florence

small bone. In this area, the hammer bone has a diameter of approximately 1 mm and a total length of about 4/5 mm (see Figs. 4.89, 4.90, 4.91, 4.92).

In the course of human evolution, significant changes occurred in the structures and relationships between the temporomandibular joint (TMJ) and the ossicular chain. Bipedal locomotion and the reduction in canine tooth size have long been considered as hallmarks of humanity, as they appear in the earliest human fossils

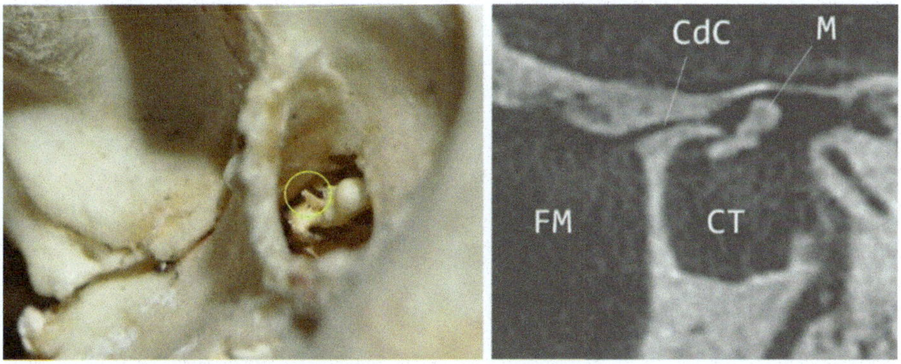

Fig. 4.90 Left: highlighted in yellow is the anterior process of the malleus; right: cone beam CT scan in sagittal section (FM mandibular fossa, CT tympanic cavity, CdC Civinini canal, M malleus). Specimen from the Museum of Anthropology at the University of Florence

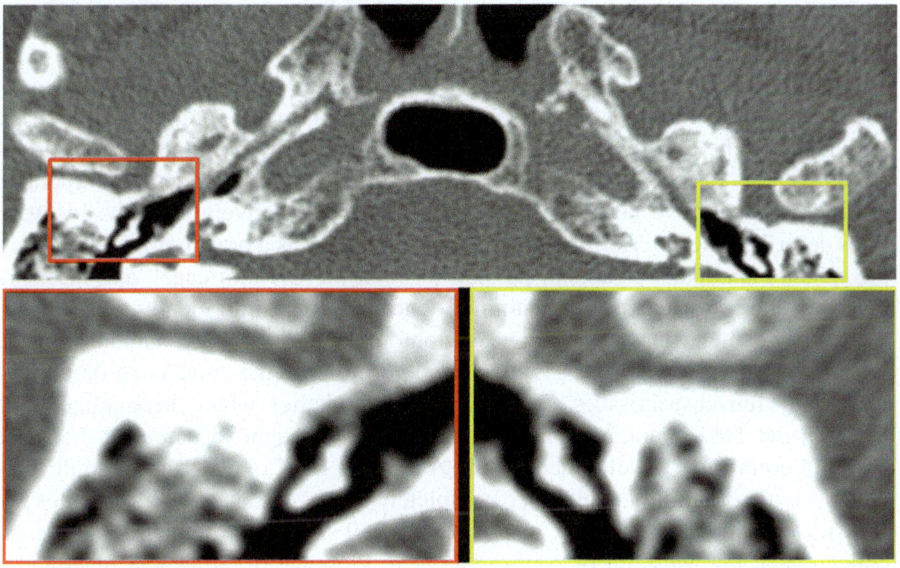

Fig. 4.91 Cone beam CT scan, horizontal section: highlighting the malleus and its ligamentous relationship with the Civinini canal

discovered to date. However, this list of characteristics should be expanded to include changes in the hammer bone. In the early hominids (*Paranthropus robustus, Australopithecus africanus*), the hammer bone clearly resembles the human hammer bone in terms of the proportions of the handle and body, while the anvil and stirrup bear a closer resemblance to those of large African and Asian primates. This suggests that the anvil and stirrup evolved towards their current human form later in our evolutionary history (Quam et al., 2013).

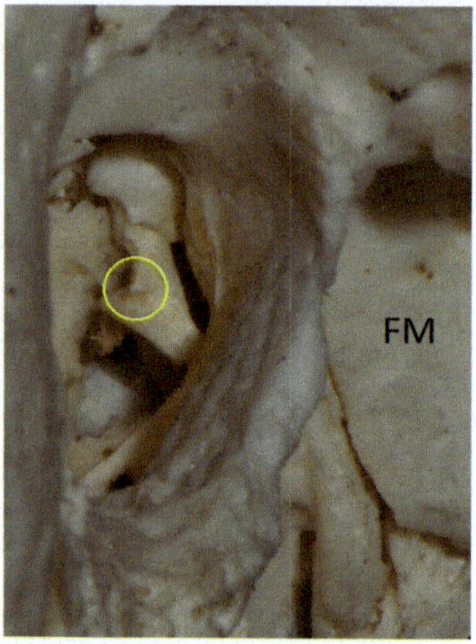

Fig. 4.92 Through the external auditory meatus, the malleus is visible with its anterior process highlighted in yellow (FM: mandibular fossa). Specimen from the Museum of Anthropology at the University of Florence

Within the Civinini canal, the chorda tympani nerve also runs, along with a connection between small venous vessels that connect from the anterior region of the middle ear to the retrodiscal venous plexus, ultimately draining into the retromandibular venous system (Rodriguez Vazquez et al., 1993). The ligamentous structures within the canal have been identified and described relatively recently. In 1954, Rees described the anterior malleolar ligament (AML) as a branch originating from the sphenomandibular ligament, passing through the Civinini canal, and inserting onto the anterior process of the hammer bone. In 1962, Pinto identified the disco-malleolar ligament (DML), which inserts onto the anterior process and neck of the hammer bone, traversing the Civinini canal and joining the connective tissue network of the retrodiscal tissues. Contrary to previous belief (where these connections between the TMJ, mandible, and middle ear were thought to be only a transient phase in embryonic development), both AML and DML persist in the adult individual, as the Civinini canal never completely closes (Rodriguez Vazquez et al., 1998). In an unpublished study conducted on 219 skulls from the anthropological collection of the Museum of Anthropology in Florence (totaling 438 mandibular fossae), the Civinini canal remained patent in over 95% of the samples.

The canal can be divided medio-laterally into three portions, determined by the paths of the chorda tympani, AML, and DML. The bony portions of the canal mirror the anatomy of the ligaments that pass through it. For the DML, it is wide and flat, taking on a triangular shape, while the canal for the AML is narrow and deep with a linear form (Mérida-Velasco et al., 2012) (see Fig. 4.93).

The pathways of the DML and AML are often separated by a bony spur or septum, and they run independently (Mérida-Velasco et al., 2012) (see Figs. 4.94, 4.95, 4.96, 4.97).

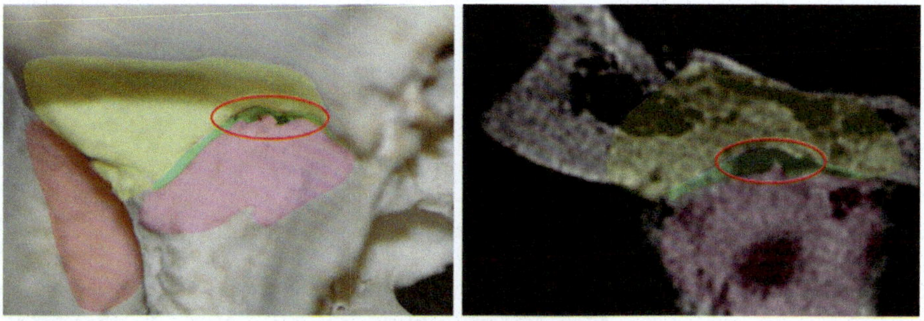

Fig. 4.93 Mandibular fossa: red, external auditory meatus; yellow, anterior portion of the mandibular fossa; green, area of the petro-tympanic fissure; highlighted in red, the opening of the Civinini canal with the thin bony spur that frequently divides it lateromedially. Left: antero-inferior view; right: cone beam CT scan, coronal section. Specimen from the Museum of Anthropology at the University of Florence

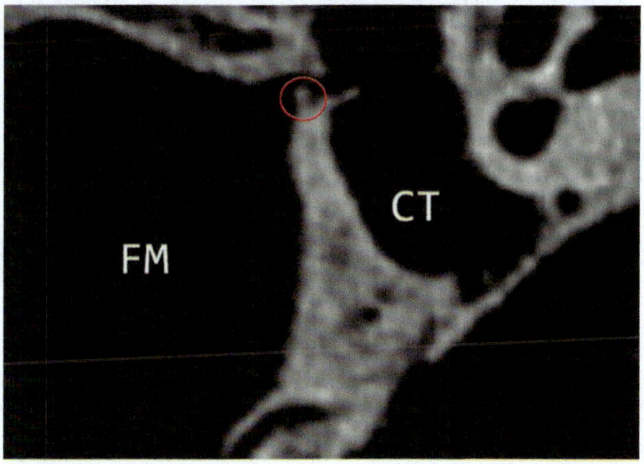

Fig. 4.94 Cone beam CT scan in sagittal section of the posterior portion of the mandibular fossa: FM, mandibular fossa; CT, tympanic cavity; highlighted in red, the thin bony spur dividing the Civinini canal. Specimen from the Museum of Anthropology at the University of Florence

The DML appears as a flat, triangular-shaped band of connective tissue lying on the horizontal plane, crossing the Civinini canal laterally to the AML and medially to the chorda tympani (Rowicki & Zakrzewska, 2006). The base of this triangle is continuous with the posteromedial portion of the disk and the joint capsule. The DML contains elastic fibers (Anastasi et al., 2020), which are absent in the AML (Sencimen et al., 2009), and two proteins, fibronectin and tenascin C (an elastic protein that connects to fibronectin), which are mostly arranged along the collagen fibers of the DML. This suggests that the DML is an elastic structure with tensile capabilities, capable of biomechanically integrating with the stretching of retrodiscal tissues during mandibular movements (Arai & Sato, 2011).

In humans, during the course of the ligaments within the Civinini canal, a contingent of fibers originating from the retrodiscal region attaches to the walls of the petrotympanic fissure and the canal walls. This adhesion is not considered capable

Fig. 4.95 Posterior wall of the left mandibular fossa. Civinini canal divided by the thin bony spur that frequently separates the course of the two ligaments. Specimen from the Museum of Anthropology at the University of Florence

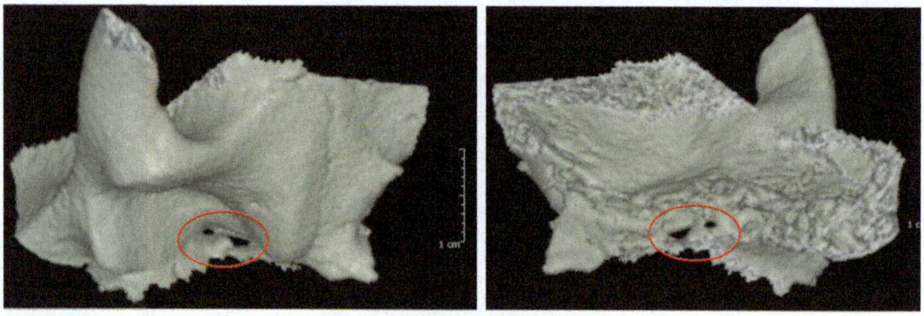

A B

Fig. 4.96 Cone beam CT of the posterior wall of the mandibular fossa; showing the Civinini canal in its medial portion. (**b**) Posterior view of the same anatomical volume as in (**a**); the access to the canal is observed as if from inside the middle ear cavity. Specimen from the Museum of Anthropology, University of Florence

of preventing nanometric transmission movements caused by the traction of retrodiscal tissues associated with anterior movement of the condyle (Eckerdal, 1991).

The AML has a different structure and arrangement compared to the DML. It possesses a more rounded morphology and a more medial course. It originates from the sphenomandibular ligament, which runs from the spine of the sphenoid bone to the upper edge of the mandibular foramen on the inner surface of the ascending branch of the mandible.

In the literature, there are numerous studies that link temporomandibular joint (TMJ) disorders to those of the middle ear. The action of the ligaments connecting the hammer bone to the TMJ and the mandible is crucial because abnormal tension in the DML at the attachment point of the hammer bone can lead to issues within the hammer-anvil-stirrup chain (Loughner et al., 1989).

TMJ disorders can result in increased stiffness of the middle ear (Riga et al., 2010), and there are significant correlations between ipsilateral tinnitus (ringing in

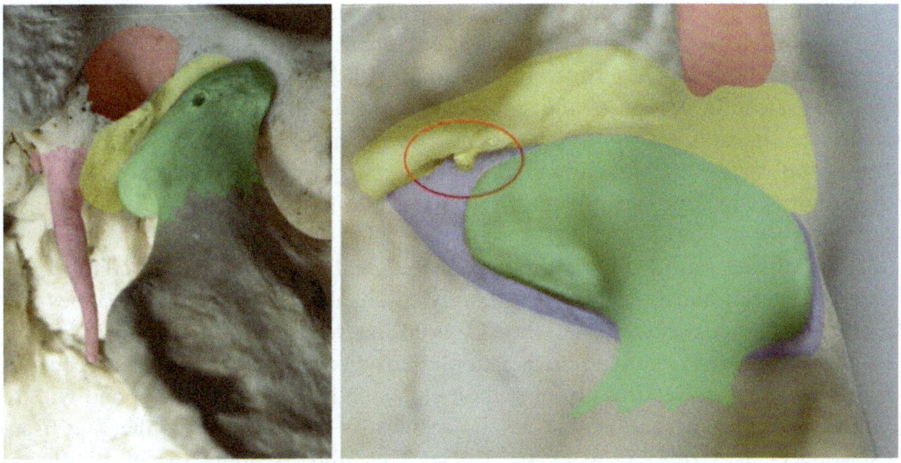

Fig. 4.97 Relationships between condyle, mandibular fossa, and acoustic meatus. Left: infero-posterior view; right: inferior view. In red: acoustic meatus; in pink: styloid process; in green: condylar process; in yellow: posterior portion of the mandibular fossa; in blue: anterior portion of the mandibular fossa; highlighted in red, the emergence of the condylar process with the bony spur dividing it. Specimen from the Museum of Anthropology, University of Florence

the ear on the same side) and TMJ disorders related to disc displacement (Ren & Isberg, 1995). Tinnitus or acufene (the perception of sound in the absence of an actual auditory stimulus) is a complex issue with multiple causes and is often reported in patients with TMJ disorders. The origin of tinnitus can be multifactorial (Haider, 2015; San Juan & Basura, 2019), and it can be useful for clinicians to try and determine if there is a genuine connection between the TMJ disorder and tinnitus in a particular individual.

For this purpose, since this phantom perception can change in quality and intensity during extreme jaw movements, patients are asked to perform maximum mouth opening and maximum lateral movements without occlusal contact in a noise-free environment, after paying close attention to the quality and intensity of their tinnitus. If the perceived sound changes with these jaw movements, it can reasonably suggest a correlation, perhaps partial, between the two disorders.

The DML can undergo biomechanical tension due to functional disturbances caused by disc displacement or inflammatory edema. Disc displacement and an increase in intra-articular pressure from edema can create tension on the AML. TMJ disorders, disc dislocations, edema, and inflammatory disorders can all lead to tension on the hammer bone through the Civinini canal (Isberg et al., 1986; Myers, 1988; Williamson, 1990).

In TMJ disorders, attention is often primarily focused on anterior disc displacement, which generates the most noticeable symptoms, such as clicking or joint locking. However, a dislocated disc (typically anteriorly) corresponds to a condyle that changes its position within the fossa, shifting to a more posterior position.

In individuals with painful disc dislocation in the joint, histological changes in the retrodiscal tissues have been observed. These changes include a significant increase in fibroblasts, arterial vessels that are either obliterated or reduced in caliber, and a similarly significant reduction in elastic fibers (Pereira et al., 1996). Retrodiscal tissues are highly elastic structures capable of expanding during forward condylar movement, allowing the rich venous plexus to fill with blood, which then returns when the condyle returns to its physiological position within the fossa (see Fig. 4.98).

The delicate retrodiscal tissues are histologically predisposed to adapt to the antero-posterior movements of the condyle. Thanks to their elasticity, they can increase in volume and return to their original size when the condyle moves back to its position in maximum intercuspation. Retrodiscal tissues, when subjected to repeated compressive stress from pathological condylar retrusion, respond with an inflammatory-type reaction that results in fibrosis of the loose connective tissue (Stasolla et al., 2004; Yang et al., 1992). This fibrosis alters the elasticity and dynamic response of the retroarticular tissues, potentially affecting the dynamics of the ligaments that connect this structure to the hammer bone.

It's worth noting that the degree of tension these ligaments can exert on the hammer bone depends on various anatomical characteristics, including the length and diameter of the canal, as well as the level of insertion of their fibers into the walls of the Civinini canal (Cheynet et al., 2003; Şencimen et al., 2008; Vázquez et al., 1992).

While the anatomical relationships between the Civinini canal and the ligaments that run through it have been precisely defined, the function of these two elements remains less clear. They connect two anatomically complex and refined areas with entirely different activities and functions. At present, the limited knowledge regarding the purposes of these ligaments can be effectively represented by this

Fig. 4.98 Dissection of a human TMJ: TR retrodiscal tissue, D disc, C mandibular condyle

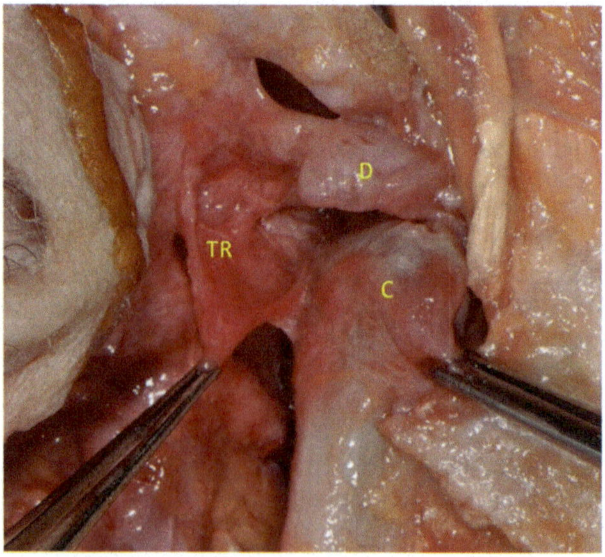

disheartening statement regarding the DML: "[the disco-malleolar ligament] has no physiological role but can be responsible for the dislocation of the ossicular chain of the ear after temporomandibular dislocation or trauma" (Cheynet et al., 2003). Similarly, for the anterior process of the hammer bone, most studies do not provide any theories regarding its function. It is suggested that the anterior malleolar process, in combination with the AML, acts as the pivot point for the rotation axis of the hammer/anvil complex, also serving as a cushion for movements stimulated by low frequencies (Proops et al., 1984). Furthermore, the biomechanical differences between the two ligaments, driven by both anatomical and histological distinctions, should be highlighted. The DML, being more elastic, comes under tension during the stretching of retrodiscal tissues during antero-posterior and antero-medial movements of the condyles. In contrast, the AML, anatomically originating from the sphenomandibular ligament, appears to be stiffer and undergoes tension and relaxation with mouth opening and closing movements (Baragar & Osborn, 1984; Burch, 1970). Therefore, the two ligaments come under tension in different ways during mandibular dynamics, suggesting the hypothesis of a proprioceptive regulation function through the ossicular chain and the vestibular system in managing the mandibulo-cranial relationship (see Fig. 4.99).

In humans, the proximity and connection between the TMJ and the middle ear may still need to be considered, given the possible and frequent negative repercussions of this contiguity. It may be an evolutionary mismatch, a price to pay for the cranial anatomical adaptations induced by bipedalism and encephalization.

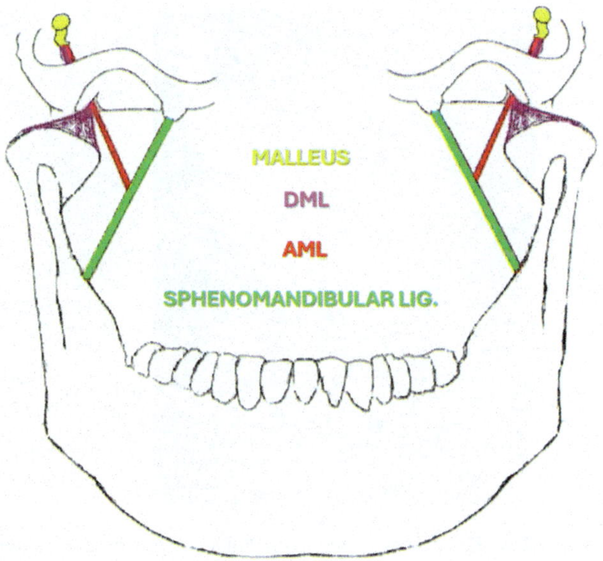

Fig. 4.99 Ligaments crossing through the Civinini canal, connecting the malleus to the TMJ and mandible

4.6 Tilt of the Glenoid Fossa Plane

In hominids, the mandibular fossa is located on the lateral edge of the cranial base and is primarily formed by the squamous portion of the temporal bone. The posterior wall of the fossa is formed by the junction of the anterior surface of the tympanic plate and the lower projection of the squamous portion of the temporal bone, which together create a reduced postglenoid process. Medially, the fossa is closed off by the angular spine of the sphenoid bone, from which the sphenomandibular ligament originates (see Figs. 4.100, 4.101, 4.102).

The petrotympanic fissure (also known as the 'Glaserian fissure') divides the mandibular fossa into two parts: the anterior part belonging to the squamous portion of the temporal bone and the posterior part belonging to the tympanic plate (Durband, 2008).

Among primates, the shape of the articular eminence, although highly variable, is often relatively flat, undefined, and has a slight inclination towards the anterior margin of the glenoid fossa (Terhune, 2011) (see Fig. 4.103).

This 'open' morphology is in sharp contrast to what is observed in humans. One of the differences between the human TMJ and that of great apes is that in humans, the mandibular fossa is characteristically deep (Sergi & Ascenzi, 1974), with the tympanic portion positioned vertically (Lockwood et al., 2002) and with a steep pre-glenoid area (Ashton & Zuckerman, 1954; Kimbel et al., 1986; Weidenreich, 1943) (see Fig. 4.104).

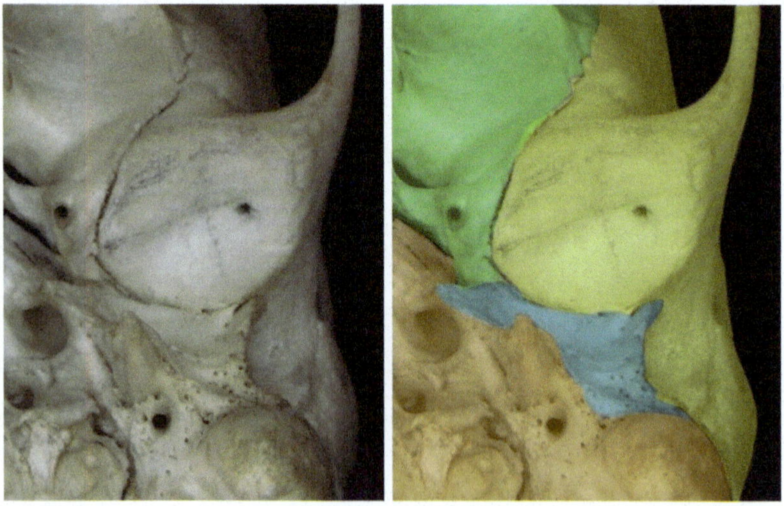

Fig. 4.100 Basal view of the mandibular fossa, with explanatory colorations: yellow, squamous portion of the temporal bone; light blue, tympanic bone; orange, petrous portion of the temporal bone; green, sphenoid bone. Skull specimen 327 from the Museum of Anthropology, University of Florence

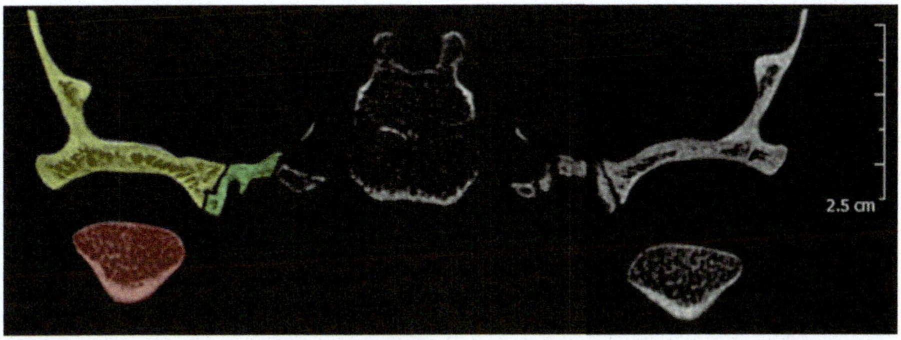

Fig. 4.101 Cone beam CT coronal section: yellow, squamous portion of the temporal bone; green, sphenoid bone; red, condylar head. Specimen from the Museum of Anthropology, University of Florence

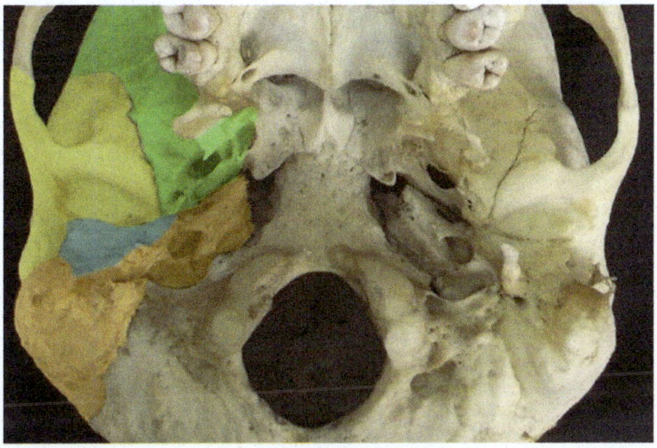

Fig. 4.102 Basal view of the mandibular fossa. Left: various anatomical components colored as follows: yellow, squamous portion of the temporal bone; light blue, tympanic bone; orange, petrous portion of the temporal bone; green, sphenoid bone. Skull specimen 2538 from the Museum of Anthropology, University of Florence

Under normal conditions, the inclination of the roof of the mandibular fossa follows a symmetric growth pattern, with no statistically significant difference in development between the right and left sides (Katsavrias, 2002). Its inclination is calculated in degrees relative to the so-called 'Frankfurt plane' (a plane that runs from the highest point of the external auditory meatus to the lowest point of the orbit). Statistically, this inclination ranges between 30° and 60° (see Fig. 4.105).

The function of the articular eminence of the TMJ in humans does not differ from its function in other primates (Terhune, 2011). This can be observed by examining the inclination of the roof of the mandibular fossa in a chimpanzee. With the skull in its natural posture on the cervical spine, the anterior-superior portion of the mandibular fossa exhibits a steepness similar to that of humans. Typically, the dry skull of a primate is described when observed with the same orientation used for a human skull, with the Frankfurt plane parallel to the horizontal plane. However, in the natural clinograde posture of anthropoid apes, the Frankfurt plane is not parallel to the horizontal plane but inclined forward compared to humans (see Fig. 4.106).

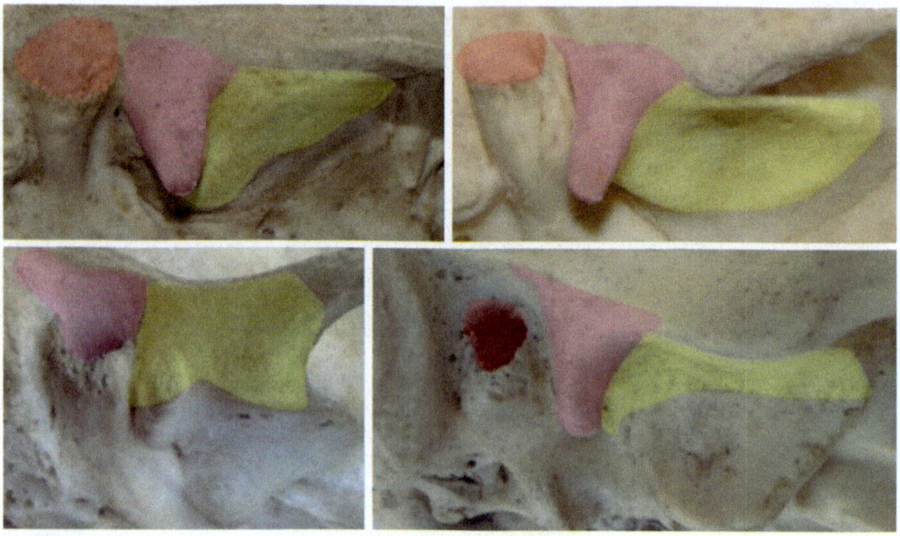

Fig. 4.103 Lateral-inferior view of the roof of the mandibular fossa in several primates: red, external acoustic meatus; pink, postglenoid process; yellow, roof of the mandibular fossa. Clockwise from top left: macaque (*Macaca radiata*); vervet monkey (*Cercopithecus albogularis*); chimpanzee (*Pan troglodytes*); gorilla (*Gorilla gorilla*). Specimens from the Museum of Zoology, University of Florence

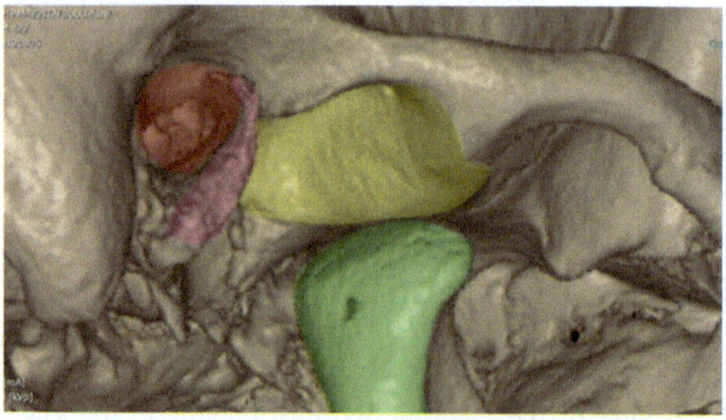

Fig. 4.104 Human mandibular fossa, postero-lateral-inferior view: yellow, glenoid roof of the mandibular fossa; pink, posterior portion of the mandibular fossa related to the tympanic bone (note its distinct verticality); red, external acoustic meatus; green, condylar process. Cone beam CT from a specimen at the Museum of Anthropology, University of Florence

In this position, the steepness of the glenoid fossa roof effectively increases compared to what is observed in the usual examination of a dry skull when placed on a horizontal plane (see Fig. 4.107).

The glenoid fossa in *Australopithecus afarensis* is wide and shallow, and even in *Australopithecus africanus*, it is broad but with a more pronounced anterior articular eminence. In *Homo sapiens*, however, the mandibular fossa, when viewed from

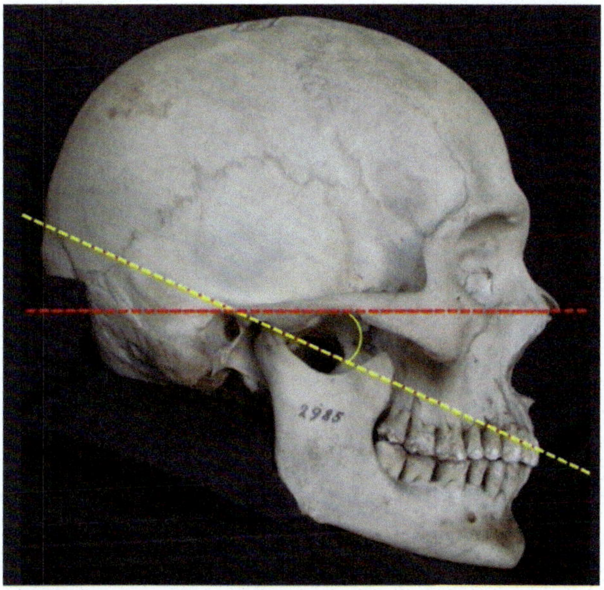

Fig. 4.105 Skull (H. sapiens), lateral norma: inclination of the mandibular fossa roof (yellow line) relative to the Frankfurt plane (red line). Specimen from the Museum of Anthropology, University of Florence

the side, is generally markedly deeper (Aiello & Dean, 1990). With the acquisition of bipedalism in an upright posture, there was a modification in the cranial alignment, with an extension (posterior rotation in the sagittal plane) of the cranial base. However, the masticatory function remained not too dissimilar from that of our ancestors. In compensatory adaptation to this extension, which would have led to a relative flattening of the glenoid fossa roof of the TMJ in humans, there occurred a deepening of the mandibular fossa (Kieser, 1997) (see Fig. 4.108).

The remodeling in this area has led to several changes, including the proximity between the middle ear and the mandibular fossa, the slender proximity that separates these two structures with a steeper glenoid fossa roof, the appearance of the Civinini canal, and the two ligaments that connect the hammer bone to the retrodiscal tissues and the mandible. Additionally, the vectorial forces that are distributed on the articular surfaces contribute to making this structure susceptible to joint disorders. The imbalances between the condyle and the meniscus that frequently afflict the TMJ can be considered as issues resulting from evolutionary adaptations of pre-existing architectures.

The temporomandibular joint (TMJ) is not only one of the most recent but also one of the most complex, perhaps even the most complex, among the joints in the human body. Its anatomical and functional delicacy originates from its imperfect constitution, in which evolution has acted like a bricoleur, creating adapted but sometimes imperfect structures from what was available (Pievani, 2019). Because the inclination of the glenoid fossa roof of the mandibular fossa is a cranial parameter, it is evident that it is influenced by head posture. While it may vary daily within a certain range of positions in each individual, it always maintains a characteristic inclination of the species.

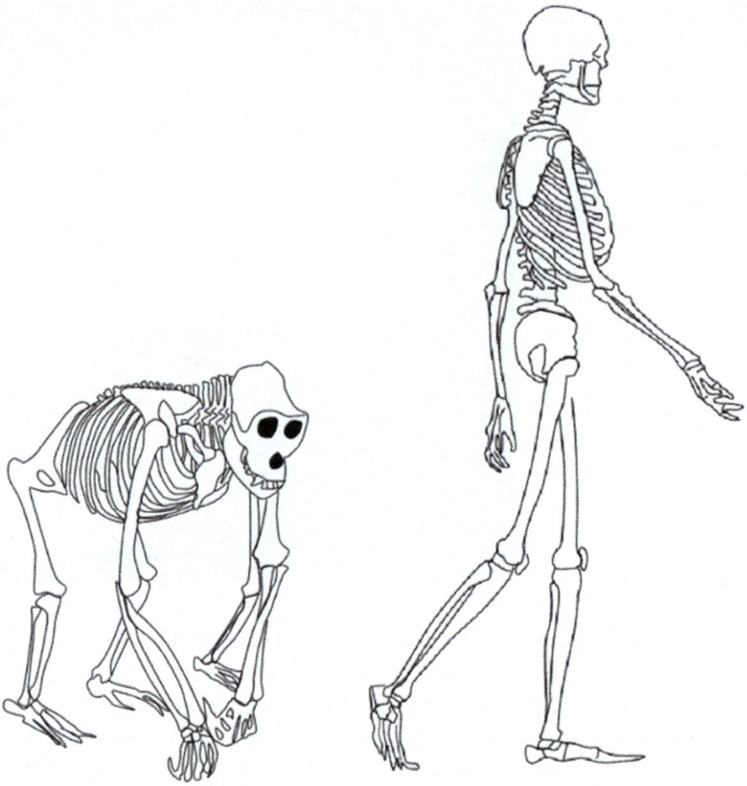

Fig. 4.106 Different cranial inclination between the clinograde posture of anthropoid apes and the erect posture of humans

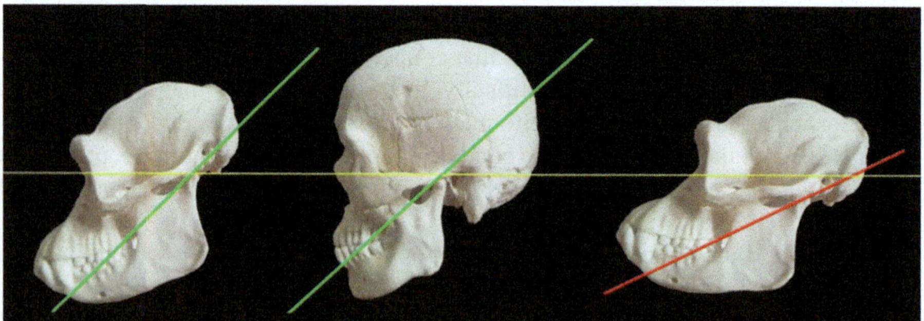

Fig. 4.107 Highlighted in green are the inclinations of the glenoid roofs in natural posture for chimpanzees (*Pan troglodytes*) and humans (*Homo sapiens*). In red is the inclination of the glenoid plane in chimpanzees, not accounting for the natural cranial posture in primates

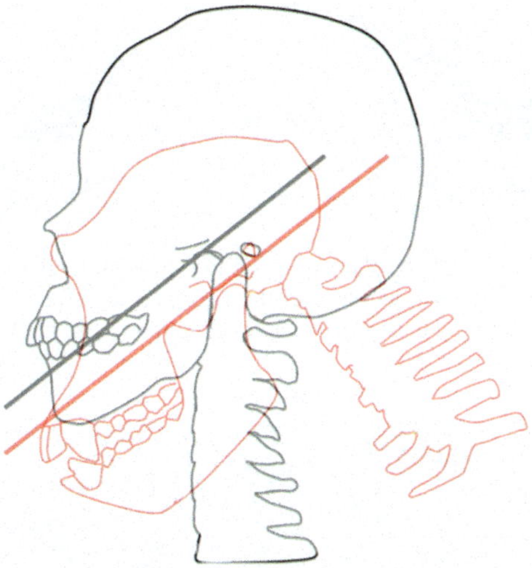

Fig. 4.108 Overlay of a human skull *(Homo sapiens*, in black) with a chimpanzee skull (*Pan troglodytes*, in red): with contours placed in their natural postural positions, the inclinations of the glenoid roofs become nearly overlapping. The figure aims to demonstrate not an exact overlap between human and chimpanzee glenoid roofs, but rather to show that in these two primate species, the inclinations are not markedly different, similar to their masticatory dynamics

The glenoid fossa roof of the mandibular fossa and its overall morphology can partly change according to the main directions of active occlusion forces (Sergi & Ascenzi, 1974). The steepness of the mandibular fossa roof is considered a defining characteristic of the human TMJ and is one of the factors regulating masticatory dynamics. In our stomatognathic system, a close connection exists between the inclination of the glenoid fossa roof and the relationships between occlusal surfaces. It is evident that tooth attrition and loss can influence the height of the articular fossa (Aiello & Dean, 1990). In regulating the excursive movements of protrusion and lateralization in the horizontal plane, there is a close and continuous anatomical-functional interdependence between the inclination of the glenoid fossa roof (posterior guidance) and the overbite or incisal overjet (anterior guidance) (Mack, 1984).

At birth, the articular eminence of the mandibular fossa is flat and develops with the eruption of the first deciduous teeth (Katsavrias, 2002). It rapidly increases in its vertical dimension following the eruption of the first deciduous molars (Ito et al., 1995).

The continuous succession of teeth in the dental arch during dental eruption results in variations in both the depth of the mandibular fossa and the inclination of the glenoid fossa roof. This occurs in response to masticatory patterns generated throughout life and in relation to functional stresses on the dentition (Hinton, 1981; Owen et al., 1992).

However, the characterization of the mandibular fossa roof is not limited to an increase in its inclination in response to dental eruption. The physiological alternation of the chewing side also shapes the articular portion of the TMJ with a unique

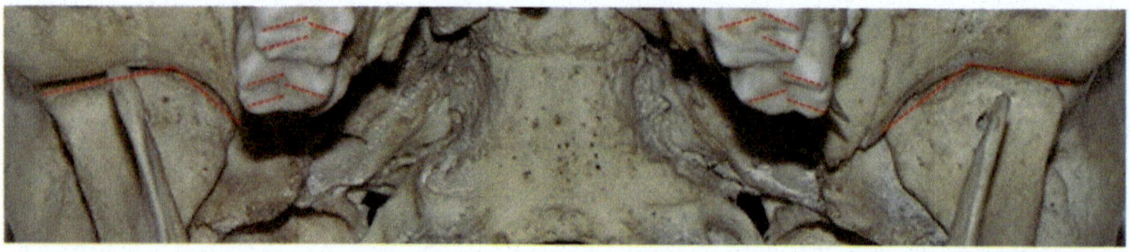

Fig. 4.109 Articular rim of the TMJ shaped by the two characteristic facets from a physiological alternated unilateral masticatory function: note the wear facets on the surfaces of the molars, both on the entry and exit sides of the masticatory cycle. Specimen from the Museum of Anthropology, University of Florence

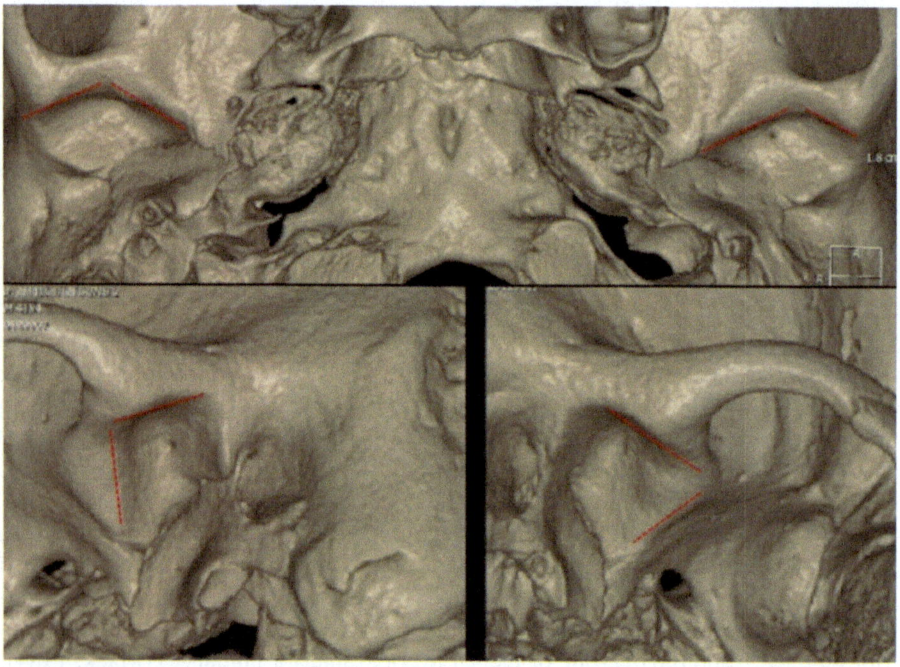

Fig. 4.110 Glenoid roof of the TMJ: top image shows view from below; bottom image shows infero-posterior view. Dashed red lines highlight the two facets shaped by a physiological alternated unilateral masticatory function. Cone beam CT from a specimen at the Museum of Anthropology, University of Florence

form, in which both the anterior-superior surface of the head of the mandibular condyle and the glenoid fossa roof of the mandibular fossa exhibit two typical facets (see Fig. 4.109, 4.110, 4.111).

Dental occlusal surfaces strongly reflect dietary profiles, just as the shape of the TMJs reflects masticatory forces and the range of mandibular movements. Overall, joint morphology and dental form vary in predictable ways according to diet, influencing each other. The triturating occlusal surface, glenoid fossa, mandibular and cranial morphology collectively reflect feeding strategies (Terhune et al., 2015).

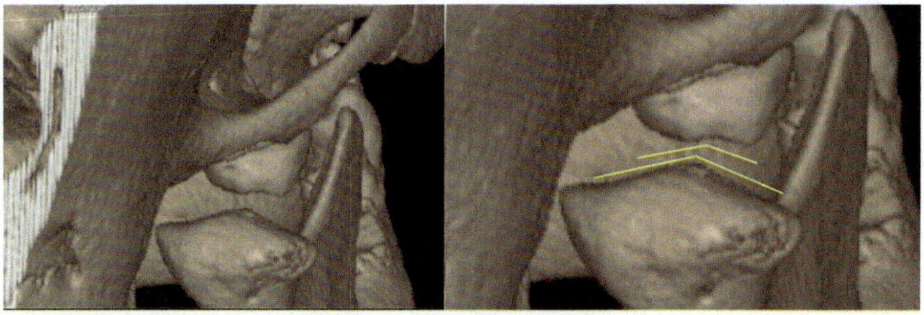

Fig. 4.111 Mandibular condyles, postero-lateral view: the yellow lines highlight the correspondence between the inclined planes of the occlusal molar slopes, which regulate the entry and exit of the chewing cycle, and the condylar facets shaped by a physiological alternating unilateral mastication. Cone beam CT of a specimen from the Museum of Anthropology, University of Florence

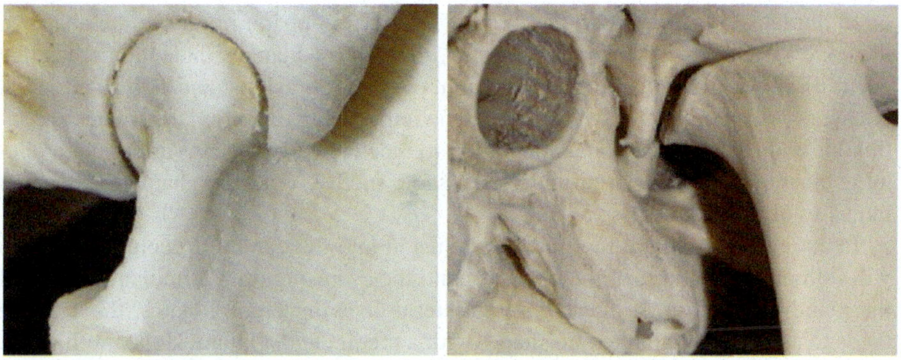

A B

Fig. 4.112 (**a**) Joint space of the temporomandibular joint (TMJ) of a badger (*Meles meles*); (**b**) Joint space of a waterbuck or water antelope (*Kobus ellipsiprymnus*). Specimens from the Museum of Zoology, University of Florence

In carnivorous mammals, the joint dynamics of the TMJ are closely tied to the relationships between their articular surfaces and condyles, which are so constrained to rotation within the fossa that they resemble the hinges of a door: a cylindrical condyle within a cylindrical articular cavity (Fig. 4.112a) with essentially rotary movement and minimal axial translational component. On the other hand, in herbivorous mammals, the articular surfaces of the TMJ tend to be flattened (Fig. 4.112b), with the condylar surface almost concave, adapted to extensive excursive movements in the horizontal plane (Fig. 4.113).

The occlusal morphological evolution of the tribosphenic molar and its differentiation in various species adapted to different evolutionary niches is also concurrent with functional changes in the force vectors of the superficial masseter, medial pterygoid, and temporal muscles (Grossnickle, 2017).

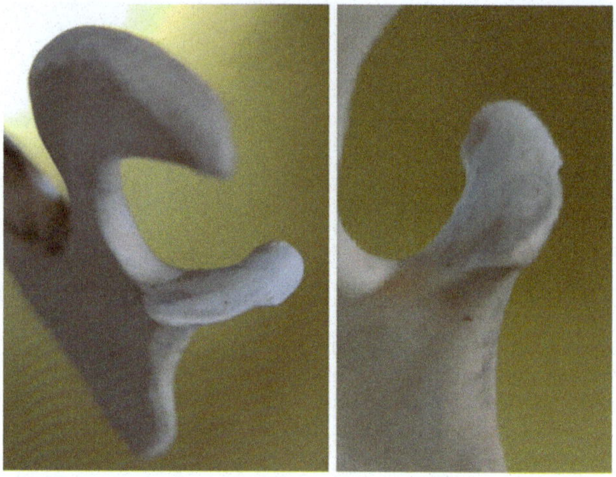

Fig. 4.113 Condyle of a roe deer (*Capreolus capreolus*): the surface appears slightly concave

In humans, too, the extent and type of occlusal abrasion influence the shape of the condyle and fossa, which are strongly affected by changes in dental arches that occur throughout life due to their function, occlusal abrasions, and/or loss of dental elements (Mongini, 1972, 1975, 1977). The remodeling of condylar surfaces is continuous and reflects functional adaptation to mechanical stresses: variations in dental occlusion that occur in an individual's life modify both the characteristics and intensity of these stresses.

In a single individual, the type of occlusion and the extent of occlusal abrasion will influence the shape of the condyles and the corresponding articular surfaces of the mandibular fossa. If alternating unilateral chewing predominates, the condyle tends to assume a characteristic shape, with a medial and a lateral slope, both steep, symmetrical, and evenly developed (Mongini, 1984) (Figs. 4.114 and 4.115).

The anterior surface of the mandibular fossa, on which the condyle slides with the interposition of the articular disc, will also be remodeled by masticatory function and tend to form two surfaces that are consistent with those of the condylar surface (Fig. 4.116).

In an unpublished study conducted by the author using *cone beam computed tomography* (CBCT) scans, the shape of the mandibular condyles in 207 skulls from the anthropological museum in Florence was evaluated (Fig. 4.117). The skulls belonged to individuals from two populations, one from the Syracuse area and the other from the Florence area, who died in the second half of the 1800s, and whose gender, age, and cause of death are known. Given the era in which they lived, we can assume that these individuals represent a population similar to the present population.

It should be noted that in the Florence skull collection, virtually all skulls show distinct functional abrasions of the occlusal surfaces, always related to the age of the individual. Occlusal wear is distributed on both working and balancing sides (indicating a physiological pattern of alternating unilateral chewing) and is practically ubiquitous and consistent across different and numerous populations from around the world. It was therefore possible to identify a natural and physiological prevalent

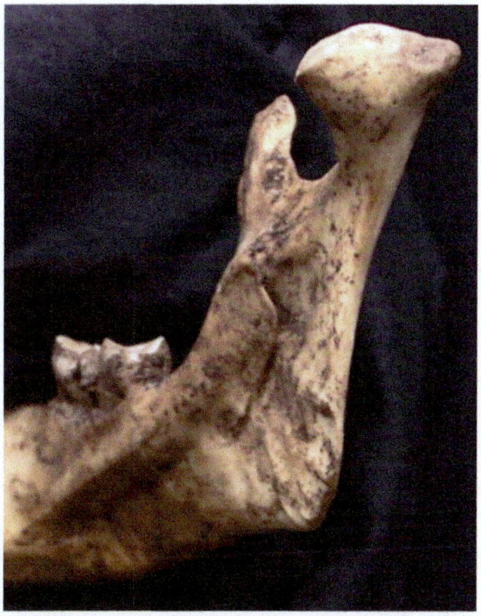

Fig. 4.114 Condylar surfaces shaped by the alternating unilateral function, in the two characteristic functional facets. Specimen from the Museum of Anthropology, University of Florence

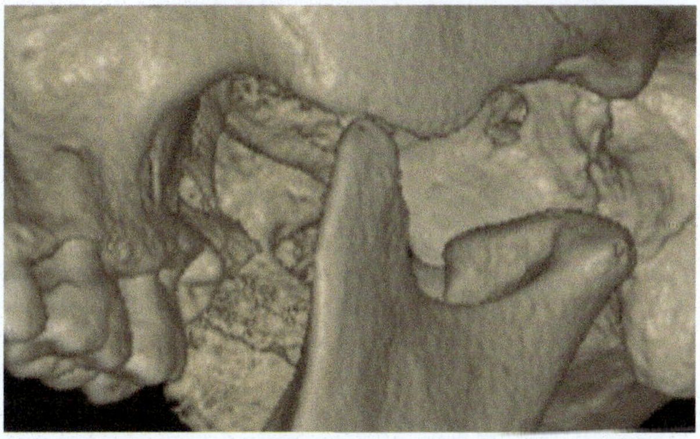

Fig. 4.115 Condylar articular facets: note the correspondence between the inclinations of the occlusal surfaces of the molars and those of the condylar surfaces. Cone beam CT of a specimen from the Museum of Anthropology, University of Florence

shape of the condylar surfaces and glenoid fossa roofs, which exhibit two articular facets shaped by function, characteristic of alternating unilateral chewing (Figs. 4.118, 4.119, 4.120).

The results of the study indicate that in the skulls studied, 83.7% of the observed TMJs have condyles and glenoid fossa roofs with a double surface characterized by two facets, one anteromedial and one anterolateral (Fig. 4.121).

The predominance of articular surfaces characterized by two facets (Fig. 4.122), essentially mirror images between the two condyles, highlights that the prevalent masticatory function during periods when food was not modified by the food industry was physiological alternating unilateral chewing. This is typical of symmetrical masticatory dynamics guided by correct occlusal surfaces for working and balancing.

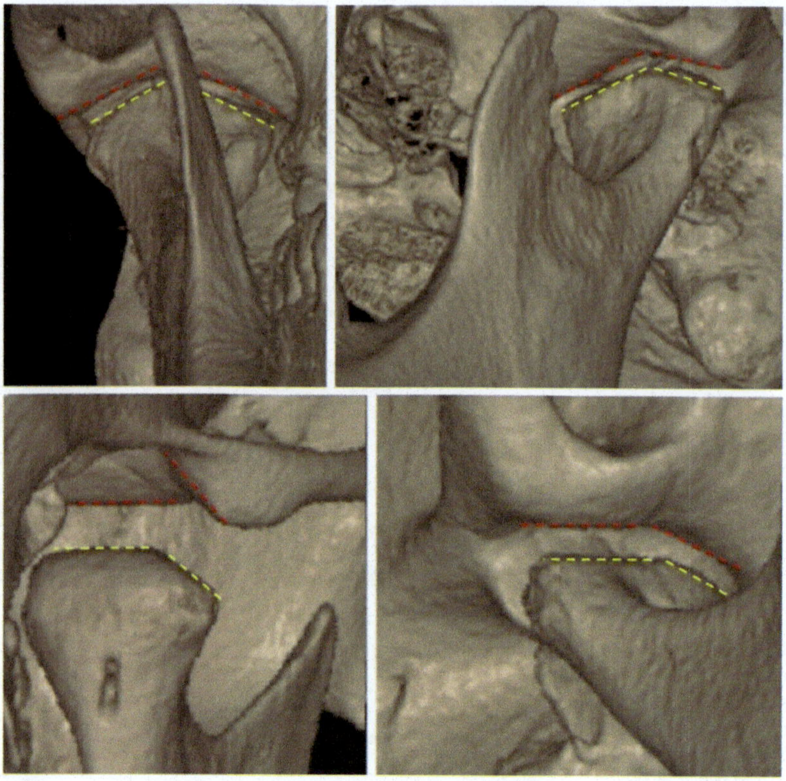

Fig. 4.116 Functional facets present on the articular surfaces of the condyle (yellow) and the mandibular fossa (red): top left shows anterior view, top right shows antero-latero-inferior view; bottom left shows postero-latero-superior view, bottom right shows latero-inferior view. Cone beam CT of a specimen from the Museum of Anthropology, University of Florence

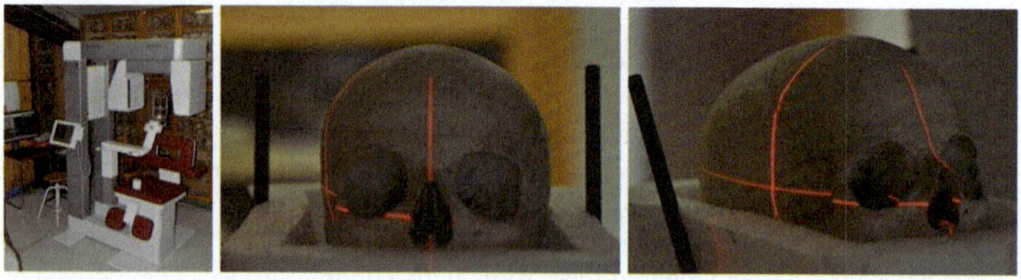

Fig. 4.117 CBCT Scanora Soredex device, temporarily installed in the skull collection of the Museum of Anthropology, University of Florence, specifically for the study of the skull collection

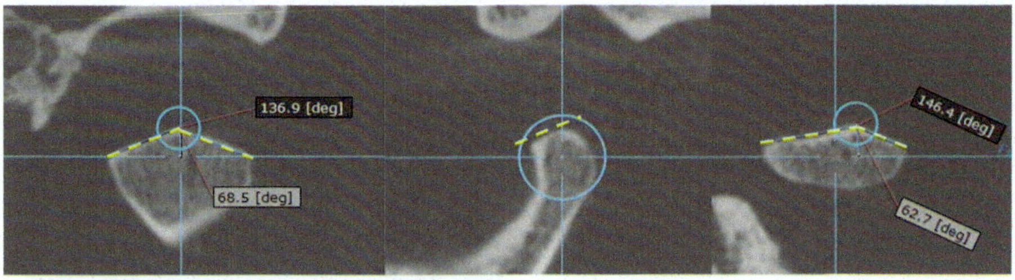

Fig. 4.118 Physiological articular surface of the condyle, developed on two distinct functional facets: left shows coronal section; center shows sagittal section; right shows horizontal section. Cone beam CT of a specimen from the Museum of Anthropology, University of Florence

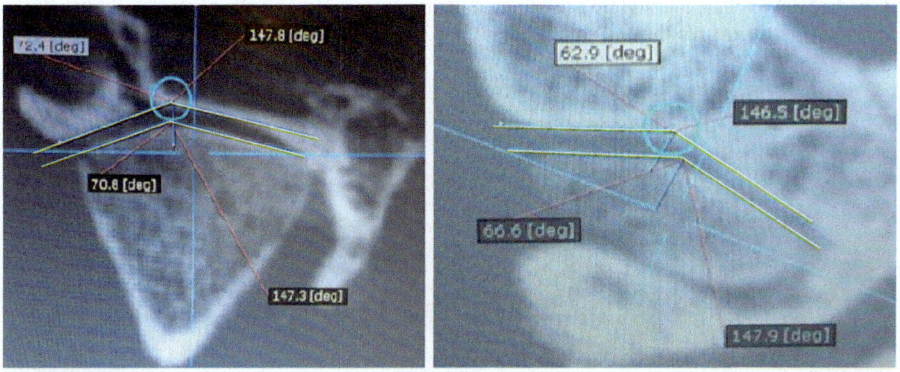

Fig. 4.119 Physiological articular surface of the glenoid roof, developed on two distinct functional facets. Cone beam CT of a specimen from the Museum of Anthropology, University of Florence

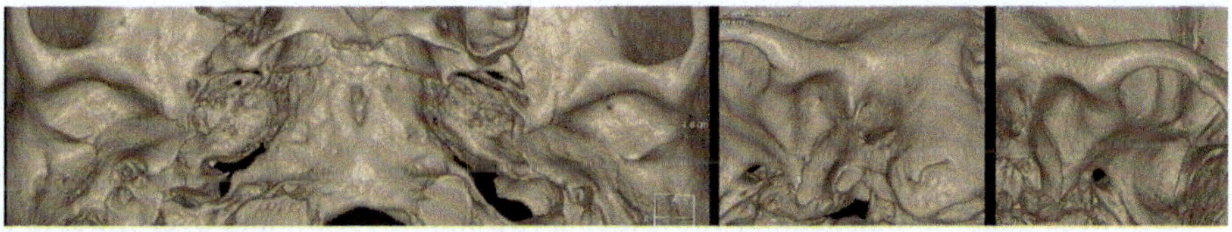

Fig. 4.120 Relationship between the condyle and the mandibular fossa: left shows a coronal section, right shows a horizontal section of the mandibular fossa and the condyle; note the two distinct functional facets and the correspondence between the opposing articular surfaces. Cone beam CT of a specimen from the Museum of Anthropology, University of Florence

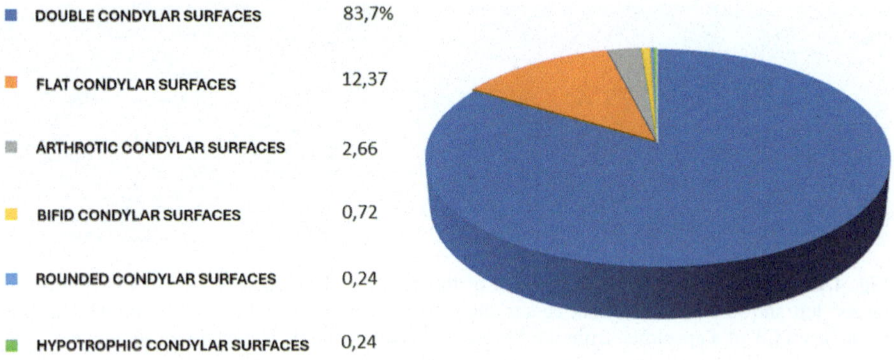

Fig. 4.121 Pie chart highlighting the predominant presence of articular surfaces with two functional facets, antero-medial and antero-lateral

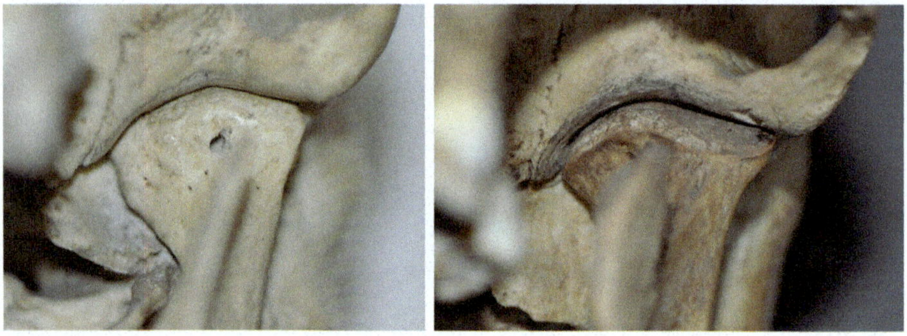

Fig. 4.122 Articular surfaces with two functional facets, antero-medial and antero-lateral. Specimen from the Museum of Anthropology, University of Florence

As discussed in Chap. 3, the shift to mass-produced, processed foods has brought about a profound change in masticatory dynamics, resulting in reduced functional stimuli required for the proper maturation of the form and function of various components of the stomatognathic apparatus, including the articular surfaces of the TMJ. These adapt to masticatory movements (Mongini, 1980), and over time, in a subject with a fixed unilateral chewing pattern, condylar and glenoid fossa roof morphology will adapt, acquiring different anatomical characteristics (Fig. 4.123) compared to the contralateral joint.

The glenoid roofs and functional surfaces of the corresponding condyles can have different inclinations in a single individual based on their 'functional history' (Wall, 1999). For example, individuals whose diet consists of less consistent foods may have smaller condylar dimensions compared to others who consume tougher foods (Scheidegger et al., 2018). In fixed unilateral chewing, on the dominant chewing side, the working condyle will gradually become more globular, and the mandibular fossa will become deeper. On the opposite side, the balancing condyle will have a flatter and more elusive shape, corresponding to a shallower mandibular fossa (Dettori et al., 2001).

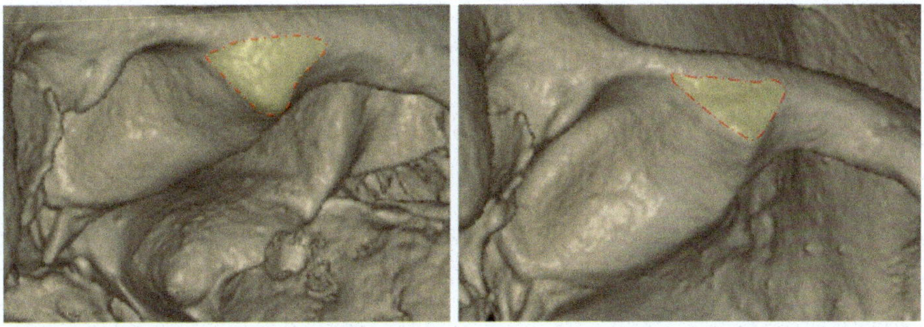

Fig. 4.123 Different shapes of the mandibular condyles (green) and different inclinations of the glenoid roof of the mandibular fossa (yellow dashed lines), between the right and left TMJ: in a subject with fixed unilateral mastication, the working side shows a more globular condyle and a steeper glenoid roof

As discussed extensively earlier, from an evolutionary perspective, fixed unilateral chewing cannot and should not be considered physiological. Mammals chew on one side at a time but alternate the work and distribute stress on both sides, and all primates, including humans, deviate from this pattern. Fixed unilateral chewing creates a potentially pathological dysfunctional situation that can lead to the so-called 'habitual chewing side syndrome.' In this syndrome, the dominant chewing side exhibits the following characteristics: reduced lateral guidance, a higher incidence of condyle-meniscus incoordination, clicking during opening, chronic pain, habitual chewing on the painful side (Mora et al., 2015), as well as asymmetric wear, different occlusal planes, different compensatory curves, etc. (see Fig. 3.86).

4.7 The Structuring of the Temporo-Mandibular Ligament

In the articular capsule of anthropoid apes, there exists a broad and resistant external oblique ligament with an antigravity function that maintains close contact between the condyle, disc, and articular eminence, both in static conditions and during opening and closing movements (Dean, 1990).

The temporomandibular ligament (LTM), which is notably reduced or absent in primates, is a unique anatomical feature in humans. The condylar dynamics in humans confirm the principle that accessory ligaments have evolved around the temporomandibular joint (TMJ) to reduce the need for neuromuscular control by providing mechanical support. The LTM is fully developed in the *Homo* genus, and the external surface of the articular tubercle serves as its point of insertion. The size of the articular tubercle is likely proportional to the size of the ligament itself (Wall, 1999). The LTM originates from the external portion of the articular tubercle and the lower margin of the zygomatic arch (Fig. 4.124), and it can be rightfully considered the conductor of TMJ movements because its configuration influences multiple aspects of the dynamics of this joint (Osborn, 1989).

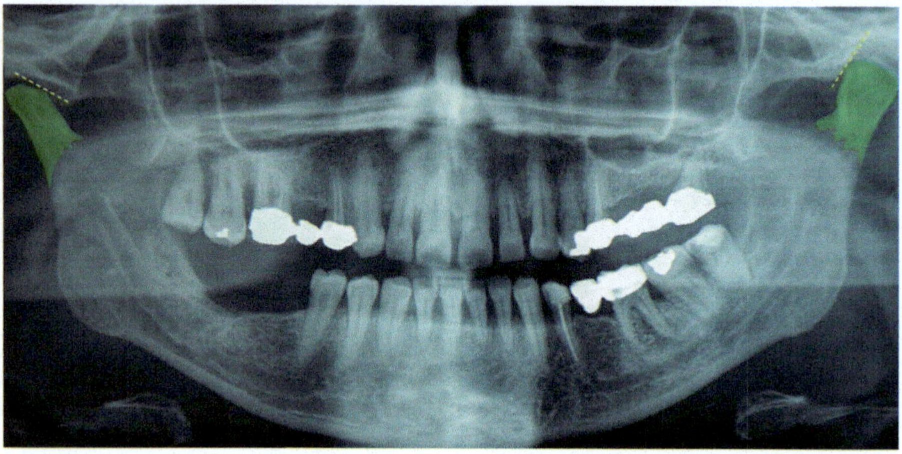

Fig. 4.124 Different shape of the mandibular condyles (green) and different inclination of the glenoid roof of the mandibular fossa (dashed yellow lines), between the right and left TMJs. In a subject with fixed unilateral chewing, the working side has a more globular condyle and a steeper glenoid roof

Fig. 4.125 Different orientation of the fibers of the temporomandibular ligament (TML) running inside the articular capsule: red, fibers of the oblique portion; yellow, fibers of the horizontal portion; green, articular disc. Specimen from the Museum of Anthropology, University of Florence

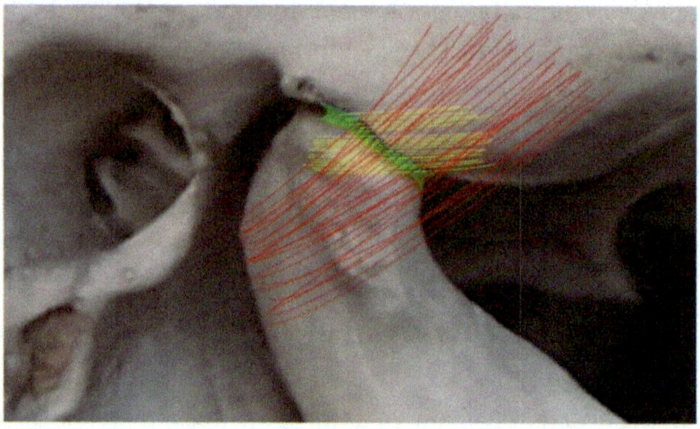

Its fibers insert externally and posteriorly onto the neck of the mandibular condyle (Fig. 4.125).

Classical gnathology describes condylar movement during mouth opening as initially involving rotation due to the action of the superior head of the external pterygoid muscle, resulting in an opening of approximately 10–15°. However, pure condylar rotation at the start of mouth opening is primarily an experimental observation (Rees, 1954). In reality, the condyle undergoes an immediate rotation-translation motion, not solely in response to the vectorial forces of the muscles acting on the mandible, but rather due to the constraints imposed by the shape of the joint and the presence of the LTM (Osborn, 1989). A body subject to constraints does not necessarily move in the direction of the applied force. For example, a door, regardless of the direction of the force acting upon it, can only swing open or closed along the axis defined by its hinges.

There are several anatomic and functional aspects that cast doubt on the notion of pure condylar rotation at the onset of mouth opening:

1. Early studies on TMJ movement were conducted by manually inducing movements in cadavers placed in a supine position, without the influence of neuromuscular control, and with the mandible held in a closed position (Osborn, 1989).
2. Due to its "antigravity" structure, the LTM is naturally taut and contributes to maintaining contact between the articular surfaces even during mandibular movements (Cuccia et al., 2011).
3. When the opening movement begins, the condylar head, being non-spherical, increases its radius of rotation, further accentuating the tension of the oblique fascia. This occurs regardless of the depth of the articular eminence: a deep and steep articular eminence only results in increased tension compared to a shallow and less inclined eminence (Wall, 1999). The eccentric rotation of the condyle immediately causes the LTM to act like the chain of a swing, with the mandible rotating around the lower insertion point of the LTM and swinging around the posterior insertion of the articular tubercle. During the condylar rotation on the sagittal plane, the articular eminence and the LTM irresistibly induce the condyle to translate, moving downward and forward (Osborn, 1989).
4. During the opening movement of the mandible, the lower portion of the external pterygoid muscle is active and, in synergy with the suprahyoid muscle group, assists in the downward translation of the condylar head. In this phase, the upper portion of the external pterygoid muscle, responsible for positioning and stabilizing the condylar head and the disc against the articular eminence during mandibular closing movements, is not active. Therefore, the different components of the external pterygoid muscle can be considered as functionally distinct muscle bands (McNamara Jr. & Moyers, 1973).

The natural opening movement is a simultaneous combination of rotation and translation, where the rotational component is slightly more significant in the initial phase compared to the later phase of opening (Chen, 1998). In normal subjects, mouth opening and closing are determined more by mandibular rotation than translation. However, in no case has pure rotation been found (Mapelli et al., 2009) because the already-taut oblique fascia immediately triggers the translational movement, resulting in an immediate condylar rotation-translation. Ligaments stabilize joints by assisting and guiding them through their normal range of motion. The stability of the TMJ is primarily attributed to ligaments that protect the joint from biomechanical stresses. The temporomandibular ligament (LTM) serves as a limiting and containing element of the joint dynamics and is characterized by two distinct portions: a deeper, horizontal portion and an outer, oblique portion, which behave differently in controlling various functional movements (Fig. 4.126).

Just like the external pterygoid muscle, even though the LTM is considered a single structure, it has two parts that serve different functions and needs: the anti-gravitational function of the oblique band of the ligament is more closely related to the acquisition of an upright posture, while the anti-retrusive function of the horizontal band seems to be more related to cranial remodeling processes resulting from encephalization.

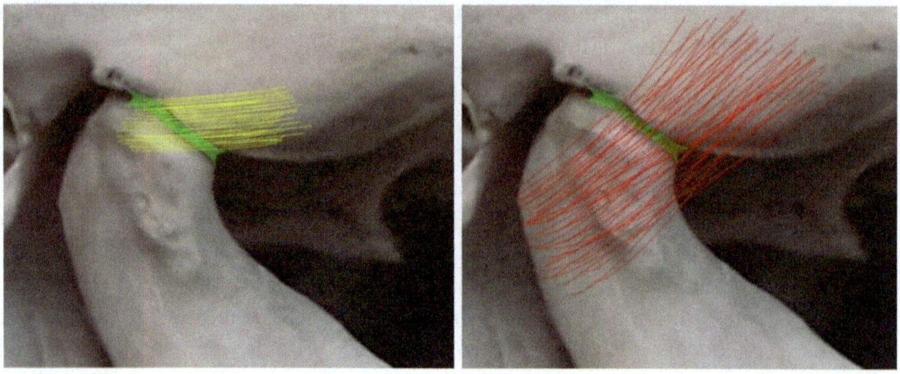

Fig. 4.126 TML: fibers of the oblique portion (in red) and fibers of the horizontal portion (in yellow); in green, the articular disc. Specimen from the Museum of Anthropology, University of Florence

4.7.1 Oblique Portion of the Temporomandibular Ligament

The oblique ligamentous fibers ensure the maintenance of mutual contact between the articular surfaces and, in humans, are functionally related to posture due to the antigravitational action they exert in opposing the separation of the articular surfaces. The oblique band controls the initial phase of mouth opening by limiting the rotational component of the condyle and initiating the translational component (Osborn, 1989).

The tension in the oblique band of the LTM is activated due to both the morphology of the condyle (its rotation immediately increases the range of movement) and the cranial extension movement associated with the opening phase. As we have seen, the oblique band acts as a swing chain that induces the rotation of the mandible around the ligament's insertions on the condyle and around the posterior insertion on the articular tubercle (Osborn, 1993a, 1993b) (Fig. 4.127).

The condylar translation initiates movements of the mandible on the horizontal plane. In opening and closing the mouth, the sliding of the condyle on the glenoid fossa is necessary to make the chewing cycle efficient (Smith, 1985). The cutting and grinding of food require continuous adjustments of movements on the horizontal plane to avoid damaging impacts (interferences). These continuous adaptations of mandibular dynamics are made possible by the oblique band of the LTM, which allows the condyles to slide forward and downward against the articular eminence (Katsavrias, 2002). Furthermore, during mouth opening, the oblique band of the LTM acts as a biomechanical constraint, by translating the condyle forward, reducing the posterior movement of the region near the mandibular angle that would occur with just mandibular rotation alone (Fig. 4.128).

The upright posture led to the verticalization of the spinal column, the forward displacement of the occipital foramen below the cranial box, and the retrusion of the

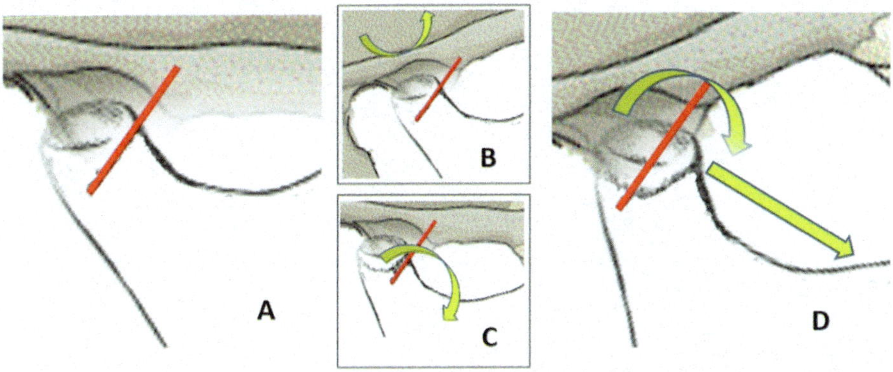

Fig. 4.127 (**a**) Position of the condyle with the arches in maximum intercuspation and the oblique band of the TML in anti-gravity tension; (**b**) cranial extension; (**c**) mandibular rotation; (**d**) both these movements, however, occur simultaneously, accentuating the tension of the TML, which, like a chain, causes the condyle and the disc to translate over the glenoid roof

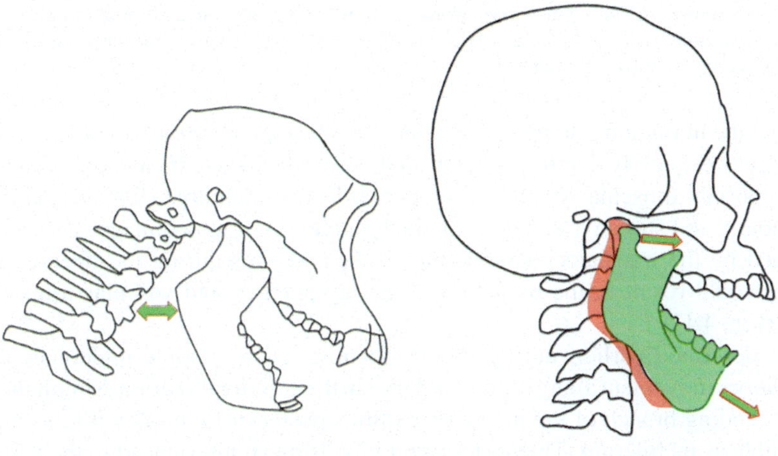

Fig. 4.128 The TML in humans is structured as a constraint related to the upright posture, necessary to maintain the patency of the airways that could be hindered by excessive retraction of the mandibular angle and the connected soft tissues. Left: mandibular outline of a chimpanzee; note in anthropomorphic apes how the clinograde posture determines a spatial relationship between the mandible, cervical spine, and interposed airways that allows greater patency of the latter when the mouth is maximally opened; right: mandibular outline in *Homo sapiens*, with the red area showing the opening due to rotation alone, and the green area showing the opening in rototranslation: note, with equal opening, the decreased posterior positioning of the posterior portion of the mandible due to the translational movement

facial structure. These factors, combined with significant global encephalization, resulted in a reduction of the space between the posterior part of the mandible and the anterior wall of the cervical spinal column. This space accommodates the alimentary canal, the upper airways, and the large arterial and venous vessels of the

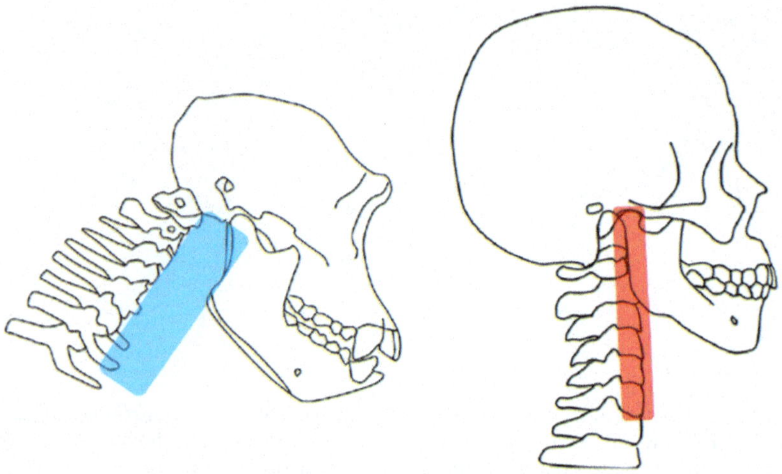

Fig. 4.129 With the upright posture, the vertebral column becomes vertical, reducing the retro-mandibular space, as seen in the outlines: the figure in the left shows a chimpanzee (*Pan troglodytes*); the figure on the right shows *Homo sapiens*. Note how in anthropomorphic apes, the clinograde posture, with the characteristic cranial flexion, results in less anterior prominence of the facial mass on the horizontal plane

neck. If the maximum mouth opening and the subsequent retraction of the mandibular angle were solely due to condylar rotation, it would also lead to the retraction of the important muscular structures supported by the mandible (lingual, the floor of the mouth, and suprahyoid muscles). Such retraction, if not compensated and counteracted by the translatory component of the condyles during the opening movement, would result in the reduction of airway patency and vascular trunks in the neck (Fig. 4.129).

As described earlier, during mouth opening, each condyle undergoes a roto-translatory movement along the glenoid roof in the anterior direction. Simultaneously, the ascending branch of the mandible exhibits posterior rotation, with the condylar mandible as its fulcrum. The posterior rotation is most pronounced in the part of the ascending branch farthest from the center of rotation: the region of the mandibular angle (Fig. 4.130). During the opening movement of the mandible, the oblique band of the LTM serves as a crucial biomechanical constraint for mandibular translation, necessary to reduce the posteriorization of the gonial angle.

Simple condylar rotation without roto-translatory movement would result in increased posteriorization of the gonial angle, leading to a further invasion of the retro-mandibular spaces (Fig. 4.131).

The pharyngeal airway space is surrounded by soft tissues enclosed by bony structures such as the mandible and cervical vertebrae: a flexible tube enveloped by soft material bounded by rigid structures. The positions of the skull and mandible influence the patency of the upper airways and the activity of the primary dilator muscle of these airways, the genioglossus. Head extension reduces upper airway obstruction (a maneuver commonly used to maintain airway patency in

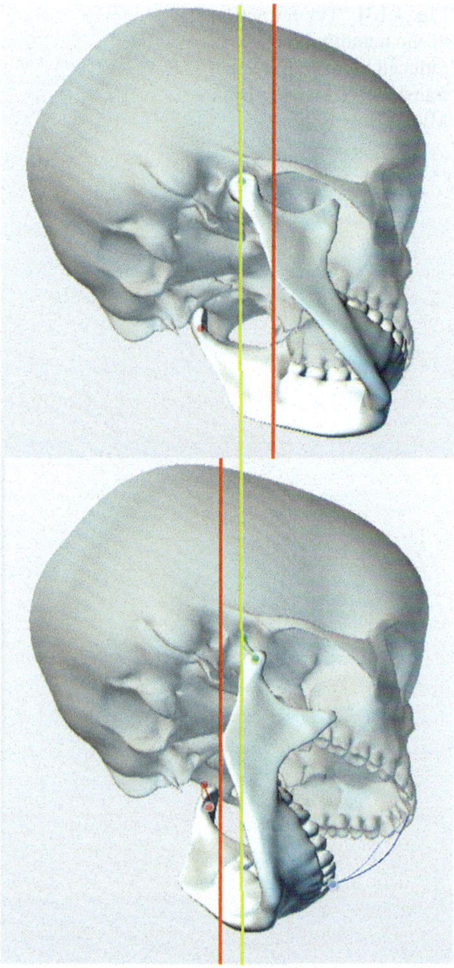

Fig. 4.130 Mouth opening with retraction of the mandibular angle: yellow line indicates the position of the condyle in the fossa in maximum intercuspation; red line indicates the position of the mandibular angle in maximum intercuspation (above) and in maximum opening (below)

resuscitation and anesthesia); conversely, extreme head flexion reduces pharyngeal space and airflow (Walsh et al., 2008). It is known that mandibular protrusion improves airway patency, and different head positions can provide mechanical advantages or disadvantages in stabilizing and dilating the pharyngeal spaces (Cai et al., 2016).

The flexion-extension of the neck and mouth opening influence airway patency both at the level of the soft palate and oropharynx. The tongue muscles, which originate and are enclosed by the mandible, change their position according to that of the lower jaw. Therefore, it is not surprising that the airways can undergo significant changes in patency due to changes in mandibular position. Soft palate tissues, on the other hand, are influenced not so much by anatomical connections with tongue muscles but by their close proximity to the dorsal surface of the tongue. Changes in mandibular position can alter the relationships between the tongue and velopharynx, leading to modifications in airway patency (Isono et al., 2004).

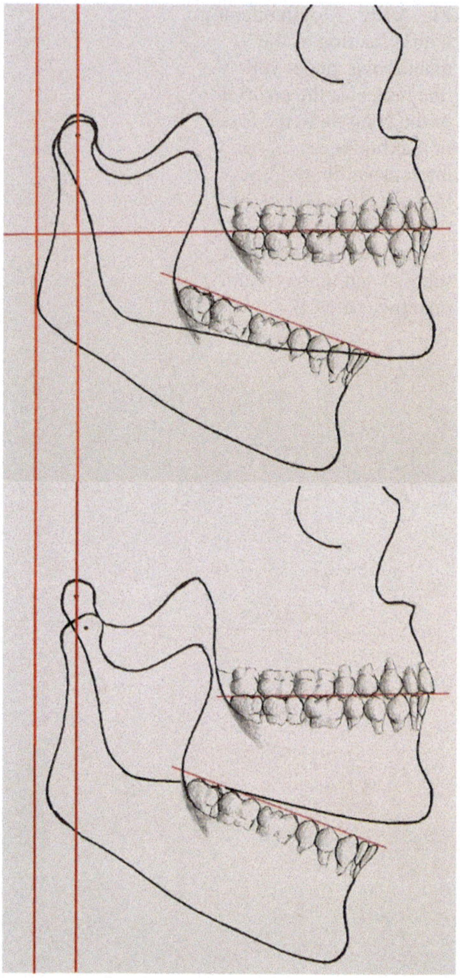

Fig. 4.131 The retraction of the mandibular angle is reduced by the condylar translation (Drawing by Marco Boggioni)

Mandibular advancement leads to an expansion of the lateral airways through a direct connection between the mandibular ramus and the lateral walls of the pharynx (Brown et al., 2013), making the upper airways less collapsible. Forward movement of the tongue is reduced during head extension and rotation.

Mouth opening, due to the downward and backward movement of the mandible, results in a reduction of the retro-lingual air space, causing a narrowing of the pharyngeal lumen until it is compressed by the soft tissues surrounding the airways (Ayuse et al., 2004). During maximum opening, excessive posteriorization of the mandibular angle and the associated soft tissues tend to encroach on the space occupied by vascular and nervous structures. Condylar translation prevents the tongue and mandible from compressing the airways and large neck vessels (Craddock, 1948; Du Brul, 1964; Sicher, 1937, 1951; Smith, 1985) (Fig. 4.132).

It is important to remember that mouth opening is not solely due to mandibular movement but also involves cranial extension with associated flexion of the cervical

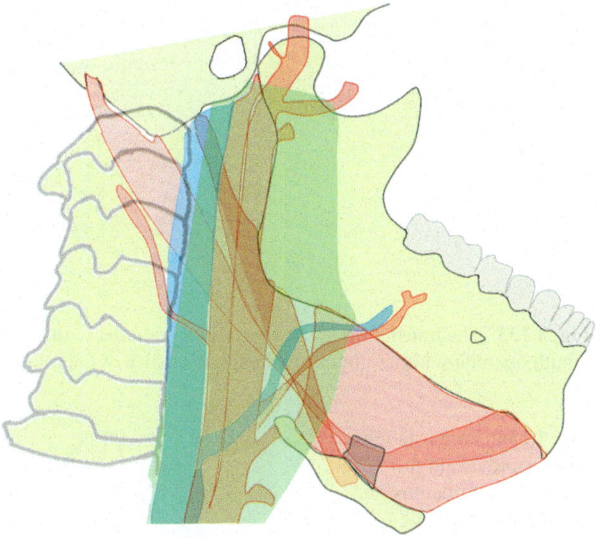

Fig. 4.132 Retromandibular space: the carotid and jugular vascular trunks pass through the narrow space located between the ascending ramus of the mandible and the cervical vertebral column

spine. There is a functional relationship between the temporomandibular system and the cranio-cervical system, in which mouth opening is the result of the simultaneous activation of these two regions, involving concurrent movements of both the TMJ and the atlanto-occipital joint and cervical joints (Eriksson et al., 2000; Kohno et al. 2001, 2001).

This mandibulo-cranio-cervical dynamic aims to achieve maximum mouth opening without compressing vital neck organs (Dean, 1990) while maintaining the effectiveness of the masseter and internal pterygoid muscles. Condylar translation enhances the contraction of the masseter and internal pterygoid muscles, which would otherwise be stretched to the point where they can no longer provide effective force for closing the mouth (Smith, 1985). It also helps to avoid excessive tension on the neurovascular bundle of the inferior alveolar nerve, which would be exerted by pure rotation.

In the movement of maximum mouth opening, head extension reduces the possibility of compression of vital structures behind the mandible (Dean, 1990). This extension contributes to reaching the extreme limit of movement, reducing the posteriorization of the mandibular angle (Fig. 4.133).

In humans, due to the limited bony constraints that the mandibular fossa offers to the condyle, the regulation of mandibular movements, especially in the horizontal plane, is primarily controlled by the restraining action of the LTM. Condylar translation regulated by the LTM allows the lower incisors to engage with the palatal surface of the upper incisors during protrusion, thereby initiating their guiding function. In lateral movements, it enables the lower canines to guide movements in the horizontal plane, which are essential for the chewing cycle. Pure rotation would only allow point contact between teeth, without any incisal and canine guidance during chewing (Fig. 4.134).

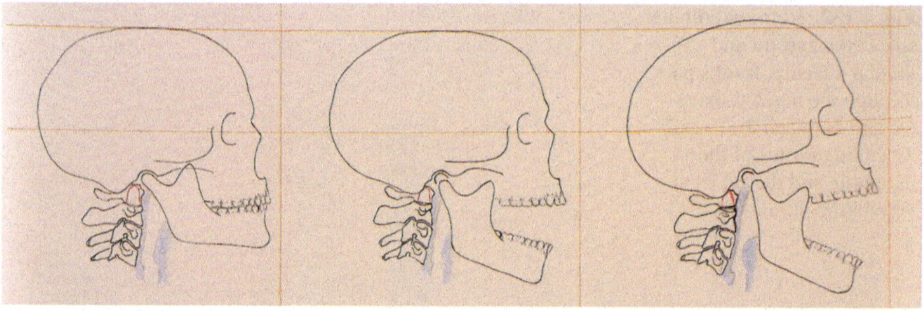

Fig. 4.133 The translation of the condyles associated with cranial extension allows for maximum mouth opening without impairing the functionality of the masseter and internal pterygoid muscles

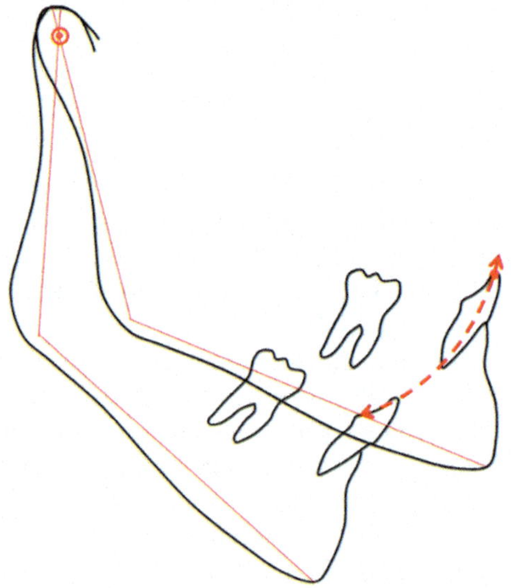

Fig. 4.134 Pure rotation of the condyles in the movement of opening and closing the mandible would not allow for any horizontal movement, making it impossible to implement the functional guides of the incisors in protrusion and the canines in lateral movements

4.7.2 Horizontal Portion of the Temporomandibular Ligament

The horizontal portion of the LTM originates from the external surface of the articular eminence and inserts onto the lateral surface of the condylar pole. The function of this small and deep band is to firmly resist excessive mandibular retrusion, which would lead to traumatic compression of the soft tissues behind the TMJ.

With the dental arches in maximum intercuspation, the condyle should ideally be supported by the vertical dimension of occlusion, creating a state of 'homeostatic floating' within the joint. From this position, during functional movements, the condyles are capable of only limited retrusive movement. Excessive posteriorization of the condyle would lead to stress on the delicate tissues at the posterior portion of the TMJ. In humans, this thin but extremely resilient horizontal band serves as a surrogate for the function of the postglenoid bony process, which in other primates acts as a bony barrier protecting retrodiscal structures. The horizontal portion of the LTM compensates for the volumetric reduction of the postglenoid process that has occurred in our species.

In monkeys, the postglenoid process articulates with the condyle. Indeed, on the posterior-superior condylar surface of large primates, there is a clear imprint of the postglenoid process, confirming its role in limiting condylar retrusion. In anthropoid apes, the LTM is rudimentary or absent. In addition to the postglenoid processes (Fig. 4.135), the large canines in anthropoid apes also play a role in preventing mandibular retrusion (Dean, 1990) (Fig. 4.136).

In anthropoid apes, the bony processes that border the TMJ posteriorly and medially serve as constraints that limit condylar movements, effectively 'boxing in' the

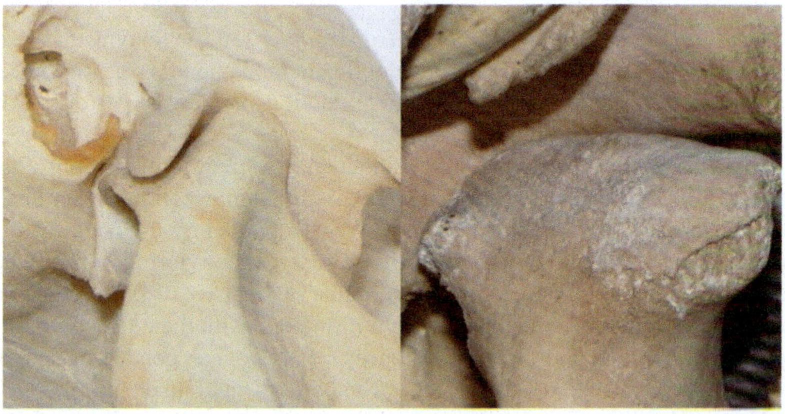

Fig. 4.135 Postglenoid process. Right: Guianan howler monkey *(Alouatta macconnelli)*; left: rhesus macaque (*Macaca mulatta*). Specimens from the Museum of Zoology, University of Florence

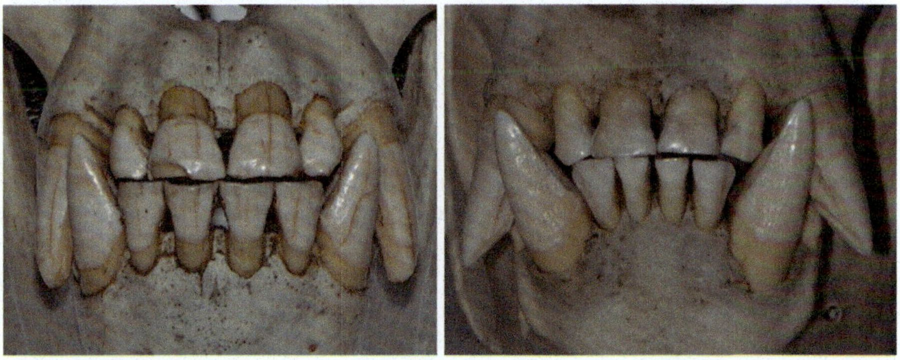

Fig. 4.136 The occlusal interlocking of the canines, which extend significantly beyond the occlusal plane, causes the upper canine to limit the retrusion of the mandible. Left: Bornean orangutan (*Pongo pygmaeus*); right: gorilla (*Gorilla gorilla*). Specimens from the Museum of Zoology, University of Florence

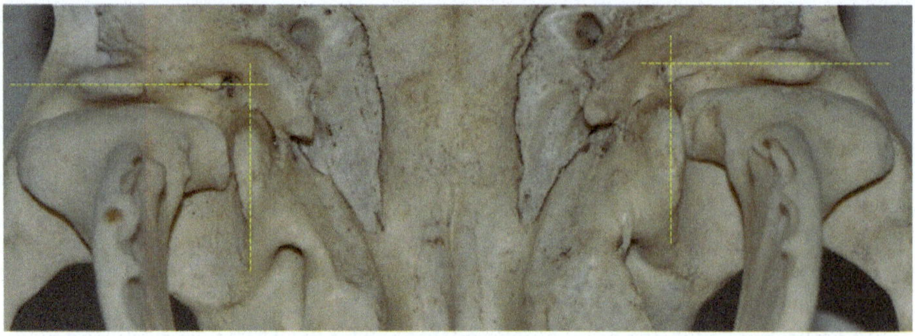

Fig. 4.137 In great apes, the post- and entoglenoid processes (yellow dashed lines) 'box in' the condyle and its movements. Specimen from the Museum of Zoology, University of Florence

condyle within the fossa (Fig. 4.137). In *Homo sapiens*, these bony walls are drastically reduced, and within the TMJ capsule, ligamentous bundles are structured that partially take over the dynamic constraint function previously performed by the postglenoid and entoglenoid processes.

Condylar retrusion occurs within a limited range, representing a physiological movement of the mandible starting from maximum intercuspation and ending in the position of maximum retrusion. This movement is guided by the occlusal surfaces of the posterior teeth, although it is not typically realized during normal chewing cycles. From maximum intercuspation, the value of retrusive displacement can vary from a few tenths of a millimeter to a few millimeters, with the most common value being 1.5 mm, without significant differences related to age, gender, or dental class (Foglio-Bonda et al., 2006). Therefore, in maximum intercuspation, the condyles should be in their homeostatic position within the fossa, with the fibers of the horizontal portion of the LTM not under tension, allowing the condyles themselves to retract—a physiological joint freedom described graphically by Ulf Posselt (Posselt, 1962). When the mandible retracts, the horizontal band comes under tension, halting the condylar retrusion and protecting the retrodiscal tissues and structures from harmful compressions (Fig. 4.138).

In addition to protecting delicate retrodiscal tissues, the LTM also plays a crucial role on the working side of mastication. During the chewing cycle, the working condyle does not move much from its position when the arches are in maximum intercuspation. It is not restricted from retrusion but is constrained to orbit and slide outward (Dean, 1990; Fanucci et al., 2008), with a slight lateral shift known as 'Bennett movement' (Fig. 4.139).

In the resting position of the mandible, the horizontal band is relaxed, and when the working condyle rotates along its vertical axis, this band comes under tension, preventing condylar retrusion. The rotational movement of the working condyle is limited by the action of the horizontal portion of the LTM, which allows for a slight lateral movement but opposes the posterior traction of the temporal muscle (Fig. 4.140).

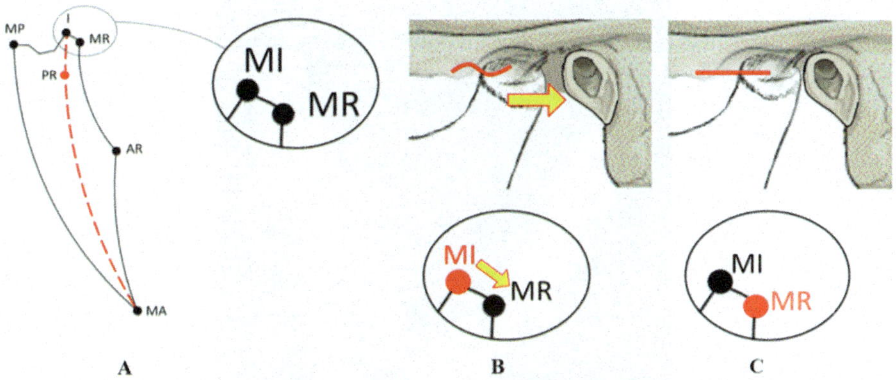

Fig. 4.138 (**a**) Posselt diagram (MI maximum intercuspation, MP maximum protrusion, MA maximum opening, PR rest position, AR maximum retruded opening, MR maximum retrusion); (**b**) in maximum intercuspation the horizontal band of the TML is not in tension; (**c**) with the posterior displacement of the condyle, the horizontal band becomes tense and stops the movement at maximum retrusion

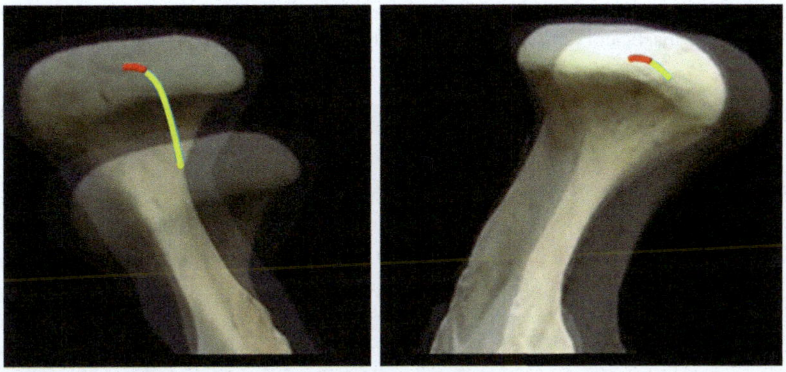

Fig. 4.139 Mandibular condyles, top view. Left lateral movement of the mandible: Bennett movement is divided into two parts: an initial part (red line), called immediate Bennett; and a subsequent part (yellow line), called progressive Bennett

The opposing actions of the horizontal band of the temporomandibular ligament, which resists condylar retrusion, and the temporal muscle, which exerts posterior traction, work together to strongly stabilize the working condyle and determine its rotation on its vertical axis (Aiello & Dean, 1990) (Fig. 4.141).

In the resting mandibular position, the horizontal band of the LTM is relaxed, but during chewing, at the beginning of the power stroke from the working side, the condyle rotates—as previously mentioned—along its vertical axis. This causes the backward movement of its lateral pole, putting the horizontal band under tension. This activation of the horizontal portion prevents further condylar retrusion, forcing it into a slight lateral oscillation. During the phase described earlier, when the bolus

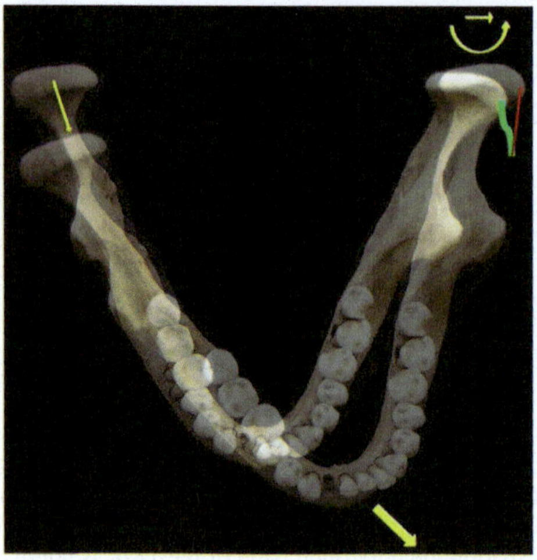

Fig. 4.140 Left lateral movement of the mandible: the right balancing condyle moves forward and medially, while the left working condyle rotates around a vertical axis and translates laterally under the constraint of the horizontal band of the TML, which becomes tense. Note on the working condyle the two representations of the horizontal band of the TML: in red, the band is tense at the beginning of the masticatory stroke; in green, the band is relaxed in maximum intercuspation

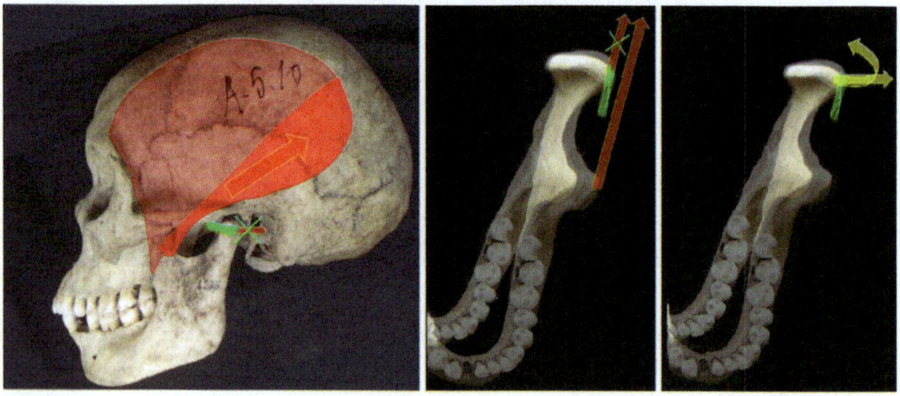

Fig. 4.141 The vector of muscle contraction of the temporalis (large red arrow) inserting on the coronoid process tends to posteriorize the condyle (small red arrow), while the horizontal band of the TML, which inserts on the lateral pole, opposes this movement (green band): their synergy generates a couple of forces that leads to the rototranslation of the working condyle (yellow arrows). Specimen from the Museum of Anthropology, University of Florence

placed between the arches is about to be crushed, specific condylar positions can be identified to illustrate this movement (Fig. 4.142).

The horizontal band not only protects against accidental condylar retrusion but also, and more importantly, against excessive posterior condylar movement during the power stroke of chewing when the forces from the adductor muscles of the

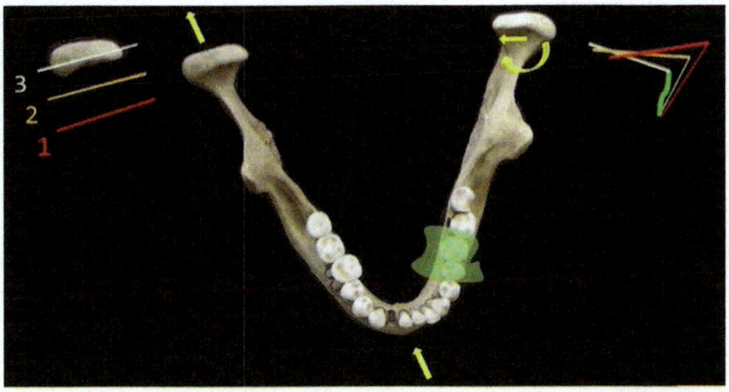

Fig. 4.142 (*1*) Red. Beginning of the power stroke on the left side: centripetal lateral movement; after the mouth opens and the mandible shifts sideways, the horizontal band of the TML is tense to prevent the lateral pole from experiencing a posteriorizing force during the maximum masticatory load: the balancing condyle is moved forward, downward, and medially. (*2*) Yellow. Intermediate phase of the power stroke: the condyles approach the position they will occupy in maximum intercuspation; the working condyle rotates around a vertical axis and moves slightly inward (side shift); the balancing condyle translates laterally, posteriorly, and upward along the articular eminence. (*3*) White. End of the power stroke: both the working and balancing condyles reach the position they occupy in the fossa with the arches in maximum intercuspation (Del Palomar & Doblare, 2006; Hylander, 2006)

mandible are discharged with their full power. In this phase, the condyle is pushed (with the interposition of the disc) against the glenoid fossa roof, which, due to its inclination, tends to make the condyle slide posteriorly. The relationship between the articular surfaces of the TMJ occurs on a plane that is inclined posteriorly and upwards, where the muscle vectors of the jaw elevators exert their force with a resultant that tends to retract the condyle into the fossa (Fig. 4.143).

The possibility of tensioning the horizontal band of the LTM manually has led to the false belief that one can find a repeatable point of registration for the mandibulo-cranial relationship in guided (but actually forced) mandibular retrusion. It is undeniable that this position allows for some repeatability, but it is equally clear that this maneuver induces a mandibulo-cranial relationship in which the condyles are placed in a non-physiological position of maximum retrusion. This position puts the horizontal band of the LTM under tension, which, physiologically, with the dental arches in maximum intercuspation, should not be taut. Therefore, manual maneuvers that arbitrarily induce a posterior position of the condyles in the fossa not only result in a repeatable position but also, and more importantly, in a repeatable error.

Condylar retrusion is a potentially harmful movement, whether it occurs at low intensity (the retrodiscal tissues are not anatomically and histologically suitable for compression) or with violent mandibular retrusion, caused, for example, by a traumatic event. A violent posterior push of the condyles could, if not blocked, result in severe and irreversible damage to the delicate organ of balance and hearing due to an invasion of the spaces of the middle ear (Fig. 4.144).

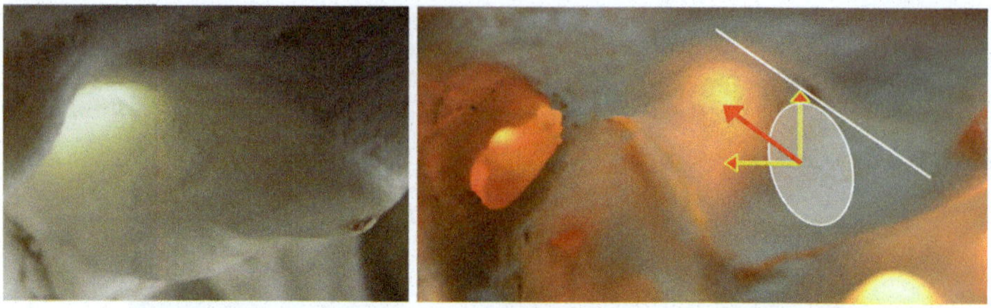

Fig. 4.143 Mandibular fossa, latero-inferior view: during the power stroke, the inclined plane of the glenoid roof tends to slide the condyle in an anteroposterior direction. Specimen from the Museum of Anthropology, University of Florence

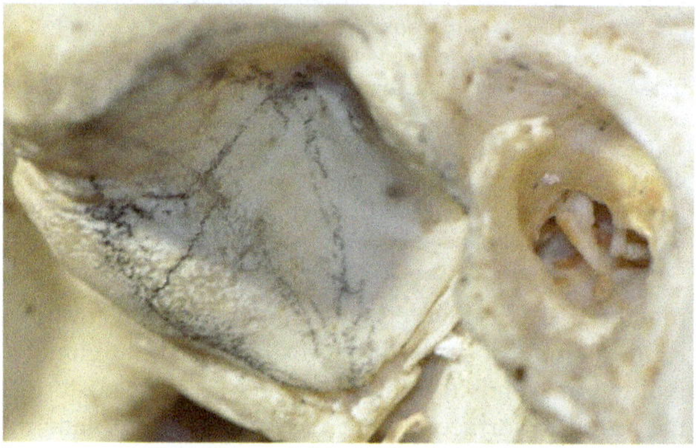

Fig. 4.144 Mandibular fossa, with the two characteristic articular slopes and, posteriorly, the auditory meatus: note in the foreground within the tympanic cavity the malleus and, behind, the articulation between the incus and stapes. Skull 1397, specimen from the Museum of Anthropology, University of Florence

4.8 Modification of Dental Volumes and Disappearance of the C/P3 *Honing Complex*

In mammals, mastication has been the driving force behind adaptive modifications of the stomatognathic apparatus and splanchnocranium. The shapes and volumes of teeth have been the functional elements primarily involved in changes related to the trophic niches of individual species.

4.8.1 Modification of Dental Volumes

It is typical for mammals to have three types of teeth that serve distinct functions, such as cutting, piercing, and grinding. Even 250 million years ago, the teeth of cynodonts, from which the class of mammals originated, were grouped into three distinct morphological types: incisors, canines, premolars-molars, unlike other reptiles. The longevity of this dental formula testifies to its undeniable adaptive advantage, which has been maintained by historical and modern mammals, including humans (Benoît & Granat, 2010).

In anthropoid apes and many other primates, incisors are relatively more developed than premolars and molars. This adaptation is functional for an "arboreal" diet consisting mainly of fruits to be bitten and peeled, but which does not require intense masticatory effort (Figs. 4.145 and 4.146).

Compared to anthropoid apes, australopithecines, especially *Paranthropus boisei*, showed a reversal of this trend with a reduction in the volumes of incisors and canines, which barely exceeded the occlusal plane, while the posterior teeth increased in size. As Fabio Di Vincenzo writes: "... in australopithecines, there were marked modifications of the stomatognathic apparatus and teeth; around 4 million years ago, the precursor of all subsequent humanity, *Australopithecus anamensis*, appeared, capable of moving on the ground with a stable bipedal gait.

In males of *A. anamensis*, the canines were still relatively large but had an incisor-like (trapezoidal) morphology with apical wear typical of later hominins. This was different from the conical canines characteristic of anthropoid apes and

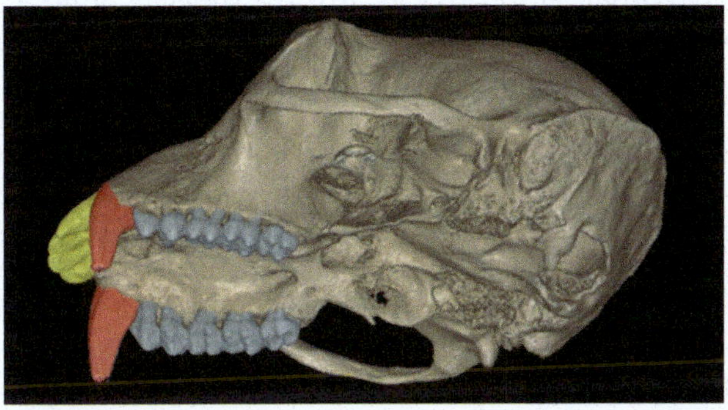

Fig. 4.145 Mammalian teeth have specialized in specific functions: the incisor group (yellow) for cutting and peeling, the canines (red) for biting, piercing, and tearing, and the premolar and molar group (blue) for shredding and grinding. Here they are highlighted in a skull of a colobus monkey (*Colobus guereza*): note the volumetric predominance of the incisor group compared to the premolars and molars. Specimen from the Museum of Zoology, University of Florence

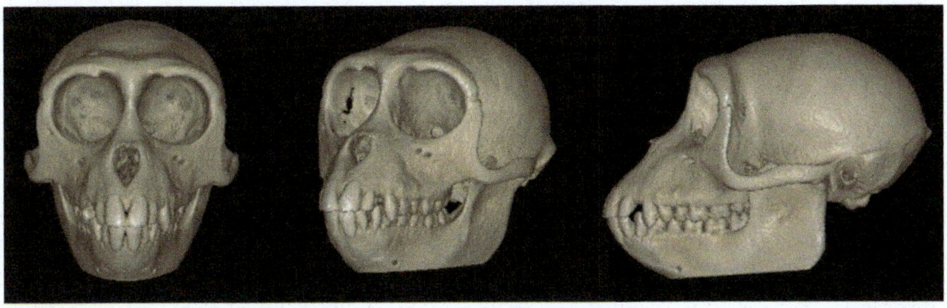

Fig. 4.146 Skull of a macaque (*Macaca sinica*). Premolars and molars are less developed compared to the incisor group. Specimen from the Museum of Zoology, University of Florence

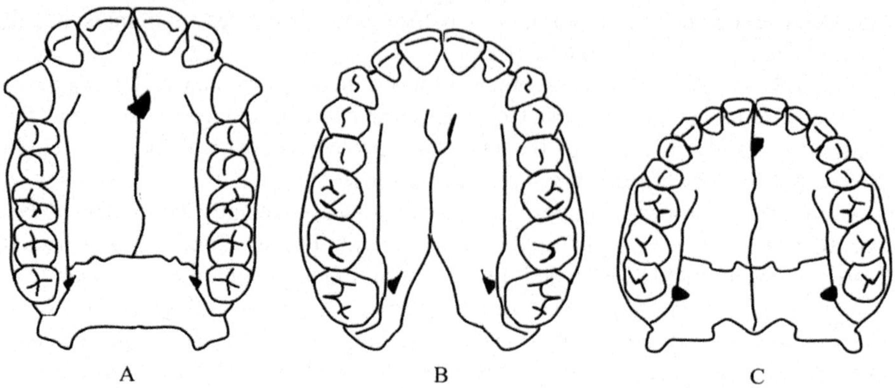

Fig. 4.147 (**a**) Chimpanzee (*Pan troglodytes*); (**b**) *Australopithecus afarensis*; (**c**) *Homo sapiens*. In the genus *Homo*, the profile of the canine crown recedes and the characteristic canine diastemas are lost, with a reduction in the size of the zygomatic arches

early hominins" (Di Vincenzo, 2011, p. 70). The structure of the arches remained similar to that of anthropoid apes, i.e., U-shaped with rows of parallel teeth, unlike the parabolic morphology typical of other australopithecines and the genus *Homo*. In *A. anamensis*, the facial bones had not yet begun to retract due to the reduced importance and size of the anterior dentition, which was better suited for cutting and chewing soft foods like leaves and fruit, compared to the posterior dentition, which was more useful for consuming tougher plant materials (seeds, corms, and tubers), more readily available in an open grassland environment (Fig. 4.147).

A. anamensis, the diastema between the lateral incisor and canine of the upper arch, which accommodated the lower canine, was lost. The maxillary canines, while large in absolute size, did not extend much beyond the occlusal plane. The posterior dentition, on the other hand, was more markedly developed, a new evolutionary characteristic that identifies *A. anamensis* as the ancestor of the australopithecines and subsequent humans. Compared to earlier hominins, the molars were larger, and the premolars began to develop roots and accessory cusps. The molar cusps became low and rounded, better suited for breaking and crushing hard and resistant materials. The thickness of their enamel increased compared to that in the molars of earlier primates and even current anthropoids (chimpanzees and gorillas). This thicker occlusal enamel provided greater protection against excessive wear due to prolonged chewing.

A. anamensis, despite sharing the same forest ecology as previous hominins like *Sahelanthropus, Orrorin*, and *Ardipithecus*, was better adapted than these species to consume tough, fibrous, and resistant foods primarily found in open savanna environments. Chewing these types of foods required intense and prolonged effort. The increase in the size of the posterior dentition and the complex of craniofacial and musculature structures related to chewing would characterize the subsequent evolution of the australopithecines (currently known to include slightly less than ten species) and the immediate predecessors of the genus *Homo*.

Between 3 and 2.5 million years ago, there was a sudden aridification in the tropical regions of Africa inhabited by our ancestors. This led to a retreat of the forest and an expansion of grasslands, causing a drastic change in the available food sources. This ecosystem modification forced the australopithecines to adapt their masticatory apparatus to the new dietary resources.

Starting around 2.5 million years ago, the so-called gracile australopithecines were "replaced" by the paranthropines, a new group of more specialized hominins. Although they had unchanged stature and weight, their posterior dentition became enormous. These "robust" australopithecines, named for their extreme development of the masticatory apparatus, further emphasized the "molarized" pattern, with an expansion of molar chewing surfaces accompanied by additional enamel thickening. Premolars became "molarized" by developing broad occlusal areas, resembling small grinding stones, and the height of alveolar bones increased to accommodate deeper and more extensive dental roots. Both the body and ramus of the mandible became more robust.

The action of such large teeth and jaws required extremely powerful and developed muscle bundles. This caused a prominent sagittal crest to form on top of the skulls of the robust australopithecines. Similar to modern gorillas, the temporal muscle surrounded the small cranial box from both sides, exerting pressure such that the skull of these hominins, when viewed from above, appears "squeezed" around the temporal fossae (Fig. 4.148).

The canines of the paranthropines approached the form of incisors, never exceeding the occlusal plane. The anterior dentition significantly reduced in size following the posterior displacement of the entire maxillary complex. They developed an extremely robust, broad, and flat facial skeleton, optimizing the powerful muscle levers and directing the loads of masticatory effort almost exclusively to the posterior portion of the dentition, specifically the premolars and first molars.

However, the paranthropines were not the only hominins walking on two legs in the savanna at that time. Starting from one of the species classified under the genus *Australopithecus*, a new and more versatile form of hominin emerged. This hominin survived by consuming a more omnivorous diet, including a significant propensity

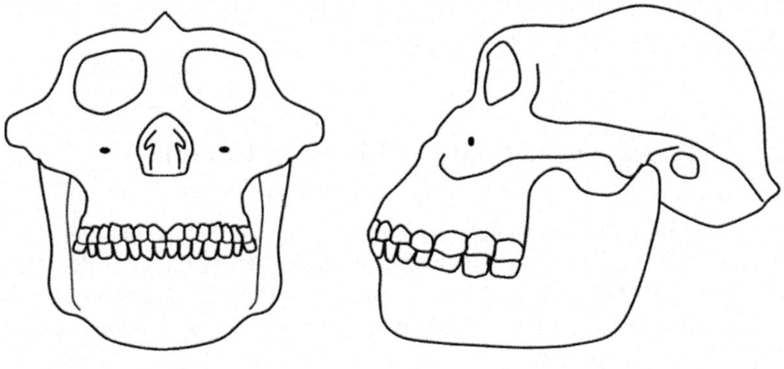

Fig. 4.148 Skull of *Paranthropus boisei*

for meat consumption. These were the first members of the genus *Homo*. Therefore, for a very long time, two distinct genera of hominins coexisted in East Africa. Likely due to competitive exclusion, the paranthropines specialized further in consuming both tender and tough plant materials successfully. On the other hand, the early representatives of the genus *Homo* further developed opportunistic eating habits. In addition to plant materials available in the environment, they also exploited a stable and relatively abundant protein source in the savanna ecosystem—carcasses of herbivores abandoned by predators, primarily large felines.

The early *Homo* individuals were characterized by an omnivorous feeding behavior with an increased tendency to scavenge carrion and carcasses, putting them in direct competition with other scavengers such as jackals, vultures, and even large felines guarding their prey.

This significant trophic resource reduced the selective pressure that had driven the australopithecines to increase dental sizes and masticatory structures. Instead, it favored mutations that led to a relative reduction in the mass of masticatory muscles and all dental and craniofacial structures around the cheekbones and mandibles. This resulted in a reduction of facial prognathism for reasons opposite to those that occurred in the paranthropines. This trend has persisted in all subsequent human species, with dental volumes subject to a scaling effect, i.e., relative reduction compared to the progressive increase in overall body size.

When determining the evolutionary path of hominins, adaptations related to diet have by far been the most important factor, whether they directly affected dental morphology or other aspects of complex human physiology such as brain development. However, while the increasing significance of posterior dentition has represented the "engine" of australopithecine evolution, a different and more revolutionary biological model, centered around the almost exponential development of brain volumes and the cognitive abilities associated with it, has characterized human evolution since the emergence of the first representatives of the *Homo* genus.

Nevertheless, even in this case, diet has set the rules of the game, and the final result has risked not being as straightforward as it may presumptuously appear when looking back at our history (Di Vincenzo, 2011). Nutrition, especially chewing, has been and continues to be so crucial in shaping cranial structures that we could paraphrase the expression of the German philosopher Ludwig Feuerbach "We are what we eat" as "We are how we chew."

4.8.2 *Disappearance of the C/P3* Honing Complex

In human dentition, the first molar is considered the king of mastication, and we can easily define the canine as the prince of proprioception. In humans, despite having lost their original size and functions, canines have become fundamental for their proprioceptive role in guiding mandibular movements, and for this reason, they deserve an in-depth discussion.

The canine is a unique tooth in the dental formula, playing a central role during mastication and differing from others in terms of shape, function, phylogeny, and embryogenesis. It constitutes a transition between the anterior incisors and the posterior grinding teeth, morphologically displaying intermediate characteristics between the shape of the anterior incisors in front of it and that of the posterior grinding teeth. The crown of the canine is the only one among the front teeth to end with a cusp.

In cynodont dinosaurs, the canine tooth differentiated early from the homogeneity of their homodontic dentition. It embarked on the path towards heterodonty, increasing in size and becoming particularly robust, with a conical and pointed shape (Fig. 4.149). Despite this differentiation, the canine tooth in mammals would retain the closest resemblance to the archaic reptilian form: a conical and pointed tooth designed to quickly immobilize and kill prey, much like a dagger.

In early mammals, the canine tooth retained the essential characteristics it had in cynodonts, despite significant variations in shape and size related to changes in dietary habits (Anirudh et al., 2018). For example, carnivorous mammals have an extremely developed root to withstand the intense stress associated with predatory behavior (Figs. 4.150 and 4.151).

The interspecific differences in the shape of canines also reflect variations in muscle strength and predatory and feeding behavior. For instance, felids have a bite that drives their canines deep, while canids tend to use them for shallow and cutting bites. This shallow action results in less stress compared to the penetrating bites of felids. If felids have relatively stronger bites than canids, they also possess a larger jaw musculature and relatively stronger canines capable of resisting greater forces and stresses.

In canids, the canines tend to have an elliptical cross-section and are flattened medio-laterally, unlike those of felids and hyaenids, which have a more rounded cross-section. This shape helps them better resist both bending in all directions and sudden, unpredictable impacts with the prey's internal bones (Valkenburgh & Ruff, 1987).

In herbivorous species, on the other hand, canines tend to be absent (Granat et al., 2008), and there is a large diastema between the front and rear teeth, which acts as a sort of storage space for chewing food (Fig. 4.152).

Fig. 4.149 Skull of a cynodont. In cynodonts—mammal-like reptiles with a structure that foreshadowed that of mammals—the canine is the first tooth to differentiate from the others

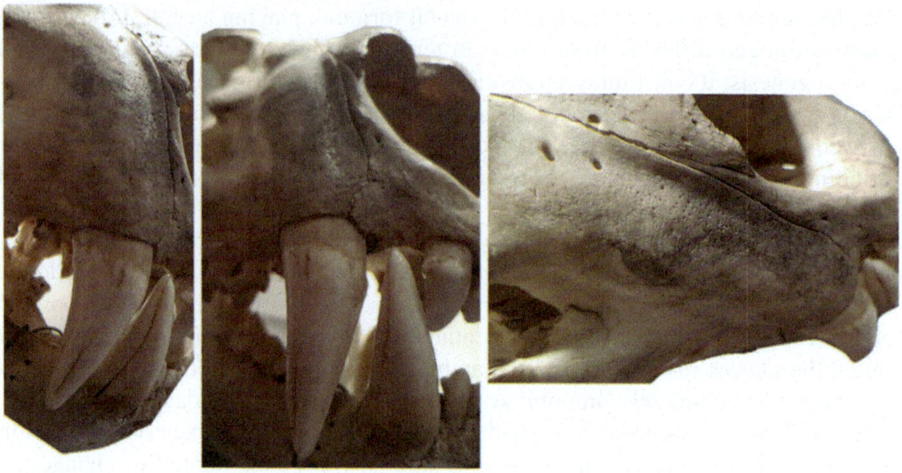

Fig. 4.150 Lion *(Panthera leo)*. Note the significant prominence of the alveolar bone that encloses an impressive root. Specimen from the Museum of Zoology, University of Florence

Fig. 4.151 Canines of a cave bear (*Ursus spelaeus*). Note the significant development not only in length but also in the maximum diameter of the root, designed to withstand the intense forces applied to the canine

In primates, elongated, conical canines with extremely developed roots remain (Fig. 4.153). However, while in cynodont dinosaurs, early mammals, and carnivorous mammals, canines primarily served and continue to serve the function of piercing prey to immobilize and injure it as deeply and effectively as possible, in primates, the canine tooth also takes on a secondary sexual role consistent with sexual dimorphism. In male primates, this tooth is more developed than in females, becoming a sexual display flag, as well as a deterrent and, if necessary, a weapon for use in combat with other males.

Fig. 4.152 Mandible of a roe deer (*Capreolus capreolus*). Note the wide diastema between the grinding teeth and the incisor group, serving as a reserve space for food to be chewed before being swallowed

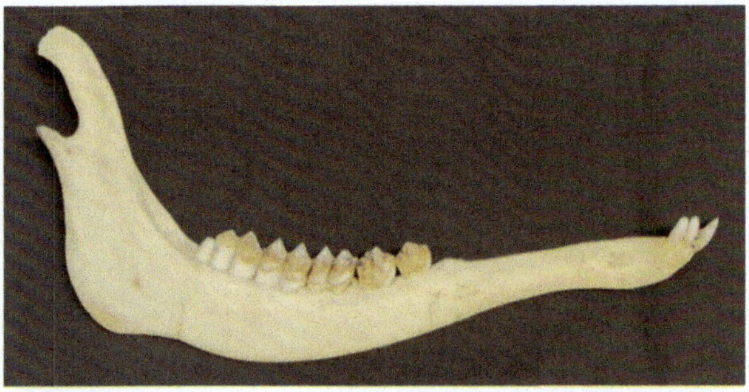

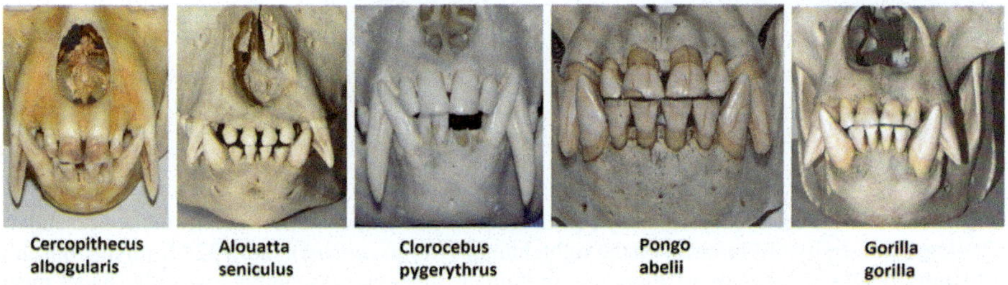

Fig. 4.153 Incisor groups of anthropomorphic and non-anthropomorphic monkeys. Note the extreme development of both the crowns and roots of the canines. Specimens from the Museum of Zoology, University of Florence

In anthropoid apes and primates in general, the large size of canines primarily serves as signals that these primates, being social animals, use for complex interactions within their groups (Manzi, 2013). In most species, female primates have much smaller canines than males (McNulty, 2010), and sexual dimorphism (especially in canine crown length) is an important morphological correlation with the social behavior of anthropoid primates, where canines function as weapons in intraspecific combat. Canine dimorphism is the result, particularly in species characterized by polygamous social groups, of intense competition among males to gain greater mating opportunities and reproductive success. Sexual dimorphism arises from competition between individuals of the same sex to possess, by demonstrating their strength, a harem of individuals of the opposite sex. The competition aims to exclude other members of the same sex looking for partners through deterrent weapons like horns and/or canines. The canines of anthropoid apes, compared to humans, are not only larger but also more elongated and pointed, whereas in humans, canines are wider, flatter, and spatula-shaped (Fig. 4.154).

The incisors of anthropoid apes and many other primates are relatively larger than premolars and molars because they are more functional for an 'arboreal' diet, primarily consisting of fruits to be bitten and peeled but not requiring strong masticatory effort. Canines, on the other hand, appear to respond to a different type of adaptation: especially upper canines, and particularly in males, protrude well

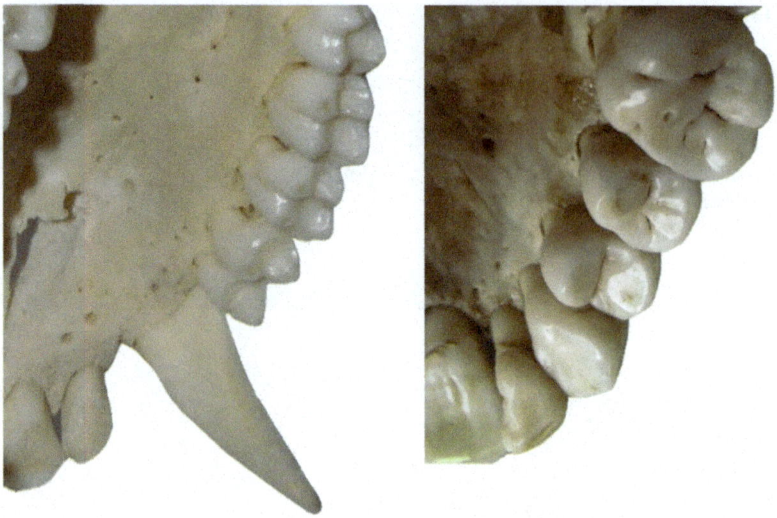

Fig. 4.154 On the left, vervet monkey (*Chlorocebus pygerythrus*). Specimen from the Museum of Zoology, University of Florence; on the right, human (*Homo sapiens*). Skull 327 from the Museum of Anthropology, University of Florence. In humans, the canine acquires a spatula shape, much less pointed and not protruding from the occlusal curves

beyond the occlusal plane, as seen in carnivorous predators where they serve defensive and offensive functions (Fig. 4.155).

In monkeys, the important canines used by males as combat weapons must also be supported by powerful neck muscles (Washburn & Moore, 1973). This involvement of neck musculature suggests that in anthropoid apes, the development of strong neck muscles depends not only on their characteristic posture but also on the intense demands placed on them during the use of canines. As we have observed, neck musculature has always been closely associated with mouth opening, and in carnivores, after the bite is closed, there is often an activation of neck muscles to shake and tear apart the prey.

However, in the evolution of carnivorous mammals, there exists a mechanical trade-off between maximum mouth opening and the force of its closure: an increase in the maximum opening capacity results in a loss of mechanical advantage and a consequent reduction in biting force. The closing force of the mouth decreases significantly as the mouth's opening width increases (Dumont & Herrel, 2003). A significant example of how canine length influences bite force and its relation to neck muscle power is provided by the mammal that embodies the most formidable example of large canine teeth in the collective imagination: the saber-toothed tiger or Smilodon. This feline, which lived from 2.5 million years ago to 500,000 years ago in North America, ranged in size from that of a modern jaguar to even larger than a Bengal tiger. Due to the length of its maxillary canines, it required a wide mouth opening, and it is unlikely that the mandibular adductors could generate sufficient force for an effective and lethal bite to the prey's throat.

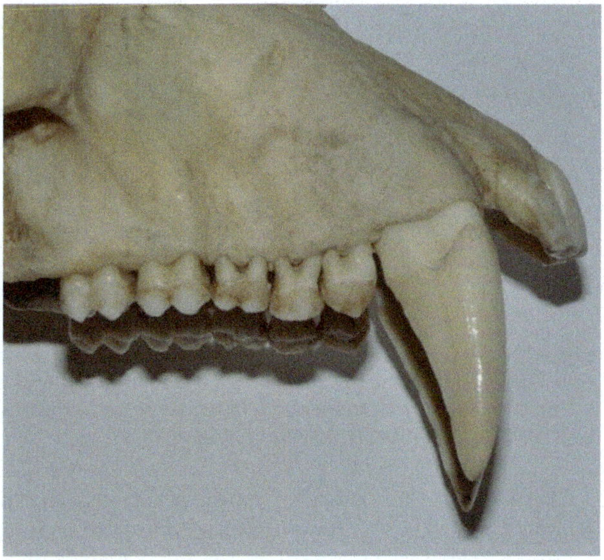

Fig. 4.155 Macaque (*Macaca nemestrina*). Note how the premolars and molars have significantly reduced volumes compared to the incisor group. Specimen from the Museum of Zoology, University of Florence

However, the Smilodon possessed powerful forelimbs to immobilize its prey, allowing it to effectively exert the considerable force of its neck flexor muscles, which rotated the skull by flexing it relative to the atlanto-occipital joint. The lower canines anchored into the prey, the forelimbs provided strong support, and the neck musculature flexed the head in a way that allowed the maxillary canines to penetrate entirely into the captured animal's neck, compressing its structures, suffocating it, and/or occluding the cerebral blood flow (Brown, 2014) (Fig. 4.156).

Compared to anthropoid apes, in the evolutionary lineage of hominids, the canines are smaller and show reduced sexual dimorphism. This is one of the main features that positions *Sahelanthropus tchadensis* as an ancestor of our evolutionary lineage, while some exclude *Orrorin* because it possesses large canines more similar to those of apes. When comparing them to the chimpanzee's dentition, what appears reduced in australopithecines (especially in *Paranthropus boisei*) are the front teeth, particularly the canines, which protrude very little from the occlusal plane, while the posterior teeth, premolars, and molars have developed more significantly (Manzi, 2006, 2007). The early australopithecines, therefore, retained a slight sexual dimorphism in their canines, a trait that was also maintained by the paranthropines, while in the genus *Homo*, sexual dimorphism in canines has nearly disappeared. The short canines and the weaker dimorphism in early hominids might suggest that among males in these species, there was reduced physically violent sexual competition, possibly due to the formation of long-term monogamous pair bonds. However, it seems imprudent to advance this socio-behavioral hypothesis (which is also heavily influenced by the cultural conceptions of the observer), as it would be contradicted by the moderate to high degree of sexual dimorphism relative to body mass observed in early hominids.

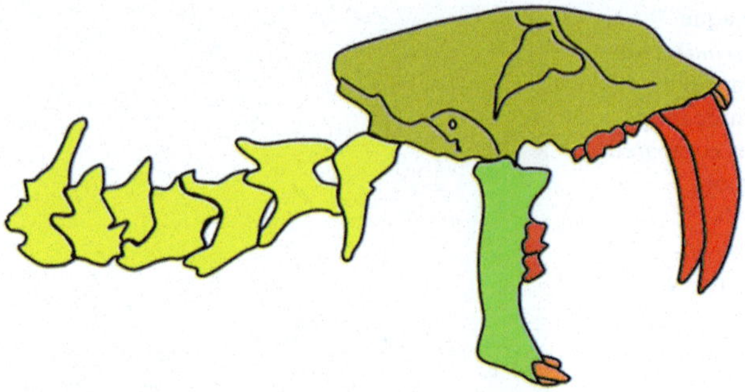

Fig. 4.156 Cervico-cranial structure of the Smilodon. Its formidable canines sank into the prey's neck thanks to the cranial flexion exerted by the powerful neck muscles. The reduced strength of the mouth-closing muscles can be inferred from the limited extension of the coronoid process, which is generally strongly developed in carnivores

Throughout the history of biology, various hypotheses and explanations have emerged regarding the progressive reduction of canines in the evolutionary path of hominids. In 1871, Darwin suggested that the functional significance of the reduction in canine crown size in humans was related to the use of tools and artifacts in food preparation. Other hypotheses involve bipedalism or an increase in social cooperation and, as mentioned earlier, the reduction of male competition for mates (Harcourt-Smith, 2010). However, despite these ongoing efforts, there is currently no complete consensus on the advantages that the reduction in canine length may have provided to early humans. Another hypothesis in this regard is that if canines were excessively long, they would hinder both effective mouth opening and chewing efficiency: long canines would require greater jaw rotation to achieve a wide mouth opening, and there would be a need for retraction of the mouth-closing muscles.

This would make these muscles less efficient for chewing, which is why shorter canines are associated with greater mechanical efficiency in chewing (Hylander, 2013).

Therefore, in the author's opinion, among the various explanations for the reduction in canine size, the most convincing one is increased chewing efficiency, linked to a functional adaptation resulting from dietary changes. In fact, the canines of hominids are not simply smaller versions of those found in apes. Their incisor-like morphology is decidedly unusual among anthropoid primates, and this particular morphology developed alongside cranial and facial evolutionary changes, including short and wide dental arches and an intercondylar space that favors an emphasis on the lateral-medial component of the chewing cycle. The strong selective pressure that australopithecines faced, triggered by changes in food sources, led to a reduction in the crown height of the canines and a change in their shape, necessary for transitioning to a mechanically more demanding diet. This reduction was driven by the need to maintain efficient chewing and proper biting force, both essential for the dietary shift. Chewing efficiency can explain the shape changes induced by a new function (Scott, 2010). Alongside the reduction in crown volume, which became shorter (and proportionally wider), and a resilient pyramidal shape, a powerful root structure of the canine was maintained. These structural modifications allowed it to withstand the strong transverse loads associated with this new function, guiding the entry phase of particularly developed chewing cycles, especially in their lateral component.

The paranthropines developed an impressive masticatory apparatus with a small intercanine arch, small incisors and canines, and very large, strong premolars and molars. The other evolutionary branch of australopithecines, that of the genus *Homo*, reversed this trend, with an expanded intercanine arch and a volumetric regression of premolars and molars. Thanks to its increasingly incisor-like morphology and its powerful root, in humans, the canine is still capable of tearing meat, whereas neither carnivores nor monkeys ever use their canines for chewing.

Bipedalism and the reduction of canines, along with the consequent loss of the C/P3 *honing complex* (the *honing complex* between canines and premolars C/P3: "to hone" means to sharpen), distinguish the earliest hominids from monkeys. Humans are characterized by small canines and short rows of post-canine teeth, while monkeys have large canines and long U-shaped dental arches (Stelzer et al., 2017). The C/P3 *honing complex*, almost ubiquitous in non-human anthropoid dentition, is a functional relationship that occurs between the upper canine and the lower canine and first premolar (Delezène, 2015; Puech et al., 1989). This occlusal relationship is achieved through canine diastemas, which are characteristic in all primates except for humans. With the teeth in maximum intercuspation, the upper canine is positioned in a space between the lower canine and the lower first premolar, while the lower canine fits between the upper canine and the upper lateral incisor (Fig. 4.157).

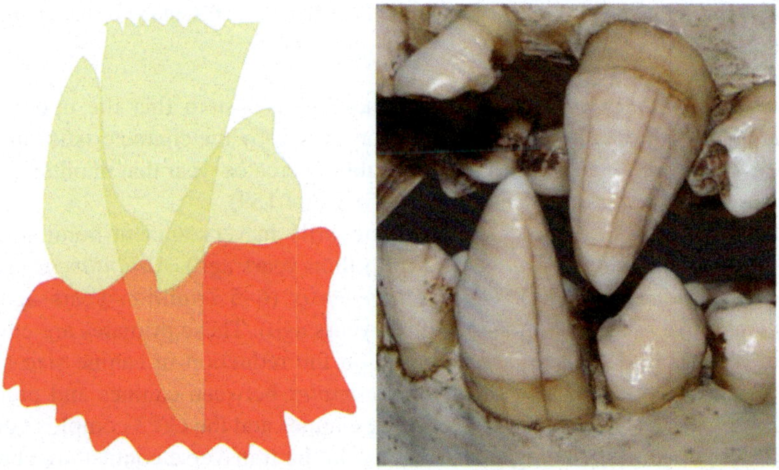

Fig. 4.157 C/P3 honing complex: thanks to the abrasive rubbing with the lower premolar, the upper canine continually maintains its sharp edges to keep its offensive potential efficient; Sumatran orangutan *(Pongo abelii)*. Specimen from the Museum of Zoology of the University of Florence

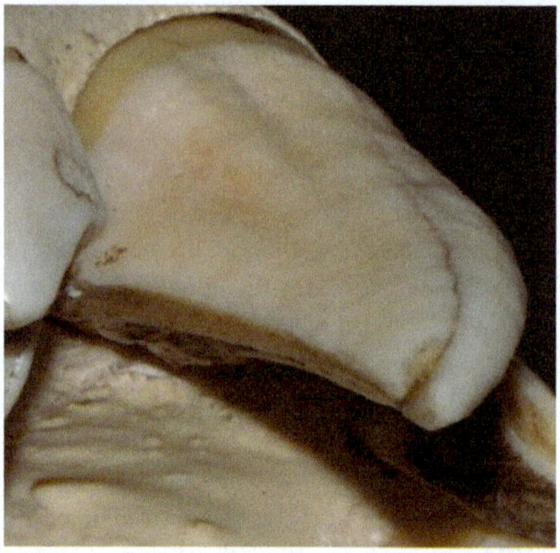

Fig. 4.158 Sharp edge of the upper canine: Sumatran orangutan (*Pongo abelii*). Specimen from the Museum of Zoology of the University of Florence

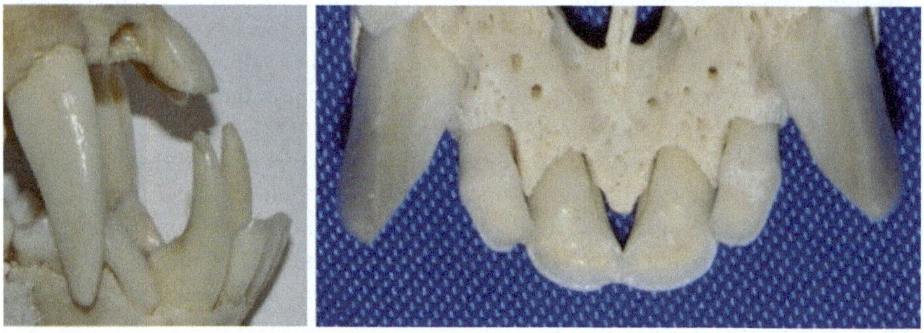

Fig. 4.159 C/P3 honing complex and wear facets on the upper canines: vervet monkey (*Chlorocebus pygerythrus*). Specimen 4894 from the Museum of Zoology of the University of Florence

The C/P3 *honing complex* is a self-sharpening system that the upper canine achieves through the physiological abrasion caused by the characteristic surface of the lower first premolar. This continuous rubbing creates wear that maintains a facet with very sharp and clear edges (Figs. 4.158 and 4.159).

The reduction of canines began evolutionarily in very ancient hominid forms, such as *Sahelanthropus tchadensis* (7–6 million years ago), *Ardipithecus kadabba* (5.8–5.2 million years ago), *Orrorin tugenensis* (6–5.7 million years ago), and *Ardipithecus ramidus* (around 4.4 million years ago). These primates are classified within the hominid lineage largely based on the reduction in canine size and the progressive disappearance of the *honing complex* between canines and premolars (C/P3) (Su, 2013). The process of reducing canines and the C/P3 complex persisted in australopithecines, leading to an increase in the transverse contact area between maxillary and mandibular teeth, with greater use of canines and premolars in food processing and chewing (Ward et al., 2010).

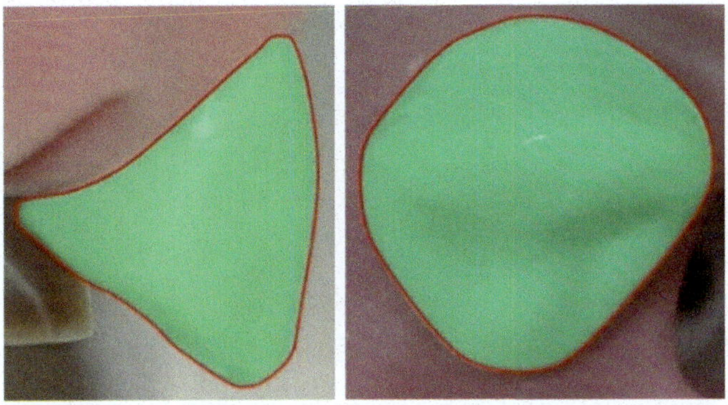

Fig. 4.160 The coronal structure of the canine gives this element formidable resistance to stress

With the genus *Homo*, the distinctive canine diastema of primates was lost. The selective force that had shaped the reduction in canine size was part of the same pressures that had modified the mandibular and maxillary geometry (Teaford & Ungar, 2000). In humans, the crown of the canine is structured with a prominent median cusp, conically pointed and convex on all its surfaces, essentially pyramidal in lateral-lateral section and rhomboidal in horizontal section (Fig. 4.160).

In humans, the canine root is single and longer than all the others, and its mesial and distal surfaces often have a central longitudinal groove for an antirotational effect (Benoît & Granat, 2010).

The upper first premolar acquired an angle similar to the other teeth and assumed a grinding function, aided by the metaconid, the inner cusp that became much more prominent. With the loss of the C/P3 complex, the diastema between the upper lateral and canine teeth also disappeared, and the crown of the upper first premolar conformed to the morphology of the second premolar. In monkeys, the lower first premolar presents as a single large cusp, the protoconid, rotated compared to the other teeth in the arch and primarily serving a cutting function (Fig. 4.161), while the metaconid is a cusp rarely found in anthropoids and, when present, has an insignificant volume.

The loss of the C/P3 complex, in addition to being related to dietary changes, is a secondary effect of bipedalism, in which the integration between the canine and premolar "breaks" (which is very strong in almost all other primates), resulting in an increase in bite functional efficiency. This condition occurred early in the evolution of hominids, leading to "cascade effects" that followed a series of different adaptations throughout the history of the human clade. With the loss of the C/P3 *honing complex*, the canine and premolar, no longer bound together, were free to evolve in a "plastic" manner in response to rapid environmental changes (Raia et al., 2018).

It's striking that in human canines, despite such a drastic change in crown shape and volume, there isn't a similarly marked reduction in their roots. Differences in root dimensions (surface, volume, length, height of bifurcation, crown base area) in

Fig. 4.161 Characteristic shape of the first lower premolar, adapted to create a wear surface with continually sharp edges on the upper canine: vervet monkey (*Chlorocebus pygerythrus*). Specimen 4894 from the Museum of Zoology of the University of Florence

Fig. 4.162 On the left, the canine of a cave bear (*Ursus spelaeus*); on the right, a human canine (*Homo sapiens*). Note how in the bear the root, besides having a considerable diameter, is about three times the length of the crown; even in humans, the root, which is quite long, is almost twice the length of the crown (yellow dashed line: separation between crown and root)

primates as well as in carnivores are strongly correlated with the differential use of teeth. Dental elements that must support significant occlusal forces typically have larger root surfaces and volumes (Fig. 4.162).

As mentioned earlier, in human canines, there is a predominance of root volume over the crown. The reason for this lies in their transition from teeth that needed to withstand intense predatory and combat forces (as well as serving as secondary sexual characteristics to impress females) to teeth that play a crucial role in regulating and guiding the phases of the chewing cycle. From the moment they erupt into the arch, our upper canines tend to intercept the lateral movements of the mandible through dynamic contact with their lower counterparts, assuming a guiding function that can subject their crowns and roots to potentially damaging transverse forces. Withstanding such forces for millions of chewing cycles requires a powerful periodontal support system, which explains its oversizing relative to the crown.

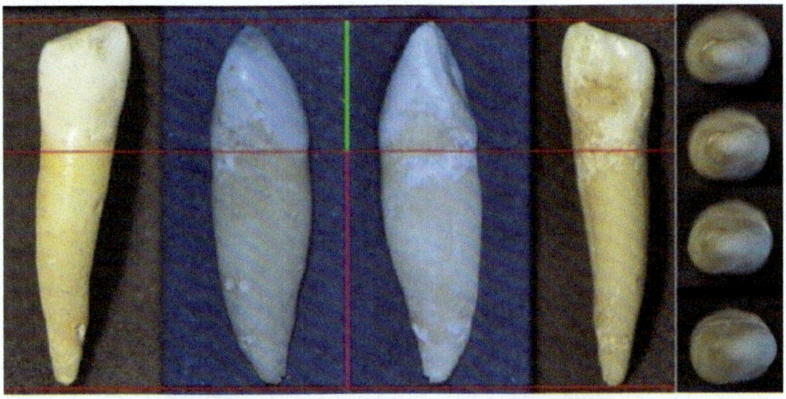

Fig. 4.163 Human lower canine: vestibular, distal, mesial, lingual views; the red horizontal lines delimit the length of the crown and the root. Note the relationship between the crown and the root, which is almost twice as long; it is also compressed at all levels, making it more resistant to dislocation—as dentists well know when they have to extract an elliptically shaped root: on the right, the root seen from its apex, focused at different planes; note its elliptical shape

Furthermore, to increase resistance, the human canine root has a flattened shape with an elliptical section, allowing it to more effectively oppose twisting forces (Fig. 4.163).

Thanks to its refined proprioceptive system capable of transmitting neural information that sets the mandibular trajectories, the canine plays the role of an extremely sensitive guide in regulating the lateral movements of the mandible (Casteyde, 2008). Sensory information is provided by the presence of numerous periodontal proprioceptive receptors, densely distributed in the apical area, which is the most subject to stress and the most delicate due to the passage of the vasculo-neural bundle through the apical foramen (Cash & Linden, 1982). The periodontal ligament is richly innervated by mechanoreceptors whose cell bodies are located in the trigeminal ganglion or the mesencephalic trigeminal nucleus. Neurons in both ganglia are sensitive to the stretching of periodontal fibers induced by tooth movement, but their activation threshold, central connections, and functional significance differ (Byers & Dong, 1989). The regulatory role of the masticatory function performed by the canine is confirmed by the fact that in patients without canine contacts in lateral movements, the posterior teeth show four times greater wear facets compared to patients with canine guidance (Kinzey, 1971).

The canine tooth is part of a unique proprioceptive functional context influenced, as we have seen, by obligatory bipedalism in an upright posture and pronounced encephalization. Its eruption, position, and orientation are influenced in each of us by orthodontic forces related to chewing (Fig. 4.164).

The chewing cycle is a characteristic of all mammals, and its lateral-medial component (horizontal extension) varies significantly among different species based on their diets. The relationship between the mesio-palatal surface of the upper canine and the disto-vestibular surface of the lower canine constitutes the primary guide

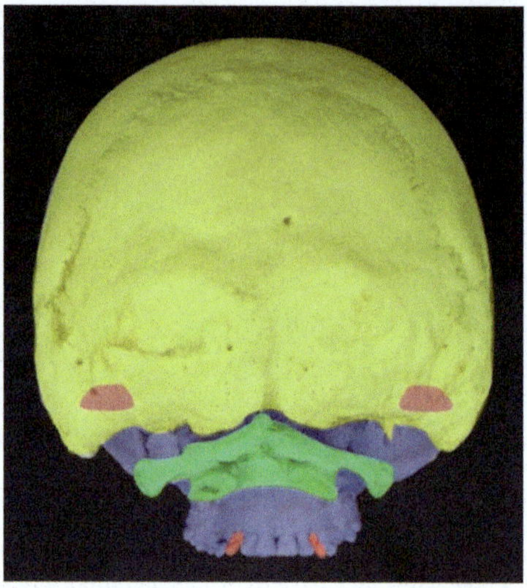

Fig. 4.164 The anatomo-functional modifications of the canine teeth in the genus Homo were not only a consequence of changed dietary habits and different social behaviors, but they also fit into a biomechanical context of modifications of the skull and facial structure, triggered first by bipedalism and then by encephalization: yellow, neurocranium; purple, splanchnocranium; pink, perspective of the mandibular fossae; green, atlas; red, canines. Specimen from the Museum of Anthropology of the University of Florence

during the phase of the cycle in which the force of the jaw adductor muscles is expressed most intensely (Fig. 4.165).

A prominently protruding canine beyond the occlusal plane would create interferences if there were an increase in the horizontal component of mandibular lateral movements during chewing cycles (Fig. 4.166).

In primates, canines establish reciprocal relationships along the coronal sides (Fig. 4.167), while in humans, the relationship involves the mesio-palatal surface of the upper canine with the disto-vestibular surface of the lower canine (Johanson et al., 1990) (Fig. 4.168), as previously illustrated in Fig. 4.165.

In humans, the canine, thanks to its unique position in the dental arch and its crown-root morphology, assumed a different and new functional role as a transitional tooth between the incisor group and the posterior teeth. Like a cornerstone situated at a critical point in the structure, it unites and supports the various components, constituting the pivotal element in the horizontal management of masticatory dynamics. Both phylogenetically and ontogenetically, its position, the orientation of its crown, and its functional facets—resulting from individual dynamic equilibria—are functional markers that do not represent the cause of the function but rather its consequence, being not so much the origin but rather the result (Picq, 2010).

Among mammals, canine teeth, although not directly involved in chewing, possess special characteristics and highly variable adaptive significance (Picq, 2010). In humans, canines can both tear food, preparing it for posterior occlusal crushing,

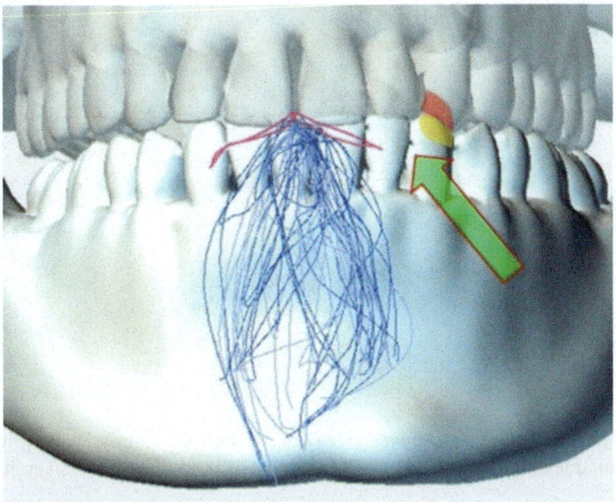

Fig. 4.165 Phase of entry of the left chewing cycle. In orange and red, the canine guides, functional surfaces that manage the centripetal sliding of the mandible during the forceful chewing stroke; a phase in which the bolus interposed between the teeth is crushed until it reaches the minimum vertical dimension: note, in fuchsia overlaid at the interincisal point, the two lateralization traces, which converge at the point of maximum intercuspation; from here the chewing cycles depart, which, in the traces of alternate unilateral chewing, form an 'entanglement' equally distributed on both sides

Fig. 4.166 Even in primates that are phylogenetically closest to us, the canines decisively exceed the occlusal plane; despite the diastemas, the mandibular dynamics are forced, in order to surpass the elongated crowns of the canines, to a greater verticality of lateral movements: Sumatran orangutan (*Pongo abelii*). Specimen from the Museum of Zoology of the University of Florence

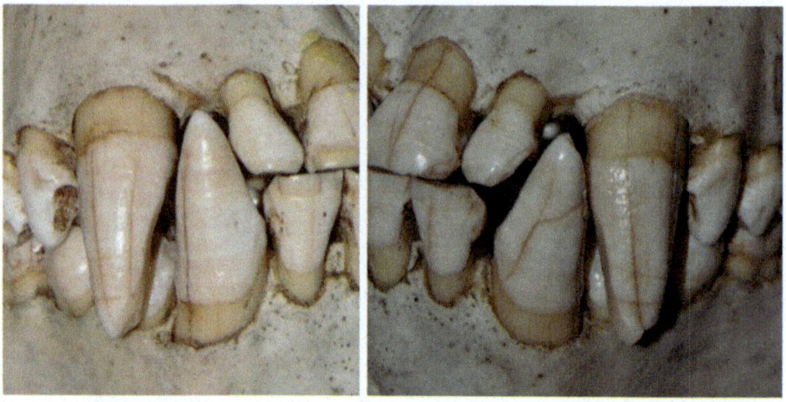

Fig. 4.167 Sumatran orangutan (*Pongo abelii*). C/P3 honing complex: reciprocal relationships along the coronal sides. Specimen from the Museum of Zoology of the University of Florence

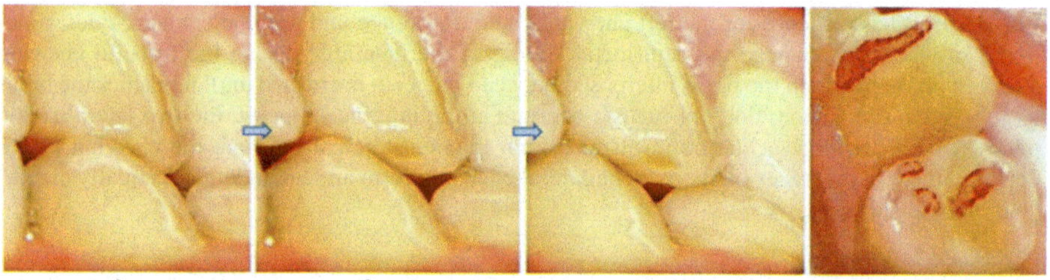

Fig. 4.168 Intercanine relationships in the entry phase of the chewing cycle: (**1**) The incisal edges contact, initiating the occlusal relationship; (**2**) The disto-vestibular surface of the lower canine slides along the mesio-palatal surface of the upper canine; (**3**) End of the entry phase of the chewing cycle, achieved upon reaching the minimum vertical dimension in maximum intercuspation; (**4**) Path of the upper canine's lateral guide highlighted with a 200-micron articulating paper

and play the crucial role of guiding horizontal mandibular movements during chewing (Fig. 4.169).

In humans, the proper management of lateral movements by the functional surfaces of the canines guides mandibular gliding along trajectories that define the gothic arch: an area that encompasses all the anterior-posterior and lateral-lateral sliding positions of the mandible against the skull on the horizontal plane. It represents the expression of the countless positions that the refined modulation of the external pterygoid muscles is capable of managing (Fig. 4.170).

From the moment the canine tooth positions itself in the dental arch, usually between the ages of 9 and 12, it immediately begins its role as a guide for lateral movements, excluding the other occlusal surfaces through a pure canine guidance. Pure canine guidance is physiological at this stage of life, but in a properly maturing masticatory system, all repeated occlusal contacts are naturally and inevitably destined to wear over time. The wear of canine guides leads to reduced verticality in the

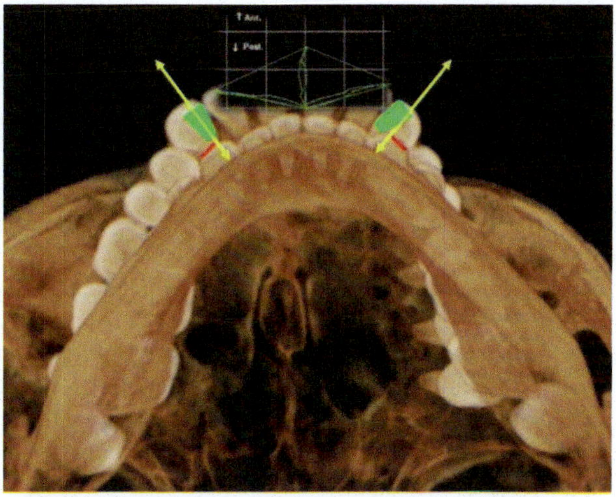

Fig. 4.169 Occlusal relationships between the mandible and the upper jaw, viewed from below: functional surfaces (in green for the upper canines and in red for the lower canines) that regulate and guide the intercanine relationship; superimposed above, the gothic arch, i.e., the boundary paths of lateral and protrusive movements on the horizontal plane

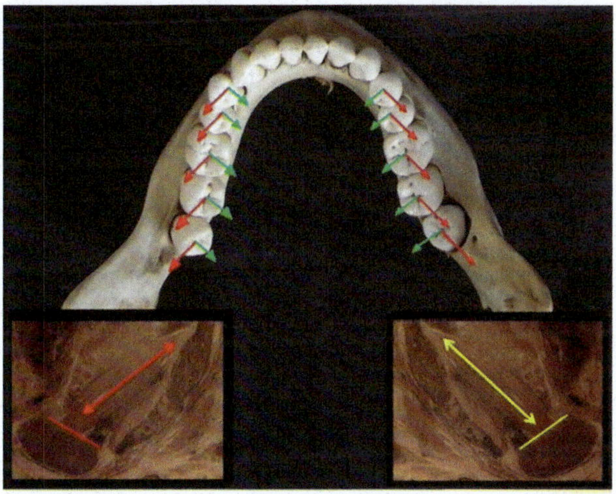

Fig. 4.170 The contraction of the external pterygoid muscles determines the opposite lateral movement: bottom left, horizontal plane section of the left condyle and left external pterygoid muscle (red arrow); bottom right, horizontal plane section of the right condyle and right external pterygoid muscle (yellow arrow)

entry phase of the chewing cycle, gradually involving the posterior teeth, thus establishing group guidance: the vestibular cusps of premolars and molars come into contact at the beginning of the entry phase of the chewing cycle, increasing grinding efficiency and making bolus processing much more effective (Fig. 4.171).

Canine guidance is therefore a characteristic of younger individuals, and its frequency decreases with age, while group function is more common in mature individuals. That's why canine-protected occlusion seems to be the most suitable model for planned orthodontic and prosthetic rehabilitation in younger patients, while group function occlusion can be a good model for prosthetic rehabilitation in mature subjects (Panek et al., 2008). In a physiologic alternation of the chewing side, these maturation phenomena of occlusal contacts occur almost symmetrically on both sides, promoting both anatomical and functional symmetry of the maxillaries. The

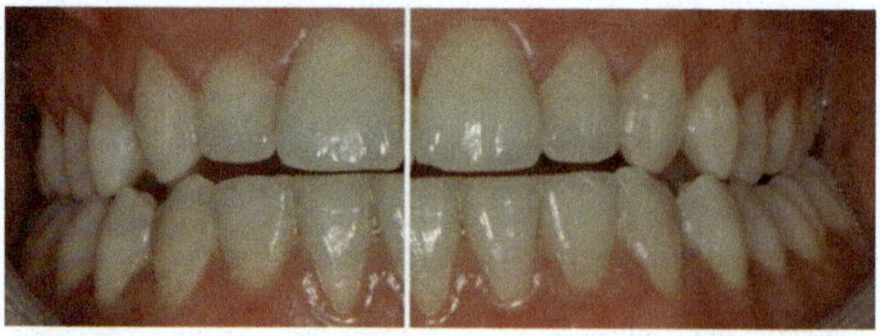

Fig. 4.171 Group relationship: on the left, entry phase of the right chewing cycle; on the right, entry phase of the left chewing cycle

absence of the physiological alternation of the chewing side, along with an asymmetric development of the bony bases, will result in asymmetric wear and maturation of the occlusal surfaces between the right and left sides, causing (especially during development) asymmetry in the dental arches, dental compensation curves, and their lateral guides. Chewing will predominantly occur on the side where movements develop with minimal vertical dimension (Planas, 1994) (Figs. 4.172 and 4.173).

The proprioceptive information from the canine is transmitted to the trigeminal nucleus and is then analyzed by the sub-cortical nervous system. The canine on the working side plays a special role in the entry phase of the chewing cycle, allowing very precise cortical control of the mandible and temporomandibular joints (TMJs) in all three planes of space. The canine on the balancing side limits the width of the exit movement during chewing. In the case of a missing canine, there is almost systematic shifting of chewing to the side where the canine is present: premolars are not suitable for the role of canine guidance, so it is sought on the opposite side (Casteyde, 1977, 2008).

The canine also differs from other dental elements at the embryonic level. The entire facial skeleton derives from the notochord that characterizes vertebrates (Le Douarin, 2004). Around the 4th week of intrauterine life, between the cranial bulge from which the head will develop and the cardiac bud, the four branchial arches or pharyngeal arches appear. The first arch is formed by two lateral swellings that split into two tubercles, one upper and one lower. The upper tubercle gives rise to the lateral processes of the maxilla, while the lower tubercle gives rise to the two mandibular processes (right and left), which rapidly grow toward the midline and meet to form the mandibular arch (Fig. 4.174).

The upper jaw (maxilla) has three distinct embryonic components: the right maxillary process, the left maxillary process, which originate from the upper tubercle of the first pharyngeal arch, and anteriorly interposed, the premaxillary or nasofrontal process, which forms the front portion of the palate and develops from the cranial bulge. These three components come together at the level of the canines/premolars, forming the upper jaw (maxilla) (Fig. 4.175).

The premaxilla is also known as the "incisive bone," "intermaxillary bone," "premaxillary bone," or even "Goethe's bone," to whom its discovery is romantically but erroneously attributed. The formation of the premaxilla internally stabilizes the facial skeleton and is closely related to the development of the human face. Due to its position and relationships, the premaxilla can be ideally compared to the

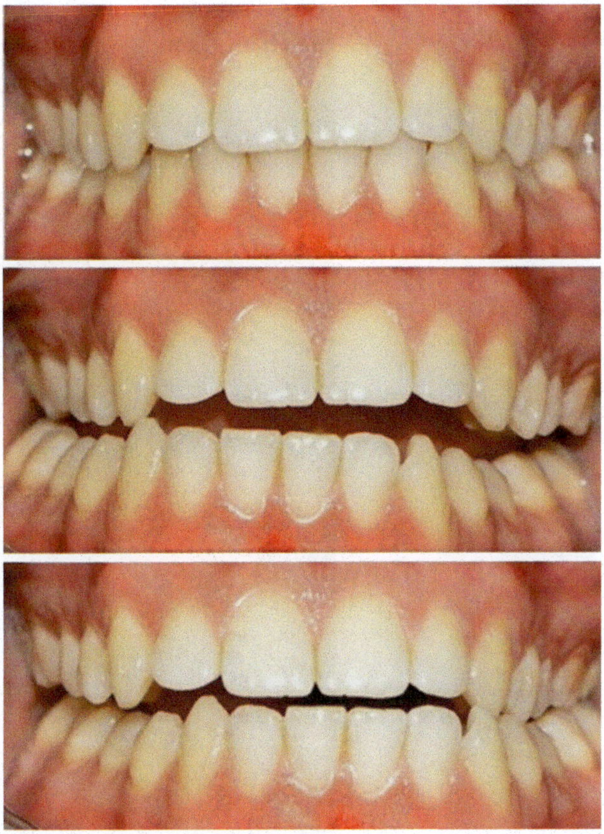

Fig. 4.172 Asymmetrical development of the maxillae, and of the von Spee and Wilson curves: at the top, maximum intercuspation; in the middle, right lateral movement; at the bottom, left lateral movement. Note how the right lateral movement engages only the canines, completely discluding the dental arches (pure canine guidance) with a greater vertical dimension at the start of the chewing cycle: at this functional moment, the posterior teeth do not contact, losing efficiency in processing the bolus; the left lateral movement, instead, develops a smaller vertical dimension at the start of the cycle, with all dental elements collaborating in the food comminution, as well as in guiding the entry phase, precisely when the forceful masticatory stroke and the maximum intensity of the mandibular adductor muscles' force are expressed

keystone of an arch (Trevizan & Consolaro, 2017). In each maxilla, the premaxilla is delimited by the suture that runs through the palate from the incisive foramen to the alveolar margin of the canine region (Fig. 4.176), then extending to the lower portion of the piriform aperture. This suture remains patent in the palatal region throughout early childhood, and patency can sometimes persist into adulthood. The premaxilla is a somewhat independent bony component within the maxillary complex, and it is the presence of the sutures that connect it to the maxillary processes that explains the success of antero-posterior expansions to stimulate midfacial growth.

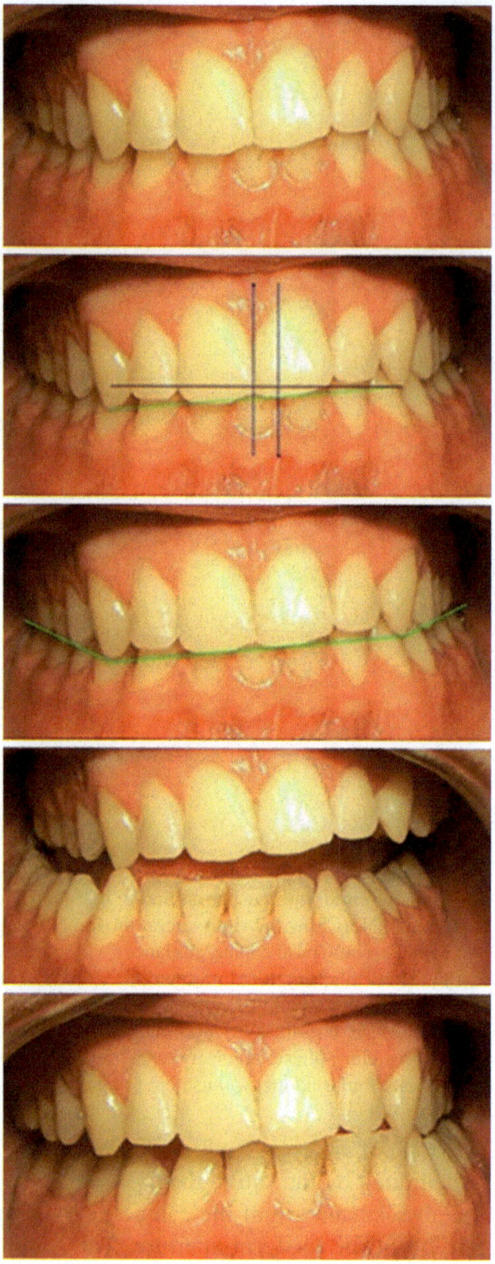

Fig. 4.173 Five images, from top to bottom. Maximum intercuspation: left mandibular laterodeviation; note that both the interincisal line and the labial frenula are shifted to the left. Misalignment of the interincisal lines (vertical blue lines) and horizontal leveling of the canines. Planas' Z occlusal plane. Right lateral movement with canine guidance. Left lateral movement with group function

A bone suture is a connection between two distinct elements, consisting of a narrow band of dense fibrous connective tissue, forming an immovable joint. Even within the maxillary complex, sutures, continuously remodeled by biomechanical stimuli, play an important role in growth. The maxillary complex consists of highly

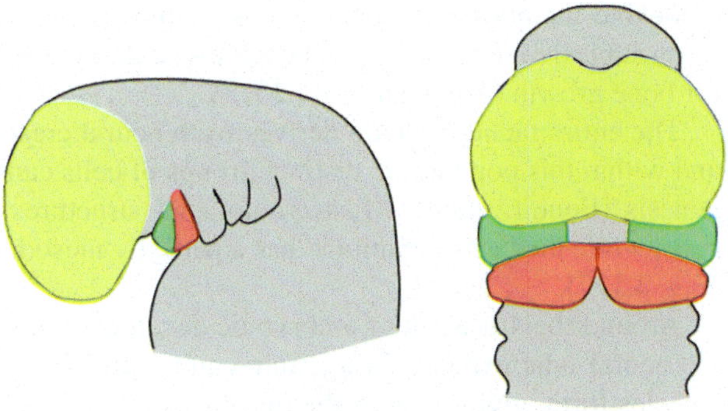

Fig. 4.174 Human embryo around the 4th week: yellow, frontonasal process; green, maxillary process; red, mandibular process

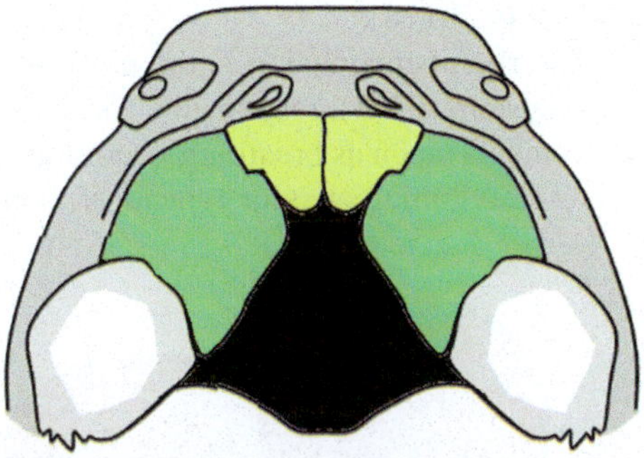

Fig. 4.175 Formation of the palate in the human embryo: yellow: premaxilla derived from the nasofrontal process; green: maxillary processes

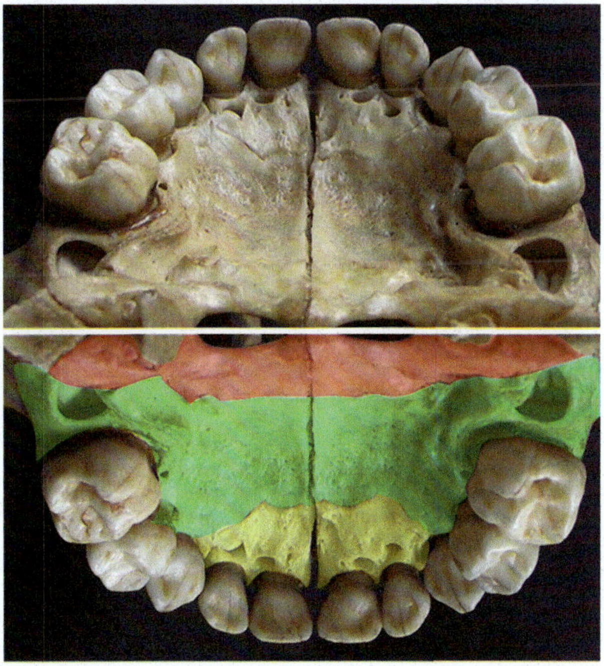

Fig. 4.176 Various palatal bone components: yellow, premaxilla; green, palatine process of the maxilla; red, horizontal plate of the palatine bone. Skull 2857 from the Museum of Anthropology of the University of Florence

malleable membranous bones, whose sutures act as growth sites that proliferate when subjected to tension. In a neonate's skull, the sutures are widely open to allow for bone growth (Trevizan et al., 2018).

The entire facial skeleton derives from neural crest cells (Jugessur et al., 2009), and within this population, distinct groups of cells can be identified during odontogenesis (Benoît, 2004). Most craniofacial structures originate from neural crest cells, while the "neurocranium" has a partially mesodermal origin (Kuratani, 2005) (Fig. 4.177).

Around the 4th week of embryonic development, there is a ventral folding, and the neural tube expands in a cranio-caudal direction. In the cranial portion, three vesicles form, giving rise to the prosencephalon (forebrain), mesencephalon (midbrain), and rhombencephalon (hindbrain), with the spinal cord located more caudally (Fig. 4.178).

The rhombencephalon is the most ancient and primitive part of the central nervous system (CNS), responsible for vital aspects by executing various "automatic" movements (such as breathing, locomotion, swallowing, digestion, sleep, and chewing—a rhythmic behavior typical of mammals). In the rhombencephalon, you can

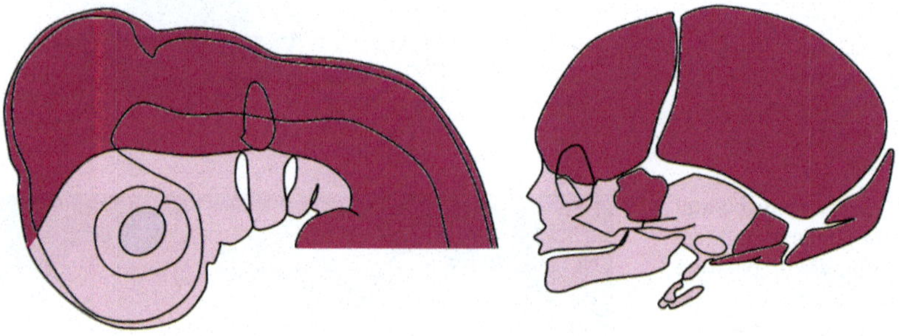

Fig. 4.177 Different embryonic origins of the neurocranium and splanchnocranium

Fig. 4.178 Different embryonic origins of the neurocranium and splanchnocranium

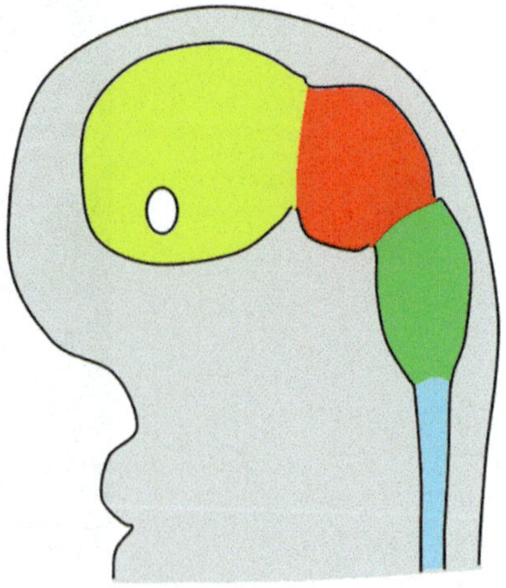

find the CPGs (central pattern generators), neural networks that generate the basic rhythmic pattern of movements associated with functions typically not controlled by our conscious part. The motor command for the basic chewing movements is generated by a neural network located in the brainstem's medulla and pons (Morquette et al., 2012; Nakamura & Katakura, 1995).

CPGs—these interconnected sets of neurons capable of generating basic motor output patterns—can be divided into "constitutive," active throughout the entire lifespan (such as the CPGs for respiration or digestion), or "conditional," which control and regulate episodic movements like locomotion or chewing (Arshavsky et al., 2015). The generation of a masticatory rhythmic pattern is a highly complex process, involving three distinct phases: the genesis of the masticatory rhythm, the creation of a pattern of muscular activity (involving the jaw, facial muscles, and tongue), and the coordination of the activities of all the muscles involved (Nakamura & Katakura, 1995). Nevertheless, masticatory cycles vary and are always adapted to the current situation: peripheral afferents are of fundamental importance in interacting with CPG activity depending on the type of food introduced into the oral cavity, and masticatory movements vary significantly, both due to the variable consistency of food and because it changes during the chewing process (Schwenk, 2000). In mammals, the masticatory cycles generated by CPGs vary depending on the species and their diet, but they all fall within a dynamic cyclic structure that applies to all mammals (Hiiemae, 1984).

To the ancient rhombencephalon, the more recent mesencephalon joined, which we share with many animals. It controls balance, oversees visual and auditory reflexes, and hosts essential functions for the rapid and automatic processing of external stimuli. Additionally, it connects the spinal cord to the cerebral hemispheres for controlling expression, phonation, writing, upright posture, and all other bodily attitudes.

The prosencephalon (composed of the telencephalon and diencephalon) is the most recent phylogenetic part and represents about 80% of the human brain. It processes the most refined and complex actions, hosts the highest cognitive functions, and the capacity for consciousness. In more primitive vertebrates, such as fish, the telencephalon, while serving as an integrative center for various stimuli, has a much less complex structure, and its activity is more modest. In less primitive forms, the prosencephalon develops progressively, eventually dominating the activities of the posterior centers (Padoa, 1991).

Returning to the embryological development of facial structures, on each side of the neural tube, clusters of cell populations migrate through the facial mesenchyme, forming the ectodermal and odontogenic skeletal mesenchyme and differentiating into numerous and different types of tissues within the craniofacial region (Benoît & Granat, 2010; Mossey et al., 2009). The bony skull of adults is formed from a complex set of embryological precursors of multiple tissues (German & Palmer, 2006). Groups of cells derived from the neural crest of the prosencephalon migrate to form the nasofrontal bud (Lieberman, 2011), from which the incisor teeth will develop. The upper frontal group, therefore, derives from a cell population that will give rise to neural structures capable of sophisticated proprioceptive discrimination.

Canines, on the other hand, originate from mesencephalic cell clusters that migrate to the anterior part of the maxillary area, initiating the development of these dental elements (Benoît & Granat, 2010). They originate from cells that will form the mesencephalon, a set of neural structures responsible for controlling and connecting the basic automatic functions of the spinal cord with the organizing and more refined functions of the cerebral hemispheres, allowing for rapid, automatic, and appropriate responses to external stimuli.

The canine has numerous periodontal receptors that send afferent signals to the mesencephalic nucleus of the trigeminal nerve. The number and density of these receptors confirm the profound embryogenetic relationships between these two structures (Piancino et al., 2017). The mesencephalic nucleus is located in the brainstem and contains the cell bodies of primary afferents involved in proprioception of the face, mandible, and dental mechanoreceptors. The mesencephalic nucleus is found in all vertebrates with a mandible, while it does not exist or is very rudimentary in agnathans (jawless vertebrates). This suggests that its unique design is related to mandibular dynamics (Gerstner et al., 2011).

From the rhombencephalon, on the other hand, cells migrate that will form the forehead, the middle and lower parts of the face, the pharynx, and the neck (Lieberman, 2011), as well as the dental buds of premolars and molars (Benoît & Granat, 2010). Premolars and molars, therefore, originate from cell populations that are part of the more ancient portion of the central nervous system (CNS), which oversees basic functions that are not under conscious control and regulates rhythmic and stereotyped motor behaviors such as walking, swimming, breathing, and, indeed, chewing (Fig. 4.179).

So, summarizing embryologically: premolars and molars originate from cells of rhombencephalic origin, canines from cells coming from the mesencephalon, and the upper incisor group from cells originating in the prosencephalon. It is remarkable to reflect on how the tasks performed by the different tooth groups seem to coincide with the general functions exercised by the neural regions of origin: molars and premolars originate from the archaic rhombencephalon for the execution of automatic movements; canines from the mesencephalon for the processing of rapid automatic responses and connections; and the upper incisor group from the recent prosencephalon for the processing of more refined and complex actions (Fig. 4.180).

The dentition of mammals is therefore composed of different types of teeth, each with a distinct morphology of the crown and root, highly adapted to their particular masticatory function; in the embryo, the development of individual teeth is based on complicated interactions between the ectoderm of the first branchial arch and mesenchymal cells derived from the neural crest, which migrate to this region from their original site. In the early stages of tooth development, the role of the ectoderm is essential, but the mesenchymal component also acquires specific competence and dominance in stabilizing the formation of cell populations that will give rise to individual dental elements; in the first branchial arch, the local regulation between the ectoderm and neural crest-derived ectomesenchyme is crucial to establish the shape of the appropriate tooth in the specific maxillary region (Cobourne & Mitsiadis,

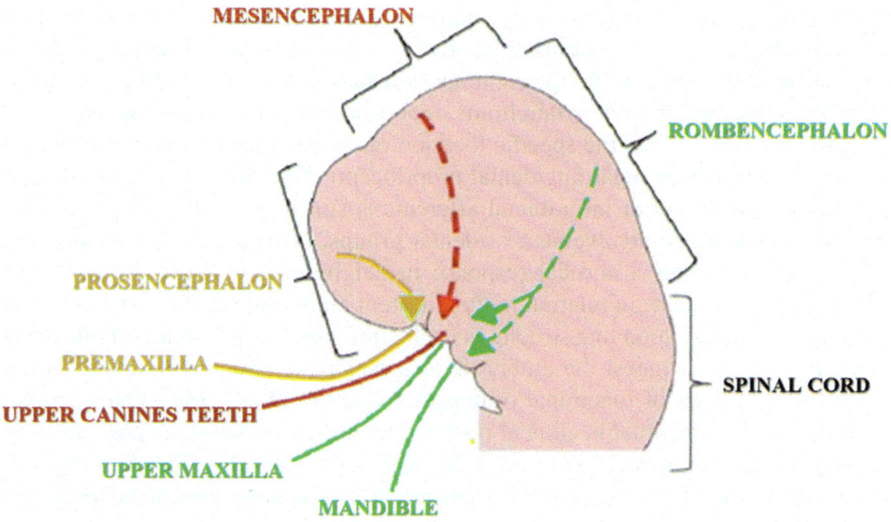

Fig. 4.179 Migration trajectories of different cell populations from distinct neural structures: rhombencephalon, mesencephalon, prosencephalon; these distinct groups of cells will also give rise to the buds of various dental elements. It is fascinating to note the correspondence between the functions performed by the three different dental groups (refined proprioception of the incisors, coordination and guidance of the canines' movements, and rhythmic and repetitive work of the grinding teeth) with the types of functions of the CNS structures from which the teeth derive

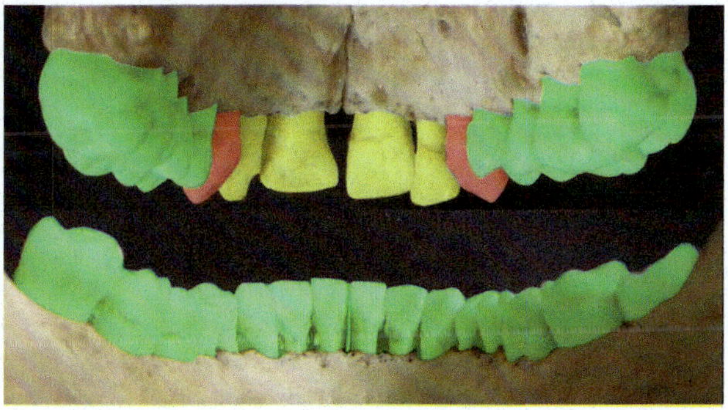

Fig. 4.180 Division of dental elements based on their embryological origin: yellow, prosencephalon; red, mesencephalon; green, rhombencephalon. Skull 372 from the Museum of Anthropology of the University of Florence

2006). In the process of cellular migration just described, the cells that will form the maxillary arch are distinguished from those that will form the mandible, having a common origin, deriving from the maxillary arch of the first branchial arch (Cerny et al., 2004).

The roots of different teeth (incisors, canines, and premolars-molars) are located in portions of the alveolar bone and innervated by different branches of the trigeminal nerve: the premaxilla area coincides with the territory of the nasopalatine nerve, the lateral areas of the maxillary bone are innervated by the right and left major palatine nerves, the maxillary lateral areas are innervated by the right and left major palatine nerves, and in the upper maxillary on the border of these areas the canines are positioned; the mandibular body and the upper teeth are innervated by the right and left inferior alveolar nerves (Fig. 4.181). The development and homeostatic adaptation of the premaxilla, the right and left maxillary areas, and the two hemimandibles respond, within certain limits, independently to functional stimuli.

Little is known about the specific location of the reference areas in the primary somatosensory cortex regarding dental proprioception because research investigating the cortical target of intra-dental afferents is still in its early stages; but what emerges is a sensory involvement of dental groups, in which the stimulation of a single tooth activates a cortical response in neighboring elements as well. This group neural response, in relation to the different embryonic areas of the jaws, and their specific innervation that structures itself simultaneously with their bone development, seems to suggest the embryological, anatomical, and neural foundations supporting the laws of functional orthopedics (Dettori et al., 2001; Planas, 1994): Pedro Planas with a notable clinical intuition identified in the upper jaw and in the mandible a developmental territoriality that realizes itself with differentiated growth responses to the biomechanical stimulations of mastication, according to the distinct embryonic origin in the different sections of the jaws.

Even if the different bony components of the jaws are innervated by fibers pertaining to the entire trigeminal nerve, the stimulation of different active territories has shown functional hemispheric laterality: functional magnetic resonance imaging has indeed shown a significant correlation of extracephalic activations

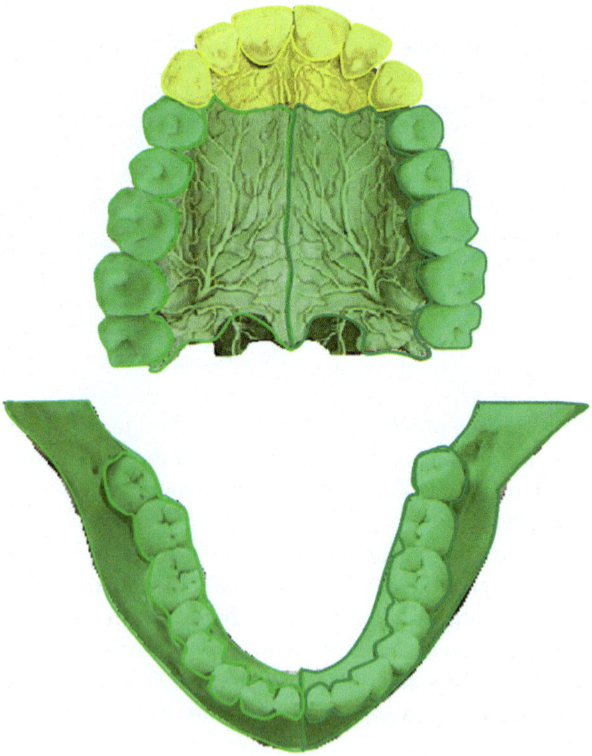

Fig. 4.181 The maxillae are innervated by distinct branches of the trigeminal nerve: the premaxilla by the nasopalatine nerve, the posterior portions of the upper maxillae by the greater palatine nerves; the upper canines are positioned at the boundary between these two innervations, as if their transitional morphology between the incisor group and the masticatory group is also supported from a neurological perspective. Each hemimandible is innervated by its own alveolar nerve

between the incisal bite and the molar bite (Yoshizawa et al., 2019). Each hemisphere preferentially responds to the stimulation of the contralateral side and in both hemispheres there is a clear somatotopic arrangement with the representation of the maxillary teeth, which differs from the cortical representation of the mandibular dental arch (Bono, 2020); cerebral circulation is activated during mastication, independently of the rhythm and voluntary control of the side of mastication (Hasegawa et al., 2011).

The skeleton has a well-developed sensory innervation, which, like in other parts of the body, serves to detect and respond to various stimuli, including mechanical loads. Sympathetic nerve endings that regulate autonomous functions related to bone tissue are more abundant in the medullary spaces and the periosteum, where the density of innervation is higher near areas of active remodeling. Only recently have the functions of this complex interaction between nervous tissue and bone tissue started to become known. The peripheral nervous system plays a role in bone physiology and its adaptation to the mechanical environment, regulating autonomous functions related to osteogenesis, including homeostatic (Heffner et al., 2017) remodeling. Even during the embryonic development of limbs and their ossification, nerve fibers play a distinct role in skeletal growth, innervating the subchondral bone and modulating vascularization and the differentiation of the endochondral matrix (Grässel, 2014). During intrauterine life, there is a supply of sensory nerve fibers in the developing bone, seemingly parallel to the increasing mineralization, indicating neural influence on skeletal development processes (Sisask et al., 1995).

In the developing skeleton of mice, functional nerve supply is early and prominently localized, especially in areas with high osteogenic activity, co-localizing with a rich capillary network (Sisask et al., 2013). Nerve fibers are most frequently found in metabolically active bone areas: in periosteal vessels, Volkmann's canals, bone marrow, the osteochondral junction of the growth plate, and the attachment of the synovial membrane. On the bone surface, periosteal nerves are more numerous in the epiphysis than in other regions (García-Castellano et al., 2000).

Although the two upper maxillae and the mandible represent opposing parts of the same functional system for food breakdown, they exhibit embryological, anatomical, and functional differences, each playing its role with distinct proprioceptive regulations. The receptive field of human periodontal mechanoreceptors often extends beyond a single tooth, and about half of the periodontal mechanoreceptors of posterior teeth respond to the stimulation of adjacent teeth (from two to four; these "receptive fields" enhance proprioceptive localization and spatial resolution of individual dental elements. During chewing, not only are the periodontal afferents that innervate the tooth where force is applied directly activated, but also those innervating adjacent teeth. This group proprioception also results from a mechanical connection between neighboring teeth, through transseptal collagen fibers or interdental contact, rather than branching of nerve endings to more than one tooth (Johnsen & Trulsson, 2003; Trulsson, 2006). A common feature of all these multiple afferents is that the involved teeth are adjacent.

Therefore, both due to their embryological origin and the territorial innervation of different nerve branches, the maxillae can be divided into well-defined functional areas: the premaxilla and upper lateral maxillae superiorly, and the two lower hemi-mandibles inferiorly.

The upper canines are located at the border of the premaxilla, in a bony territory highly sensitive to biomechanical stimuli, both due to its histological nature and embryological origin. The biomechanical stimuli applied to the upper canine are shared anteriorly by the premaxilla and posteriorly by the maxilla, with a common growth stimulus.

As for the premaxilla, functional stimulation of any tooth in the upper incisor sector is collected as a global bone growth stimulus for that sector, but it does not extend beyond the canines. Monolateral chewing promotes the growth of the alveolar process and basal bone through neural stimulation caused by occlusal forces transmitted by the periodontium. According to Planas' developmental laws, on the working side, both the maxilla and the hemi-mandible strengthen, while this does not occur on the balancing side, where the hemi-mandible is stimulated to grow anteroposteriorly (Fig. 4.182).

As we have seen, canines are very unique teeth in terms of shape and function. Both their long root and incisiform crown, while being single-pointed, enable them to withstand considerable forces (Fig. 4.183). They are the only teeth among all dental elements that perform the tearing function, which involves enormous transverse effort, with stress occurring even during the guidance of lateral movements of the mandible.

These considerations provide a new perspective on the profound implications of the unique role that canines play in the development and proprioception of the maxillary bones.

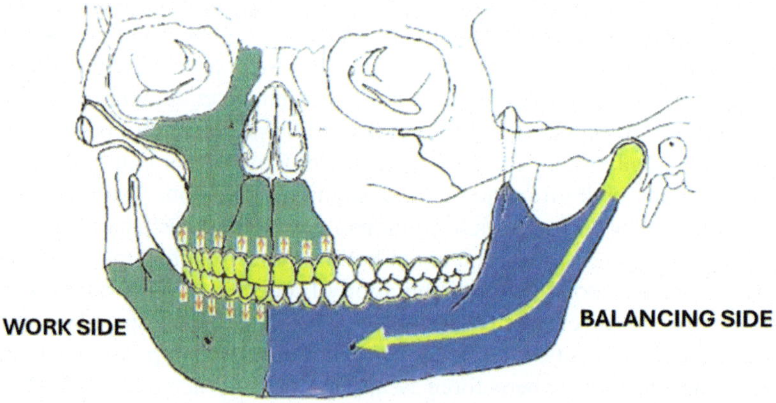

Fig. 4.182 Asymmetrical development of the maxillae in fixed unilateral mastication (Image from: "Evolution of the Functional Dynamic Relationship between Occlusion and TMJ," Dettori et al., 2001)

Fig. 4.183 The canine is frequently used instinctively, unconsciously relying on its ability to withstand intense stress

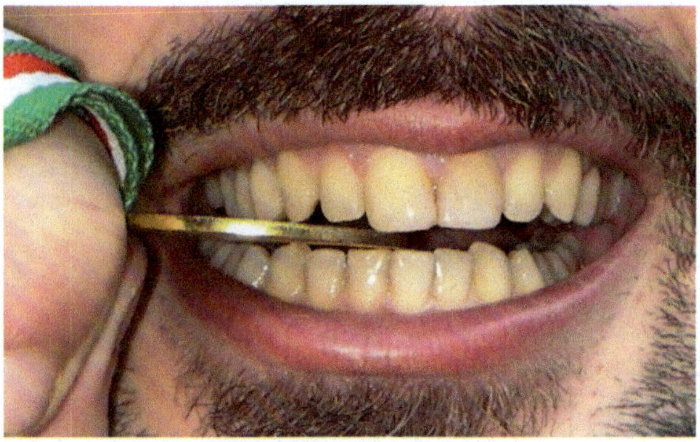

4.9 Reduction in Mandibular Length

The evolutionary development of the mandible in the *Homo* genus, particularly in *H. sapiens*, should be viewed as the result of multiple factors rather than solely as a consequence of biomechanical pressures or cranial constraints (Veneziano et al., 2018). The robustness of the mandible is described by the ratio of its width to height, an important factor in primates to determine its resistance to torsional and bending forces during chewing (Hylander, 1985). The increase in the height of the mandibular body observed in the evolution of the *Homo* genus has led to a reduction in mandibular robustness (Veneziano et al., 2018).

Australopithecines and humans are characterized by having short and deep mandibles with relatively small incisors and canines. An analysis of stress on primate mandibles suggests that in the molar region, mandibles are vertically deep as an adaptation to counter the increased sagittal bending stress on the balancing side of the mandibular body during unilateral chewing. This increase in bending stress on the balancing side is caused by the intense muscular forces activated during chewing (Hylander, 1979a, 1979b, 1979c). A widespread hypothesis suggested that this series of traits evolved in early clade members in response to changing environmental conditions and variations in available food. With the emergence of the *Homo* genus, the functional significance of mandibular shape variation was linked to technological development, particularly fire control. However, this hypothesis is contradicted by the exceptionally rapid evolution of mandibular shape in both australopithecines and humans compared to any other primate clade. The entire hominid clade stands out among primates as a unique example of mandibular shape evolution. The evolution of the masticatory apparatus has been linked to diet and food processing, but the physical connection between the neurocranium and the lower jaw underscores the role of encephalization in dental and mandibular reduction.

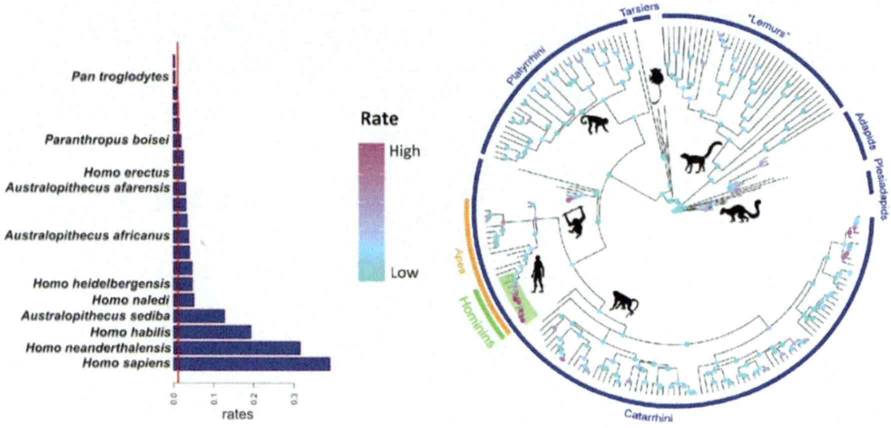

Fig. 4.184 Evolutionary rates of mandible shape in the primate tree (Image from "Unexpectedly rapid evolution of mandibular shape in hominins," Scientific Reports, 2018)

Although the neurocranium and mandible may have evolved under the influence of various factors, encephalization occurred almost simultaneously with the reduction in dental and mandibular volume. Furthermore, since australopithecines had not developed food processing techniques, we must consider that the shortening of the mandible shared with the *Homo* genus is not related to food processing but is also due to the unique locomotor adaptation represented by bipedalism (Raia et al., 2018). Bipedalism is what led to a diversification of dental conformation, with the reduction in canine volume, the loss of the canine/premolar complex, and the acquisition of new masticatory potential. This allowed us access to new trophic niches, paving the way for new diets and dietary adaptations (Manzi, 2018) (Fig. 4.184).

4.10 Expansion of the Upper Jaw

The shape of dental arches and the size of canines distinguish humans from all other primates. Humans are characterized by small canines and dental arches with short parabolas and short rows of post-canine teeth, while monkeys and apes have large canines with long and narrow U-shaped dental arches. Within the hominid family, as canine size reduced and the diastema diminished, incisors moved backward, and dental arches became parabolic (Stelzer et al., 2017) (Fig. 4.185).

Compared to the crania of more archaic species that had a more elongated shape, in *Homo sapiens*, the neurocranium developed a globular encephalization with a contraction of the anteroposterior diameter. This reduction in anteroposterior dimension "pulled" back the base of the upper jaw, which simultaneously expanded its base (Fig. 4.186).

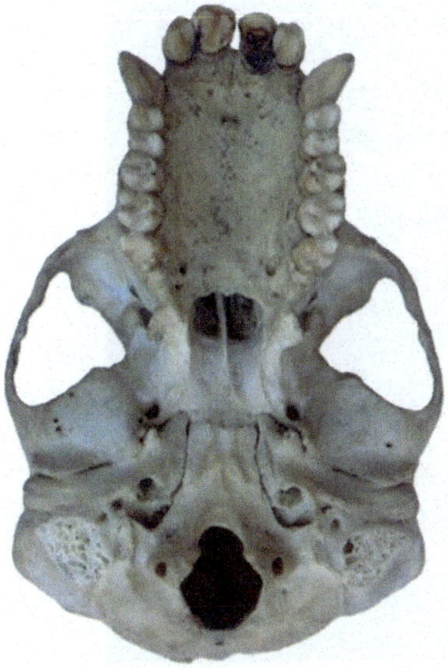

Fig. 4.185 On the left, chimpanzee (*Pan troglodytes*), specimen from the Museum of Zoology of the University of Florence; on the right, *Homo sapiens*, skull 4924 from the Museum of Anthropology of the University of Florence

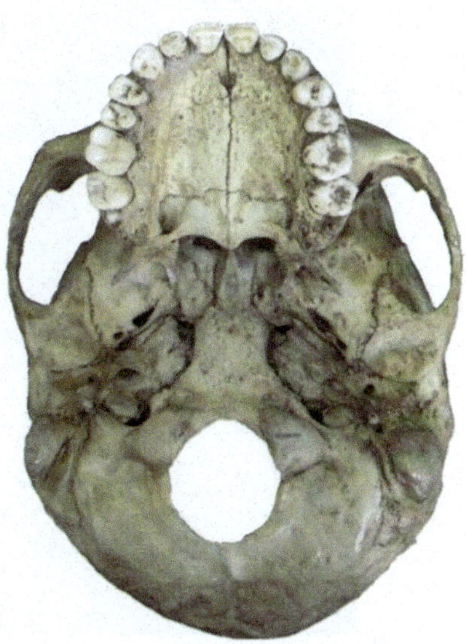

Fig. 4.186 Compared to more archaic species (green outline), in *Homo sapiens* (red outline) there was a contraction of the anteroposterior diameter of the neurocranium, which influenced the shape of the basicranium and facial structure

4.11 Modification of the Occlusal Plane

The occlusal plane is part of a more complex functional framework, the components of which influence and coordinate with each other (Dickerson et al., 2001). The modification of the geometry of various cranial regions related to bipedalism and encephalization has influenced the vectoring of masticatory muscles and, in turn, the arrangement and orientation of teeth. Muscular forces affect the position of teeth, just as the arrangement of teeth influences muscular movement, both with the inclination of individual elements and the overall orientation of all teeth. The long

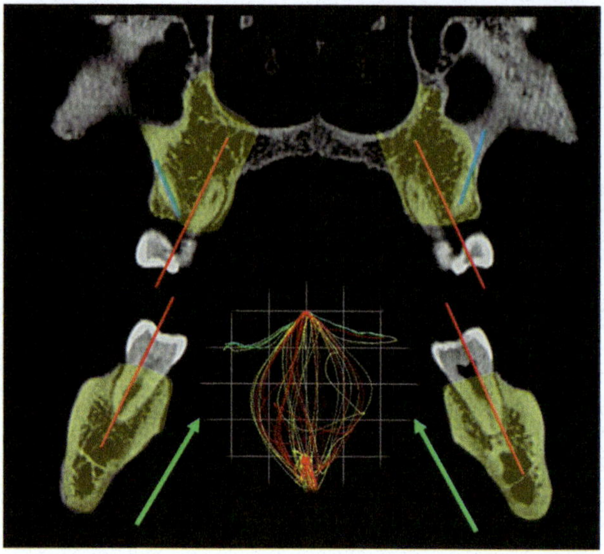

Fig. 4.187 Dental axes: the red lines indicate the axes of the teeth; the light blue lines indicate the axes of the vestibular roots of the upper molars, which are oriented to also distribute forces toward the zygomatic pillar; the green arrows indicate the vectors of the entry phase of the chewing cycle where the forceful stroke develops; in the center, a kinesiographic tracing of the chewing cycles with the superimposed lateral traces that guide the cycles. Note how the dental axes, the root axes, and the bony bases themselves orient in relation to the main vectors of mastication

axes of teeth align according to the most favorable inclination to receive pressure induced by function (Thielemann, 1958) (Fig. 4.187). A tooth is a biomechanically inert element, and its position stabilizes in bone when the resultant forces equal zero—a situation more theoretical than practical.

During chewing, muscular forces with various directions develop simultaneously, but they will have a single resultant force (Osborn, 1993a, 1993b). The learning of the motor pattern for chewing is structured based on the morphology of the occlusal plane (and modifications of other determinants of occlusion): each change leads to a new phase of adaptation and motor learning (Dickerson et al., 2001; Federici & Centofanti, 1977; Pompa et al., 2010). Since the occlusal plane is connected to other functional aspects of occlusion (inclination of the mandibular fossa, cusp shape, incisal guidance during protrusion, canine guidance during lateral movements, canine length, intercondylar distance, compensatory curves), in humans, it takes on species-specific characteristics linked to individual masticatory function and age (see Chap. 3 on dental wear) (Fig. 4.188).

The shortening of the face is a significant trend in the evolution of hominids, and an increase in the axial inclination of molars seems to be associated with the retrusion of the maxillae and the reduction of dental volumes, functionally related to cranial structure. Among the earliest hominids, *Australopithecus afarensis* already exhibits a certain degree of axial orientation of the molars (Holly Smith, 1986).

Fig. 4.188 On the left, gorilla (*Gorilla gorilla*), specimen from the Museum of Zoology of the University of Florence; on the right, *Homo sapiens*, skull 3677 from the Museum of Anthropology of the University of Florence

4.12 Formation of Overbite

In the human dentition, overbite measures the vertical overlap of the upper central incisors over the lower counterparts, with reference to the incisal ridges (Fig. 4.189).

The dentition of monkeys and apes is specialized for an 'arboreal' diet, predominantly focused on consuming fruits. They have long, wide, and prominent incisors adapted for biting, piercing, and peeling fruit skins, with an incisal ridge-to-ridge ratio that proves effective for handling this type of food (Osborn, 1993a, 1993b). In humans, the advancement of the anterior incisor group over the lower one is linked to a modification of the maxillae, attributed to the new cranial arrangement resulting from bipedal locomotion and an upright posture (Raia et al., 2018). In our evolutionary lineage, overbite appears with *Australopithecus africanus*, in which the incisal overjet is only slightly accentuated (Moggi-Cecchi et al., 1998), and it becomes a functional determinant of occlusion with *Homo sapiens* (Fig. 4.190).

The arrangement of the human incisor group doesn't directly stem from the frugivorous structure of primates but rather results from the omnivorous-carnivorous dietary specialization of australopithecines, in which overbite makes its appearance. Human incisors are short, relatively narrow, and vertically oriented, suited for cutting and separating tough pieces of cartilage and muscle. Meanwhile, the short canines, useless for combat, are stubby, bluntly rounded, and only slightly longer than the incisors, aiding in grinding and tearing operations (Dart, 1934).

It's also worth noting that until the last century, human dentition was subject to gradual wear due to intensive chewing, which progressively reduced the incisal and canine guidance. Occlusal abrasion led to a gradual loss of overjet and overbite, eventually resulting in an incisal ridge-to-ridge relationship. Proprioceptive monitoring of mandibular movements, from incisors in protrusion to canines in lateral excursions, gradually extended to encompass control over posterior teeth as wear facets spread.

In the preceding pages, we have seen how the postural factor of bipedalism, coupled with our unique globular encephalization, jointly shaped the structure of

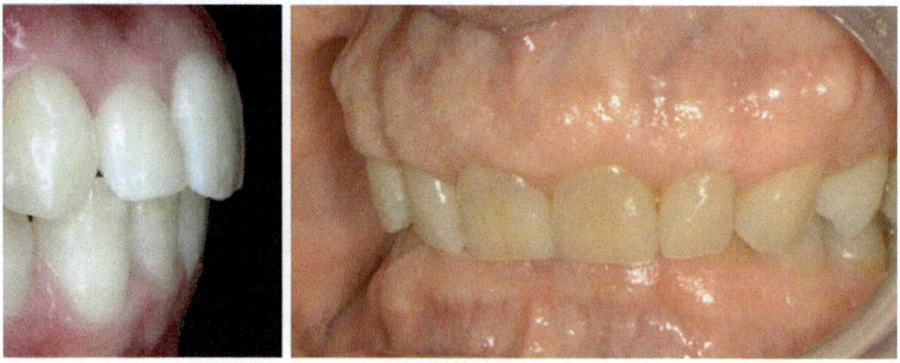

Fig. 4.189 On the left: correct overbite; on the right: deep bite

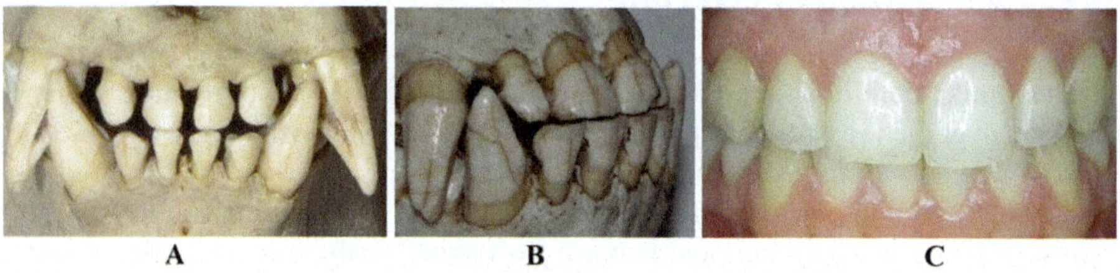

Fig. 4.190 (**a**) Red howler monkey (*Alouatta seniculus*), specimen from the Museum of Zoology of the University of Florence; (**b**) orangutan (*Pongo pygmaeus*), specimen from the Museum of Zoology of the University of Florence; (**c**) *Homo sapiens*

the stomatognathic apparatus. To deny the correlation between posture and dental occlusion means not only disregarding the embryological, anatomical, and functional relationships among the various segments of the postural chain to which the mouth belongs but also willfully ignoring the evolutionary history that inexorably links the mandible, skull, and vertebral column.

References

Abitbol, M. (1988). Effect of posture and locomotion on energy expenditure. *American Journal of Physical Anthropology, 77*(2), 191–199.

Aboitiz, F. (1994). The evolution of brain size and organization in vertebrates. A program for research. *Biological Research, 27*, 15.

Aiello, L., & Dean, C. (1990). *An introduction to human evolutionary anatomy*. Academic Press.

Alghadir, A. H., Zafar, H., & Iqbal, Z. A. (2015). Effect of three different jaw positions on postural stability during standing. *Functional Neurology, 30*(1), 53.

Alves, N. (2008). Study about the development of the temporomandibular joint in the human fetuses. *International Journal of Morphology, 26*(2), 309–312. Nov., 2011: *88* (3), 89–101.

Anastasi, M. R., et al. (2020). The discomallear ligament: Anatomical, microscopical, and radiologic analysis. *Surgical and Radiologic Anatomy*, 1–7.

Anirudh, B. V. M., Yuvaraj Babu, K., & Mohanraj, K. G. (2018). Comparing the dimensions of canine tooth in different species. *Drug Invention Today, 10*(3), 3311–3313.
Anson, B. J., Bast, T. H., & Richany, S. F. (1955). LXXVI The fetal and early postnatal development of the tympanic ring and related structures in man. *Annals of Otology, Rhinology & Laryngology, 64*(3), 802–823.
Anthwal, N., Joshi, L., & Tucker, A. S. (2013). Evolution of the mammalian middle ear and jaw: Adaptations and novel structures. *Journal of Anatomy, 222*(1), 147–160.
Arai, H., & Sato, I. (2011). Anatomical study of the human discomallear ligament using cone beam computed tomography imaging and morphological observations. *Okajimas Folia Anatomica Japonica, 88*(3), 89–101.
Arlegi, M., Gómez-Olivencia, A., Albessard, L., Martínez, I., Balzeau, A., Arsuaga, J. L., & Been, E. (2017). The role of allometry and posture in the evolution of the hominin subaxial cervical spine. *Journal of Human Evolution, 104*, 80–99.
Arlegi, M., Gómez-Robles, A., & Gómez-Olivencia, A. (2018). Morphological integration in the gorilla, chimpanzee, and human neck. *American Journal of Physical Anthropology, 166*(2), 408–416.
Arnold, P. (2020). Evolution of the mammalian neck from developmental, morpho-functional, and paleontological perspectives. *Journal of Mammalian Evolution, 28*, 173–183.
Arshavsky, I., Deliagina, T. G., & Orlovsky, G. N. (2015). Central pattern generators: Mechanisms of the activity and their role in the control of "automatic" movements. *Zhurnal vysshei nervnoi deiatelnosti imeni IP Pavlova, 65*(2), 156–187.
Ashton, E. H., & Zuckerman, S. (1954). The anatomy of the articular fossa (fossa mandibularis) in man and apes. *American Journal Of Physical Anthropology, 12*(1), 29–62.
Ayuse, T., Inazawa, T., Kurata, S., Okayasu, I., Sakamoto, E., Oi, K., Schneider, H., & Schwartz, A. R. (2004). Mouth-opening increases upper-airway collapsibility without changing resistance during midazolam sedation. *Journal of Dental Research, 83*(9), 718–722.
Baragar, F. A., & Osborn, J. W. (1984). A model relating patterns of human jaw movement to biomechanical constraints. *Journal of Biomechanics, 17*(10), 757–767.
Been, E., Gómez-Olivencia, A., Shefi, S., Soudack, M., Bastir, M., & Barash, A. (2017). Evolution of spinopelvic alignment in hominins. *The Anatomical Record, 300*(5), 900–911.
Been, E., Gómez-Olivencia, A., & Kramer, P. A. (Eds.). (2019). *Spinal evolution: Morphology, function, and pathology of the spine in hominoid evolution*. Springer.
Benoît, R. (2004). Génétique et arcades dentaires. *Orthod fr, 75*(2), 107–121.
Benoît, R., & Granat, J. (2010). The canine in the dento-alveolar system. *Journal of Dentofacial Anomalies and Orthodontics, 13*(1), 11–39.
Bono, D. (2020). *Intraoral somatosensation: A NASCENT perspective*. Doctoral dissertation, UCL (University College London).
Bramble, D. M., & Lieberman, D. E. (2004). Endurance running and the evolution of Homo. *Nature, 432*(7015), 345.
Brown, E. C., Cheng, S., McKenzie, D. K., Butler, J. E., Gandevia, S. C., & Bilston, L. E. (2013). Tongue and lateral upper airway movement with mandibular advancement. *Sleep, 36*(3), 397–404.
Brown, J. G. (2014). Jaw function in *Smilodon fatalis*: A reevaluation of the canine shear-bite and a proposal for a new forelimb-powered class 1 lever model. *PLoS One, 9*(10), e107456.
Bruner, E., Manzi, G., & Arsuaga, J. L. (2003). Encephalization and allometric trajectories in the genus *Homo*: Evidence from the Neandertal and modern lineages. *Proceedings of the National Academy of Sciences, 100*(26), 15335–15340.
Buffoli, B., Galli, M., Papini, A., Ferrari, M., Belotti, F., Labanca, M., Tschabitscher, M., & Rodella, L. F. (2014). Morphometric analysis of Huguier's canal by Cone Beam CT. *Italian Journal of Anatomy and Embryology, 118*(2), 36. ISSN 2038-5129.
Burch, J. G. (1970). Activity of the accessory ligaments of the temporomandibular joint. *The Journal of Prosthetic Dentistry, 24*(6), 621–628.

Byers, M. R., & Dong, W. K. (1989). Comparison of trigeminal receptor location and structure in the periodontal ligament of different types of teeth from the rat, cat, and monkey. *Journal of Comparative Neurology, 279*(1), 117–127.
Cai, M., Brown, E. C., Hatt, A., Cheng, S., & Bilston, L. E. (2016). Effect of head and jaw position on respiratory-related motion of the genioglossus. *Journal of Applied Physiology, 120*(7), 758–765.
Cash, R. M., & Linden, R. W. (1982). The distribution of mechanoreceptors in the periodontal ligament of the mandibular canine tooth of the cat. *The Journal of Physiology, 330*(1), 439–447.
Casteyde, J. P. (1977). Etio-pathogénie et symptomatologie des syndromes dysfonctionnels de l'appareil manducateur. *Revue d'Odonto-Stomatologie, 6*(6), 477–486.
Casteyde, J.-P. (2008). L'occlusion de la canine. Importance, options de réglages, risques et précautions. *Actualités Odonto-Stomatologiques, 244*, 355–366.
Cerny, R., Lwigale, P., Ericsson, R., Meulemans, D., Epperlein, H. H., & Bronner-Fraser, M. (2004). Developmental origins and evolution of jaws: New interpretation of "maxillary" and "mandibular". *Developmental Biology, 276*(1), 225–236.
Chen, X. (1998). The instantaneous center of rotation during human jaw opening and its significance in interpreting the functional meaning of condylar translation. *American journal of Physical Anthropology: The Official Publication of the American Association of Physical Anthropologists, 106*(1), 35–46.
Cheynet, F., Guyot, L., Richard, O., Layoun, W., & Gola, R. (2003). Discomallear and malleomandibular ligaments: Anatomical study and clinical applications. *Surgical and Radiologic Anatomy, 25*(2), 152–157.
Clark, G. T., Browne, P. A., Nakano, M., & Yang, Q. (1993). Co-activation of sternocleidomastoid muscles during maximum clenching. *Journal of Dental Research, 72*(11), 1499–1502.
Cobourne, M. T., & Mitsiadis, T. (2006). Neural crest cells and patterning of the mammalian dentition. *Journal of Experimental Zoology Part B: Molecular and Developmental Evolution, 306*(3), 251–260.
Craddock, F. W. (1948). The muscles of mastication and mandibular movements. *New Zealand Dental Journal, 44*, 233–239.
Cuccia, A. M., Caradonna, C., & Caradonna, D. (2011). Manual therapy of the mandibular accessory ligaments for the management of temporomandibular joint disorders. *The Journal of the American Osteopathic Association, 111*(2), 102–112.
Dart, R. A. (1934). The dentition of *Australopithecus africanus*. *Folia Anatomica Japonica, 12*(4), 207–221.
Dean, C. (1990). *An introduction to human evolutionary anatomy*. Academic Press.
Delezène, L. K. (2015). Modularity of the anthropoid dentition: Implications for the evolution of the hominin canine honing complex. *Journal of Human Evolution, 86*, 1–12.
Del Palomar, A. P., & Doblare, M. (2006). Finite element analysis of the temporomandibular joint during lateral excursions of the mandible. *Journal of Biomechanics, 39*(12), 2153–2163.
De Toffol, L., Cozza, P., di Girolamo, M., & Cirillo, F. (2004). Il condilio mandibolare. *Virtual Journal of Orthodontics [Serial Online], 6*(2), 1–8.
Dettori, S., Confaloni, A., & Pazzola, T. (2001). *RNO: evoluzione del rapporto dinamico funzionale tra occlusione ed ATM*. Mercurio editore.
Dickerson, W. G., Chan, C. A., & Carlson, J. (2001). The human stomatognathic system: A scientific approach to occlusion. *Dentistry Today, 20*(2), 100.
Di Vincenzo, F. (2011). Evoluzione: da *Australopithecus* a *homo*.
Di Vincenzo, F., & Manzi, G. (2013). Accademia Nazionale dei lincei Seminario sulla evoluzione biologica e i grandi problemi della biologia cervello in evoluzione (Roma, 27–28 febbraio 2013) E poi venne il cervello: evoluzione umana ed encefalizzazione.
Du Brul, E. L. (1964). Evolution of the temporomandibular joint. In B. G. Sarnat (Ed.), *The temporomandibular joint*. Charles C Thomas.
Dumont, E. R., & Herrel, A. (2003). The effects of gape angle and bite point on bite force in bats. *Journal of Experimental Biology, 206*(13), 2117–2123.

Dunbar, D. C., Badam, G. L., Hallgrímsson, B., & Vieilledent, S. (2004). Stabilization and mobility of the head and trunk in wild monkeys during terrestrial and flat-surface walks and gallops. *Journal of Experimental Biology, 207*(6), 1027–1042.

Dunbar, D. C., Macpherson, J. M., Simmons, R. W., & Zarcades, A. (2008). Stabilization and mobility of the head, neck and trunk in horses during overground locomotion: Comparisons with humans and other primates. *Journal of Experimental Biology, 211*(24), 3889–3907.

Durband, A. C. (2008). Mandibular fossa morphology in the Ngandong and Sambungmacan fossil hominids. *The Anatomical Record: Advances in Integrative Anatomy and Evolutionary Biology: Advances in Integrative Anatomy and Evolutionary Biology, 291*(10), 1212–1220.

Ebben, W. P., Flanagan, E. P., & Jensen, R. L. (2008). Jaw clenching results in concurrent activation potentiation during the countermovement jump. *The Journal of Strength & Conditioning Research, 22*(6), 1850–1854.

Eckerdal, O. (1991). The petrotympanic fissure: A link connecting the tympanic cavity and the temporomandibular joint. *CRANIO®, 9*(1), 15–22.

Eriksson, P. O., Häggman-Henriksonn, B., Nordhh, E., & Zafarr, H. (2000, June). Co-ordinated mandibular and head-neck movements during rhythmic jaw activities in man. *Journal of Dental research, 79*(6), 1378–1384. https://doi.org/10.1177/00220345000790060501

Eriksson, P. O., Zafar, H., & Nordh, E. (1998). Concomitant mandibular and head-neck movements during jaw opening–closing in man. *Journal of Oral Rehabilitation.* https://doi.org/10.1046/j.1365-2842.1998.00333.X. Article first published online: 27 October 2009.

Fanucci, E., Spera, E., Ottria, L., Barlattani Jr, A., Fusco, N., Mylonakou, I., ... & Simonetti, G. (2008). Bennett movement of mandible: A comparison between traditional methods and a 64-slices CT scanner. *ORAL & Implantology, 1*(1), 15.

Federici, E., & Centofanti, S. (1977). Occlusion according to neuromuscular function. *Mondo Odontostomatologico, 19*(4), 8.

Foglio-Bonda, P. L., Migliaretti, G., Cavallo, F., Rocchetti, V., & Bodin, C. (2006). Incisor and mandibular movement during retrusion. *Archives of Oral Biology, 51*(7), 581–586.

Fricano, E. E. I. (2018). *The primate ectotympanic tube: Correlates of structure, function, and development*. Doctoral dissertation, Johns Hopkins University.

Galis, F., Van Dooren, T. J. M., Feuth, J. D., Metzt, J. A. J., Witkam, A., Ruinard, S., Steigenga, M. J., & Wijnaendts, L. C. D. (2006). Extreme selection in humans against homeotic transformations of cervical vertebrae. *Evolution, 60*, 2643–2654.

Galway-Witham, J., & Stringer, C. (2018). How did Homo sapiens evolve? *Science, 360*(6395), 1296–1298.

Gangloff, P., & Perrin, P. P. (2002). Unilateral trigeminal anaesthesia modifies postural control in human subjects. *Neuroscience Letters, 330*(2), 179–182.

García-Castellano, J. M., Díaz-Herrera, P., & Morcuende, J. A. (2000). Is bone a target-tissue for the nervous system?: New advances on the understanding of their interactions. *The Iowa Orthopaedic Journal, 20*, 49.

German, R. Z., & Palmer, J. B. (2006). Anatomy and development of oral cavity and pharynx. *GI Motility Online*. https://doi.org/10.1038/gimo5

Gerstner, G., Madhavan, S., & Crane, E. (2011). Mammalian oral rhythms and motor control. In V. Klika (Ed.), *Biomechanics in applications*. InTech.

Goel, A., Shah, A., Kothari, M., Gaikwad, S., & Dhande, P. L. (2011). Comparative quantitative analysis of osseous anatomy of the craniovertebral junction of tiger, horse, deer, and humans. *Journal of Craniovertebral Junction and Spine, 2*(1), 32.

Gould, S. J. (2002). *I have landed*. Harvard University Press.

Granat, J., Benoît, R., & Peyre, É. (2008). La canine. Première partie: Définition et évolution durant l'Ère secondaire. *Actualités Odonto-Stomatologiques, 244*, 309–316.

Grässel, S. (2014). The role of peripheral nerve fibers and their neurotransmitters in cartilage and bone physiology and pathophysiology. *Arthritis Research & Therapy, 16*(6), 1–13.

Gregersen, G. G., & Lucas, D. B. (1967). An in vivo study of the axial rotation of the human thoracolumbar spine. *JBJS, 49*(2), 247–262.

Grieznis, L., Apse, P., & Blumfelds, L. (2010). Passive tactile sensibility of teeth and osseointegrated dental implants in the maxilla. *Stomatologija, 12*(3), 80–86.
Grossnickle, D. M. (2017). The evolutionary origin of jaw yaw in mammals. *Scientific Reports, 7*, 45094.
Häggman-Henrikson, B., & Eriksson, P.-O. (2004). Head movements during chewing: Relation to size and texture of bolus. *Journal of Dental Research, 83*(11), 864–868.
Häggman-Henrikson, B., et al. (2006). Head immobilization can impair jaw function. *Journal of Dental Research, 85*(11), 1001–1005.
Haider, H. (2015). Management of tinnitus patients. *Journal of Hearing Science, 5*(4).
Hanau, R. L. (1926). Articulation defined, analyzed and formulated. *Journal of the American Dental Association, 13*(12), 1694–1709.
Hannam, A. G., et al. (2008). A dynamic model of jaw and hyoid biomechanics during chewing. *Journal of Biomechanics, 41*(5), 1069–1076.
Harcourt-Smith, W. H. E. (2010). The first hominins and the origins of bipedalism. *Evolution: Education and Outreach, 3*(3), 333.
Harvey, P. H., & Pagel, M. D. (1992). Scaling of brain mass on body mass. *Neuroscience Year-Supplement, 2*, 139–141.
Hasegawa, Y., Ono, T., Sakagami, J., Hori, K., Maeda, Y., Hamasaki, T., & Nokubi, T. (2011, February). Influence of voluntary control of masticatory side and rhythm on cerebral hemodynamics. *Clinical Oral Investigations, 15*(1), 113–118.
Hashimoto, T., Ojiri, H., & Kawai, Y. (2011). The foramen of Huschke: Age and gender specific features after childhood. *International Journal of Oral and Maxillofacial Surgery, 40*(7), 743–746. https://doi.org/10.1016/j.ijom.2011.03.017. Epub 2011 May 6. PMID: 21549561.
Heffner, M. A., Genetos, D. C., & Christiansen, B. A. (2017). Bone adaptation to mechanical loading in a mouse model of reduced peripheral sensory nerve function. *PLoS One, 12*(10), e0187354.
Hiiemae, K. (1984). Functional aspects of primate jaw morphology. In D. J. Chivers, B. A. Wood, & A. Bilsborough (Eds.), *Food acquisition and processing in primates* (pp. 257–281). Springer.
Hinton, R. J. (1981). Changes in articular eminence morphology with dental function. *American Journal of Physical Anthropology, 54*(4), 439–455.
Hogervorst, T., Bouma, H. W., & De Vos, J. (2009). Evolution of the hip and pelvis. *Acta orthopaedica, 80*(suppl 336), 1–39.
Hosoda, M., Masuda, T., Isozaki, K., Takayanagi, K., Sakata, K., Takakuda, K., ... & Morita, S. (2007). Effect of occlusion status on the time required for initiation of recovery in response to external disturbances in the standing position. *Clinical Biomechanics, 22*(3), 369–373. 2002, 179–182.
Humphrey, T. (1968, April). The development of mouth opening and related reflexes involving the oral area of human fetuses. *Alabama Journal of Medical Sciences, 5*(2), 126–157.
Hylander, W. L. (1979a). The functional significance of primate mandibular form. *Journal of Morphology, 160*(2), 223–239.
Hylander, W. L. (1979b). Mandibular function in *Galago crassicaudatus* and *Macaca fascicularis*: An in vivo approach to stress analysis of the mandible. *Journal of Morphology, 159*(2), 253–296.
Hylander, W. L. (1979c). An experimental analysis of temporomandibular joint reaction force in macaques. *American Journal of Physical Anthropology, 51*(3), 433–456.
Hylander, W. L. (1985). Mandibular function and biomechanical stress and scaling. *American Zoologist, 25*, 315–330.
Hylander, W. L. (2006). Functional anatomy and biomechanics of the masticatory apparatus. In *Temporomandibular disorders: An evidenced approach to diagnosis and treatment*. Quintessence.
Hylander, W. L. (2013). Functional links between canine height and jaw gape in catarrhines with special reference to early hominins. *American Journal of Physical Anthropology, 150*(2), 247–259.

Igarashi, N., Yamamura, K., Yamada, Y., & Kohno, S. (2000). Head movements and neck muscle activities associated with the jaw movement during mastication in the rabbit authors. *Brain Research, 871*(1), 151–155.

Isberg, A., et al. (1986). Hyperplastic soft-tissue formation in the temporomandibular joint associated with internal derangement: A radiographic and histologic study. *Oral Surgery, Oral Medicine, Oral Pathology, 61*(1), 32–38.

Isono, S., Tanaka, A., Tagaito, Y., Ishikawa, T., & Nishino, T. (2004). Influences of head positions and bite opening on collapsibility of the passive pharynx. *Journal of Applied Physiology, 97*(1), 339–346.

Ito, H., Su, M. J., Saitoh, H., Tamatsu, Y., Abe, T., Hori, H., Fujimura, A., & Nozaka, Y. (1995). Studies on the development of the articular part of the temporal bone with special reference to the postglenoid process. *The Bulletin of Tokyo Dental College, 36*(3), 145–158.

Izzo, R., Guarnieri, G., Guglielmi, G., & Muto, M. (2013). Biomechanics of the spine. Part I: Spinal stability. *European Journal of Radiology, 82*(1), 118–126.

Janssen, M. M., Vincken, K. L., Kemp, B., Obradov, M., De Kleuver, M., Viergever, M. A., ... & Bartels, L. W. (2010). Pre-existent vertebral rotation in the human spine is influenced by body position. *European Spine Journal, 19*(10), 1728–1734.

Jenkins, F. A. (1969). The evolution and development of the dens of the mammalian axis. *The Anatomical Record, 164*(2), 173–184.

Johanson, D., Edey, M., & Edey, M. A. (1990). *Lucy: The beginnings of humankind* (p. 363). Simon and Schuster.

Johnsen, S. E., & Trulsson, M. (2003). Receptive field properties of human periodontal afferents responding to loading of premolar and molar teeth. *Journal of Neurophysiology, 89*(3), 1478–1487.

Jugessur, A., Farlie, P. G., & Kilpatrick, N. (2009). The genetics of isolated orofacial clefts: From genotypes to subphenotypes. *Oral Diseases, 15*(7), 437–453.

Kardong, K. V. (1995). Vertebrates: Comparative anatomy, function, evolution. No. QL 805. K35 1995.

Katsavrias, E. G. (2002). Changes in articular eminence inclination during the craniofacial growth period. *The Angle Orthodontist, 72*(3), 258–264.

Kieser, J. A. (1997). Basicranial flexion, facial reduction and temporomandibular joint dysfunction. *Medical Hypotheses, 49*(5), 409–411.

Kimbel, W. H., White, T. D., Johanson, D. C., Yaroch, L. A., Skelton, R. R., & McHenry, H. M. (1986). On the phylogenetic analysis of early hominids. *Current Anthropology, 27*(4), 361–365.

Kinzey, W. G. (1971). Evolution of the human canine tooth 1. *American Anthropologist, 73*(3), 680–694.

Klineberg, I., & Murray, G. (1999). Osseoperception: Sensory function and proprioception. *Advances in Dental Research, 13*(1), 120–129.

Kohno, S., Kohno, T., & Medina, R. U. (2001, August). Rotational head motion concurrent to rhythmical mandibular opening movements. *Journal of Oral Rehabilitation, 28*(8), 740–747.

Kohno, S., Matsuyama, T., Medina, R. U., & Arai, Y. (2001). Functional-rhythmical coupling of head and mandibular movements. *Journal of Oral Rehabilitation, 28*(2), 161–167.

Koolstra, J. H., & Van Eijden, T. M. (1997). Dynamics of the human masticatory muscles during a jaw open-close movement. *Journal of Biomechanics, 30*(9), 883–889.

Kuratani, S. (2005). Craniofacial development and the evolution of the vertebrates: The old problems on a new background. *Zoological Science, 22*(1), 1–19.

Le Douarin, N. M. (2004). The avian embryo as a model to study the development of the neural crest: A long and still ongoing story. *Mechanisms of Development, 121*(9), 1089–1102.

Le Huec, J. C., et al. (2011). Equilibrium of the human body and the gravity line: The basics. *European Spine Journal, 20*(5), 558.

Lieberman, D. (2011). *The evolution of the human head.* Harvard University Press.

Lieberman, D. E., McBratney, B. M., & Krovitz, G. (2002). The evolution and development of cranial form in Homo sapiens. *Proceedings of the National Academy of Sciences, 99*(3), 1134–1139.
Lockwood, C. A., Lynch, J. M., & Kimbel, W. H. (2002). Quantifying temporal bone morphology of great apes and humans: An approach using geometric morphometrics. *Journal of Anatomy, 201*(6), 447–464.
Loughner, B. A., Larkin, L. H., & Mahan, P. E. (1989). Discomalleolar and anterior malleolar ligaments: Possible causes of middle ear damage during temporomandibular joint surgery. *Oral Surgery, Oral Medicine, Oral Pathology, 68*(1), 14–22.
Lovejoy, C. O. (2005). The natural history of human gait and posture: Part 1. Spine and pelvis. *Gait & Posture, 21*(1), 95–112.
Lumsden, R. M., II, & Morris, J. M. (1968). An in vivo study of axial rotation and immobilization at the lumbosacral joint. *JBJS, 50*(8), 1591–1602.
Mack, P. J. (1984). A functional explanation for the morphology of the temporomandibular joint of man. *Journal of Dentistry, 12*, 225–230.
Manzi, G. (2006). *Homo sapiens. Breve storia naturale della nostra specie*. Il Mulino.
Manzi, G. (2007). *L'evoluzione umana: ominidi e uomini prima di Homo sapiens*. Il mulino.
Manzi, G. (2013). *Il grande racconto dell'evoluzione umana*. Società Editrice il Mulino.
Manzi, G. (2017). *Ultime notizie sull'evoluzione umana*. Il mulino.
Manzi, G. (2018, Giugno). *Le scienze*.
Mao, J. J., & Nah, H.-D. (2004). Growth and development: Hereditary and mechanical modulations. *American Journal of Orthodontics and Dentofacial Orthopedics, 125*(6), 676–689.
Mapelli, A., Galante, D., Lovecchio, N., Sforza, C., & Ferrario, V. F. (2009). Translation and rotation movements of the mandible during mouth opening and closing. *Clinical Anatomy: The Official Journal of the American Association of Clinical Anatomists and the British Association of Clinical Anatomists, 22*(3), 311–318.
Marchi, D. (2016). *Il mistero di Homo naledi*. Ed Mondadori.
McCollum, M. A., Rosenman, B. A., Suwa, G., Meindl, R. S., & Lovejoy, C. O. (2010). The vertebral formula of the last common ancestor of African apes and humans. *Journal of Experimental Zoology Part B: Molecular and Developmental Evolution, 314*(2), 123–134.
McNamara, J. A., Jr., & Moyers, R. E. (1973). Electromyography of the oral phase of deglutition in the rhesus monkey (*Macaca mulatta*). *Archives of Oral Biology, 18*(8), 995–1002.
McNulty, K. P. (2010). Apes and tricksters: The evolution and diversification of humans' closest relatives. *Evolution: Education and Outreach, 3*(3), 322.
Mérida Velasco, J. R., Rodriguez Vazquez, J. F., & Jimenez Collado, J. (1993). The relationships between the TMJ disc and related masticatory muscles in humans. *Maxillofacial Surgery, 51*(4), 390–395. Discussion 395–6.
Mérida Velasco, J. R., Rodriguez Vàsquez, J. F., Merida Velasco, J. A., Sànchez-Montesinos, I., Espìn Ferrara, J., & Jimenez-Collado, J. M. (1999, May 1). Development of the human temporomandibular joint. *Anatomical Record, 255*(1), 20–33.
Mérida-Velasco, J. R., de la Cuadra-Blanco, C., Pozo Kreilinger, J. J., & Mérida-Velasco, J. A. (2012). Histological study of the extratympanic portion of the discomallear ligament in adult humans: A functional hypothesis. *Journal of Anatomy, 220*(1), 86–91.
Miyahara, T. (1991). Modulation of soleus H-reflex by teeth clenching. *Kokubyo Gakkai zasshi. The Journal of the Stomatological Society, Japan, 58*(4), 670–686.
Moggi-Cecchi, J. (2017). Human evolution: The hard evidence of an ever-changing bush. *PARADIGMI*. https://doi.org/10.3280/PARA2017-002007
Moggi-Cecchi, J., Tobias, P. V., & Beynon, A. D. (1998). The mixed dentition and associated skull fragments of a juvenile fossil hominid from Sterkfontein, South Africa. *American Journal of Physical Anthropology: The Official Publication of the American Association of Physical Anthropologists, 106*(4), 425–465.
Molina, M., Magnano, G., Pisa, W., & Valsesia, C. (1980). *Concetti fondamentali di gnatologia moderna*. Ilic'.

Mongini, F. (1972). Remodelling of the mandibular condyle in the adult and its relationship to the condition of the dental arches. *Cells Tissues Organs, 82*(3), 437–453.
Mongini, F. (1975). Dental abrasion as a factor in remodeling of the temporomandibular joint and occlusion. *Acta Anatomica, 92*(2), 292–300.
Mongini, F. (1977). Anatomic and clinical evaluation of the relationship between the temporomandibular joint and occlusion. *The Journal of Prosthetic Dentistry, 38*(5), 539–551.
Mongini, F. (1980). Rimodellamento condilare dopo terapia occlusale. *The Journal of Prosthetic Dentistry, 43*(5), 568–577.
Mongini, F. (1984). *L'Apparato stomatognatico: funzione disfunzione e riabilitazione*. Scienza e tecnica dentistica edizioni internazionali.
Mora, U. S. (2015). *The habitual chewing side syndrome* (Doctoral dissertation, Universidade de Santiago de Compostela).
Morquette, P., Lavoie, R., Fhima, M. D., Lamoureux, X., Verdier, D., & Kolta, A. (2012). Generation of the masticatory central pattern and its modulation by sensory feedback. *Progress in Neurobiology, 96*(3), 340–355.
Mossey, P. A., Little, J., Munger, R. G., Dixon, M. J., & Shaw, W. C. (2009). Cleft lip and palate. *The Lancet, 374*(9703), 1773–1785.
Motoyoshi, M., Shimazaki, T., Jun, N. A. K. A. J. I. M. A., & Akira, N. A. M. U. R. A. (2000). Biomechanical influences on the upper vertebrae during mastication, an examination using the finite element method. *Orthodontic Waves: Journal of the Japanese Orthodontic Society: 日本矯正歯科学会雑誌, 59*(3), 183–190.
Mudry, A. (2015). Glaser fissure, Huguier canal, and Civinini canal: A confused eponymical imbroglio. *Otology & Neurotology, 36*(6), 1115–1120.
Myers, L. J. (1988). Possible inflammatory pathways relating temporomandibular joint dysfunction to otic symptoms: An anatomical hypothesis. *CRANIO®, 6*(1), 64–70.
Nakamura, Y., & Katakura, N. (1995). Generation of masticatory rhythm in the brainstem. *Neuroscience Research, 23*(1), 1–19.
Osborn, J. W. (1989). The temporomandibular ligament and the articular eminence as constraints during jaw opening. *Journal of Oral Rehabilitation, 16*(4), 323–333.
Osborn, J. W. (1993a). A model to describe how ligaments may control symmetrical jaw opening movements in man. *Journal of Oral Rehabilitation, 20*(6), 585–604.
Osborn, J. W. (1993b). Orientation of the masseter muscle and the curve of Spee in relation to crushing forces on the molar teeth of primates. *American Journal of Physical Anthropology, 92*(1), 99–106.
Owen, C. P., Wilding, R. J. C., & Adams, L. P. (1992). Dimensions of the temporal glenoid fossa and tooth wear in prehistoric human skeletons. *Archives of Oral Biology, 37*(1), 63–67.
Padoa, E. (1991). *Manuale di anatomia comparata dei vertebrati* (Vol. 4). Feltrinelli Editore.
Panek, H., Matthews-Brzozowska, T., Nowakowska, D., Panek, B., Bielicki, G., Makacewicz, S., & Mankiewicz, M. (2008). Dynamic occlusions in natural permanent dentition. *Quintessence International, 39*(4).
Pang, D., & Li, V. (2004). Atlantoaxial rotatory fixation: Part 1—Biomechanics of normal rotation at the atlantoaxial joint in children. *Neurosurgery, 55*(3), 614–626.
Pękala, J. R., Pękala, P. A., Satapathy, B., Henry, B. M., Skinningsrud, B., Paziewski, M., ... & Tomaszewski, K. A. (2018). Incidence of foramen Tympanicum (of Huschke): Comparing cadaveric and radiologic studies. *Journal of Craniofacial Surgery, 29*(8), 2348–2352.
Pereira, F. J., Lundh, H., Eriksson, L., & Westesson, P. L. (1996). Microscopic changes in the retrodiscal tissues of painful temporomandibular joints. *Journal of Oral and Maxillofacial Surgery, 54*(4), 461–468.
Piancino, M. G., Isola, G., Cannavale, R., Cutroneo, G., Vermiglio, G., Bracco, P., & Anastasi, G. P. (2017). From periodontal mechanoreceptors to chewing motor control: A systematic review. *Archives of Oral Biology, 78*, 109–121.
Picq, P. (2010). The human canine-Its evolution and adaptive significance. *Journal of Dentofacial Anomalies and Orthodontics, 13*(1), 4–10.

Pievani, T. (2019). *Imperfezione: una storia naturale*. Raffaello Cortina Editore.
Planas, P. (1994). *Rehabilitación neuro-oclusal*. Editorial Salvat.
Plomp, K. A., Viðarsdóttir, U. S., Weston, D. A., Dobney, K., & Collard, M. (2015). The ancestral shape hypothesis: An evolutionary explanation for the occurrence of intervertebral disc herniation in humans. *BMC Evolutionary Biology, 15*, 68. Published 27 April 2015. https://doi.org/10.1186/s12862-015-0336-y
Pompa, G., Giovannetti, A., Gentile, T., & Di Carlo, S. (2010). Control factors in removable complete dentures: From the articulation quintet to kinetic contact. *Annali di stomatologia, 1*(2), 14.
Posselt, U. (1962). *Physiology of occlusion and rehabilitation*. F.A. Davis. 41.
Powell, V., Esteve-Altava, B., Molnar, J., Villmoare, B., Pettit, A., & Diogo, R. (2018). Primate modularity and evolution: First anatomical network analysis of primate head and neck musculoskeletal system. *Scientific Reports, 8*(1), 1–10.
Proops, D., Hawke, M., Berger, G., & MacKay, A. (1984). The anterior process of the malleus. *The Journal of Otolaryngology, 13*(1), 39–43.
Puech, P.-F., Cianfarani, F., & Ribot I Trafi, F. (1989). Maxillary canine microwear in *Dryopithecus* from Spain. *American Journal of Physical Anthropology, 80*(3), 305–312.
Quam, R. M., de Ruiter, D. J., Masali, M., Arsuaga, J. L., Martínez, I., & Moggi-Cecchi, J. (2013). Early hominin auditory ossicles from South Africa. *Proceedings of the National Academy of Sciences, 110*(22), 8847–8851.
Raia, P., Boggioni, M., Carotenuto, F., Castiglione, S., Di Febbraro, M., Di Vincenzo, F., Melchionna, M., Mondanaro, A., Papini, A., Profico, A., Serio, C., Veneziano, A., Vero, V. A., Rook, L., Meloro, C., & Manzi, G. (2018). Unexpectedly rapid evolution of mandibular shape in hominins. *Scientific Reports, 8*(1), 7340.
Rees, L. A. (1954). The structure and function of the mandibular joint. *British Dental Journal, 96*, 125–133.
Ren, Y.-F., & Isberg, A. (1995). Tinnitus in patients with temporomandibular joint internal derangement. *CRANIO®, 13*(2), 75–80.
Riga, M., et al. (2010). Aural symptoms in patients with temporomandibular joint disorders: Multiple frequency tympanometry provides objective evidence of changes in middle ear impedance. *Otology & Neurotology, 31*(9), 1359–1364.
Rodriguez Vazquez, J. F., Merida Velasco, J. R., & Jiménez Collado, J. (1993). Relationships between the temporomandibular joint and the middle ear in human fetuses. *Journal of Dental Research, 72*(1), 62–66.
Rodriguez Vazquez, J. F., Merida-Velasco, J. R., Merida-Velasco, J. A., & Collado, J. (1998). Anatomical consideration on the discomalleolar ligament. *Journal of Anatomy, 192*, 617–621.
Rowicki, T., & Zakrzewska, J. (2006). A study of the discomalleolar ligament in the adult human. *Folia Morphologica, 65*(2), 121–125.
Samman, T. (2006). Craniocervical functional morphology of several North American coelurosaurian dinosaurs.
San Juan, J., & Basura, G. J. (2019). Tinnitus management in lateral skull base lesions. *Journal of Neurological Surgery Part B: Skull Base, 80*(02), 125–131.
Scheidegger, R., Koletsi, D., & Eliades, T. (2018). The impact of dietary consistency on structural craniofacial components: Temporomandibular joint/condyle, condylar cartilage, alveolar bone and periodontal ligament. A systematic review and meta-analysis in experimental in vivo research. *Archives of Oral Biology, 94*, 33–47.
Schindler, H. J., Lenz, J., Türp, J. C., Schweizerhof, K., & Rues, S. (2010, May). Influence of neck rotation and neck lateroflexion on mandibular equilibrium. *Journal of Oral Rehabilitation, 37*(5), 329–335. https://doi.org/10.1111/j.1365-2842201002064.X
Schwenk, K. (Ed.). (2000). *Feeding: Form, function and evolution in tetrapod vertebrates*. Elsevier.
Scott, J. E. (2010). *Nonsocial influences on canine size in anthropoid primates*. Arizona State University.
Şencimen, M., Yalçin, B., Doğan, N., Varol, A. L. T. A. N., Okçu, K. M., Ozan, H. A. S. A. N., & Aydintuğ, Y. S. (2008). Anatomical and functional aspects of ligaments between the malleus and the temporomandibular joint. *International Journal of Oral and Maxillofacial Surgery, 37*(10), 943–947.

Sencimen, M., Varol, A., Baykal, B., Altug, H. A., Dogan, N., Sahin, S., ... & Yalcin, B. (2009). Histological characteristics of ligaments between middle ear and temporomandibular joint. *European Journal of Dentistry, 3*(4), 280–284.

Senter, P. (2010). Were Australopithecines ape human intermediates or just apes? A test of both hypotheses using the "Lucy" skeleton. *The American Biology Teacher, 72*(2), 70–76.

Sergi, S., & Ascenzi, A. (1974). *Il cranio neandertaliano del Monte Circeo*. Accademia Nazionale dei Lincei.

Sforza, C., Tartaglia, G. M., Solimene, U., Morgun, V., Kaspranskiy, R. R., & Ferrario, V. F. (2006). Occlusion, sternocleidomastoid muscle activity, and body sway: A pilot study in male astronauts. *CRANIO®, 24*(1), 43–49.

Shimazaki, T., Motoyoshi, M., Hosoi, K., & Namura, S. (2003). The effect of occlusal alteration and masticatory imbalance on the cervical spine. *The European Journal of Orthodontics, 25*(5), 457–463.

Sicher, H. (1937). Phylogenesis of human temporomandibular articulation (Abstr.). *Journal of Dental Research, 16*, 339–340.

Sicher, H. (1951). Functional anatomy of the temporomandibular articulation. *The Australian Journal of Dentistry, 55*(2), 73–85.

Sisask, G., Bjurholm, A., Ahmed, M., & Kreicbergs, A. (1995). Ontogeny of sensory nerves in the developing skeleton. *The Anatomical Record, 243*(2), 234–240.

Sisask, G., Silfverswärd, C. J., Bjurholm, A., & Nilsson, O. (2013). Ontogeny of sensory and autonomic nerves in the developing mouse skeleton. *Autonomic Neuroscience, 177*(2), 237–243.

Smith, B. H. (1986). Development and evolution of the helicoidal plane of dental occlusion. *American Journal of Physical Anthropology, 69*(1), 21–35.

Smith, R. J. (1985). Functions of condylar translation in human mandibular movement. *American Journal of Orthodontics, 88*(3), 191–202.

Sockol, M. D., Raichlen, D. A., & Pontzer, H. (2007). Chimpanzee locomotor energetics and the origin of human bipedalism. *Proceedings of the National Academy of Sciences, 104*(30), 12265–12269.

Sperber, G. H. (1975). The temporomandibular joint. In D. D. Ded (Ed.), *Craniofacial embryology* (2nd ed., pp. 121–149). Wright.

Spoor, F. (1997). Basicranial architecture and relative brain size of Sts 5. *South African journal of science, 93*(4), 182–186.

Stasolla, D., Mazza, A., Zaher, K., Maccioni, F., & Marini, M. (2004). Valutazione conRM delle alterazioni morfo-strutturali del tessuto retrodiscale nell'incoordinazione condilo-discale dell'ATM. *La Radiologia Medica, 107*, 261–268.

Stedman, H. H., Kozyak, B. W., Nelson, A., Thesier, D. M., Su, L. T., Low, D. W., ... & Mitchell, M. A. (2004). Myosin gene mutation correlates with anatomical changes in the human lineage. *Nature, 428*(6981), 415–418.

Stelzer, S., Gunz, P., Neubauer, S., & Spoor, F. (2017). Hominoid arcade shape: Pattern and magnitude of covariation. *Journal of Human Evolution, 107*, 71–85.

Stevens-Sparks, C., & Strain, G. M. (2014). The canine jaw-ear connection: The malleomandibular and tympanomandibular ligaments. *The Anatomical Record, 297*(5), 876–891.

Su, D. F. (2013). The earliest hominins: *Sahelanthropus*, *Orrorin*, and *Ardipithecus*. *Nature Education Knowledge, 4*(4), 11.

Takada, Y., Miyahara, T., Tanaka, T., Ohyama, T., & Nakamura, Y. (2000). Modulation of H reflex of pretibial muscles and reciprocal Ia inhibition of soleus muscle during voluntary teeth clenching in humans. *Journal of Neurophysiology, 83*(4), 2063–2070.

Tardieu, C., Hasegawa, K., & Haeusler, M. (2017). How did the pelvis and vertebral column become a functional unit during the transition from occasional to permanent bipedalism? *The Anatomical Record, 300*(5), 912–931.

Teaford, M. F., & Ungar, P. S. (2000). Diet and the evolution of the earliest human ancestors. *Proceedings of the National Academy of Sciences, 97*(25), 13506–13511.

Terhune, C. E. (2011). Modeling the biomechanics of articular eminence function in anthropoid primates. *Journal of Anatomy, 219*(5), 551–564.
Terhune, C. E., Cooke, S. B., & Otárola-Castillo, E. (2015). Form and function in the platyrrhine skull: A three-dimensional analysis of dental and TMJ morphology. *The Anatomical Record, 298*(1), 29–47.
Thielemann, K. (1958). Biomécanique de la parodontose. L'équilibre de l'articulé par le meulage sélectif. Traduction J autissier ed. J prelat.
Torisu, T., Yamabe, Y., Hashimoto, N., Yoshimatsu, T., & Fujii, H. (2001). Head movement properties during voluntary rapid jaw movement in humans. *Journal of Oral Rehabilitation, 28*(12), 1144–1152.
Trevizan, M., & Consolaro, A. (2017). Premaxilla: An independent bone that can base therapeutics for middle third growth! *Dental Press Journal of Orthodontics, 22*(2), 21–26.
Trevizan, M., Nelson Filho, P., Franzolin, S. D. O. B., & Consolaro, A. (2018). Premaxilla: Up to which age it remains separated from the maxilla by a suture, how often it occurs in children and adults, and possible clinical and therapeutic implications: Study of 1,138 human skulls. *Dental Press Journal of Orthodontics, 23*(6), 16–29.
Trulsson, M. (2006). Sensory-motor function of human periodontal mechanoreceptors. *Journal of Oral Rehabilitation, 33*(4), 262–273.
Tucker, A. S. (2017). Major evolutionary transitions and innovations: The tympanic middle ear. *Philosophical Transactions of the Royal Society B: Biological Sciences, 372*(1713), 20150483.
Valkenburgh, B. V., & Ruff, C. B. (1987). Canine tooth strength and killing behaviour in large carnivores. *Journal of Zoology, 212*(3), 379–397.
Vázquez, J. F. R., Merida Velasco, J. R., & Jimenez Collado, J. (1992). Development of the human sphenomandibular ligament. *The Anatomical Record, 233*(3), 453–460.
Veneziano, A., Meloro, C., Irish, J. D., Stringer, C., Profico, A., & De Groote, I. (2018). Neuromandibular integration in humans and chimpanzees: Implications for dental and mandibular reduction in Homo. *American Journal of Physical Anthropology, 167*(1), 84–96.
Walker, W. F., & Liem, K. F. (1994). *Functional anatomy of the vertebrates: an evolutionary perspective.*
Wall, C. E. (1997). The expanded mandibular condyle of the Megaladapidae. *American Journal of Physical Anthropology: The Official Publication of the American Association of Physical Anthropologists, 103*(2), 263–276.
Wall, C. E. (1999). A model of temporomandibular joint function in anthropoid primates based on condylar movements during mastication. *American Journal of Physical Anthropology, 109*(1), 67–88.
Walsh, J. H., Maddison, K. J., Platt, P. R., Hillman, D. R., & Eastwood, P. R. (2008). Influence of head extension, flexion, and rotation on collapsibility of the passive upper airway. *Sleep, 31*(10), 1440–1447.
Ward, C. V., Plavcan, J. M., & Manthi, F. K. (2010). Anterior dental evolution in the *Australopithecus anamensis–afarensis* lineage. *Philosophical Transactions of the Royal Society of London B: Biological Sciences, 365*(1556), 3333–3344.
Washburn, S. L., & Moore, R. E. (1973). *Ape into man: A study of human evolution.* Little, Brown.
Weidenreich, F. (1943). *The skull of Sinanthropus pekinensis: A comparative study on a primitive hominid skull.* Geological Survey of China.
Williams, S. A., Middleton, E. R., Villamil, C. I., & Shattuck, M. R. (2016). Vertebral numbers and human evolution. *American Journal of Physical Anthropology, 159*, 19–36.
Williams, S. A., & Russo, G. A. (2015). Evolution of the hominoid vertebral column: The long and the short of it. *Evolutionary Anthropology: Issues, News, and Reviews, 24*(1), 15–32.
Williamson, E. H. (1990). The interrelationship of internal derangements of the temporomandibular joint, headache, vertigo, and tinnitus: A survey of 25 patients. *CRANIO®, 8*(4), 301–306.
Wolpoff, M. H. (1971). Interstitial wear. *American Journal of Physical Anthropology, 34*(2), 205–227.

Yang, L., Wang, H., Wang, M., Ohta, Y., & Suwa, F. (1992). Development of collagen fibers and vasculature of the fetal TMJ. *Okajimas Folia Anatomica Japonica, 69*(4), 145–155.

Yoshizawa, H., Miyamoto, J. J., Hanakawa, T., Shitara, H., Honda, M., & Moriyama, K. (2019). Reciprocal cortical activation patterns during incisal and molar biting correlated with bite force levels: An fMRI study. *Scientific Reports, 9*(1), 1–10.

Zafar, H. (2000). Integrated jaw and neck function in man. Studies of mandibular and head-neck movements during jaw opening-closing tasks. *Swedish Dental Journal. Supplement, 143*, 1–41.

Zafar, H., Alghadir, A. H., & Iqbal, Z. A. (2019). Effect of jaw functional status on neck muscle endurance. *Archives of Oral Biology, 101*, 30–33.

Zeredo, J. L., Toda, K., & Soma, K. (2003). Nature of neck motor unit activities evoked by different trigeminal inputs in rats. *Journal of Dental Research, 82*(5), 402–405.

Chapter 5
Masticatory Function

Since you are what you eat, it well

— Selling England by the pound, Genesis. 1973

Abstract This chapter offers an in-depth exploration of masticatory function from evolutionary, biomechanical, and physiological perspectives. It begins by defining functional occlusion as the balance among teeth, temporomandibular joints (TMJs), masticatory muscles, and the nervous system. Occlusal stability, both static (maximum intercuspation) and dynamic (during chewing), is described as an expression of physiological adaptation.

The TMJ is presented as one of the most complex and sensitive joints in the body, with movements finely regulated by musculature and proprioceptive stimuli. Mastication, essential for survival, is also viewed as a plastic function, capable of adapting to various biomechanical conditions. Physiological tooth wear is interpreted as an integral part of occlusal maturation, shaped by diet and chewing habits, and capable of progressively modifying the form and relationships between occlusal surfaces.

The chapter discusses different masticatory guidance models (canine guidance, group function, and balancing contact), showing how they evolve over time with wear and contribute to efficient force dissipation. The importance of alternating unilateral chewing, functional symmetry, and the central role of the TMJ in cranio-cervical posture and balance are emphasized.

The text integrates concepts from functional anatomy, anthropology, neurology, and biomechanics to present a comprehensive view of occlusal dynamics, highlighting how proper masticatory function not only preserves oral health but also has significant cognitive and systemic effects.

A. Papini, *On the Origin of Human Masticatory Function*,
https://doi.org/10.1007/978-3-031-98664-2_5

In the previous chapters, we have discussed the evolutionary continuity of the anatomy of our masticatory apparatus. Now, we need to explore its correct functioning, both in statics (maximum intercuspation) and dynamics (during feeding), in an attempt to provide at least a partial answer to the implicit question in the phrase: "In any field, it is fundamental to understand the concept of physiological, as it is based on this concept that pathologies or abnormalities are diagnosed".

First and foremost, we need to define the concept of dental occlusion in maximum intercuspation, which coincides with the stability of the mandibulo-cranial relationship, meaning the reciprocal positioning that the skull and mandible (and cervical spine) assume in the three planes of space every time the dental arches come together.

Occlusion is the dynamic, morphological, and functional relationship that exists among all the components of the stomatognathic system. This includes not only the teeth but also the supporting tissues, the neuromuscular system, the temporomandibular joints (TMJ), and the craniofacial bones. Within the various tissues of the system, there exists a homeostatic functional balance in which biological processes and local environmental factors are in equilibrium. The stresses to which the teeth are subjected are typically dissipated without issues because the stress is counterbalanced by the adaptive capacity of the supporting tissues, the muscles of the masticatory apparatus, and the TMJ. A physiological occlusion can exhibit numerous morphological variations, but it is always characterized by acceptable functional relationships in which the tissues of the system have developed and maintain a structural and functional equilibrium. Variations that are free from tissue pathologies are, in practice, expressions of adaptation to a set of intrinsic and extrinsic factors. Physiological variations are thus quite common and represent the norm. Pre-established treatment schemes are based on ideal models related to the morphological structure of the system rather than its functionality and have little to do with considerations of health (McNeill, 1999).

The condition of homeostasis in all the tissues contributing to the function of the stomatognathic system becomes both the guiding principle and the goal of our dental interventions. So, what are the occlusal patterns that lead to the proper dissipation of the intense forces generated during the closure of the bite and during feeding? Although the definition of 'normal occlusion' was first proposed by Angle in 1899, to define the ideal to be achieved in the treatment of malocclusions, the characteristics that constitute an 'ideal' functional occlusion have not been definitively established, and deficiencies have been highlighted in this research area. There are no long-term studies to measure the impact of non-ideal occlusal patterns on dentition.

The anthropological perspective has had very little influence on various gnathological schools that approach occlusion from different perspectives and with equally different conclusions. However, it should be firmly stated that distinct viewpoints can lead to different narratives but need not necessarily be contradictory. While bearing in mind individual variations, the physiology of proper human masticatory function falls within a range common to the class of mammals, then to the order of primates, the family of hominids, and finally, the genus *Homo*.

The temporomandibular joint (TMJ) is perhaps the most complex joint in the human body and is certainly the most frequently used. Mandibular mobility is controlled by at least 18 muscle groups, which generate often asymmetrical but carefully managed movements, to be precise at sub-millimetric levels, especially in the vicinity of occlusal contact. The TMJ is part of a multitasking mechanical system that performs the functions of biting, chewing, swallowing, breathing, expressing emotions, and speaking (Scarr & Harrison, 2017).

Dentists, in every one of their interventions, often (and sometimes unknowingly) enter into the regulation of this highly sensitive system. The fortunate aspect for the dental profession is that the adaptability of this system is so vast that it almost always overcomes the proprioceptive disturbances resulting from incongruous interventions. The refined proprioception of the system activates for thicknesses of just a few microns and far exceeds the maximum precision that we dentists can achieve with our work. Dental adaptation is a behavioral response that aims to restore and maintain homeostatic conditions. The adaptability of the patient is crucial for the success of therapy, not only in prosthodontics (Anochin & Agafonow, 1961).

Feeding is an essential function for the life of an animal, a primary need because it provides assimilable nutrients necessary for energy, maintenance, growth, and reproduction (Reilly et al., 2001). Among all the functions of the stomatognathic system, feeding (which includes chewing and swallowing) is indispensable for life. In theory, we could breathe only through the nose, we could stop speaking and cease all facial expressions and manage to survive (miserably), but we cannot in any way stop chewing and swallowing what we have chewed. Teeth and oral structures also play a further biological role in many aspects of animal life, such as grooming, parental care, digging, defense, aggressive interactions, and material transport. Therefore, while the oral apparatus must primarily be explained in terms of food reduction, its significant multipurpose, selective potential for other functional influences should not be forgotten (Gans et al., 1978).

Among all the functions performed by the mouth, feeding (swallowing first and then chewing) is phylogenetically the most ancient, and other functions have been added later. Although the masticatory function plays the primary role in characterizing and specializing the oral apparatus of mammals, we must also consider the other biological roles that contribute to the development of the structures we observe in the mouth (Fig. 5.1).

The stomatognathic system must, therefore, be capable of adapting to chew and swallow in almost every situation, under the penalty of death by starvation for the animal. This is why redundant mandibular dynamics have developed, capable of adapting and fulfilling the nutritional needs even in non-ideal situations.

The immense adaptability of chewing means that its movements, while remaining within a species-specific and individual pattern, change from moment to moment, adapting to the biomechanical needs of each chewing stroke to achieve maximum triturating efficiency while avoiding harmful forces to the integrity of the teeth. Both totally repetitive patterns due to a lack of freedom of movement and overload of structures (unilateral chewing should be considered potentially

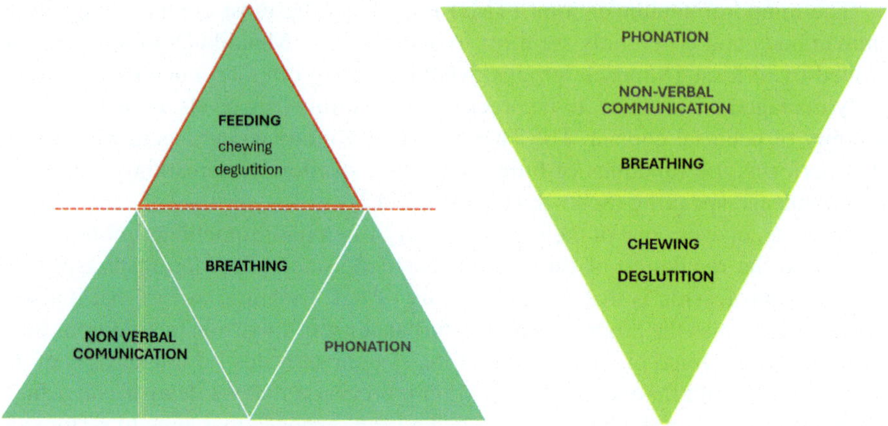

Fig. 5.1 Among the various functions that the mouth performs in humans, chewing and swallowing are essential for life; in the pyramid diagram on the left, they are highlighted with a red border and are located above a hypothetical survival line; on the right, in the inverted pyramid, the chronological succession in the development of the functions of the oral apparatus is represented from bottom to top

pathological precisely because it reduces the range of possible movements), and totally variable patterns due to a lack of neuromuscular memory should be considered pathological (Piancino & Kyrkanides, 2018).

The dynamic potentials of the individual patterns is generated by muscles and regulated by occlusal surfaces: an occlusion that is harmonized with the muscular and articular components will allow the myriad variations in movement necessary to effectively cope with the complexity of an omnivorous diet.

Most mammals have a kinematically redundant masticatory system: to generate a specific jaw movement, theoretically, multiple motor patterns can be generated (Langenbach & van Eijden, 2001). The human temporomandibular joint (TMJ) is a typical example of a kinematically and mechanically indeterminate joint system, where mandibular fossae and condyles articulate incongruously (Koolstra & van Eijden, 1999). This lack of morphological correspondence between the articular surfaces of the TMJ is the foundation of its freedom of movement. This freedom allows for the development of countless dynamic microvariations adapted by our highly diversified omnivorous diet, which varies in the types and consistencies of foods. This enormous potential given by the combination of muscular components can produce resultant forces that allow both TMJs movements in six degrees of freedom (Koolstra & van Eijden, 1996), and each muscle is capable of influencing more than one (Pileicikiene & Surna, 2004).

Therefore, in the functional triad TMJ/masticatory muscles/teeth, the latter represent the interface through which dentists daily interact (consciously or unconsciously) with the system, either restoring, maintaining, or disrupting the physiological homeostatic harmony.

After these general considerations, it remains to examine the basic principles on which to establish a homeostatic occlusion.

5.1 General Criteria for a Homeostatic Occlusion

As we have seen, the principles that characterize the physiology of our way of chewing are deeply rooted in our evolutionary history and are common first to the class of mammals, then to the order of primates, passing through all evolutionary branches that have led from the last common ancestor through the genus *Homo* to us. The extreme adaptability of the masticatory apparatus also allows for our pronounced individual phenotypic diversity. This ability to function, regardless of adhering to a supposed ideal, makes the criteria defining a state of occlusal health less obvious. The general principles of a physiologically homeostatic occlusion must first be sought in two functionally distinct moments: the first in a 'static' situation with occlusal contacts occurring in maximum intercuspation when the arches come into the position of minimal verticality, and the other in a 'dynamic' context with contacts occurring during chewing movements.

5.1.1 Static Occlusion

The position of maximum intercuspation is reached with the full engagement of the dental arches, which occurs both during swallowing—approximately 2000 times in 24 h, with an average daily force of over 2000 kg—(Molina et al., 1980), and at the end of the chewing cycle when the bolus has been finely shredded.

During maximum intercuspation, occlusal support should mainly rely on premolar and molar teeth. These teeth, with their crown-root anatomy, are better suited to withstand the intense forces generated by the closing muscles of the mouth. The incisor group should ideally only be involved in light contacts, as these dental elements are intended to perform functions such as cutting food or proprioceptive sensitivity rather than dissipating the forces of the adductor muscles.

5.1.1.1 Uniform Distribution of Contacts on the Supporting Cusps

Each individual tooth plays a significant role in occlusal control because, through its contacts with opposing structures, it becomes a fine regulator of the mandibulo-cranio-cervical relationship.

Maximum intercuspation is not only a static point of arrest and stabilization of the mandibulo-cranio-cervical relationship but also influences and diversifies the pattern of dynamic occlusion (Al-Nimri et al., 2010). The occlusal surfaces suitable for bearing the load of maximum intercuspation are primarily the supporting cusps (an ideal evolutionary continuation of the archaic protocone). Occlusal forces should be distributed evenly across all cuspidated elements to achieve uniform distribution without overloading individual teeth and their supporting structures. Static relationships between dental arches should be distributed with simultaneous contacts of all posterior teeth on both sides of the dental arch (Fig. 5.2).

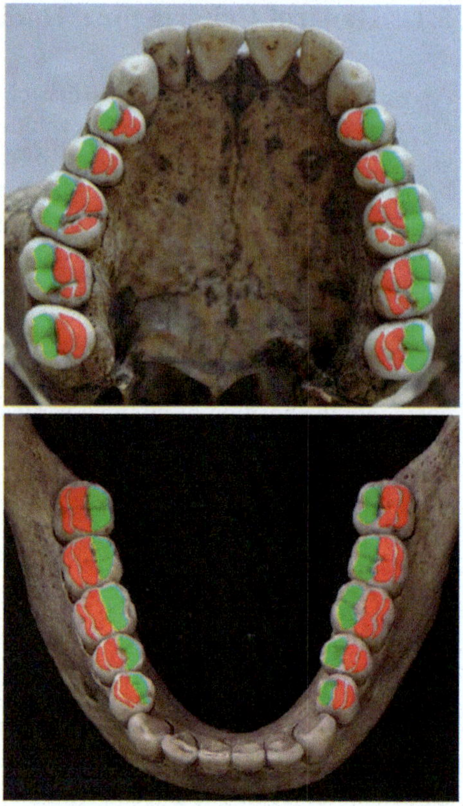

Fig. 5.2 In red: functional surfaces of the supporting cusps; in green: functional surfaces of the cutting cusps. Specimens from the Museum of Anthropology of the University of Florence

The theory of point contacts on cusps loses validity when we consider that physiological wear is an inevitable phenomenon that affects every contact area of enamel surfaces. As discussed in the chapter on wear, for the health of the stomatognathic system, it is necessary for each contact "point" to become over time a progressively larger area.

Furthermore, the theory of tripod contacts implies the three-dimensional immutability of each tooth, which is diametrically opposed to the wonderful capacity of enamel to wear and the periodontium to continuously adapt the position of the tooth to changing biomechanical environments.

5.1.1.2 Incisal Contact

The relationship of static and dynamic incisal contacts between maxillary and mandibular incisors plays an essential role because it guides mandibular dynamics by providing functional guidance. The overbite or incisal overlap, functionally defined as anterior guidance, interacts with the posterior guidance provided by the inclination of the roof of the mandibular fossa and is the primary factor in regulating the mandible in the horizontal plane. Anomalies in anterior guidance, such as an open bite or a deep bite, represent dysfunctional factors because they either force other teeth to compensate for the guiding function of the incisors (open bite) or create occlusal interferences and subsequent adaptive mandibular movements that stress muscles and ligaments.

In prehistoric populations, although wear could vary depending on the observed population and the age of the individual, overbite tended to decrease and disappear as individuals aged. This type of occlusal maturation is common not only to all archaic human species but also to all populations of *Homo sapiens* who, due to isolation, have maintained their diet unchanged. This demonstrates that the progressive loss of overbite should not be considered an exclusive characteristic of a primitive condition, but rather it characterizes and defines human occlusal maturation. This maturation has evolved over the last 2 million years and has been altered in the last century and a half due to the drastic change in diets. In past populations, intense wear of dental surfaces was almost the norm, leading to the gradual reduction of anterior guidance to a point where it reached a head-to-head relationship (Laplanche et al., 2010). This phenomenon was part of a maturation of the stomatognathic apparatus that involved a modification of all occlusal determinants:

- Decrease in overjet.
- Decrease in overbite.
- Decrease in cusp height.
- Flattening of compensatory curves.
- Decrease in the slope of the roof of the mandibular fossa.
- Transformation of interproximal dental contact points into contact areas.
- Subsequent mesialization of arches and incisal head-to-head relationship.

The latter phenomenon, called "mesial drift," refers to the mesial migration of individual teeth within the alveolar bone, not the movement of the entire dentition as part of skeletal growth processes. The result of this process is the absence of interdental spaces, even in individuals with advanced interproximal wear, and a reduction in the length of dental arches, characteristics present in the dentition of a wide variety of fossil and living hominids and pongids (Kaifu et al., 2003). Therefore, with the progressive physiological wear of occlusal surfaces, the sharp incisal edge also transforms into a flat surface (Figs. 5.3 and 5.4).

Fig. 5.3 Worn incisal edges. Skull 5049 from the Museum of Anthropology of the University of Florence

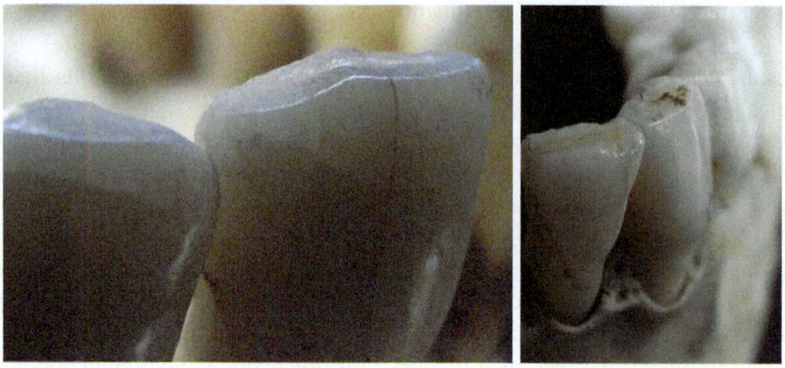

Fig. 5.4 Worn edges. Skull 5049 from the Museum of Anthropology of the University of Florence

The craniofacial arrangement, both phylogenetic and ontogenetic, influences occlusal patterns: the analysis of the incisal relationship in human skulls from different historical periods suggests that maintaining the anterior guidance into adulthood is a recent phenomenon, which rapidly appeared during the twentieth century. The lack of wear on the anterior guidance has occurred mainly in recent generations, in combination with an increase in occlusal alterations, dental crowding, and class II malocclusions. The rate of class II malocclusions has progressively increased and has become a common feature in modern "industrialized" populations (Guichard et al., 2001). Changes in diet toward softer and pre-packaged foods reduce the masticatory forces required to break down food and decrease abrasive wear on the teeth. The diet of a modern man living in an industrialized nation leads to teeth no longer being used/worn to mature an incisal head-to-head relationship. Dentitions of past populations underwent significant wear, constantly modifying and dynamically adapting teeth throughout life, including mesial migration as a compensatory mechanism. All contemporary humans inherit these compensatory mechanisms, but the recent reduction in wear intensity has led to an increase in various pathologies of the stomatognathic system in modern societies (Kaifu et al., 2003).

The definition of these evolutionary concepts is the first step towards an anthropological analysis of occlusion and the definition of what should be considered a "normal" occlusion (Laplanche et al., 2010).

The function performed by the incisor group that corresponds to the physiological overbite can be compared to that of the blades of scissors; it serves to cut and nibble food. In maximum intercuspation, the entire incisor group should have a contact of light touch, which is much lighter than that supported by the posterior teeth. Several factors make an incisor tooth potentially more delicate than a molar, giving it a structure that is generally unsuitable for bearing heavy loads: the tapered geometry of the dental crown, a root that is generally thinner compared to a molar, often non-coincident orientation between the dental axis and the vectors of forces that occur during movements, a frequently thin alveolar process, as well as a higher threshold for load sensitivity compared to premolars and molars.

In mammals, periodontal mechanoreceptors share a fundamental element: sensitivity to stretching, even though the microanatomical ultrastructure, neural connections, and functions can differ among different types of teeth and species. For example, in rabbits, contact on molars rapidly excites the masseter muscles, and the same happens for the canines of cats. In contrast, in humans, a touch on a central incisor elicits rapid inhibition of these muscles. Similarly to incisors, during lateral movements of the mandible in humans, contacts occurring on canines inhibit the activity of closing muscles (Manns et al., 1987). Mechanical stimulation of the first molar, however, immediately causes excitation (Brinkworth et al., 2004). In mammals, the inhibitory action of the three most anterior maxillary teeth may be considered an apomorphy, meaning a new character that has evolved to serve as a sensor for both chewing and swallowing. It inhibits the temporal and masseter muscles at the beginning of the tongue's swallowing action (Vaahtoniemi, 2020).

In humans, receptors in the molar region exhibit physiological characteristics distinct from those in the incisor region (Türker et al., 1997). Primary reflex responses have shown different behaviors between these two groups of teeth. This difference is likely due to more extensive and complex root structures, allowing for a more solid connection with the alveolar bone and, consequently, reduced root movement. Additionally, there is a reduced representation of periodontal mechanoreceptors in the posterior teeth (Brinkworth et al., 2004). The root surface of incisors is smaller compared to premolars and molars, and the periodontal structures are less robust. A precontact or premature contact in the anterior region triggers muscle contractions that reduce the closing force of the mandible. This serves to protect the more delicate incisor group from traumatic contacts (Biavati, 2019). Studies on periodontal mechanoreceptors in the incisor group have shown that the maximum forces during voluntary clenching drop abruptly when the anterior teeth come into contact (Gerstner et al., 2011). A rapidly increasing stimulus on the incisor group often induces the inhibition of the contraction of masticatory muscles. This reflex mechanism effectively protects the anterior teeth and their supporting tissues from harmful forces when the applied stimulus develops too quickly, such as the inadvertent crushing of a small hard object within soft food (Yang & Turker, 1999).

For these reasons, dentists must pay close attention to controlling and eliminating excessive contacts of the incisor group in maximum intercuspation. Such contacts can trigger reflexive and protective alterations in mandibular posture, often involving increased activity of the external pterygoid muscles, which attempt to retract the mandible to avoid anterior trauma. Both in maximum intercuspation and during protrusive movements, contacts occurring between the lower incisal edge and the upper palatal surface do not engage the entire incisal edge in the same manner. In particular, the outer edge of the lower incisal edge should be carefully evaluated for preservation, as it contributes to determining the vertical dimension of the bite when contacting the upper incisors (Fig. 5.5).

On the other hand, the lingual portion of the incisal edge is more frequently involved in dynamic interferences during excursive movements on the horizontal plane of the mandible. This is why there are greater possibilities for occlusal adjustments in this area (Fig. 5.6).

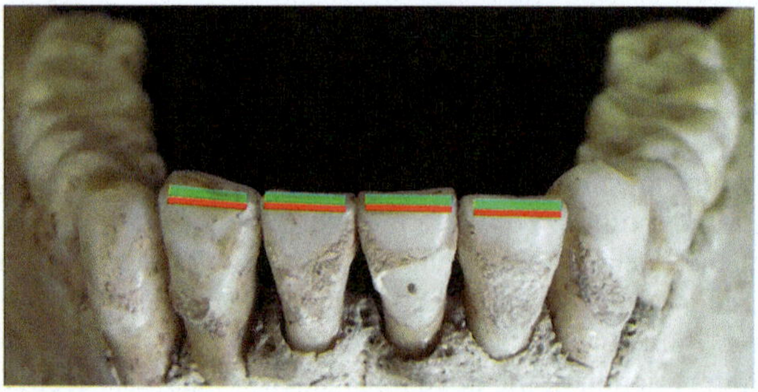

Fig. 5.5 The incisal edge presents functionally diversified zones: the vestibular margin of the incisal edge (red lines) in maximum intercuspation primarily contacts the palatal surface of the upper incisors, helping to maintain the vertical dimension; in green are the surfaces of the incisal area less involved in the contact of the anterior teeth during maximum intercuspation. Skull 3895 from the Museum of Anthropology of the University of Florence

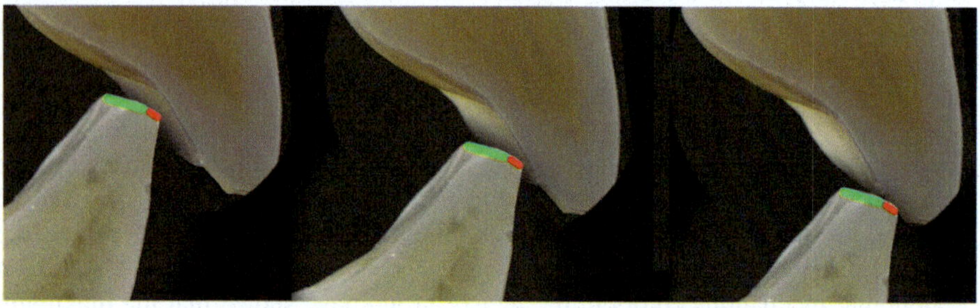

Fig. 5.6 Even during protrusive and lateral movements, the incisal edge has functionally different areas

Just as human hands can grasp an object with power grip (where all the fingers of a hand participate in gripping an object against the palm) or precision grip (where the applied force is lower and contact with the object is limited to the fingertips of the index and thumb), the two main types of teeth, molars and incisors, also differ in terms of function: molars play a role in crushing and grinding food through powerful chewing, while incisors play a key role in fine motor control. The sensitivity of incisors allows for the delicacy with which mammals can perform "precision grip" behaviors, both with their front limbs and their mouths. Mothers can gently grasp their offspring with their anterior teeth, and in primates, when an animal grasps an object either with both hands or with its mouth, the same portion of the premotor cortex is activated. Cortically, the sensitivity of the hands is comparable to that of the oral cavity, as demonstrated by Penfield's homunculus, in which the size of different parts of the body is related to the extent of dedicated motor brain tissue (Fig. 5.7).

During the developmental period, the overbite also serves to regulate the mandibulo-cranial relationship: in situations where the mandible tends to develop excessively forward, the overbite represents proprioceptive control that tends to

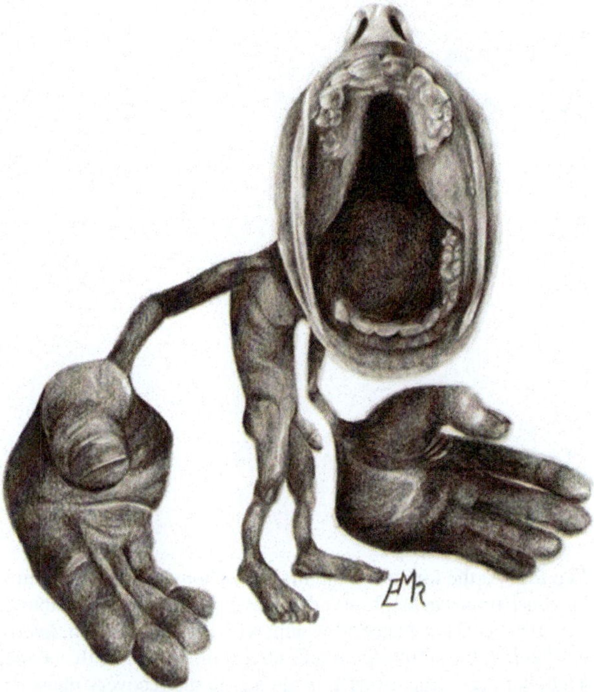

Fig. 5.7 Penfield's homunculus revisited by the author: in the drawing, an attempt has been made to highlight the stomatognathic system in relation to its extreme proprioceptive sensitivity

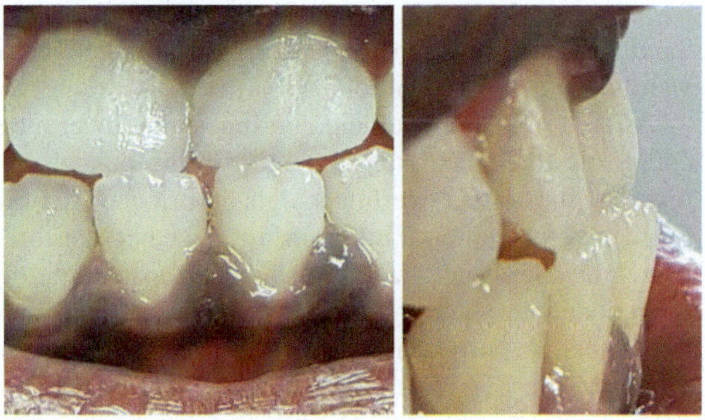

Fig. 5.8 Bite inversion: the patient presented to the clinic when it was already too late to attempt the recovery of a correct overbite with composite resin dental reconstruction ramps

limit the inversion of the incisal relationship. The loss of this anatomical and functional "brake" can contribute to excessive mandibular growth (Fig. 5.8).

When the edges of the permanent central incisors approach each other during eruption, if a situation of inversion of normal incisal relationships is anticipated, one can initially try to prevent the lower incisor edge from overlapping the upper one. In a single session and with some dental composite, the upper incisor edges can be extended vestibularly at an angle, thus creating, even in a rudimentary way, "tracks" that intercept and overtake the lower edge. With these temporary composite "slides," the lower incisor edges are intercepted during the closing movement as it approaches

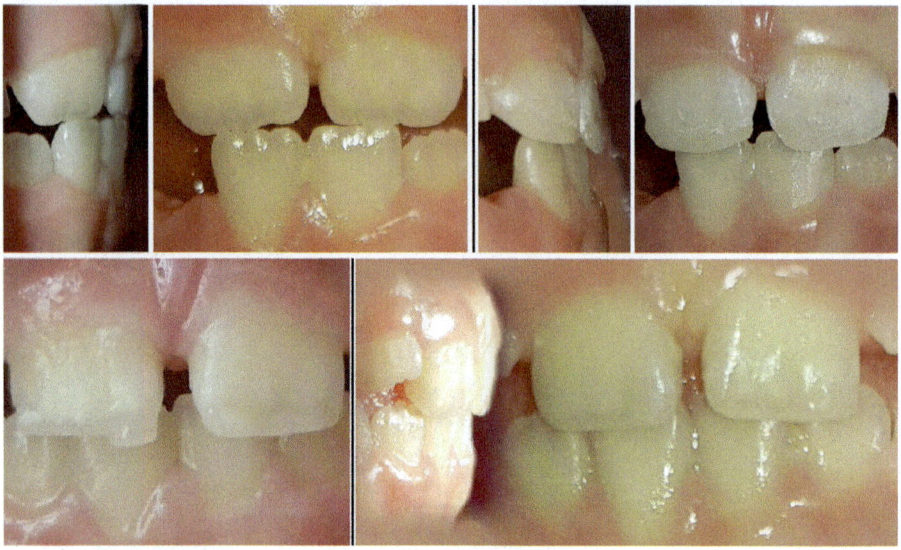

Fig. 5.9 At the top left in the first two photos, the relationships of the incisal edges in maximum intercuspation can be observed: an inversion of the bite is about to occur; at the top right, composite ramps have been applied which, with their inclination, intercept the upper incisal edges. At the bottom left, the composite tracks after 5 months (note that some parts of the ramp have been damaged, but the correct overlap is already in the recovery phase); at the bottom right, the completed recovery of the overlap with the removal of the composite

maximum intercuspation, recreating proprioceptive contact and a limit to the forward sliding of the mandible.

As the eruptive movement continues, the lower incisors will tend to slide on the composite tracks and subsequently on the palatal surfaces of the upper incisors, restoring the correct overbite and its proprioceptive brake on mandibular protrusion. Once the correct incisal relationships have been restored, the composite will be easily removable. It goes without saying that this type of solution makes sense as an immediate intervention aimed at simplifying the subsequent necessary orthodontic treatments. Its purpose is to prevent a reversal of the bite and uncontrolled mandibular protrusion (Fig. 5.9).

5.1.2 Dynamic Occlusion (Chewing)

Although there are exceptions, most mammals chew alternately on one side at a time (Popowics & Herring, 2006), and spontaneous changing of the side of bolus treatment is the first foundation of correct masticatory function. A fixed monolateral chewing pattern is a disruption of the alternation of the chewing cycle and can be considered a dysfunction in itself, even though, thanks to the extreme adaptability of the masticatory system, it may remain clinically silent for a long time.

Mammalian chewing arises from the interaction between an intrinsic rhythmic neural pattern and sensory feedback generated by the interaction of the system (muscles, bones, joints, teeth, soft tissues) with food. The mandible is subject to control mechanisms that act in real-time, determining variations in force and direction of movement that adapt to functional needs at each moment. The output of the central pattern generator (CPG) is modified by inputs from higher brain centers and feedback from sensory receptors: intraoral tactile receptors, muscle spindles, and mechanoreceptors specialized in the periodontal ligament have particularly powerful effects on movement parameters (Lund & Kolta, 2006). Proprioceptive feedback signals play a critical role in shaping basic locomotor patterns, modulating the extent and timing of muscle activity guided by the CPG to ensure correct movement execution. For example, when an unexpected disruption of the chewing cycle occurs during chewing, proprioception modifies muscle activation by correcting muscle dynamics (Tuthill & Azim, 2018); increasing evidence suggests that the dorsal part of the main sensory nucleus of the trigeminal nerve contains a significant portion of the chewing CPG (Kolta et al., 2000; Morquette et al., 2012; Tsuboi et al., 2003).

Exercise promotes axonal expansion and synaptogenesis (Chen, 1998), and in humans, neural regulation of masticatory activity is closely connected to multiple central nervous system (CNS) activities. Proper chewing activates the contralateral medial temporal lobe (important for declarative memory), the hippocampus, the entorhinal cortex, and the para-hippocampal cortex. Activation of the medial temporal lobe is believed to protect against age-related memory loss or mitigate memory impairment in patients with brain injuries (Choi et al., 2017). Reduced chewing diminishes the flow of information from the oral area to the CNS (Deitch & Rubel, 1989) and reduces the activation of all related brain areas. Research has shown a correlation between the absence of molars and a decline in learning ability, which significantly improved when missing molars were restored with artificial crowns (Watanabe et al., 2002). Impaired masticatory stimulation caused by a soft food diet leads to learning deficits (Yamamoto & Hirayama, 2001). Conversely, proper chewing increases blood flow and activates the somatosensory cortex, the striatum, the thalamus, and the cerebellum. Chewing immediately before a cognitive task increases blood oxygen levels, thereby improving the performance of the prefrontal cortex and the hippocampus, structures involved in learning and memory (Onozuka et al., 2008). Dysfunctional chewing affects cognitive function, and in the elderly, masticatory deficits contribute to a decrease in quality of life and senile dementia. The systemic effects of tooth loss are a risk factor for Alzheimer's disease (Kondo et al., 1995). Masticatory dysfunction resulting from tooth loss, a soft diet, or occlusal disharmony induces pathological changes in the hippocampus and memory and learning deficits (Onozuka et al., 2008). Therefore, we can say that proper masticatory function is important for maintaining cognitive function.

Teeth, with their occlusal surfaces, guide the masticatory function, and if the occlusion does not allow proper functioning, the system must still chew (and swallow) by adopting adaptive, asymmetric dynamics to avoid occlusal obstacles. To perform alternate unilateral chewing, the mandible must be able to execute centripetal lateral movements (entry into the masticatory cycle) with the same ease and smoothness to both the right and left sides.

The relationship between occlusal surfaces "trains" an individual to learn more efficient chewing movements (Osborn, 1982), and in almost all mammals, the occlusal surfaces of newly erupted premolars and molars are not as complementary as the micrometric precision required for efficient chewing. Since the perfect reciprocity of occlusal surfaces, which is so important, is only achieved through physiological wear, some mammals grind their teeth even in utero so that they are functionally adapted and ready for use from birth (Teaford & Walker, 1983).

Most of the morphological features of the occlusal surface are adaptive evolutions designed to process food introduced by diet. During evolution, the masticatory system has developed a number of structural elements designed not only to break down food but also to stabilize dental arches and resist powerful chewing forces (Pileicikiene & Surna, 2004). Teeth are complex tools that have evolved to create fractures (Lucas, 2004), and their architecture reflects the adaptive responses to withstand the strong stresses generated during chewing (Benazzi, Kullmer, et al., 2013; Benazzi, Nguyen, Kullmer, et al., 2013; Benazzi, Nguyen, Schulz, et al., 2013). The function of a tooth as a tool is to fracture food by penetrating it with what is referred to as "forced crack propagation." This propagation occurs through both pointed surfaces and sharp blades (Lucas, 1979). Dental morphology responds to the geometric and functional needs of chewing, and its resemblance to ideal functional forms highlights how teeth and their components have the best shape for their function (Evans & Sanson, 2003).

Comparing the limited available movement options (vertical and horizontal) with the limited available shape options (points and blades) and considering the variety of foods to which mammals have adapted over evolution, one can understand the enormous diversity in dental forms. Mammalian teeth are incredibly efficient tools, with cusps that pierce and act as pestles, crests that cut like blades, and pits that serve as mortars, allowing for the processing of the bolus immediately. One of the key characteristics that allows a tooth-tool to break down hard food is its ability to penetrate through the food itself, and the two main morphological attributes of a pointed tool that serve this purpose are the tip and the blade.

- **Tip**: The stress required to initiate a crack will vary depending on the contact area between the "tool" and the food. The smaller the radius of curvature of the tip, the smaller the contact area, and thus, the greater its penetration capacity and the stress in the food (Lucas, 1982). Once an initial crack is made in the food, the tip must continually push further inside to sustain crack propagation. Greater sharpness of the cusp reduces the force and energy required for a tooth to pass through hard food (Fig. 5.10).
- **Blade:** The cutting capacity is determined by both the sharpness of the edges (i.e., the curvature of the blade edge: the smaller the radius of curvature, the sharper the cutting edge; the smaller the contact area, the greater the stress in the food) and by the action defined by the upper and lower clearance angles and the rake angle with which dental blades (crests) act, reciprocally and on the food (Evans & Sanson, 2003) (Fig. 5.11).

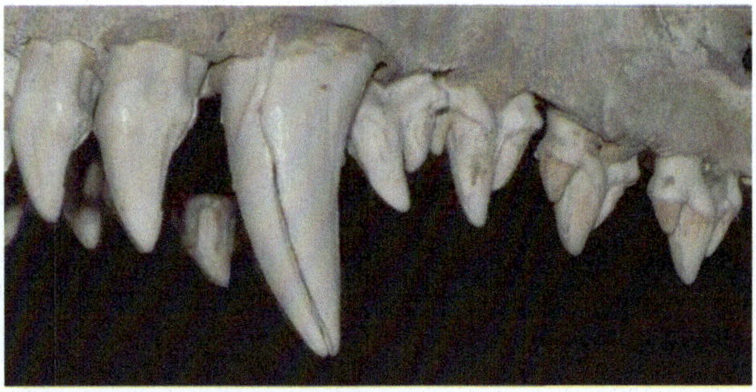

Fig. 5.10 Teeth are equipped with points and blades to create fractures in the substance of the food. Upper left dental arch of a leopard seal (*Hydrurga leptonyx*). Specimen from the Museum of Zoology of the University of Florence

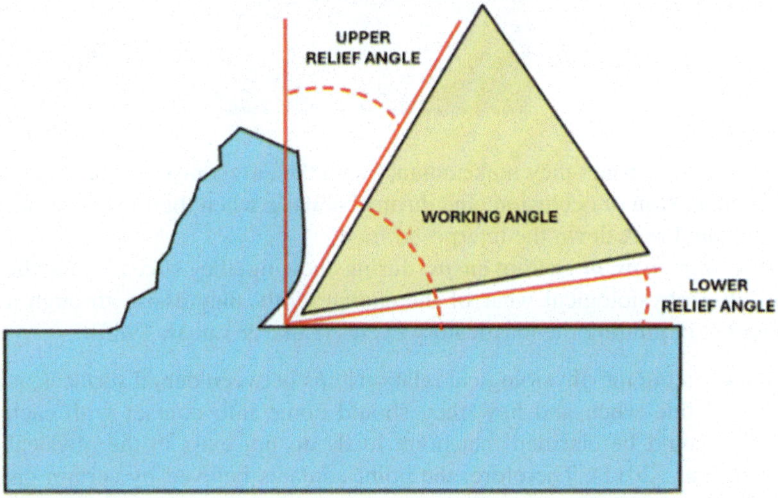

Fig. 5.11 Physical characteristics of a blade

Therefore, cusps should be considered adaptations for breaking down food and may exhibit different sharpness depending on the animal's diet. In omnivorous bunodont mammals, such as pigs and many primates, including humans, premolars and molars have low, rounded cusps that are not highly elevated on the coronal surface. On the surface of these cusp ridges, each slope plays a specific role at precise functional moments. Cusps are divided into supporting cusps, also known as working or stamp cusps, and balancing or cutting cusps (Fig. 5.12).

- ***Supporting cusps, also known as working or stamp cusps:*** Classified in anthropological nomenclature as 'protocone' for the upper and 'protoconid' for the lower, these cusps can be considered the ideal evolutionary continuation in mammals of the archaic reptilian conical tooth. They are the most voluminous and robust cusps, adapted to bear the major load of chewing and supported by an important root structure. For example, in the upper molars, which are the dental elements that perform the greatest masticatory effort, the palatal root is the largest because it supports the efforts of the supporting/working cusp (Berkovitz & Shellis, 2018). Supporting cusps have a primary role both in maximum

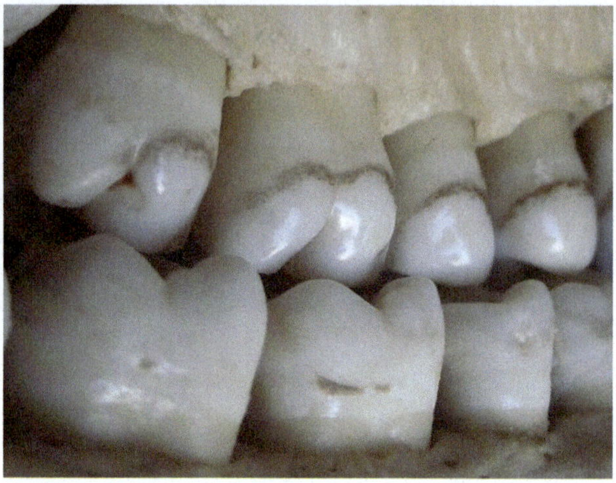

Fig. 5.12 Upper stamp cusps that fit into the corresponding antagonist fossae. Skull 318 from the Museum of Anthropology of the University of Florence

intercuspation when they make contact with the antagonist, establishing the vertical dimension of occlusion, and during chewing when their active surfaces cut, crush, and break down the interposed food.

- ***Balancing cusps, or cutting cusps***: during chewing, they slice through the bolus, and after physiological wear of the canine tooth, they assist, through a group function, in guiding the masticatory cycles (Luke & Lucas, 1983).

When delineating the physiological relationships between dental surfaces, it is necessary to define when and how they should come into contact with each other. Firstly, it should be clarified that point loads do not exist in the physical world (Benazzi et al., 2011). Therefore, the point contacts beloved by certain gnathologists are not possible in nature. Antagonistic teeth, once they have completed their eruption process, come into contact through very small areas, which we can only consider as point-like contacts in an idealized manner. These small areas bear all the force exerted by the masticatory musculature. As previously explained, the smaller the area of dental contact, the earlier its wear and the consequent formation of a functional facet (Fig. 5.13).

Tooth wear is a physiological and adaptive phenomenon (Benazzi, Kullmer, et al., 2013; Benazzi, Nguyen, Kullmer, et al., 2013; Benazzi, Nguyen, Schulz, et al., 2013) that does not affect the vertical dimension of occlusion because the dento-facial complex compensates for the potential loss of vertical dimension due to wear throughout life (Levartovsky et al., 2015) (Fig. 5.14).

5.1.2.1 Alternating Unilateral Chewing

Chewing in all mammals, including primates and humans, is characterized by being unilateral and alternating, which sets them apart from other vertebrates. This characteristic, along with the transverse movement associated with the power phase, represents two of the most important functional innovations (Wall & Smith, 2001).

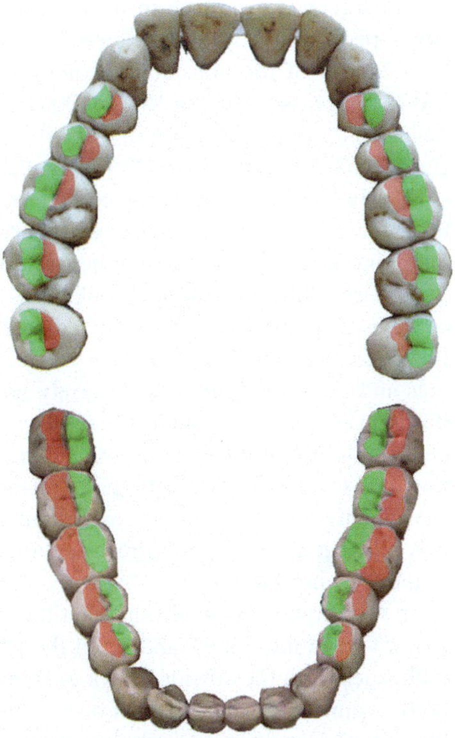

Fig. 5.13 Occlusal surfaces of the cusps: working (red), and balancing (green). Note the squared shape of the mandibular arch, a result of proper occlusal maturation (Planas, 1994)

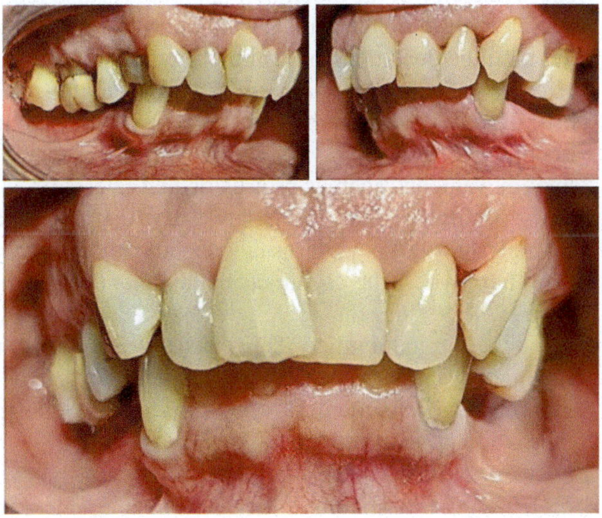

Fig. 5.14 Note how the entire upper posterior dento-periodontal-alveolar complex has migrated downward in a "desperate" search for stabilizing occlusal contact

Since their appearance in mammals, the alternating unilateral chewing pattern developed to increase the efficiency of bolus grinding through precise, differentiated control of the musculature on both sides (Gomes et al., 2010; Grossnickle, 2017; Langenbach & van Eijden, 2001). Alternating unilateral chewing allows for better control and precision of mandibular movements, increasing its effectiveness (Langenbach & van Eijden, 2001), thanks to the recruitment of musculature on both sides. Unilateral chewing promotes more precise movements that generate more effective forces and prevent excessive stress on dental, periodontal, bone, muscular, and joint structures, protecting them from overly strong stresses.

The incredible evolutionary radiation of mammals and their adaptation to diverse ecological environments has allowed them to exploit countless food sources. Within this class, individual species vary greatly in terms of dental and skeletal morphologies and chewing dynamics, a variety nearly as vast as the extraordinary range of different plumages among birds (Ungar, 2010). In mammals, alternating unilateral chewing is almost the rule, although there is a wide range of approaches to this function, and some rodents exhibit bilateral chewing. The alternating unilateral chewing movement, despite involving asymmetric mandibular dynamics (Hylander, 1979; Miyawaki et al., 2001), requires functional symmetry that allows for the alternation of the chewing side.

In vertebrates, external functionalities show a high degree of bilateral symmetry (Allard & Tabin, 2009), and often their bodies appear essentially symmetrical on both sides from the outside (Raya & Belmonte, 2006). Bilateral symmetry refers to body symmetry, evaluated with respect to a cranio-caudal axis that is found in over 99% of modern animals, including humans (Oxilia et al., 2021). This anatomofunctional symmetry of the locomotor apparatus is also realized in the masticatory apparatus through the alternation of the chewing side. The alternation of the chewing side, with the development of the characteristic functional cycle, uniquely characterizes the class of mammals, with some exceptions such as odontocetes (Hiiemae, 1978). Before the advent of mammals, in vertebrates, mandibular movements were essentially vertical, involving opening and closing. This leads us to consider horizontal movements, such as protrusion and laterality, as more recent from an evolutionary perspective. From this new muscle dynamic, already present in primitive mammals, new motor patterns developed, either reducing (in carnivores) or increasing (in herbivores) the asymmetric movement of the mandible (Langenbach & van Eijden, 2001). Chewing cycles, in their development, rarely extend significantly beyond the sagittal midline. This masticatory pattern, subject to slight variations, is fundamental for all mammals (Fish & Mendel, 1982).

In chewing, individual movement patterns are determined moment by moment by the summation of the individual vectorial contributions of the muscles involved, which establish both the direction of movement and the chewing force (Weijs, 1994). The harmonization of the occlusal component with the muscular and articular components allows for the myriad variations in movement necessary to deal with countless biomechanical situations. In a correct dynamic relationship, the inclinations of the cuspal slopes allow the teeth to slide reciprocally from eccentric positions to maximum intercuspation without interference and without introducing traumatic forces to the dento-parodontal structures (Rangarajan et al., 2015). In most primitive mammals, even the cusps of antagonistic teeth were already perfectly reciprocal, optimizing food comminution and preventing any misalignment from disturbing the entire occlusion (Crompton, 1971).

In humans, the remarkable adaptability of masticatory dynamics ensures that movements, while remaining within a species-specific and individual pattern, change from moment to moment, adjusting to the biomechanical needs of each chewing cycle, achieving maximum crushing efficiency, and avoiding uncontrolled forces that could damage the teeth. The anatomical and functional heterogeneity of individual dental groups (incisors, canines, and posterior teeth) determines their different roles in the chewing cycles. We have seen how mammalian chewing is based on the acquisition of horizontal dynamics by the mandible, and on this level, during chewing, incisors regulate movements in the antero-posterior direction, while during lateral movements, the function of guidance is assumed by the teeth that are morphologically and neurologically better suited, the canines.

5.1.2.2 Functional Chewing Guides: Pure Canine, Group, Balancing Grazing

In the dental community, there is no common agreement on the basic principles according to which dental arches should relate to each other during chewing.

Different gnathological schools have proposed various concepts regarding how dental arches should physiologically relate to each other during chewing: pure canine guidance, in which the canine excludes the entire arch without any other contacts, either on the same side or on the opposite side; group guidance, in which premolars and homolateral molars participate but not the contralateral ones; group guidance with balancing grazing, in which progressive contacts are also made on the contralateral molars.

As for canine guidance, we have seen how the canine represents a unique anatomical and functional element, differing from other dental types from an evolutionary, embryological, anatomical, and neurological perspective. When the canine erupts in the arch with its particular morphology and position on the occlusal plane, it becomes the principal element that intercepts lateral movements, temporarily becoming the exclusive guide. The increase in verticality imposed by its surfaces during this particular period of organism development creates a growth stimulus that, through the Moss matrix, contributes to completing the correct development of the bony bases. However, the mechanical stress that pure canine guides concentrate on their surfaces inevitably results in physiological wear, with a progressive decrease in the vertical dimension of the masticatory cycle entrance and the lateral pathways on the frontal plane (identified by AFMP: Planas' functional masticatory angles). The decrease in verticality due to the wear of the canine guidance involves, through the gradual recruitment of premolars and molars, the palatal surfaces of the cutting cusps of the posterior teeth, forming a group guidance on the working side (Vaahtoniemi, 2020) (Figs. 5.15, 5.16, and 5.17). Group guidance is consistently

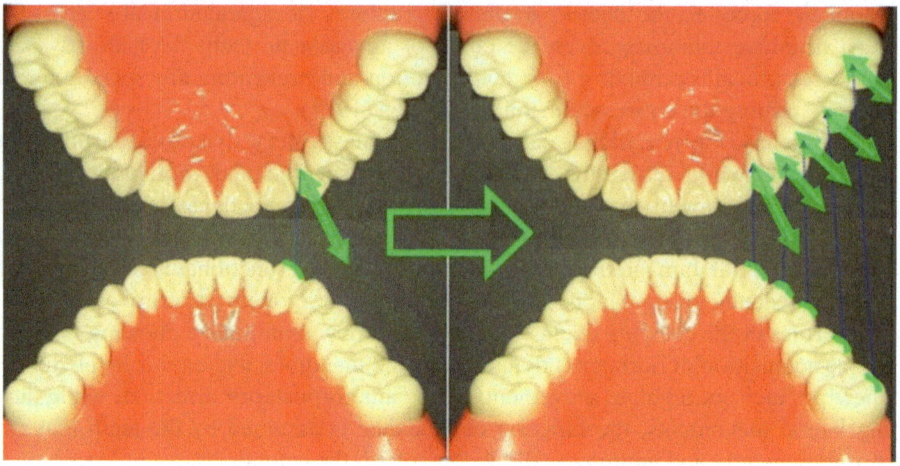

Fig. 5.15 On the left, canine guidance where, during the entry phase of the masticatory cycle, the lower dental arch initially contacts the upper one only with the canines; on the right, group guidance where the lower dental arch interfaces with the upper one through simultaneous contacts of all cusped elements

Fig. 5.16 Wear of the posterior dental surface of the upper premolars and molars involved in the chewing cycle. Note that the wear has not yet involved the third molar. Exhibit from the Museum of Anthropology at the University of Florence

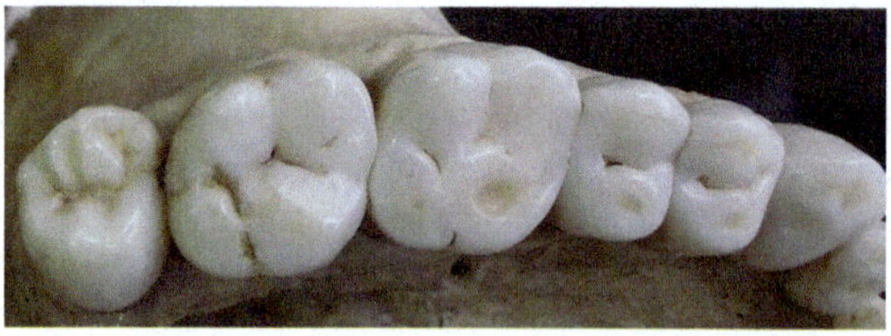

Fig. 5.17 Progressive wear on the upper posterior dental surfaces: even the palatal surfaces of the cutting cusps of premolars and molars, involved in the entry phase of the masticatory cycle, show wear; note that the wear has not yet involved the third molar. Specimen from the Museum of Anthropology of the University of Florence

found in the dentition of past populations but also in contemporary populations, albeit very rare ones, that, due to their isolation, have not adopted an 'industrialized' diet.

The same wear, associated with the maturation of occlusal curves over time, eventually leads to balancing grazing on the contralateral molars: it's the unavoidable law of wear that, together with the intimate structure of enamel, masticatory function, and the diet with which our species has evolved, matures a masticatory function guided by the canines along with the posterior group on the working side, stabilized on the opposite side by a slight balancing grazing on the lingual aspects of the vestibular cusps of the contralateral molars. Masticatory guidance on the working side, associated with contralateral grazing, not only increases grinding capacity but also allows for better dissipation of muscular forces, protecting teeth, periodontium, muscles, and the temporomandibular joint (TMJ). This balancing grazing will start on the first molar because it is the first tooth to erupt in the arch and, therefore, is the one subjected to masticatory friction and subsequent wear for the longest period (Fig. 5.18).

As the dentition is shaped and matured through physiological functional wear, balancing grazes will progressively affect all teeth posterior to the canine (Fig. 5.19).

This occlusal maturation, which also includes the progressive involvement of posterior elements in both working and balancing guidance, has been the functional pattern of occlusal relationships since the beginning of human history (Figs. 5.20 and 5.21).

Only around the end of the 19th century did occlusion maturation begin to decline due to a drastic reduction in physiological wear associated with diet; this inevitably occurred in all human groups that had adopted a different dietary regime linked to the consumption of processed foods produced by the food industry (Price, 1945).

The involvement of the musculature on the balancing side is a characteristic of mammalian chewing. In humans, the reaction forces on the balancing side are as

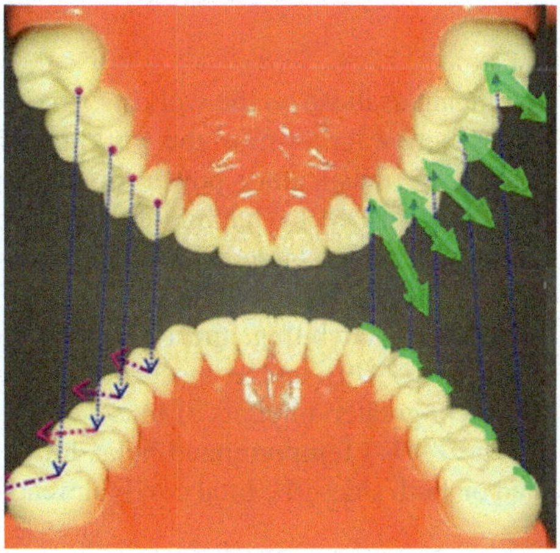

Fig. 5.18 An occlusion matured by physiological wear will present, during the entry phase of the masticatory cycle, a group guidance (green arrows), to which balancing contacts (dashed purple arrows) will be associated, stabilizing the mandibular musculo-articular system

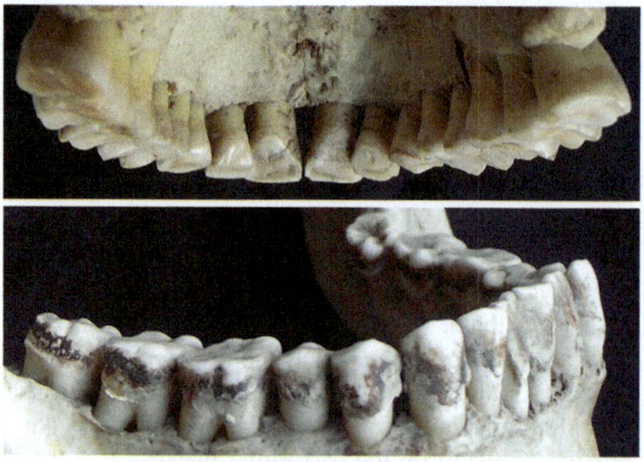

Fig. 5.19 Functionally molded occlusal surfaces: both working and balancing slopes are sculpted to allow maximum grinding efficiency as well as better force dissipation and stability. Skull 3895 from the Museum of Anthropology of the University of Florence

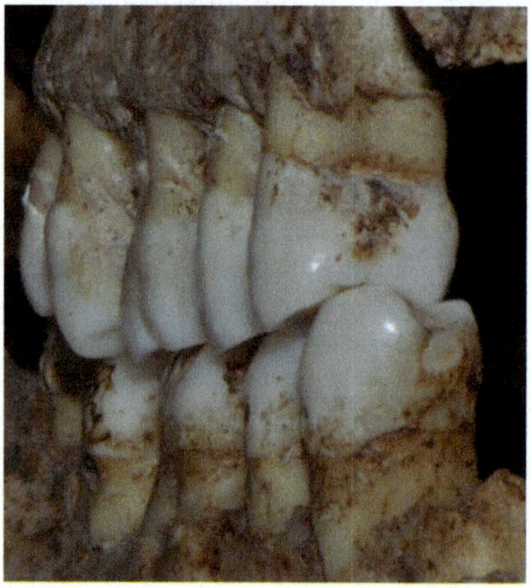

Fig. 5.20 Wear on the cuspal slopes supporting the group guidance of the grinding teeth. Skull 5049 from the Museum of Anthropology of the University of Florence

significant as those on the working side, with the rigid symphyseal area transferring the adduction forces from the working side to the opposite side (Thomason, 1997). During the masticatory power stroke, when food interposed between the dental arches is crushed by maximum muscle effort, the activity of the deep balancing masseter serves to maintain both the transverse mobility of the mandible and occlusal force (Ravosa et al., 2000). According to Hylander, during chewing and during isometric molar biting, the forces are greater on the TMJ on the balancing side (Hylander, 1979), and as masticatory force increases, most individuals tend to use relatively greater amounts of muscle force on this side (Hylander, 2006).

The TMJs (Temporomandibular Joints) also experience stresses that are distributed on both sides: the total condylar reaction force on both TMJs can be 75% of the

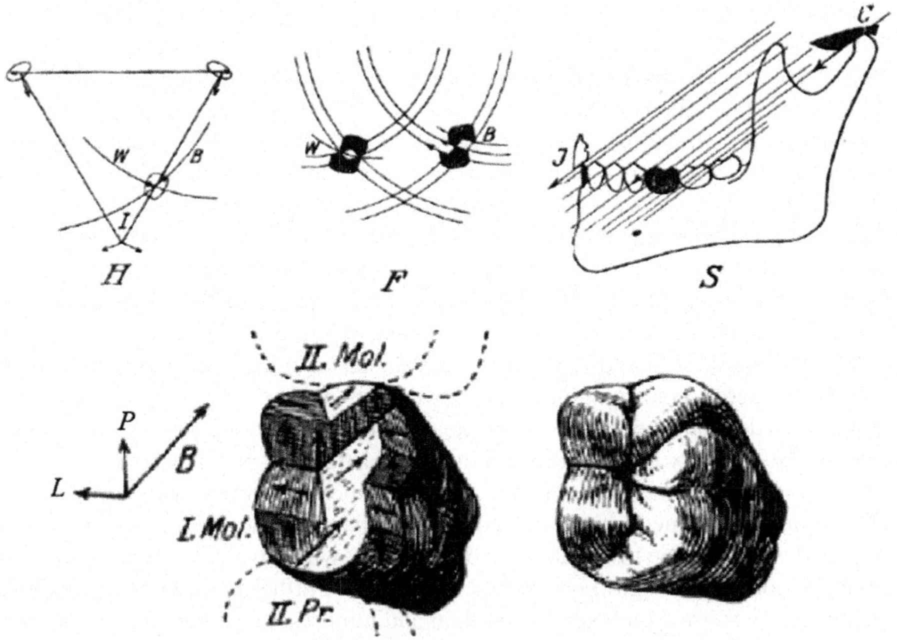

Fig. 5.21 Illustration by A. Gysi from the book "Zahnärztliche Prothetik", 1930, by Bruhn, Christian, et al., in the collection of the Fondazione Castagnola. Note how already in the early twentieth century Alfred Gysi highlighted the shaping action of mandibular movements in the three planes of space: mandibular dynamics sculpt occlusal surfaces, determining the functional slopes selectively activated in different phases of the masticatory cycle. The reader should not be misled by the fact that these anatomo-functional relationships were described over a hundred years ago: the masticatory physiology of our species and the principles guiding it have remained unchanged since its origin, approximately 200,000 years ago, until the advent of the 'industrialized' diet. The correctness of a statement, description, or discovery lies in its quality, not in the time elapsed since its enunciation—otherwise the foundations of mathematics, physics, and biology would inevitably be obsolete and no longer correct

bite force, and 70–80% of the total condylar reaction force can be borne by the balancing condyle (Smith, 1978). The differential loading of the mandibular condyles during chewing is strongly dependent on small shifts in muscle recruitment patterns: if the muscular force on the working side is roughly double that on the balancing side, the force on the working condyle is approximately 1.4 times greater than the reaction on the balancing condyle. However, if the muscular force on the working side becomes 1.5 times greater than that on the balancing side, the load is reversed, with the force on the balancing condyle becoming approximately 1.4 times greater than that on the working condyle (Hylander, 2006). Balancing contacts are a biomechanical solution that allows for the dissipation of the intense loads that develop during the power stroke of chewing.

When canines erupt in the arch, their morphology imposes an increase in the verticality of the entry movements in the chewing cycle, excluding all other dental surfaces from the function of supporting movement. The exclusive guidance of the canine during lateral movements has generated the false idea that it can be the absolute reference for understanding masticatory function and for setting up dental rehabilitations. This belief probably stems from the fact that pure canine guidance is established at a young age, presumably before the effects of life's hardships have taken their toll on a still-intact organism. Among these hardships, wear, even if physiological, is often wrongly counted, as if it were damaging the integrity of the tooth and the stomatognathic system. In this imprecise view, any other contact or grazing will always be interpreted as a pathological interference. From this arises the belief that even a grain of food on the working side is capable of separating the dental arches, eliminating any contact on the balancing side: "Enter bolus, exit balance" (Sheppard & Sheppard, 1968). This mistaken functional perspective finds equally erroneous confirmation in the movements performed by any articulator, a tool used in dental technology to replicate mandibular movements but which cannot capture the complex and wondrous dynamic elasticity of our masticatory system. An articulator, no matter how refined, can only respond to stimuli with its dynamics, which are not biomechanical but simply mechanical and therefore rigid. This response is flawed due to the absence of elasticity, which is the foundation of bone strength (completely rigid bone would be fragile). Furthermore, the inability of a mechanical system to replicate periodontal and joint resilience must be considered. During the power stroke of chewing, when a rigid bolus is interposed between the teeth of the working hemiarches, the teeth on the opposite side approach each other, up to 1 mm (Bruna et al., 2013), due to these three factors:

- Tooth intrusion related to the compressibility of the periodontal ligament (Martinez Choy et al., 2017).
- Compressibility of intra-articular tissues.
- Elasticity of the mandibular bone.

In archaic mammals and still in contemporary carnivorous mammals, despite the existence of very binding condyle-fossa relationships, the non-working hemimandible dynamically adapts to the power stroke of chewing, being able to rotate along its major axis, thanks to the adaptability of the symphyseal ligament (Crompton, 1989) (Fig. 5.22).

Even in non-human primates, the mandibular body twists both during the power stroke of chewing and during isometric biting (Hylander, 2006). In humans, the number and frequency of occlusal contacts on the balancing side vary with the distance between the mandibular and maxillary arches, significantly increasing when the arches are close to maximum intercuspation. In pure canine guidance, increased disclusion makes balancing grazes significantly less frequent than in group function (Ogawa et al., 2001).

An analysis of mandibular stress in primates suggests that deep maxillary bones in the molar region are an adaptation to counteract the increase in sagittal bending stress on the mandibular body on the balancing side during unilateral chewing, stress caused by intense balancing muscular force (Hylander, 1979). In anthropoid

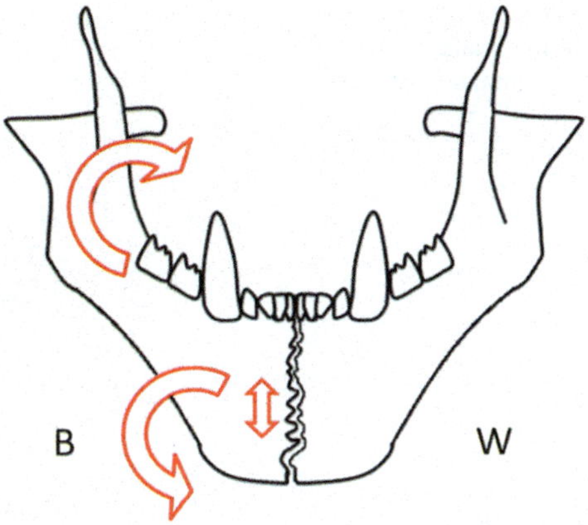

Fig. 5.22 Torsion of the balancing hemimandible in the mandible of a generic carnivorous mammal

primates, the fusion of the mandibular symphysis is an adaptive response to transfer muscular force from the balancing side to the chewing side (Hylander, 1979), increasing the efficiency of unilateral chewing (Dean, 1990; Hylander, 1979). The ossification of the symphysis strengthens the mandible against intense stresses, both bending and cutting, caused by the recruitment of adductor muscles on the balancing side, capable of generating high forces, especially during the chewing of mechanically "demanding" foods (Hylander, 1992; Hylander & Johnson, 1994; Luschei & Goodwin, 1974).

The molar region of the mandible is transversely thick as an adaptation to counteract greater torsion around the axis along the body of the hemimandible on the working side. In humans, the alveolar bone, especially in the mandible, deforms during chewing, resulting in a greater approximation of the posterior teeth on the balancing side (Pileicikiene & Surna, 2004) (Fig. 5.23).

Intense muscular forces exert stress on the mandible, causing distortions (Hylander, 1979; Korioth, 1997), including: eversion of the lower border of the mandible; inversion of the coronoid process; increased pressure of the lateral part of the mandibular condyle against the articular eminence compared to its medial part. The lateral region of the human TMJ is subjected to more intense stress and greater "wear" compared to the medial aspect. Most perforations of the articular disc occur in its lateral portion (Beek et al., 2000; Hylander, 1992; Kopp, 1976, 1978; Oberg et al. 1971; Werner et al. 1991). The mandibular body with its internal occlusal aspects also undergoes internal rotation (Fig. 5.24).

Food initially acts as a fulcrum, and the muscles on the working side tend to compress the ipsilateral joint and distract the contralateral one, while those on the balancing side perform the opposite action (Hylander, 1979). During chewing, there is a superposition/succession of multiple lever types depending on the specific conditions at that moment. The complex balance between the muscles on the working side and those on the balancing side also aims to prevent any distracting or compressive effects on the joint tissues during food grinding, maintaining intra-articular pressure within a physiological range.

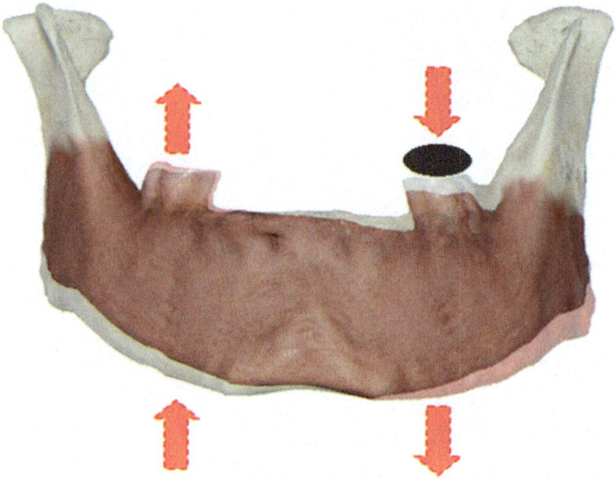

Fig. 5.23 Flexion in the frontal plane of the mandibular body

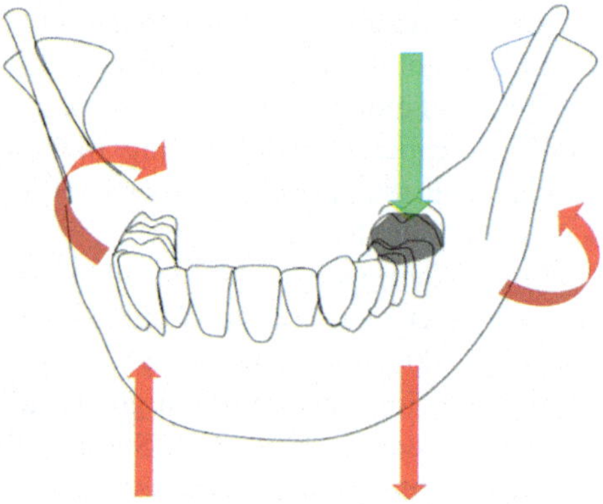

Fig. 5.24 The elasticity of the mandibular body causes it to undergo adaptive deformations during the masticatory power stroke, due to muscle forces from both the working and balancing sides

The TMJ behaves differently depending on the functional phase it is in. On the working side, the condyle functions as a hinge, essentially rotating around a vertical axis, while on the balancing side, the condyle acts as a slider, following an anterior and medial trajectory under the action of the contraction of the ipsilateral external pterygoid muscle (Fig. 5.25).

When evaluating the stresses that occur on the mandible and TMJs under various loading conditions (such as balanced occlusion, unbalanced occlusion, grinding, and clenching), it becomes evident that there is less stress on the condylar heads during balanced occlusion compared to the other measured situations (Ingawalé & Goswami, 2012).

Balancing grazing protects the TMJ from forces that tend to intrude and posteriorly displace the condyle in the fossa during the power stroke of the chewing cycle. Balancing is more effective the further posterior it is. For this reason, the disto-vestibular cusp of the lower third molar will be more effective in balancing than a balancing graze on the disto-vestibular cusp of the lower second premolar (Figs. 5.26 and 5.27).

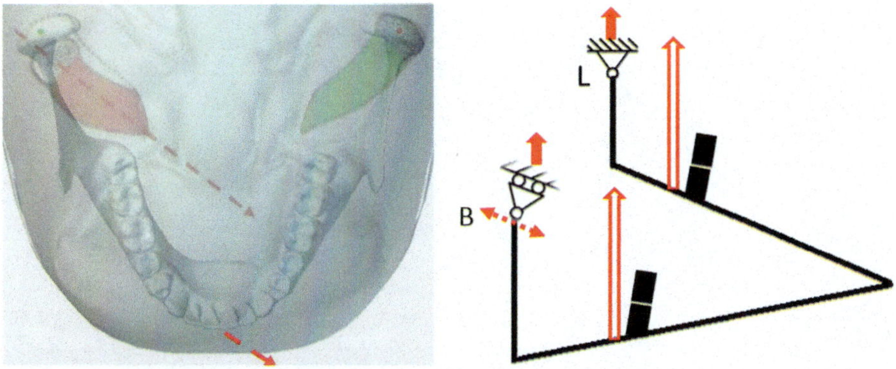

Fig. 5.25 During chewing, the working condyle essentially undergoes rotation—excluding the limited Bennett movement—and can be likened to a joint capable of rotating on its axis; whereas the balancing condyle can be seen as a carriage capable of translating anteroposteriorly and latero-medially. Proper occlusal contact on balancing surfaces supports and provides stability to the respective TMJ in its physiological movement (the absence of balancing contacts would make it less stable); however, excessive contact can cause significant muscular tensions and harmful concentrations of intra-articular forces. Below is a biomechanical diagram (black rectangles: posterior occlusal contacts; red outlined arrows: closing muscle vectors; L: working condyle with superimposed hinge constraint symbol; B: balancing condyle with superimposed carriage constraint symbol; small red arrows: intrusive tendency of condyle into fossa—sinking caused by closing muscle forces and countered by posterior occlusal contacts on both sides)

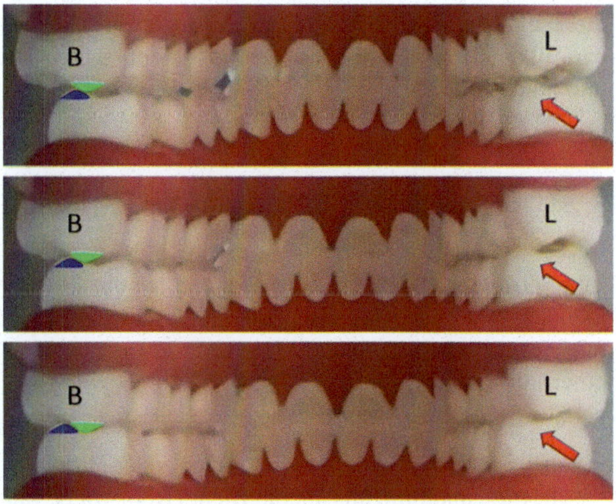

Fig. 5.26 Masticatory power stroke, right side: Note on the left side of balancing the progressive sliding of the upper molar supporting cusp (green) onto the lingual slope of the lower supporting cusp (blue)

From a biomechanical perspective, it can be concluded that the geometry of occlusal surfaces, in conjunction with a well-balanced occlusion, ensures a physiological distribution and compensation of chewing forces, thereby providing stability to the stomatognathic system and prolonged dental function (Pileicikiene & Surna, 2004).

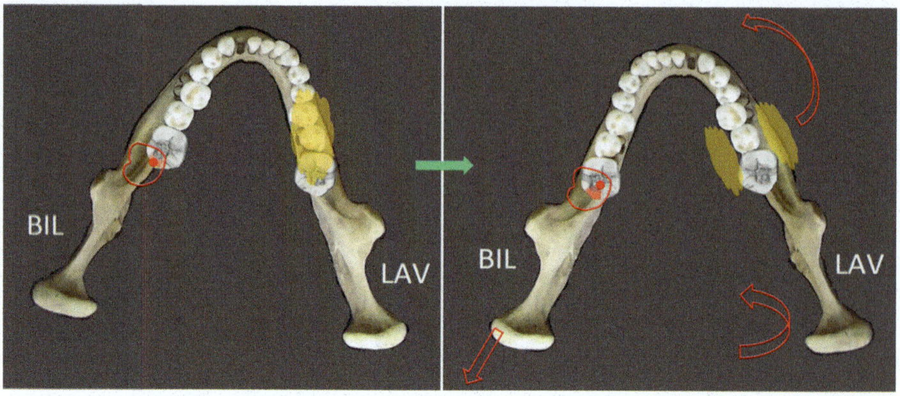

Fig. 5.27 Masticatory power stroke, right side: the hemimandible moves forward and sideways during the initial phases of the cycle, and at the beginning of the power stroke aimed at breaking down the bolus, it moves medially to achieve maximum intercuspation; note on the left balancing side the progressive sliding of the upper molar's supporting cusp (red point inside the red-bordered silhouette of the upper first molar) on the lingual slope of the lower supporting cusp. Also in red, the balancing surface of the lower first molar

If chewing forces are not effectively dissipated, they can be damaging to one or more structures of the masticatory system. For example, when there is unilateral pain in the TMJ, chewing is often done on the same side as the pain, either because balancing forces can exceed the intensity of working forces (Jacobson, 1986) or, more likely, because patients prefer to chew on the painful side to minimize the painful anterior-medial condylar movements that would occur if that side translated into balance (Hylander, 1975, 1980, 1992).

5.2 Biomechanics of Muscle Attachments

Like all joints, the TMJs are regulated by muscle levers that move the bony segments with dynamics related to force, vectoriality, and the insertion point of the muscles themselves. The TMJs have several aspects that distinguish them from other joints, and from a biomechanical, embryological, histological, and functional perspective, they exhibit unique characteristics:

- They are dynamically redundant (Scarr & Harrison, 2017).
- They are anatomically incongruent (Koolstra & van Eijden, 1999).
- They have a point of arrest in their physiological excursion, determined by extra-articular anatomical elements (the occlusal surfaces of the teeth).
- They are double, specular, and interdependent.

This last characteristic places them squarely among the joints of the axial skeleton, alongside those of the spine and the atlanto-occipital joint. The axial skeleton represents the supporting axis of our body and consists of the bones of the skull, spine, ribcage, as well as the hyoid bone and the ossicular chain of each ear (hammer, anvil, and stirrup). Its articular structures, interrelated with each other, are fundamental elements of the postural chain. Muscles and tendons insert on this part of the skeleton, using it as a fulcrum and allowing for the mobility of the appendicular skeleton connected to it (the appendicular skeleton is composed of the bones of the limb girdles and limbs) (Fig. 5.28).

Form and function are indistinguishable, complementary characteristics that mutually illuminate each other (Schwenk, 2000). Even though there are numerous asymmetries in the arrangement and placement of internal organs, vertebrates are essentially bilaterally symmetric on the outside (Raya & Belmonte, 2006). The symmetric arrangement of body parts with respect to a median plane ensures functional complementarity, optimal execution of functions, better mechanical balance structured around the gravitational field, and equal perceptual and motor capacity on both sides. External functionalities exhibit a high degree of bilateral symmetry (Allard & Tabin, 2009).

The role that the cranio-mandibular joint plays in postural balance is highlighted by the frequent association between TMJ pathology and symptoms of chronic polyarticular pain. Patients with TMJ disorders, compared to those without such

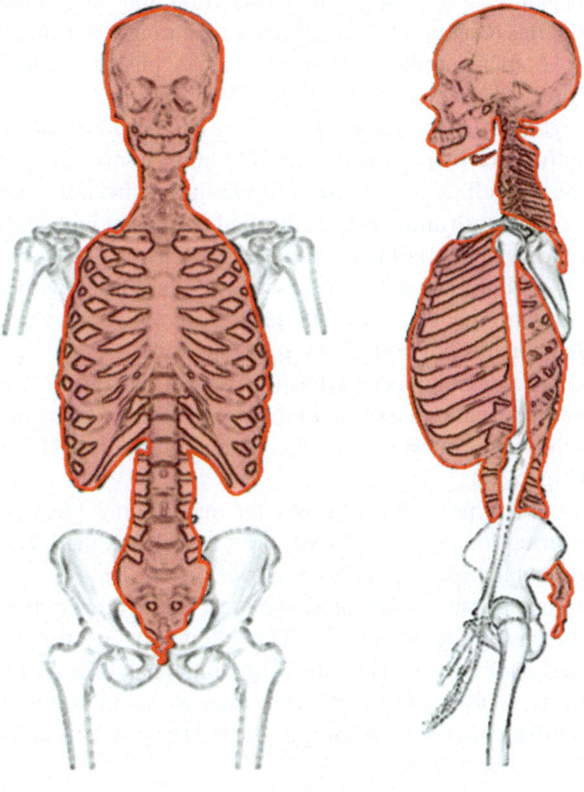

Fig. 5.28 Articulations of the axial skeleton: the cranio-mandibular joint is to be considered an integral part of the mammalian axial skeleton, as the TMJs exhibit duplicity, mirroring with respect to the midline, and dynamic interdependence that characterize the joints of this crucial part of the skeleton, so important for our posture

disorders, have a 5.5 times higher risk of having pain in other joints. This association is stronger in older age, and knee pain is the most frequently associated comorbidity (Bonato et al., 2017). The correlation between the masticatory and cervical musculature is important: in cranial and mandibular movements, these two joints interact, regulating each other. The presence of TMJ pathology correlates with the angle between C7 and the horizontal plane: the head position is more advanced in patients with TMJ disorders compared to a control group (Lee et al., 1995). In patients with joint pathology in the lower limbs, modifying the cranio-mandibular relationship, for example, using a bite, can normalize the range of motion (ROM) of the hip joint during walking (Chung et al., 2016).

However, despite these unique characteristics, we must not forget that the rules governing joint structures in arthrology, physiology, and pathology also apply to TMJs. The concept of an articulation between two bony segments inherently includes the idea of reciprocal movement permitted by the joint, allowing the joint to move freely throughout its range of motion without any hindrance or pain. From a biomechanical perspective, the dynamics of the TMJs are regulated not only by the sophisticated and redundant vectoriality of contraction that the masticatory muscles can express but also by their insertion geometry. It is interesting to note how the need for intense forces and refined precision in chewing is supported by precise muscle insertions that satisfy both requirements. The performance of a muscle also depends on its point of bony insertion and, therefore, on the lever system created: the insertion point also influences the range of motion (ROM) of the joint itself.

A mere application of Newtonian physics is sufficient to effectively express force: all skeletal muscles develop rotational forces with movements of the bony segments to which they attach. The development and shape of bones are also greatly influenced by the force exerted by muscle insertions, and this applies to the mandible as well. A muscle that inserts farther from the joint fulcrum is better suited to exert intense and highly controlled forces but with less range, speed, and immediacy (explosiveness) of movement (Fig. 5.29).

With the same level of contraction, a muscle with an insertion closer to the joint fulcrum, due to its specific lever arm, tends to produce larger and more explosive movements but with less force (Fig. 5.30).

The most important masticatory muscles of the mandible (temporalis, masseter, and internal and external pterygoids) exert a third-class lever action, in which the force (P) is interposed between the joint fulcrum (F) and the occlusal resistance (R) (Fig. 5.31).

This type of lever allows for maintaining great control and precision over the exerted force; the action of these muscles is that of a disadvantageous lever but at the same time very precise. However, these muscles can still generate considerable forces because their insertions are relatively far from the joint fulcrum, thereby increasing the power arm. This allows for exerting both intense forces and a high degree of movement control simultaneously (Fig. 5.32). This very refined control of movements is absolutely necessary to avoid unintentional traumas with potential permanent structural damage during the powerful crushing of food.

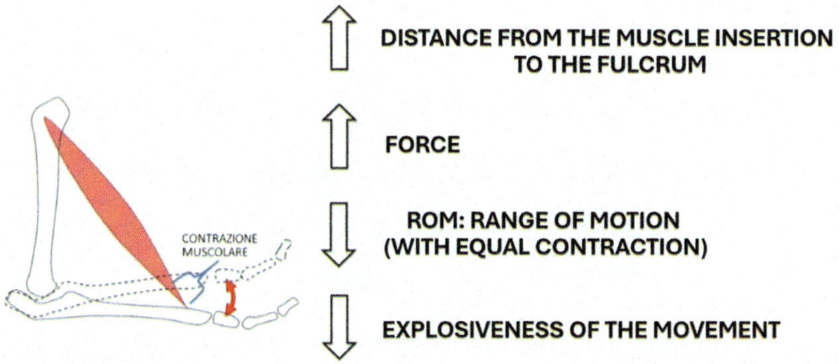

Fig. 5.29 Muscle insertion far from the joint fulcrum: increases force and movement control

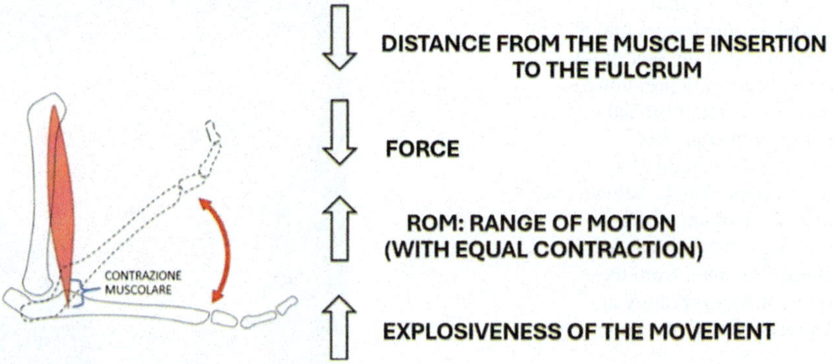

Fig. 5.30 Muscle insertion near the joint fulcrum: increases explosiveness and range of movement

Fig. 5.31 Colored areas of insertion of the mandibular adductor muscles. In this perspective, the mandible functions as a third-class lever; this type of lever is always disadvantageous because the power arm is always smaller than the resistance arm, as well illustrated by the underlying figure of the forceps, a tool that allows great precision but not great gripping force

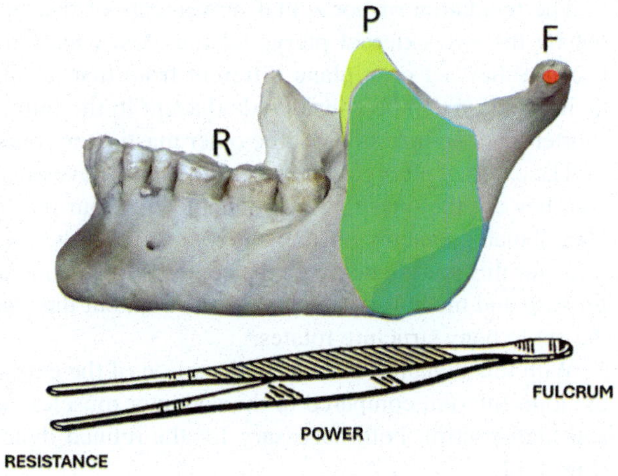

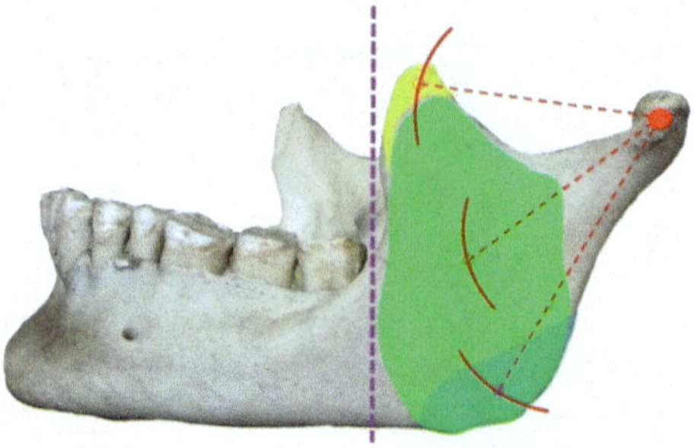

Fig. 5.32 Distances between the joint fulcrum and the central points of insertion areas of the three main masticatory muscles: note how their lever arms are developed; the extension of the insertion areas maximizes the power zone, stopping at the limit (purple line) of the occlusal resistance zone

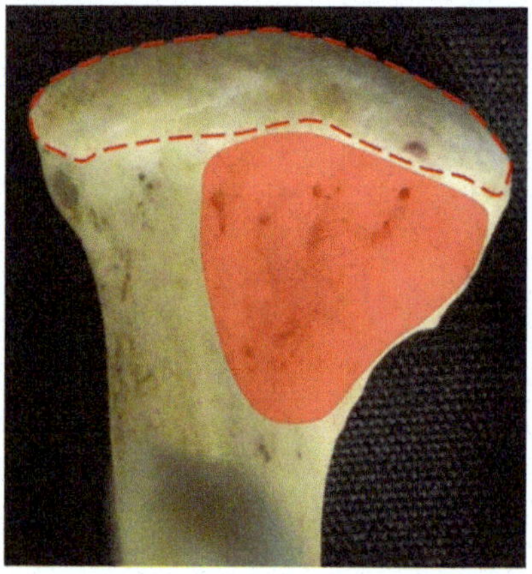

Fig. 5.33 The external pterygoid muscle inserts into a depression called the pterygoid fovea (or fossa or depression), located on the upper medial side of the condylar neck, posterior to the mandibular notch and immediately below the articular surface of the mandibular condyle (dashed red line). Specimen from the Museum of Anthropology at the University of Florence

The regulation of horizontal movements of the mandible is essentially carried out by the two external pterygoid muscles, which are unique in their own right because they act on a plane different from that of all other masticatory muscles: their insertion can be considered adjacent to the joint fulcrum, at a distance much shorter than the insertions of the other masticatory muscles (Fig. 5.33).

The external pterygoid muscles are highly developed in primates and in humans, and they are the ones that most characterize our masticatory cycle. Analyzing the mandibular geometry on the horizontal plane, we must consider the fulcrum of force for this muscle not to be the condyle to which it attaches, and which translates forward and medially under its contraction, but the contralateral one around which the entire bony structure rotates.

With this dynamic model, the insertion of the external pterygoid is farther from the joint fulcrum compared to the adductor muscles, allowing for significant force and high control, both necessary for the refined dynamic regulation of occlusion (Fig. 5.34).

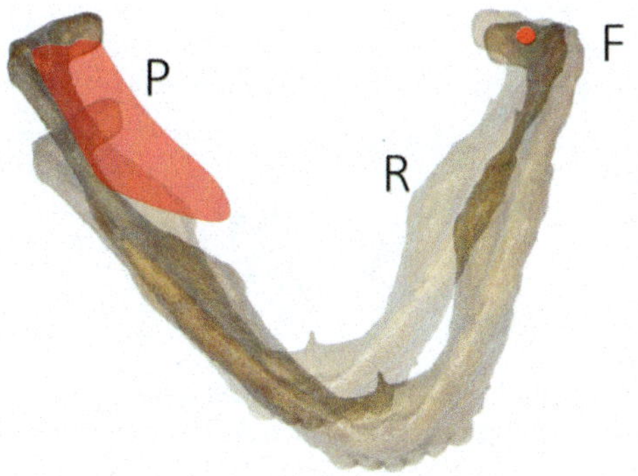

Fig. 5.34 The insertion of the external pterygoid muscle adjacent to the articular surface (unique among the masticatory muscles) provides high control, necessary for modulating the refined movements of the mandible during chewing; moreover, the power arm, greater than the resistance arm, thereby increases the developed force

Unlike the elevator muscles, the external pterygoids do not contain muscle spindles, stretch receptors that prevent excessive stretching of muscle fibers and detect any changes in the length and speed of muscle work (Rathee & Jain, 2020). When assisted by the supra- and infrahyoid muscles, the bilateral contraction of the external pterygoids contributes to mandibular depression, although this action plays a secondary role.

5.3 Mandibular Movements

Since the early days of gnathology, there have been attempts to study and reproduce the movements of the mandible through the study of the anterior midline point of the incisor group, as mandibular movements can be defined by the three-dimensional path taken by the lower central incisor point (PI) (Lewin & Ramadori, 1985). William G.A. Bonwill, Francis Balkwill, Rupert Hall, Alfred Gysi, Ulf Posselt, Pedro Planas, Bernard Jankelson are some of the fathers of gnathology who referred to this particular point in their studies.

If we start from the maximum intercuspation and have the mandible perform the widest and most extensive movements of opening, lateral movement, protrusion, and retrusion possible, the PI delineates trajectories that represent the functional limits of mandibular movement, defining a volume first described by Ulf Posselt (Posselt, 1962) (Fig. 5.35).

Within this volume, most of the movements that the mandible can normally perform during the chewing function occur. For this reason, Posselt's volume serves as a clear indicator of the health status of the two temporomandibular joints (TMJs). As previously observed, despite the unique characteristics of the TMJs, their pathophysiology falls within the principles of all joints: wide, symmetric movements

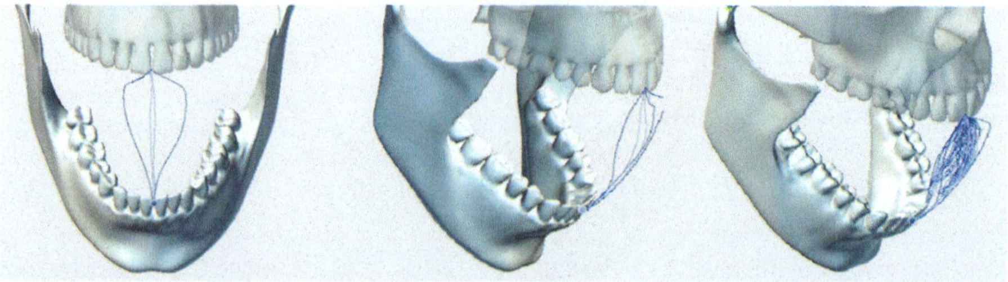

Fig. 5.35 Volume of Posselt, delineated by the extreme excursive movements of the mandible. In the lower image, the chewing movements form a tangle of lines that remain within the volume

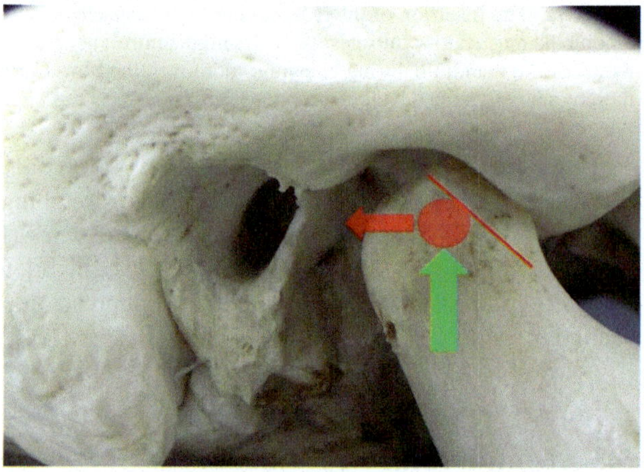

Fig. 5.36 The inclined plane of the glenoid fossa roof (red line) under the action of the closing muscles of the mouth (green arrow) tends to slide the condyle to the posterior part of the fossa (red arrow). Cranial specimen 327 from the Museum of Anthropology at the University of Florence

without interruptions or sudden changes in direction and without pain are indicative of good joint health. Except for fixed joints, every joint serves its purpose in the physiological movement of its bony ends expressed through the Range of Motion (ROM). This parameter defines the full potential of movement for any mobile or semi-mobile joint: improved joint mobility leads to greater strength, speed, ease, and fluidity in all movements. This condition results in a qualitative and quantitative improvement in movement execution, coordination and technique, motor learning processes, strength, speed, endurance, correction of muscular imbalances, postural prevention, and psychological regulation (contracted muscles are generally accompanied by psychological tension). As we have seen in previous chapters, the TMJ is characterized by a deep mandibular fossa with an articular surface inclined anteriorly and downward. The adductor muscles that close the mandible against the cranium tend to create a posteriorizing force on the condyles that articulate with the inclination of the glenoid fossa (Fig. 5.36).

The anterior articular surface of the mandibular fossa is designed to bear masticatory loads, partly due to the presence of appropriate soft tissues (fibrocartilage) and adequate bone thickness. The fibrocartilage that covers the articular ends of the TMJ, also composed of dense fibrous tissue, is more elastic and easily deformable than hyaline cartilage. When subjected to pressure, fibrocartilage deforms to adapt, creating an increase in the joint contact surface, thereby reducing stress on the individual unit of the surface.

The interposition of the articular disc also serves to prevent this characteristic from becoming a dynamic disadvantage due to excessive friction caused by a change in the microarchitecture of the articular surface (Yamada & Evans, 1970). If the joint were covered with hyaline cartilage, which is less deformable, there would be less deformation of the joint tissue, leading to greater sliding of the articular surfaces and better dynamic efficiency. However, this would result in a significantly reduced contact surface, leading to excessive concentrations of pressure stress on the joint surface. The joint forces acting on the TMJ play an important role in the gradual maturation of this region, continuing throughout adult life (Hylander, 2006). The other bony walls of the mandibular fossa, namely the superior and posterior walls, are unsuitable for bearing compressive forces (Saratti et al., 2021). The superior wall of the mandibular fossa consists of a thin bony thickness that separates it from the middle cranial fossa (Figs. 5.37, 5.38, and 5.39).

Similarly, the posterior bony wall of the glenoid fossa does not have adequate thickness to withstand compressive forces (Figs. 5.40, 5.41, and 5.42).

The posterior portion of the TMJ is a delicate area concerning its soft tissues, histologically unsuitable for withstanding compressions related to posterior forces on the condyle: the retrodiscal pad is rich in nerve endings, blood vessels, loosely packed collagen fibers, and elastic fibers. The retrodiscal tissues, located behind the articular disc of the TMJ, can be divided into three regions: the superior or temporal, intermediate, and inferior or condylar. The temporal portion attaches superiorly, posteriorly to the same bone and anteriorly to the upper part of the disc, while the

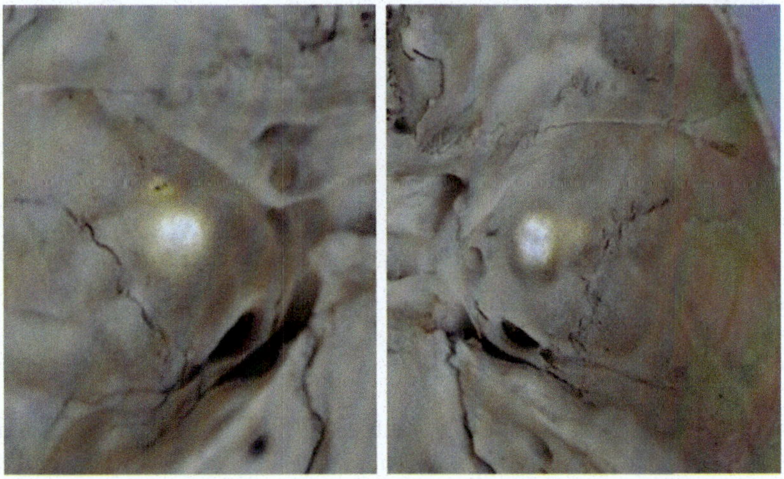

Fig. 5.37 Floor of the cranial cavity: right and left middle cranial fossae. Note the transparency of the thin bony thickness represented by the roof of the mandibular fossa through transillumination. Cranial specimen 311 from the Museum of Anthropology at the University of Florence

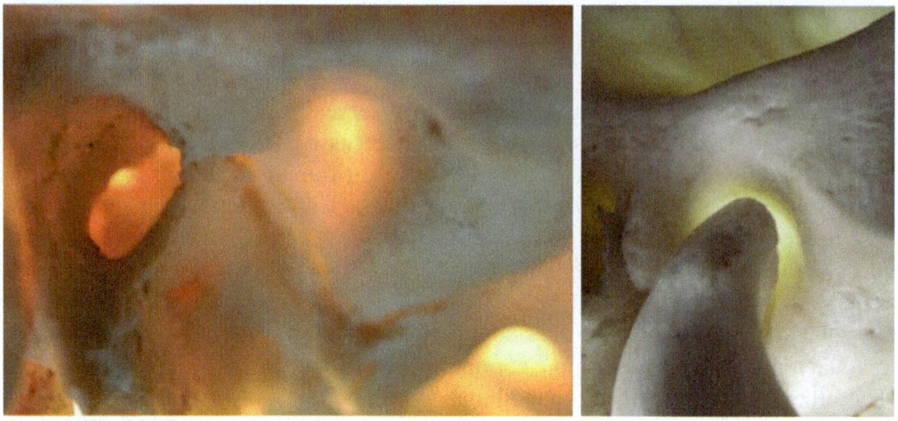

Fig. 5.38 Translucency through transillumination of the thin bony thickness of the roof of the mandibular fossa. Cranial specimen 311 from the Museum of Anthropology at the University of Florence

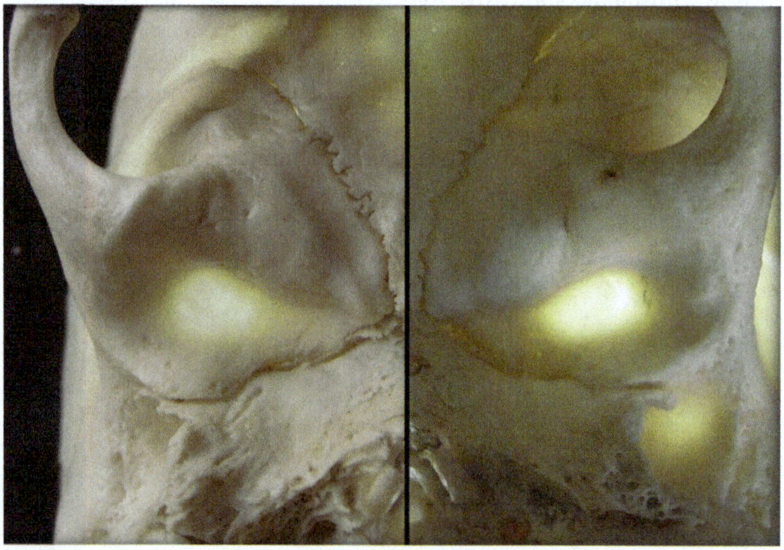

Fig. 5.39 Translucency through transillumination of the thin bony thickness of the roof of the mandibular fossa. Cranial specimen 311 from the Museum of Anthropology at the University of Florence

inferior portion attaches to the lower condylar surface of the disc. The central part is highly innervated and vascularized, primarily by an unusually rich venous plexus that acts like a hydraulic jack, controlling condylar dynamics through pressure variations. These modulate blood emptying and filling during anteroposterior movements of the condyle (Fig. 5.43).

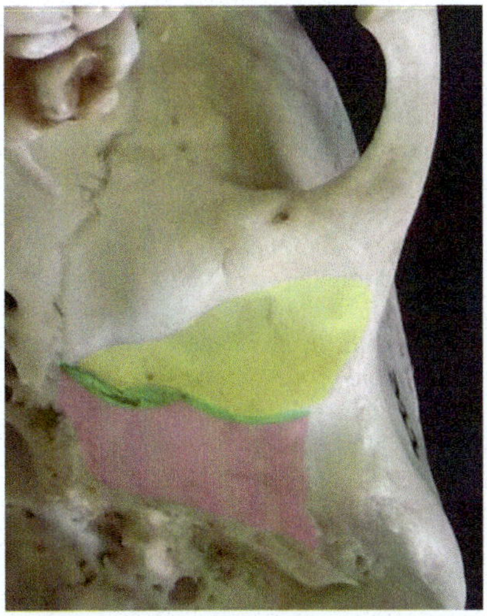

Fig. 5.40 Mandibular fossa, view from below: note—separating the anterior portion of the fossa (yellow) from the posterior portion (purple)—the area of the petrotympanic fissure (green). Cranial specimen 311 from the Museum of Anthropology at the University of Florence

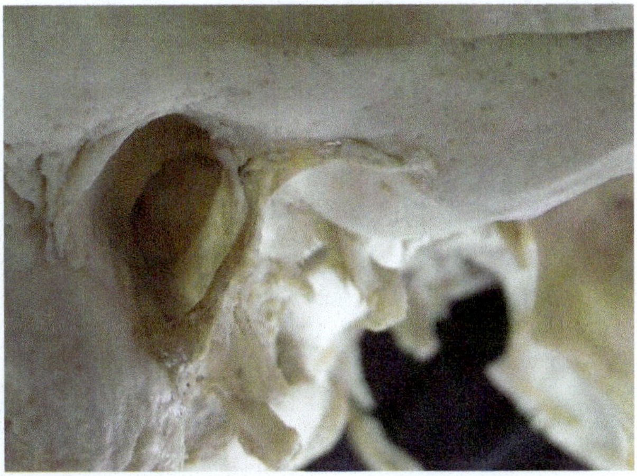

Fig. 5.41 Mandibular fossa and acoustic meatus, sagittal view: note the thin bony thickness separating the two structures. Specimen from the Museum of Anthropology at the University of Florence

The posterior area of the TMJ is highly adaptable to volume changes related to the condyle's position in the fossa, and it rearranges blood, tissue fluids, and synovial fluid internally based on this. The anatomy and histology of the retrodiscal tissues enable them to withstand and dissipate tensile forces caused by their traction. The retrodiscal vascular plexus can function like the bellows of an accordion, alternating between a state of repletion and collapse. When the condyle moves anteriorly, the retrodiscal tissues extend, increasing their volume and drawing blood inside. When the condyle returns to its resting position in the fossa, the tissues reduce their volume by emptying the accumulated blood.

The stresses imposed by mandibular movements on retrodiscal structures can be continuous, as during clenching, or dynamic and intermittent, as during chewing. For the growth, development, and homeostasis of TMJ articular tissue, physiological biomechanical stresses are required, and in general, dynamic loading leads to an

Fig. 5.42 Relationships between the condyle, mandibular fossa, and acoustic meatus: lateral-posterior-inferior view; observe the thin bony thickness separating the posterior wall of the condyle from the acoustic meatus and tympanic cavity; in the foreground, the mastoid process and behind, the styloid process. Specimen from the Museum of Anthropology at the University of Florence

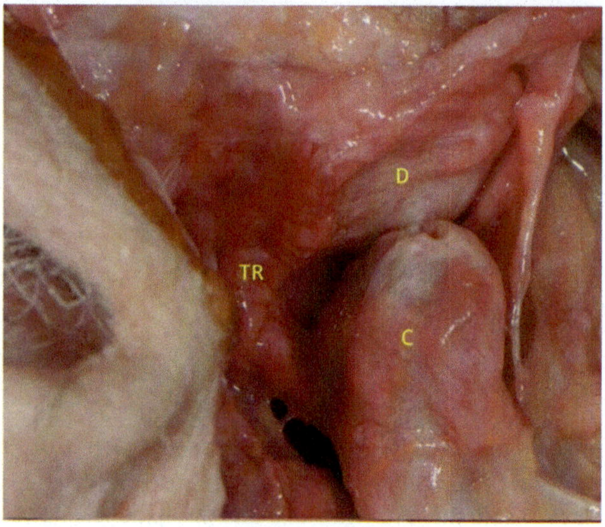

Fig. 5.43 Posterior portion of the TMJ. *TR* retrodiscal tissues; *C* condyle; *D* disc

anabolic effect on joint tissues, while static loading, if prolonged or excessive, induces a catabolic effect (Tanaka et al., 2008).

The delicate bilaminar zone of the retrodiscal tissues, if subjected to repeated traumatic compressions, triggers an inflammatory response leading to arthrogenic pain. Inflammation eventually results in fibrosis of the loose connective tissue (Stasolla et al., 2004; Yang et al., 1992), with a significant increase in fibroblast density, reduction and obliteration of arterial lumens, and a marked decrease in elastic fibers (Pereira et al., 1996). These histological changes all lead to reduced elasticity of the retrodiscal tissues, which is crucial for the proper functioning of all other joint components. Altered joint dynamics modify and reduce the circulation of synovial fluid, leading to a deficit in its lubricating and cushioning action and an increase in friction between the joint surfaces. In areas with reduced synovial fluid flow, there is an accumulation of inflammatory factors that locally accelerate the progression of joint damage (Xu et al., 2013).

Therefore, during mandibular movements, especially during the masticatory power stroke, no compression of the retrodiscal tissues should be created due to excessive retraction of the condyles from their physiological homeostatic position within the mandibular fossa.

5.4 Gothic Arch and Posselt's Scheme

Early pioneering studies in gnathology also focused on attempting to reproduce mandibular movements on the horizontal plane, aiming to reduce mandibular morphology to a triangular shape. Francis Balkwill, in 1866, was the first to describe an equilateral triangle with vertices on the occlusal surface of the lower second molars and the interincisal point (PI) (Balkwill, 1866). In 1899, William Gibson Arlington Bonwill extended this triangle to encompass the entire mandibular body, placing the two posterior vertices on the condyles and describing an equilateral triangle with sides of 100 mm, although the accuracy of these measurements has often been debated (Christensen, 1959; Nikolopoulou et al., 2019).

The two triangles relate to each other with an angle known as the Balkwill angle, which has an average of 26°, greater in males than in females (Ohm & Silness, 1982) (Fig. 5.44).

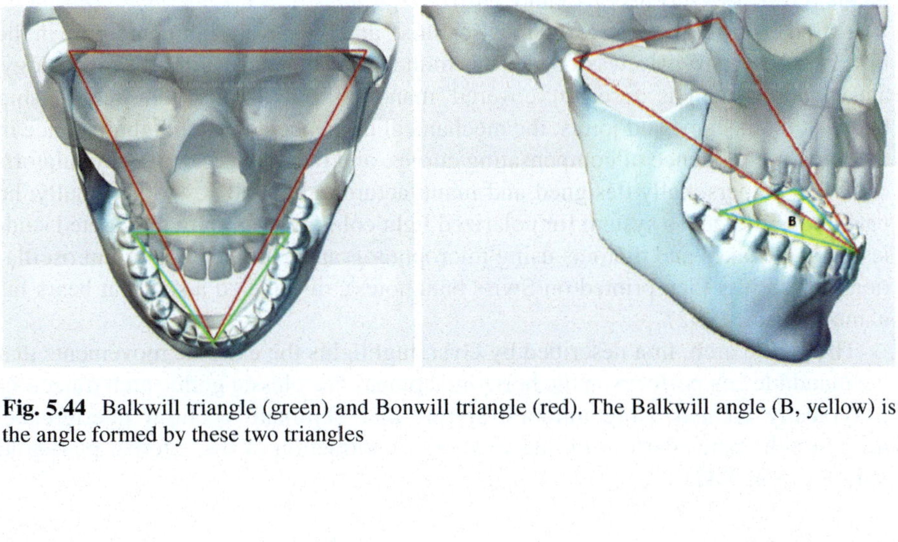

Fig. 5.44 Balkwill triangle (green) and Bonwill triangle (red). The Balkwill angle (B, yellow) is the angle formed by these two triangles

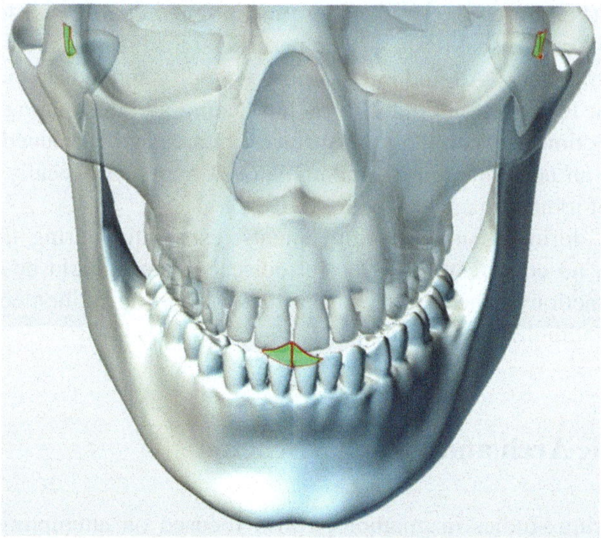

Fig. 5.45 Gothic arch (in green) outlined on the horizontal plane by the limit movements of protrusion and lateral excursion from the interincisal point. The gothic arch represents the mandible's ability to move horizontally, which is essential for masticatory dynamics: a wide and symmetrical gothic arch corresponds to equally extensive and symmetrical joint movements; conversely, a limited and asymmetrical gothic arch reflects an inability of the TMJs to perform physiological functional movements correctly, potentially triggering a pathological condition

The anterior vertex of Bonwill's triangle is used for the registration and subsequent analysis of mandibular movements. Alfred Gysi, with his recordings of mandibular movements on the horizontal plane, was the first to describe the path ("gothic arch") that the PI follows with the mandible's limiting movements in lateral and protrusive excursions, aiming to interpret and describe the resulting condylar movements (Gysi, 1930) (Figs. 5.45 and 5.46).

Alfred Gysi was a giant, a versatile scientist, and the most prominent figure in the history of dentistry. He was one of the founders of dental histology, embryology, and gnathology. His research covered mandibular dynamics, the relationship between tooth shape and joints, the mechanical influence of cusp height, balance in the joint, the influence of compensating curves, and the study of dental articulators, of which he personally designed and manufactured eleven types. Additionally, he patented a filter and a system for polarized light color photography, conducted studies on snowflakes and diatoms using microphotography, studied pendulum oscillation (which was then printed on Swiss banknotes), discovered a star that bears his name (Perrini, 1992).

The gothic arch, first described by Gysi, highlights the extreme movements that the mandible can perform on the horizontal plane. The classic gothic arch trajectory is typically created using intraoral splints that eliminate occlusal interference, revealing the paths defined by the vectors of contraction of the external pterygoid muscles (Fig. 5.47).

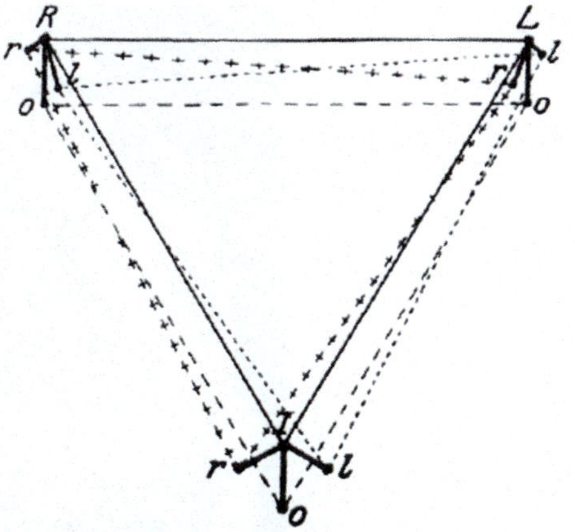

Fig. 5.46 Original drawing of the movements of the interincisal point (PI) in relation to condylar movements. Illustrations from the book "Zahnärztliche Prothetik," 1930, by Bruhn, Christian, et al., from the collection of the Fondazione Castagnola

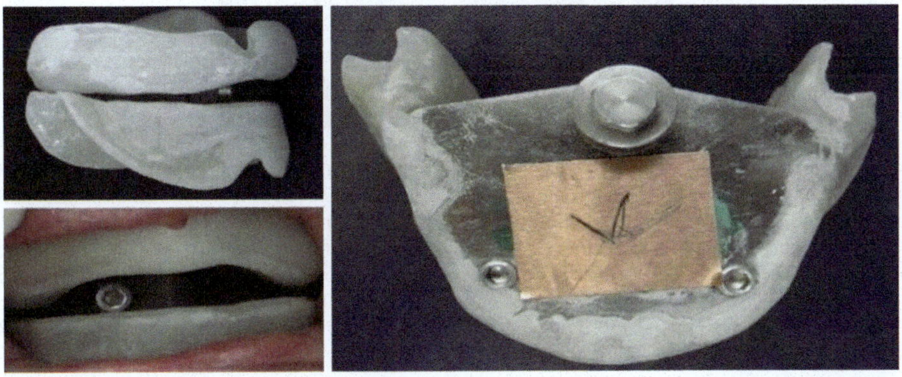

Fig. 5.47 Intraoral plates for recording the gothic arch in an edentulous patient: resin bases were made for both arches, with a writing stylus inserted into the upper base and a recording plate in the lower base. After inserting them into the patient's mouth and having them perform maximum protrusive and lateral movements, the upper stylus traced the gothic arch lines on the lower plate

By contracting the external pterygoid muscle on one side, you achieve anterior-lateral movement of the lower PI. When both muscles contract, a straight anterior protrusive movement is obtained (Fig. 5.48).

The classic devices used for recording the gothic arch exclude dental contacts (if present) and, due to their size, force the tongue into a retracted position. This modification alters the relationship between all connected structures: the mandible, skull, hyoid bone, and cervical spine. A recording of mandibular movements with minimal interference can be achieved with a kinesiograph, an instrument that can detect various mandibular movements, including those that make up the gothic arch and Posselt's volume (see Fig. 4.65 above). The kinesiograph uses a magnet placed on the vestibular surface of the lower incisors, allowing the recording of mandibular movements without occlusal interference and minimizing any potential distortions in the recording.

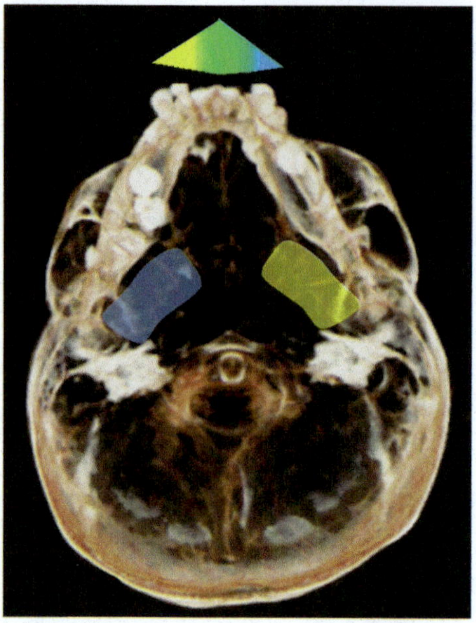

Fig. 5.48 Contraction of the external pterygoid muscles (right external pterygoid in blue, left external pterygoid in yellow) determines the opposite lateral movement; simultaneous contraction of both muscles brings the mandible forward (protrusion), and the overlapping of their contractions and their different force vectors positions the mandible at every single point within the gothic arch area. The coloration (at the top) of the gothic arch represents the sum of actions of both muscles at each point (yellow, right lateral movement caused by contraction of the left pterygoid; blue, left lateral movement caused by contraction of the right pterygoid; green, protrusion caused by simultaneous contraction of both pterygoids)

Occlusal surfaces significantly influence not only the sliding movements of the teeth but also movements that occur in the absence of dental contact, such as maximum opening. Maximum intercuspation affects and diversifies the dynamic occlusion pattern (Al-Nimri et al., 2010). The occlusal surfaces that regulate and guide lateral and protrusive movements are crucial for shaping both the gothic arch and the Posselt's volume.

The gothic arch can be interpreted as a horizontal section of Posselt's volume, just as Posselt's volume can be considered as the overlapping of infinite horizontal planes in which the gothic arch develops (Fig. 5.49). Within this volume, formed by the limit movements of opening, lateral, and protrusive, lies the dynamic potential of the mandible during mastication (Posselt, 1962).

The Gothic arch is formed by the extreme movements of lateral and protrusive excursions, which are achieved through the selective contraction of one of the two lateral pterygoid muscles (lateral excursion) or their simultaneous contraction (protrusive excursion). The contraction of the lateral pterygoid muscles generates mandibular movements on the horizontal plane, essential for chewing and characterizing the mandibular dynamics in mammals. In classic gnathology, both Posselt's volume

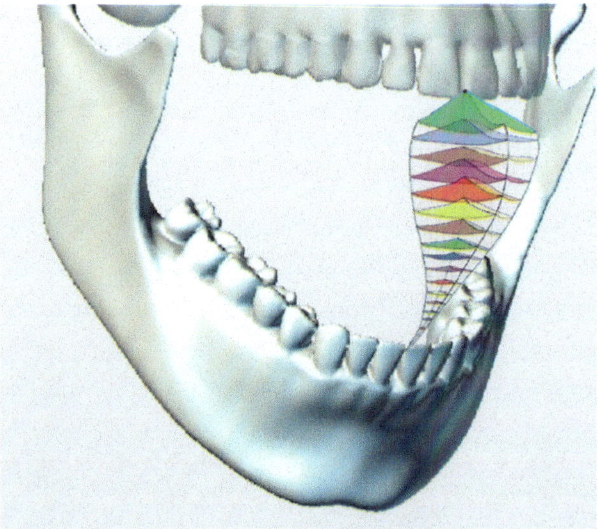

Fig. 5.49 The Posselt volume is defined by the overlap of infinite horizontal planes where maximum protrusive and lateral movements can occur, outlining the gothic arch

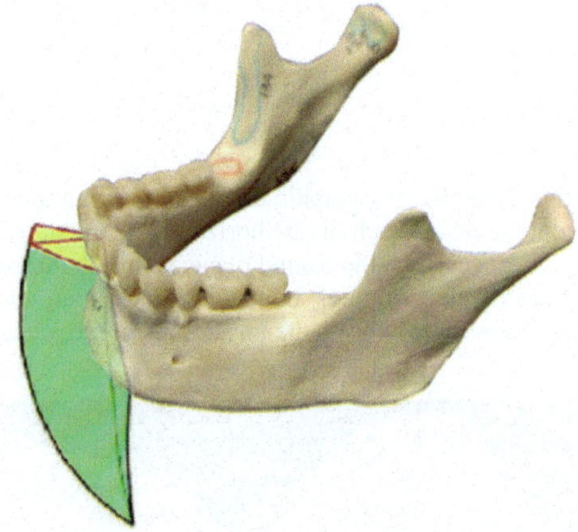

Fig. 5.50 Posselt's volume

(Fig. 5.50) and the gothic arch (Fig. 5.51) are represented in a regular and symmetrical manner, with mirrored lateral excursions and a linear protrusive path.

A physiological chewing pattern leads to selective wear patterns over time, primarily influenced by the muscle vectors of the lateral pterygoid muscles that define the gothic arch (Fig. 5.52).

The gothic arch represents the geometric expression on the horizontal plane of the ranges of motion (ROMs) of the TMJs, just as Posselt's volume represents its three-dimensional expression.

Any dental interventions aimed at widening and symmetrizing the gothic arch and Posselt's volume allow the joint system to express its physiological range and regain a normal chewing activity.

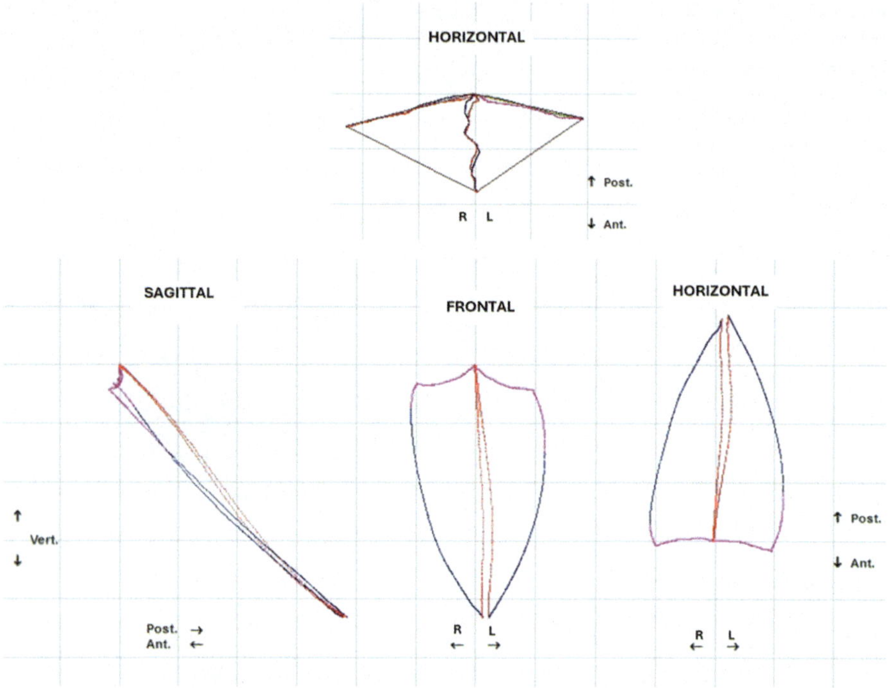

Fig. 5.51 Kinesiographic tracings depicting the interincisal point on three planes of space: at the top, the gothic arch on the horizontal plane; at the bottom, Posselt's diagram. Left: on the sagittal plane; center: on the frontal plane; right: on the horizontal plane. These tracings of the gothic arch and Posselt's volume, showing wide, regular, and symmetrical trajectories, highlight the mandible's ability to express the physiological movements of alternate unilateral mastication

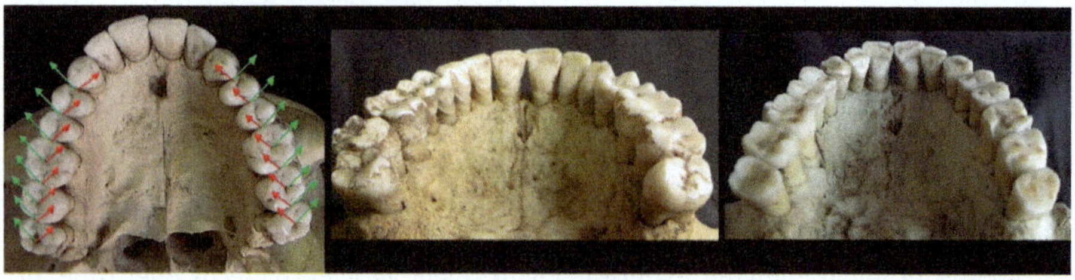

Fig. 5.52 On the left, muscle vectors of the external pterygoids. Center and right, wear patterns shaped by physiological chewing. Specimens from the Museum of Anthropology at the University of Florence

We have observed how the Range of Motion (ROM) is an indicator of the flexibility of a joint, expressed as the range of movement from the initial position to the final position. A physiological ROM contributes to maintaining correct posture, developing muscle strength properly, and preventing muscle-tendon-joint injuries. The ROM is influenced by anatomical factors such as joint structure, ligaments, joint capsule, muscles, tendons, surrounding adipose tissue, and overlying skin. Physiological factors such as gender, age, external temperature, and psychological conditions also influence joint movement.

Each joint has a physiological range of motion that depends on the balance of muscular tensions applied to it. By varying the vectors of these forces, statics, dynamics, and freedom of movement are modified. Disturbances in ROM and bite strength are associated with TMJ disorders. An extensive and symmetrical Posselt's volume indicates a broad and symmetrical freedom of movement of the mandibular muscle structures and a physiological joint ROM of the TMJs, which is essential for physiological alternating unilateral chewing (Fig. 5.53).

As an indicator of the system's ability to perform alternating chewing on either the right or left side, Pedro Planas identified the AFMP (Planas Functional Masticatory Angles), which assess the verticality of lateral movement of the interincisal point on the frontal plane. The centrifugal lateral movement (i.e., starting from maximum intercuspation and sliding the occlusal surfaces outward) is essentially the reverse of the entrance phase of the chewing cycle (Fig. 5.54).

The greater the verticality of the entrance phase of chewing cycles, the more limited they will be in their horizontal component, resulting in a decrease in the grinding capacity of the occlusal surfaces. Correct verticality of the AFMPs allows for the development of the horizontal component of chewing cycles, increasing their effectiveness. Furthermore, their symmetry allows for the spontaneous alternation of the chewing side. The further one deviates from symmetry and correct AFMP inclination, the more the system has to adapt to compensate for functional difficulties in fulfilling its task. The amplitude of each AFMP expresses the dynamics of chewing from that side: the greater the angle, the greater the verticality in the entrance phase of the chewing cycle, and the more the mandibular dynamics (and ROM of the TMJs) are penalized.

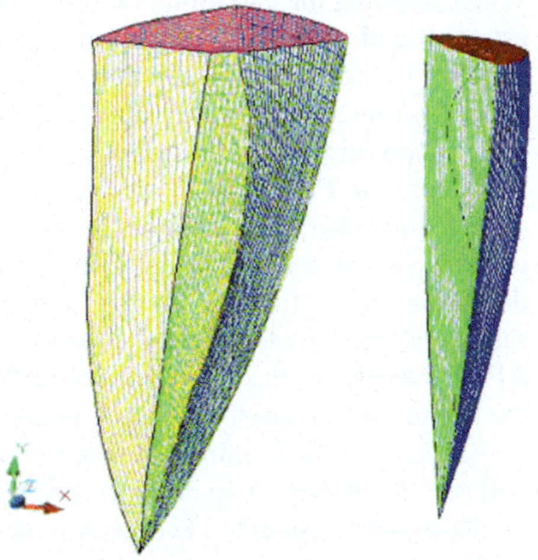

Fig. 5.53 Posselt volumes processed with CAD (computer-aided drafting): on the left, a wide and symmetrical volume, indicative of physiological masticatory movements; on the right, a reduced and asymmetrical volume, characteristic of a dysfunctional situation

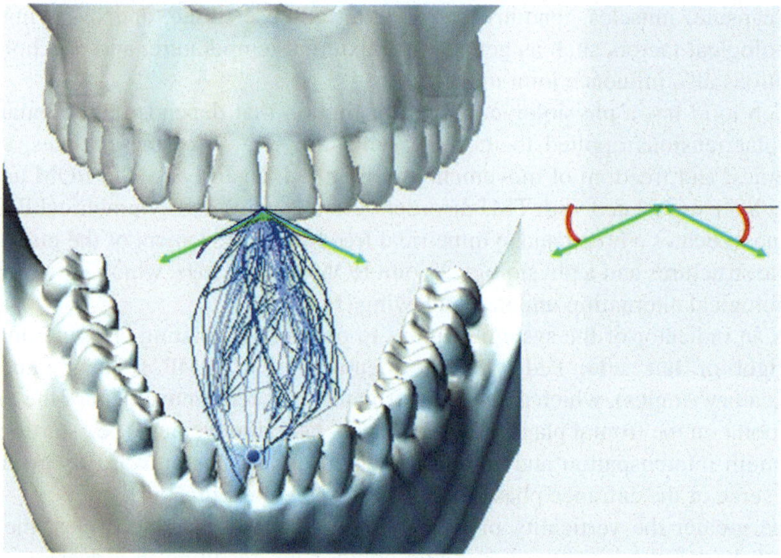

Fig. 5.54 In green: Functional masticatory angles of Planas (AFMP), superimposed on lateral and chewing cycle tracings, which on these angles find the path of entry into the occlusal phase of the masticatory force stroke

So, what is the correct inclination of the AFMPs? This question is not posed correctly because it assumes the immutability of dental surfaces and the stomatognathic system in general, as if the dentition were destined for eternal youth. Our teeth are not immutable entities but change their shape and position, adapting to their function. This adaptation over a person's lifetime should cause the inclination of the AFMPs to vary due to physiological wear, both in primary and permanent dentition. When graphing the variation of the AFMPs, you will notice that their value is highest at around 3 years, upon completion of the primary dentition, and at approximately 13 years, when the permanent dentition is complete. Following physiological wear, associated with proper function, the angles progressively decrease, reaching their minimum value at the age of 6 for primary dentition and around 60 for permanent dentition (Fig. 5.55).

In reality, every movement is never exclusively performed on a single plane but is expressed three-dimensionally on other planes. Corresponding to each AFMP on the frontal plane, there is another angle called the AFMO (horizontal masticatory functional angle) when projected onto the horizontal plane (Papini et al., 2017). AFMOs are the angles that form on the horizontal plane between the trajectories of the lateral movement of the incisal point (IP) and the frontal plane (Fig. 5.56).

AFMOs can be highlighted in kinesiographic or axiographic traces and, as we will see, are of great help in understanding the dynamics not only of the IP but also of the working condyle. They can have a physiological antero-lateral pattern or exhibit a postero-lateral pattern. However, in the case of a posterior shift of the IP, it corresponds to a potentially pathological posteriorization of the working condyle. If the right and left AFMPs have similar patterns and inclinations on the frontal plane,

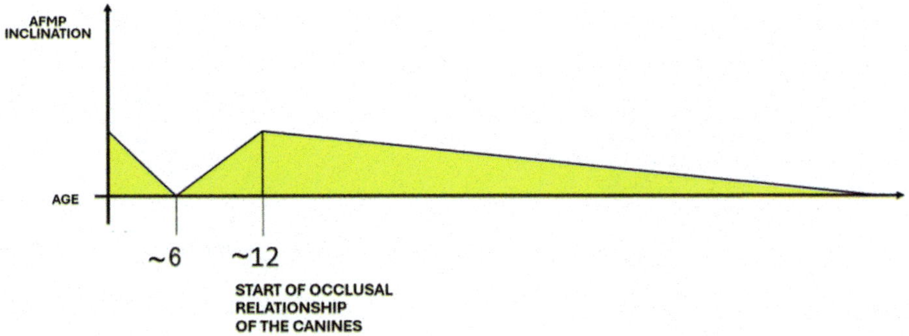

Fig. 5.55 Trend of the slope of AFMP angles during an individual's lifetime where dentition undergoes physiological wear. This graph can also be interpreted as the trend, over an individual's life, of the inevitable wear resulting from correct masticatory function

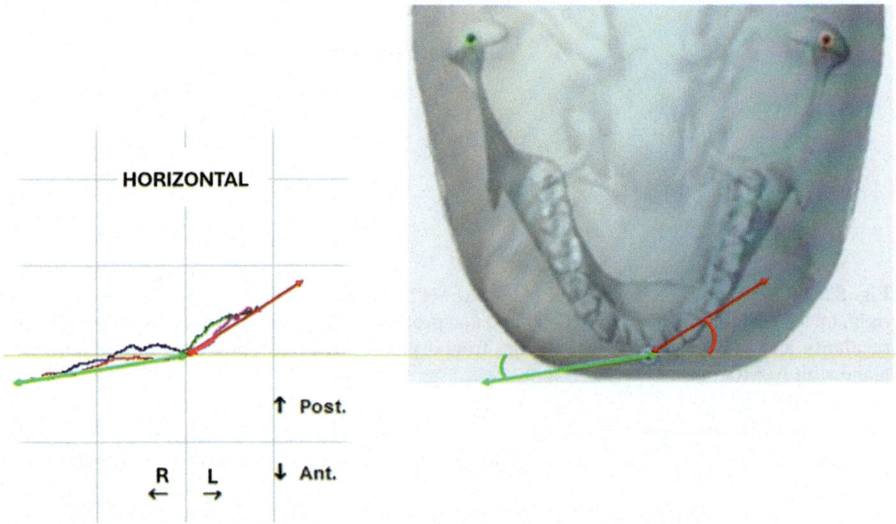

Fig. 5.56 Visualization of AFMO, horizontal functional masticatory angles: they form on the horizontal plane between the lateral paths of the PI and the frontal plane

this symmetry is also reflected on the horizontal plane, with the AFMOs on each side being equally similar and regular (Fig. 5.57).

In correct masticatory function, horizontal lateral movements should not only be symmetrical but also broad, except for certain developmental periods during which an increase in AFMPs acts as a growth stimulus—such as when the central incisors or canines erupt.

If the right and left AFMPs are equivalent, meaning that the inclination on the right side is the same as on the left, the individual chews on both sides with a physiological function of alternating monolateral chewing because the entrance phases of the chewing cycle are equivalent on both sides (Fig. 5.58).

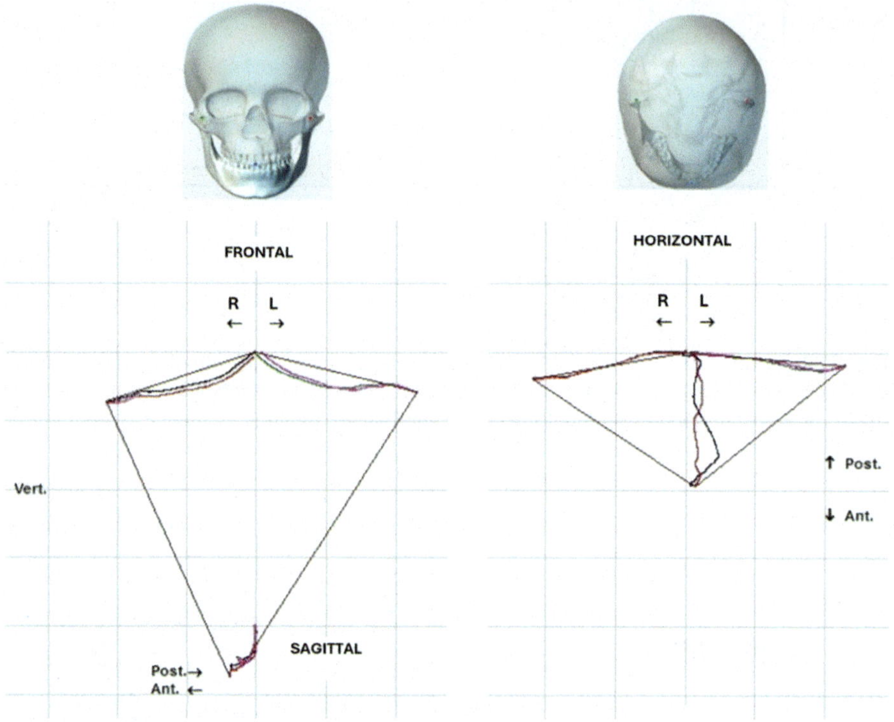

Fig. 5.57 On the left, on the frontal plane, AFMP: kinesiographic trace of lateral paths; on the right, on the horizontal plane, AFMO: lateral and protrusive trace (gothic arch). Note on the left the regularity and symmetry of AFMP on the frontal plane, which is reproduced on the horizontal plane with AFMO

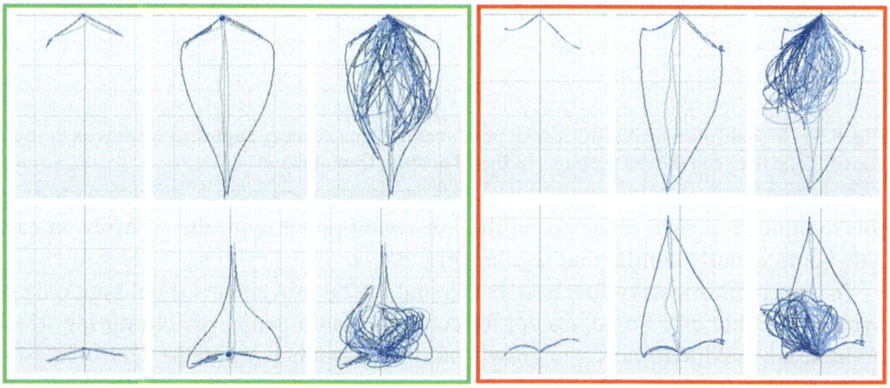

Fig. 5.58 Axiographic traces; in the green box, axiographic registration of symmetrical laterality. At the top, the three traces develop on the frontal plane; below, the same traces on the horizontal plane. AFMPs coincide and delineate the supero-anterior part of the Posselt volume; on the left, symmetrical lateral traces; in the center, the overlap of the Posselt's diagram, also wide and symmetrical; on the right, further overlap of the masticatory cycles, which draw an evenly distributed 'tangle' on both sides. In the red box below, axiographic registration of asymmetrical laterality generating asymmetrical dynamics

The Gothic arch and the Posselt diagram are, therefore, immediate references to the musculo-articular capacity that the masticatory system can express, making them direct indicators for assessing its state of health (Papini et al., 2018).

5.5 Retrusive Laterality

The functional symmetry of chewing, or the alternation of the chewing side, depends on the "ease" in the entrance phase of the chewing cycle. Nociceptive stimuli, the loss of functionality due to the absence of one or more dental elements, traumas, iatrogenic dentistry—these factors can all lead to one chewing side becoming more "comfortable" or efficient than the other. Over time, the preferred chewing side can structurally differentiate anatomically, morphologically differing from the contralateral side. This morphological differentiation occurs through modification of various anatomical areas, such as the bone bases, condyles, mandibular fossae, occlusal planes, and compensation curves (Dettori et al., 2001)

If the AFMPs are different, the increase in verticality in the entrance and exit phases of the cycle is not equal on both sides, and the individual tends to place the bolus on the side where verticality is lower and chewing is more effective. This leads to fixed unilateral chewing, to which the individual can adapt (often unconsciously). However, this falls into a dysfunctional chewing pattern that does not reflect the common masticatory alternation seen in the mammal class (Fig. 5.59).

In dysfunctional patients, chewing predominantly occurs on one side, and the musculo-articular system is unable to perform correct, wide, symmetrical, and alternating movements. The asymmetry in movement is not only evident on the frontal plane but also on the horizontal plane, resulting in deformed and asymmetric Gothic arches and Posselt diagrams (Todić et al., 2011) (Fig. 5.60).

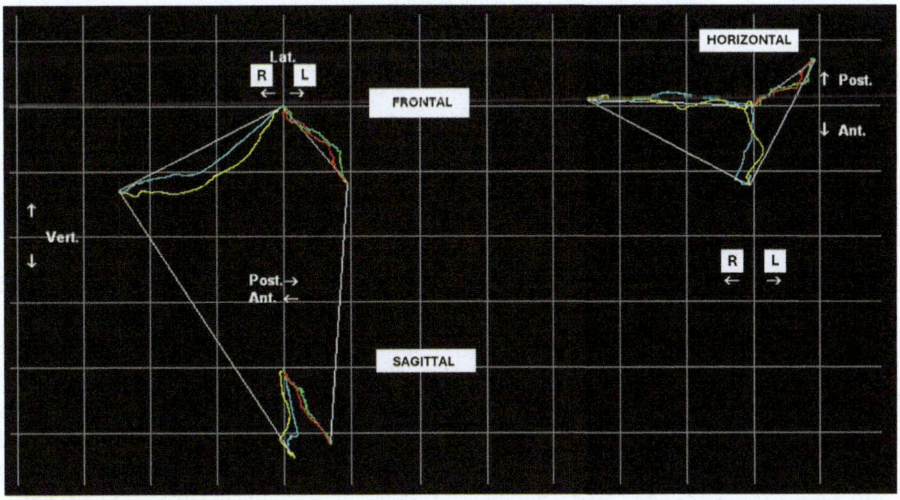

Fig. 5.59 Kinesiographic traces of laterality: note on the left the asymmetry of AFMP on the frontal plane, which is reproduced on the horizontal plane with equally asymmetrical AFMO

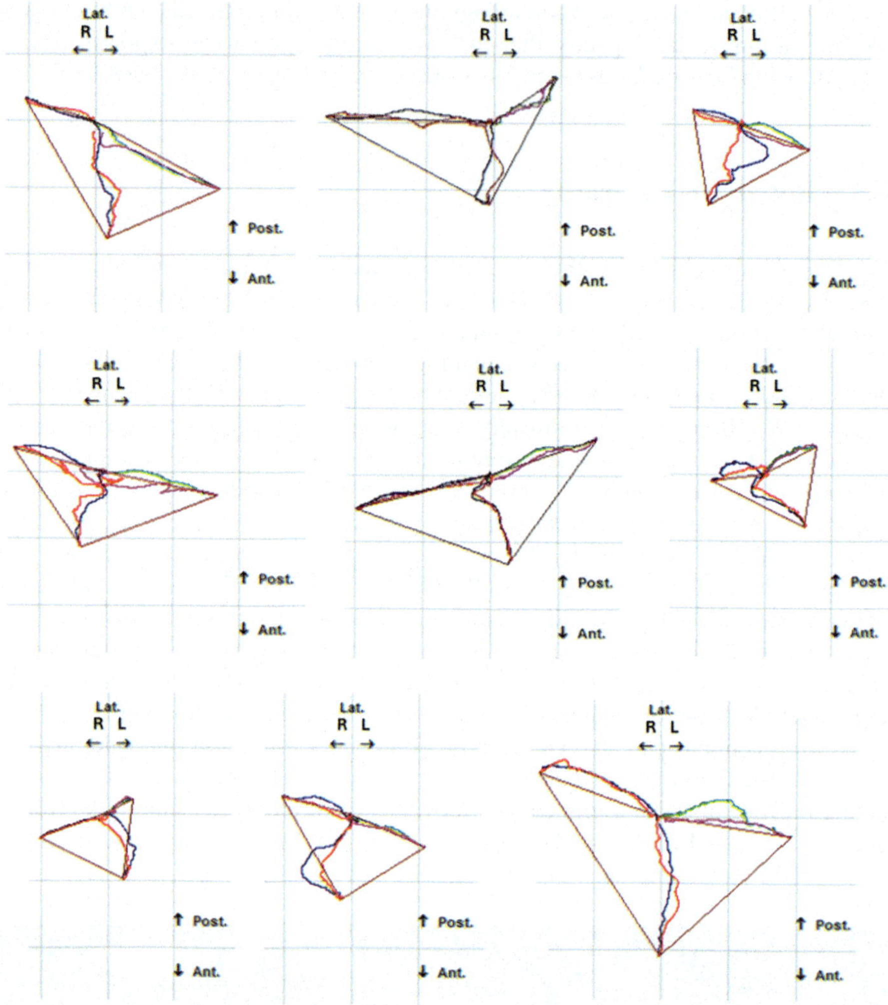

Fig. 5.60 Kinesiographic traces of laterality and protrusion on the horizontal plane in dysfunctional patients: note how the gothic arch appears deformed and how one of the lateral movements shows a retrusive pattern

This asymmetry of the Gothic arch and Posselt diagram characterizes the tracings of dysfunctional patients, which are quite different from the representations typically found in gnathology textbooks (Papini et al., 2017) (Figs. 5.61 and 5.62).

In dysfunctional patients with a predominant chewing side, functional asymmetry becomes evident with lateral movements that are also asymmetric, both on the frontal and horizontal planes. The differences between the right and left AFMPs correspond to different AFMOs on the horizontal plane. On the frontal plane, the side with the less inclined AFMP (the side of easier laterality and habitual chewing) will tend to correspond on the horizontal plane to the side with the AFMO that

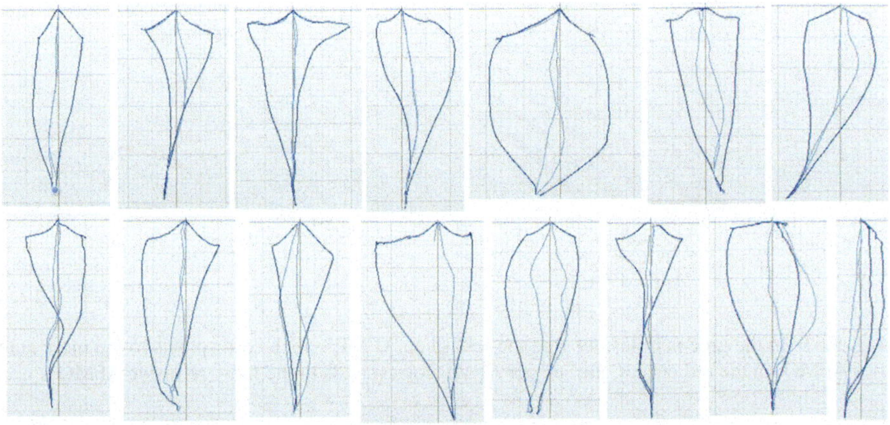

Fig. 5.61 Posselt traces on the frontal plane in dysfunctional patients: note how the lateral movements representing the 'roof' of the trace appear asymmetric, and how this asymmetry is reflected in the Posselt diagram

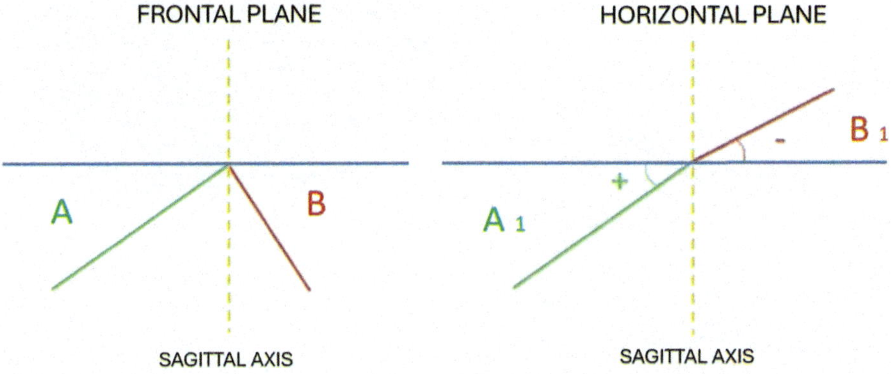

Fig. 5.62 On the frontal plane, the less steep AFMP (A) tends to correspond on the horizontal plane to the more anterior AFMO (A1); the steeper AFMP (B) tends to correspond to the more retrusive AFMO (B1)

develops anteriorly. Conversely, the side with the steeper AFMP will tend to have a more retrusive AFMO (Figs. 5.63 and 5.64).

The asymmetry on the horizontal plane frequently manifests with one of the lateral traces (and in some cases with both) showing a direction that is no longer anterior-lateral but expresses a posterior direction. The habitual chewing side, which has a chewing cycle with a smaller vertical dimension and a less inclined AFMP, tends to have an AFMO with a trace that is anterior compared to the non-chewing side. The non-chewing side, on the other hand, exhibits a chewing cycle with a greater vertical dimension and a more inclined AFMP, resulting in an AFMO with a trace that is oriented more posteriorly (Fig. 5.65).

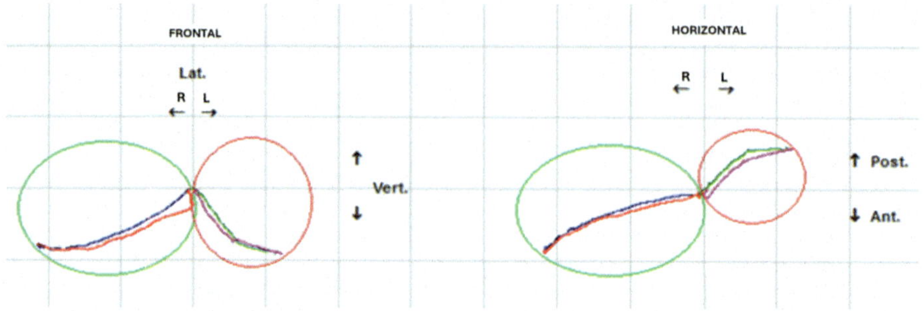

Fig. 5.63 In the green circles, we find the less steep AFMP, which corresponds to the more anterior AFMO; in the red circles, the steeper AFMP corresponds to the more retrusive AFMO

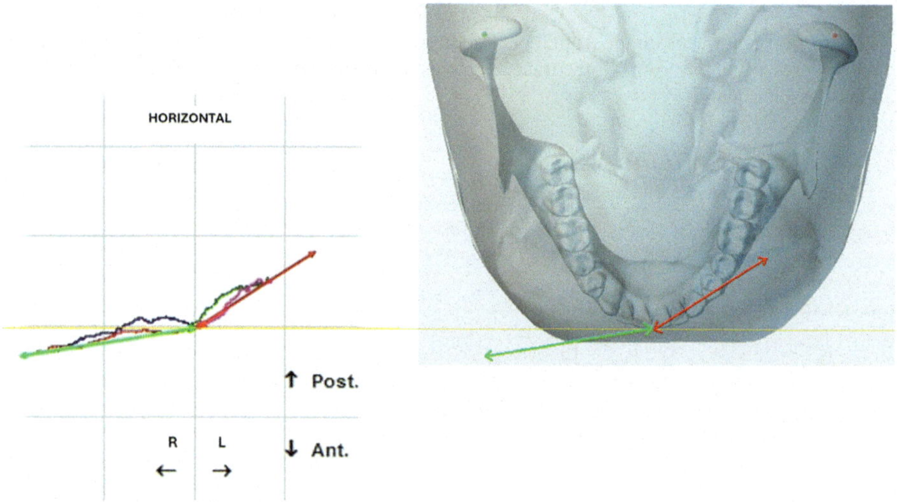

Fig. 5.64 Retrusive laterality: note how the left laterality shows a posterior direction

A study conducted by the author on 115 kinesiographic examinations performed on dysfunctional patients has shown that in 85% of the traces, there is a correspondence between the greater steepness of AFMP (Functional Mandibular Plane Angle) and the posteriorization of the AFMO (Functional Mandibular Opening Angle) trace. In accordance with Planas' law, which states that mastication occurs on the side with the less steep AFMP, on the horizontal plane, the masticatory function tends to occur on the side with the more anteriorly developed AFMO (Papini et al., 2017). This functional consideration is supported by the anatomy and physiology of the temporomandibular joint (TMJ), which, as we have seen, is entirely unsuitable for supporting compressive forces related to posterior condylar movements.

The clinical evaluation of lateral movement, performed by having the patient execute the movement from maximum intercuspation to the extreme limit of the occlusal guide, is important because it indicates how dental guides regulate the

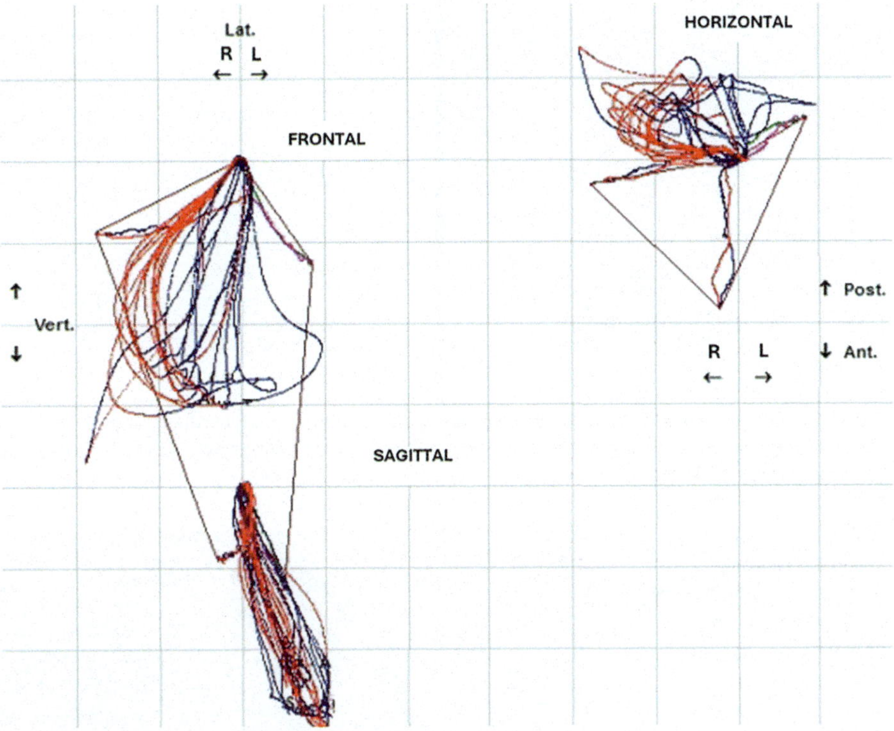

Fig. 5.65 Asymmetry of AFMP and AFMO with overlaid cycles of a unilateral chewing pattern; in each chewing cycle, the opening phase is marked in blue and the closing phase in red

mandibular movement. This lateral excursion is a convenient movement because it is easy to execute (generally) for patients. However, it should be noted that this is usually true for the habitual chewing side, while it is more difficult and sometimes impossible to execute on the opposite side. This centrifugal lateral movement is practically the same as the masticatory power stroke, but with the opposite direction—centripetal, moving from the outermost portion of the lateral guides to maximum intercuspation. It is precisely in this phase that the forces of the adductor muscles are discharged with the greatest intensity.

If we attempt to relate two concepts underlying modern gnathology on the horizontal plane, Bonwill's triangle and Gysi's gothic arch, a connection can be established between the latter and the movement of the Bonwill triangle's vertex corresponding to the working condyle. The interincisal lateral trace can be related to the movement of the condyle on the same side. The movement of the interincisal point (PI) is correlated with the movement of the working condyle. If the movement of the PI were primarily due to the balancing condyle, with only the rotation of the working condyle, its trajectory would always be anterior and lateral (Fig. 5.66).

However, the working condyle not only undergoes rotation but also lateral translation: this lateral shifting of the condyle could be considered as the residual

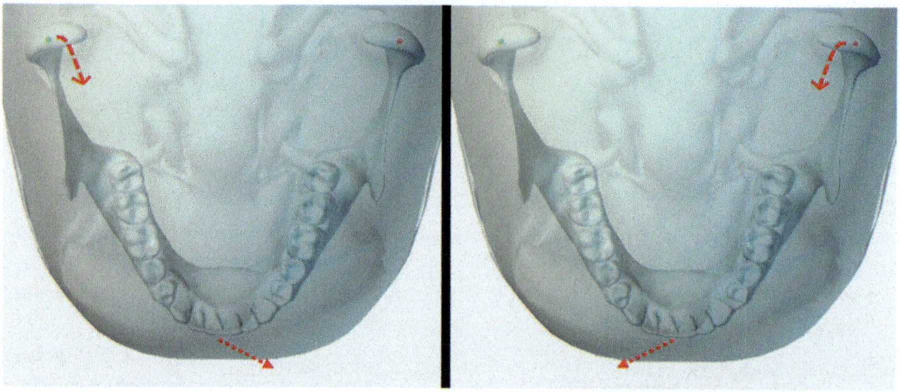

Fig. 5.66 In the figures, note how the rotation of the working condyle alone, coupled with an anterior and medial movement of the balancing condyle, would constantly lead to an anterior and lateral trajectory of the PI

Fig. 5.67 Guichet cone. Specimen from the Museum of Anthropology at the University of Florence

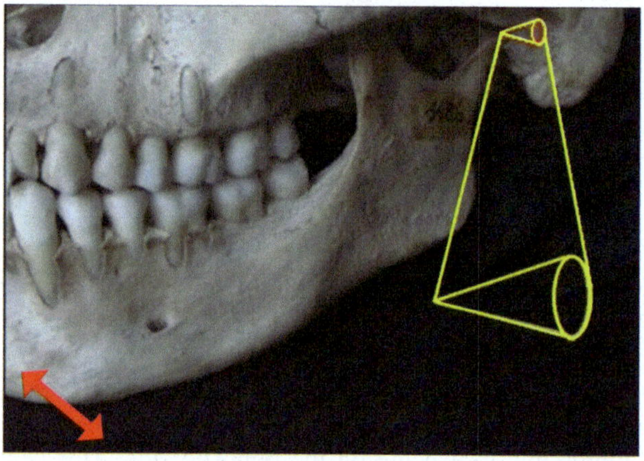

functional trace of the archaic lateral movement that allowed the lower molars on the working side to effectively interact with the upper molars, compensating for the smaller diameter of the lower arch, inscribed within the perimeter of the upper arch.

The possibility of lateral movement was described by Niles F. Guichet in 1970 as occurring within a 60-degree cone, with an axis corresponding to the intercondylar axis and the apex coinciding with the center of condylar rotation; within this cone, all possible positions of the working condyle exist (Molina et al., 1980) (Fig. 5.67).

When the mandible performs the centrifugal movement of laterotrusion, it moves according to forces and constraints that determine predictable displacements of the condyles and the PI within certain limits. On the horizontal plane:

- The balancing condyle always undergoes an anteromedial movement.
- The inter-incisal point may execute laterotrusion with an AFMO having an anterior or posterior trajectory.

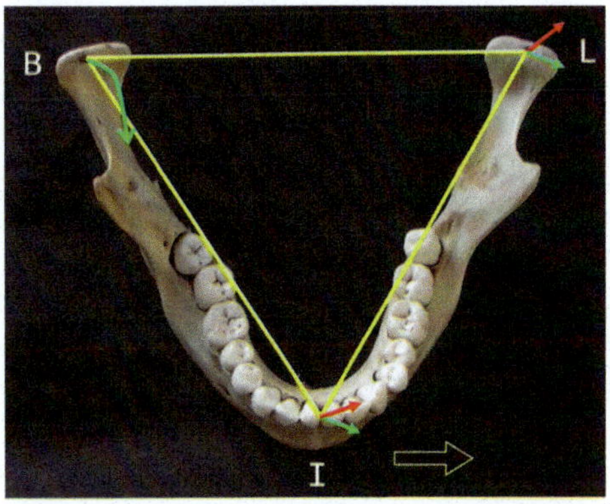

Fig. 5.68 Horizontal movement possibilities of the interincisal point (I) and the working condyle (L). Specimen from the Museum of Anthropology at the University of Florence

The working condyle may undergo an anterolateral or posterolateral movement, conditioned by the type of AFMO (Fig. 5.68).

If the condyle tends to move backward from its physiological homeostatic position in the mandibular fossa during the masticatory cycle, compressions of the delicate retrodiscal tissues against the posterior wall of the mandibular fossa can occur (Figs. 5.69 and 5.70).

If the gothic arch trace shows retrusive laterotrusion, the condyle on that side tends to move posteriorly, creating harmful compression on the delicate retrodiscal tissues.

On the horizontal plane during the laterotrusion movement, the balancing condyle—as previously observed—always undergoes a medial and forward movement, while the working condyle rotates with its lateral pole swinging outward (Aiello & Dean, 1990), restrained from sliding posteriorly by the action of the horizontal band of the temporomandibular ligament (LTM). During the power stroke, this ligamentous component plays a fundamental role by tensing due to the posterior rotation of the lateral pole of the working condyle, preventing it from sliding backward and forcing it to orbit and swing outward (Dean, 1990).

This small lateral shift (performed by the working condyle but also by the mandible as a whole) is known as Bennett movement or shift (Fig. 5.71, but also see Fig. 4.89)—which we have already discussed in Chap. 4.

Considering the path of the balancing condyle as ideally repetitive because it is always anteromedial, a direct relationship between the movement of the PI and that of the working condyle becomes evident. A slight variation in the anteroposterior position of the AFMO is reflected significantly in the posterior movement of the working condyle (Fig. 5.72).

These assessments regarding mandibular dynamics and the retrusion of the working condyle in relation to the laterality performed by the PI were already outlined by Prof. Alfred Gysi in a work from 1930 (Fig. 5.73).

Although what has been described may seem like a very theoretical and abstract topic, the evaluation of relationships on the horizontal plane between the PI and the working condyle, both in intercuspidation (Mongini, 1984) and during lateral

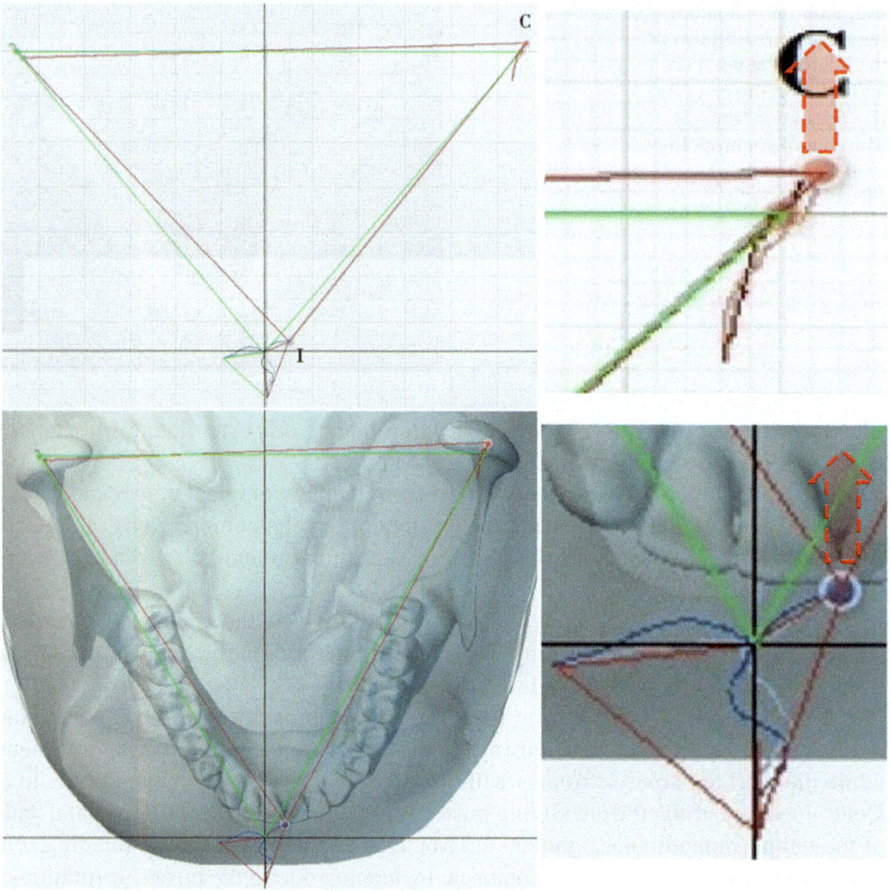

Fig. 5.69 In the trajectory of the gothic arch, retrusive laterality tends to correspond to a posterior push of the ipsilateral condyle, potentially harmful to the delicate retrodiscal tissues

Fig. 5.70 Different perspective of retrusive laterality: note how in the trajectory of the gothic arch, a posterior path of laterality tends to correspond to a posterior push of the ipsilateral condyle; this backward movement of the condyle in the mandibular fossa is—as mentioned—potentially harmful to the delicate retrodiscal tissues

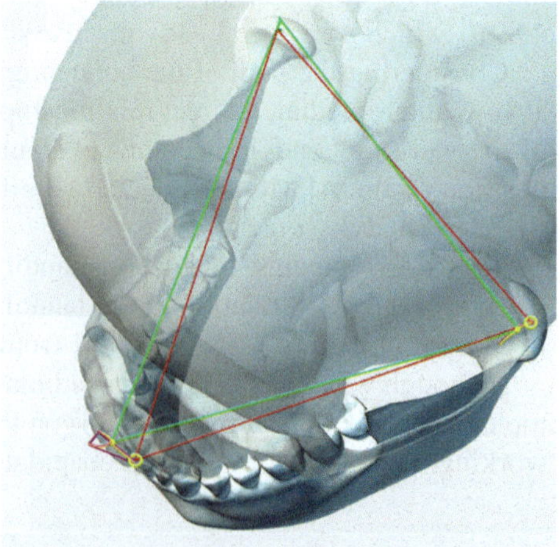

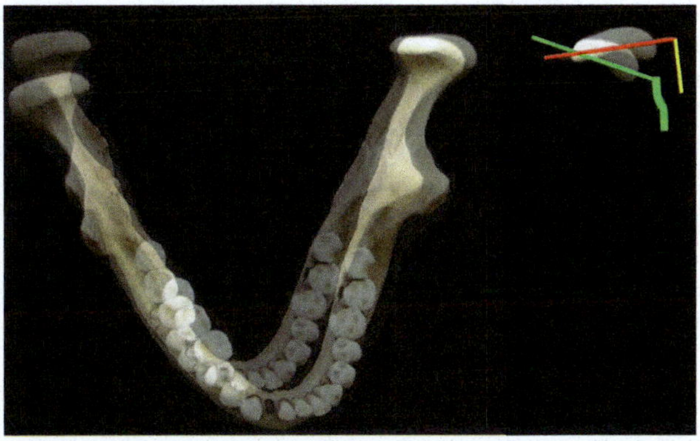

Fig. 5.71 During lateral movement on the horizontal plane, the working condyle rotates, moving its lateral pole posteriorly; this causes tension in the horizontal band of the temporomandibular ligament (LTM), which prevents the condyle from retracting into the fossa, causing it to slide outward (Bennett movement): green line, axis of the working condyle with the mandible in maximum intercuspation; green band, horizontal portion of the LTM in a stretched state; red line, axis of the working condyle with the mandible rotated in laterality; yellow band, horizontal portion of the LTM under tension

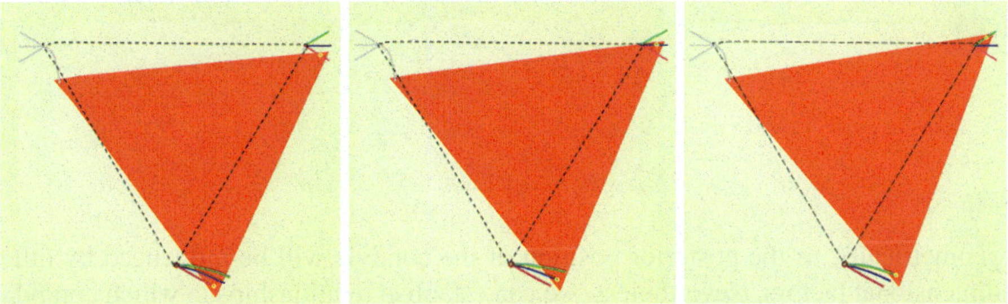

Fig. 5.72 CAD (computer-aided drafting) elaboration of the movement of a triangular body following condylar dynamics during lateral movement on the horizontal plane: there is a relationship between the retreat of the trace corresponding to the PI and the retrusion of the point corresponding to the working condyle, with a proportional increase in condylar posteriorization

movements, is crucial for the diagnosis and treatment of dysfunctional patients. If, during the lateral movement, the PI follows a retrusive trajectory (absolute or relative to the contralateral side), the corresponding working condyle tends toward retrusion. During the functional power stroke phase (a moment when, as we know, the most intense forces are generated), condylar retrusion can be traumatic for the delicate tissues in the posterior portion of the TMJ, and this potentially injurious situation can itself induce a fixed monolateral chewing pattern on the opposite side.

It should also be considered that every mandibular movement occurs simultaneously in three planes of space, and movement in one plane influences the dynamic development in the other two. For this reason, the retrusive lateral movement of the PI can indeed be related to condylar retrusion but cannot be rigorously determined

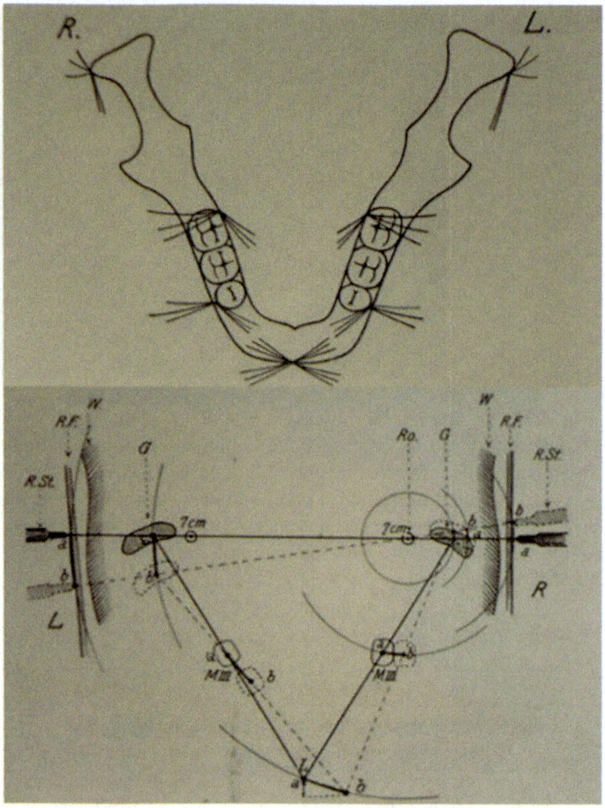

Fig. 5.73 Drawings by Prof. A. Gysi related to mandibular dynamics on the horizontal plane: in the upper figure, trajectories on the horizontal plane of different mandibular points in relation to lateral movements; in the lower figure, the retrusion of the working condyle (left) during ipsilateral laterality. Illustrations from the book "Zahnärztliche Prothetik," 1930, by Bruhn, Christian, et al., in the collection of the Castagnola Foundation

geometrically, as the posterior position of the condyle will be influenced by other dimensional factors. Nevertheless, we can establish relationships in which condylar retrusion, in relation to that of the PI, will be:

(a) Directly proportional to the posteriorization of the lateral trace (Figs. 5.74 and 5.75)

(b) Directly proportional to its extent (Figs. 5.76 and 5.77)

(c) Inversely proportional to the verticality of the PI's movement on the frontal plane (Figs. 5.78 and 5.79)

Lateral pathways, essentially overlapping with the guides of the entry phase of the masticatory cycle, are indices of the masticatory functionality on that side: they are important proprioceptive parameters upon which both masticatory cycles and their physiological alternation are based.

An occlusal interference, a precontact, an overbite, an alteration of the occlusal plane, or iatrogenic work can cause a retrusion of the entry phase of the masticatory cycle, potentially compressing the soft retrodiscal tissues, and thus lead the system

Fig. 5.74 The more the PI trace tends to retract, the more the working condyle tends to posteriorize

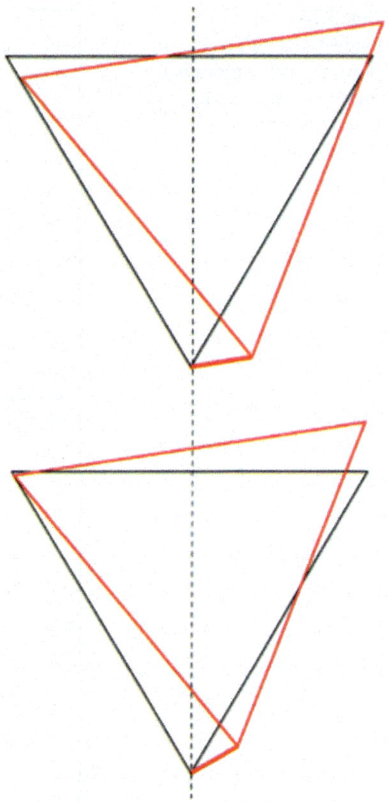

Fig. 5.75 Axial tracings of laterality on the horizontal plane: note the retrusion of the working condyle in relation to the retrusion of the PI

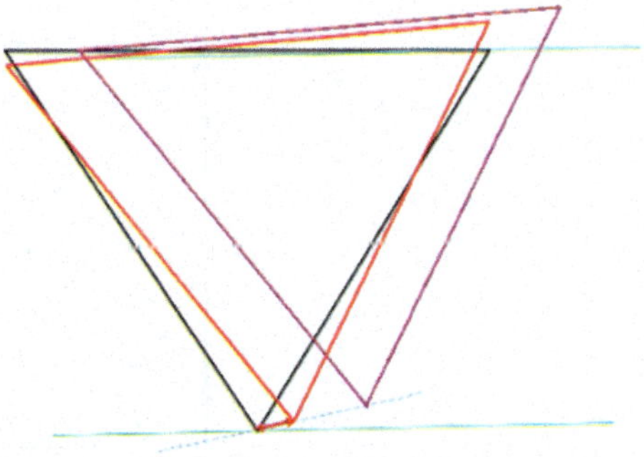

to adopt a protective pattern by shifting the chewing predominantly to the opposite side. The elimination of the functional obstacle that releases blocked horizontal movements will allow a rebalancing of the AFMP, promoting the restoration of a physiologically alternating mastication and the possibility of recovering a physiological range of musculo-articular movement of the TMJ. This functional recovery will occur not only on the frontal plane but also on the horizontal plane, with an

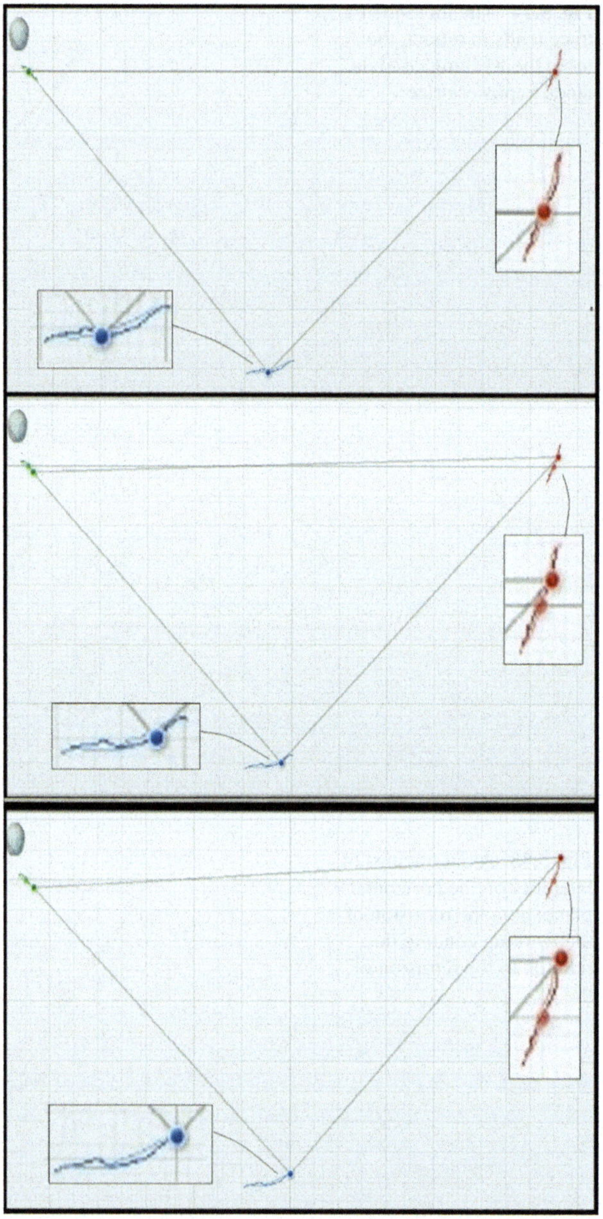

Fig. 5.76 The more extensive the PI trace, the more tendency there is for the working condyle to move posteriorly

enlargement and symmetrization of the gothic arch and, with it, the entire volume of Posselt. In a situation with asymmetric laterality and fixed monolateral chewing, we can intervene by symmetrizing the lateral pathways on the frontal plane, often reducing the excessive inclination of the steeper lateral pathway: in this way, we will tend to reduce the retrusive component of the corresponding AFMO, thus

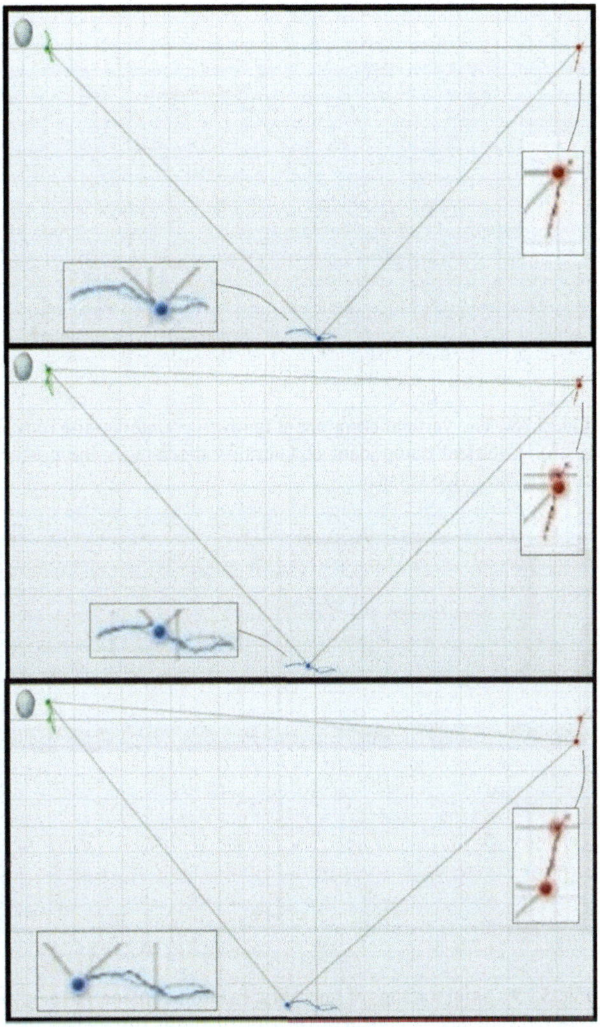

Fig. 5.77 Axial tracings of laterality on the horizontal plane: note the retrusion of the working condyle accentuated by the extent of the laterality

defusing the retrusive tendency of the working condyle (Figs. 5.80, 5.81, 5.82, and 5.83).

By symmetrizing and enlarging the AFMP, not only the AFMO but, as observed, the entire gothic arch, and consequently the entire Posselt volume, is symmetrized and enlarged. In each of these pathways, you can see the Posselt diagram before occlusal therapy (in red) and after therapy (in blue). The Posselt pathways on the sagittal plane were created without inducing mandibular retrusion, starting and ending at maximum intercuspation (Fig. 5.84).

Even in the sagittal plane, the Posselt diagram widens and lengthens, confirming an increase in musculo-articular mobility (Fig. 5.85).

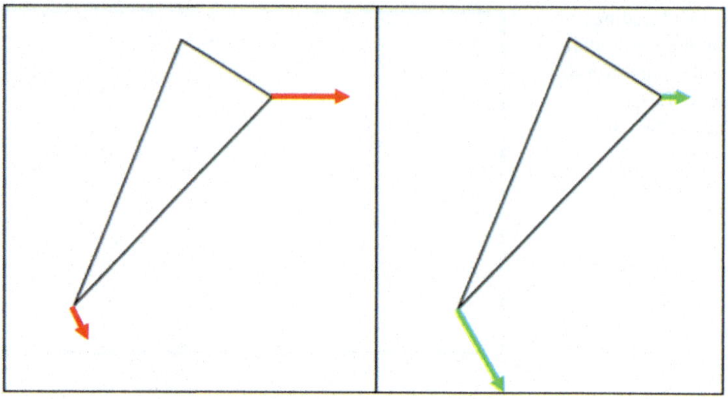

Fig. 5.78 The vertical component tends to counteract the retrusive force of the working condyle; a strong vertical component of laterality disengages the posteriorizing force of the monolateral condyle, and vice versa

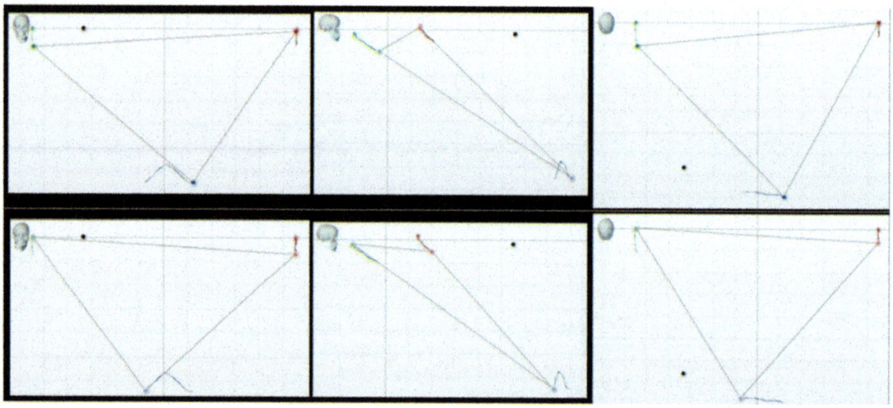

A B C

Fig. 5.79 Axial tracings of laterality: (**a**) frontal plane, (**b**) sagittal plane, (**c**) horizontal plane; note how the vertical component of laterality partially disengages the retrusion of the working condyles, which do not show any posteriorization

The study of occlusal surfaces, AFMPs, AFMOs, and their potential corrections can allow for modifications in the entry phase of the masticatory cycle, expanding the pathways and reducing asymmetry to decrease blocked laterality. In this way, the entire mandibular dynamics, including phases of the masticatory cycle away from the occlusal contact surface, can be conditioned. You can increase and regulate the volume available to the mandible, allowing it to express a freer range of movement. This enables the temporomandibular joint to recover previously blocked dynamics, increase the range of possible movements, defuse condylar retrusion, and restore alternating monolateral chewing.

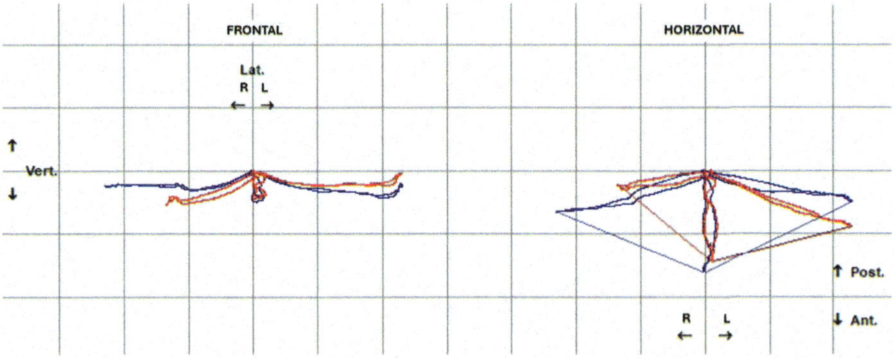

Fig. 5.80 In this and the following three figures, four different examples of how amplifying and symmetrizing AFMPs expand and symmetrize AFMOs and the Gothic arch: on the left, kinesiographic traces of laterality on the frontal plane; on the right, traces on the horizontal plane and the Gothic arch; in red, traces before occlusal correction; in blue, traces after occlusal correction

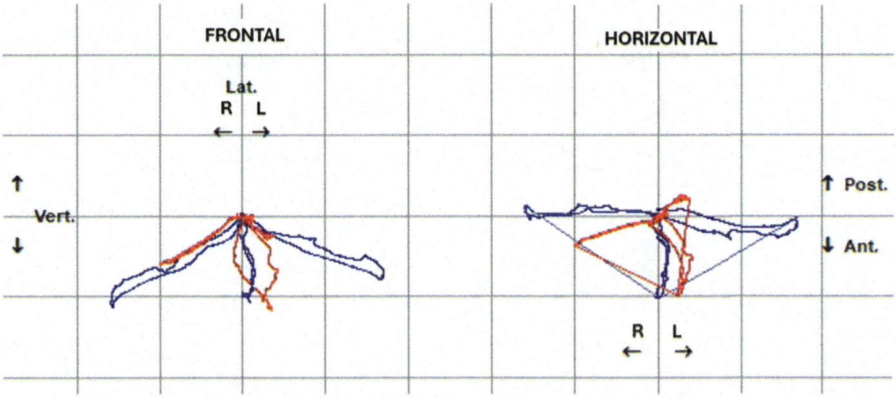

Fig. 5.81 Kinesiographic traces of laterality on the frontal plane (left) and on the horizontal plane (right): in red, traces before occlusal correction; in blue, traces after occlusal correction

A joint capable of performing wide, symmetrical, smooth movements without blockages within its physiological range is an articulation that fully expresses its physiological potential. Such an articulation distributes the force evenly across its entire surface, allowing synovial fluid to reach all areas of the joint space, preventing force concentrations that could damage its structure.

From a therapeutic perspective, restoring symmetry to the AFMPs (sometimes by increasing the inclination of the flatter one, but more often, as seen, by reducing the excessive inclination of the steeper one) leads to the recovery of physiological alternating monolateral chewing (Okeson, 2014; Mora, 2015).

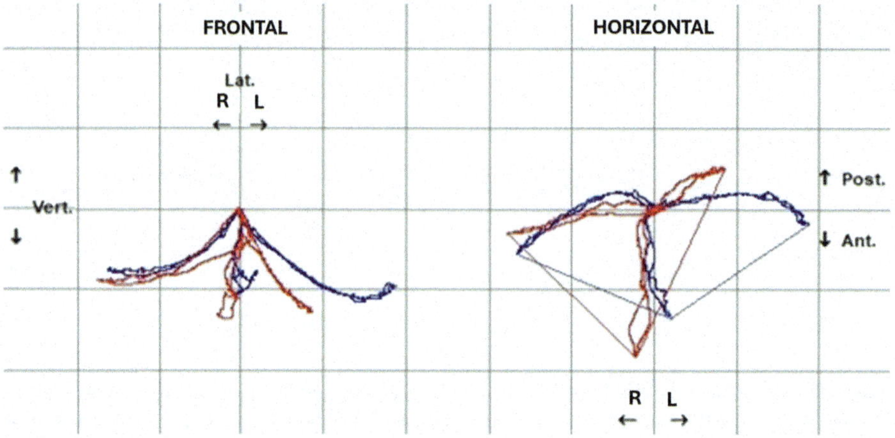

Fig. 5.82 Kinesiographic traces of laterality on the frontal plane (left) and on the horizontal plane (right): in red, traces before occlusal correction; in blue, traces after occlusal correction

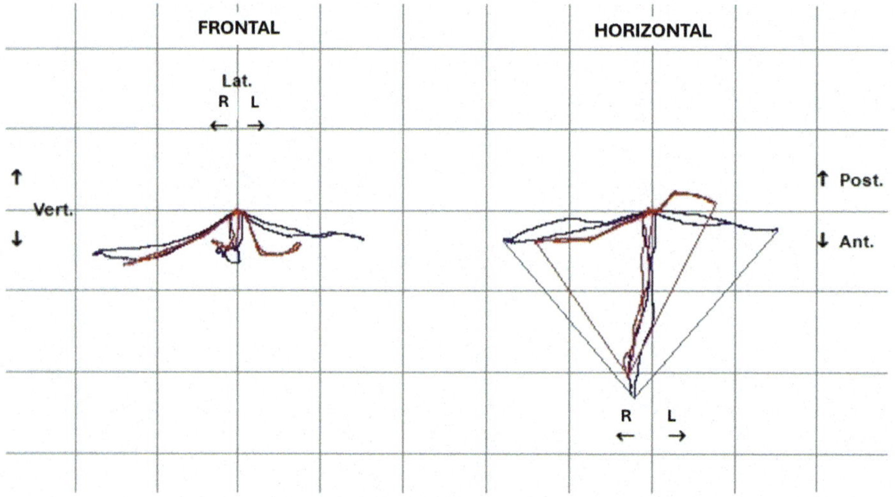

Fig. 5.83 Kinesiographic traces of laterality on the frontal plane (left) and on the horizontal plane (right): in red, traces before occlusal correction; in blue, traces after occlusal correction

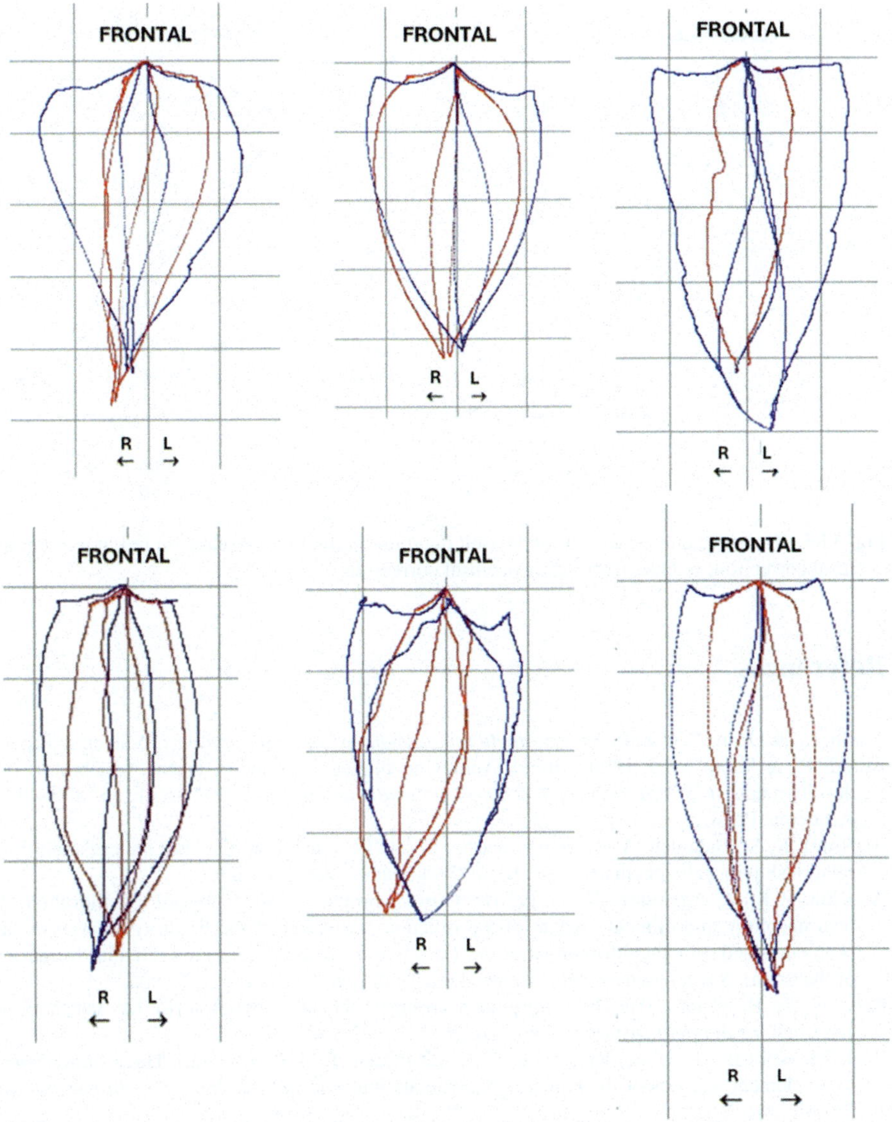

Fig. 5.84 Kinesiographic traces of the Posselt schema on the frontal plane: in red, traces before occlusal correction; in blue, traces after occlusal correction

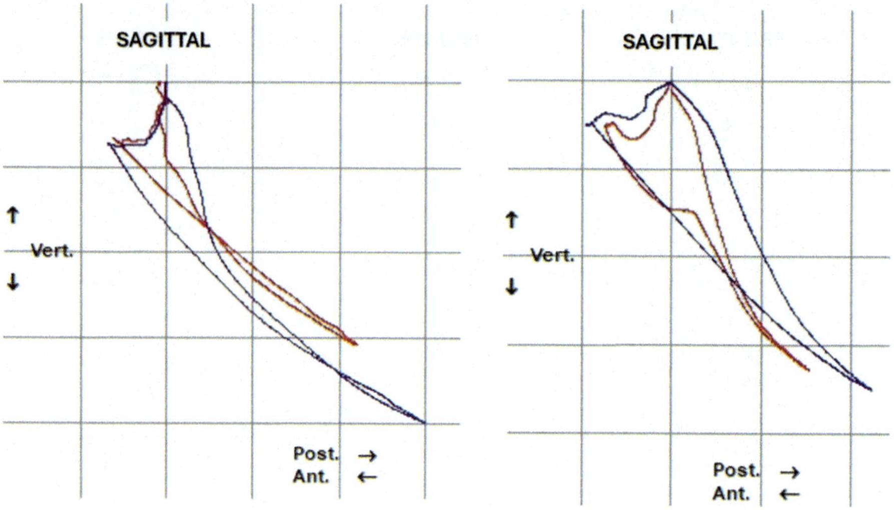

Fig. 5.85 Kinesiographic traces of the Posselt diagram on the sagittal plane: in red, traces before occlusal correction; in blue, traces after occlusal correction

References

Aiello, L., & Dean, C. (1990). *An introduction to human evolutionary anatomy*. Academic Press.

Allard, P., & Tabin, C. J. (2009, June). Achieving bilateral symmetry during vertebrate limb development. In *Seminars in cell & developmental biology* (Vol. 20, No. 4, pp. 479–484). Academic Press.

Al-Nimri, K. S., Bataineh, A. B., & Abo-Farha, S. (2010). Functional occlusal patterns and their relationship to static occlusion. *The Angle Orthodontist, 80*(1), 65–71.

Anochin, P. K., & Agafonow, W. G. (1961). Current problems in the physiology, morphology, pharmacology and clinical aspects of the reticular formation of the brain. (A survey of the problem based on the contributions of the 1st scientific conference on the reticular formation of the brain). *Sowjetwissenschaft; Naturwissenschaftliche Beitrage, 8*, 787–805.

Balkwill, F. H. (1866). The best form and arrangement of artificial teeth for mastication. *Transactions of Great Britain Odontological Society, 5*, 133–158.

Beek, M., Koolstra, J. H., van Ruijven, L. J., & van Eijden, T. M. G. J. (2000). Three-dimensional finite element analysis of the human temporomandibular joint disc. *Journal of Biomechanics, 33*, 307–316.

Benazzi, S., Kullmer, O., Grosse, I. R., & Weber, G. W. (2011). Using occlusal wear information and finite element analysis to investigate stress distributions in human molars. *Journal of Anatomy, 219*(3), 259–272.

Benazzi, S., Kullmer, O., Schulz, D., Gruppioni, G., & Weber, G. W. (2013). Individual tooth macrowear pattern guides the reconstruction of Sts 52 (*Australopithecus africanus*) dental arches. *American Journal of Physical Anthropology, 150*(2), 324–329.

Benazzi, S., Nguyen, H. N., Kullmer, O., & Hublin, J. J. (2013). Unravelling the functional biomechanics of dental features and tooth wear. *PLoS One, 8*(7), e69990.

Benazzi, S., Nguyen, H. N., Schulz, D., Grosse, I. R., Gruppioni, G., Hublin, J. J., & Kullmer, O. (2013). The evolutionary paradox of tooth wear: Simply destruction or inevitable adaptation? *PLoS One, 8*(4), e62263.

Berkovitz, B., & Shellis, R. P. (2018). *The teeth of mammalian vertebrates* (1st ed.). Academic Press. eBook ISBN: 9780128028193, Hardcover ISBN: 9780128028186, Published Date: 17th August 2018.
Biavati, P. S. (2019). *Trattamento semplificato in gnatologia: Il metodo Global Occlusion*. Edra.
Bonato, L. L., et al. (2017). Association between temporomandibular disorders and pain in other regions of the body. *Journal of Oral Rehabilitation, 44*(1), 9–15.
Brinkworth, R. S., Male, C., & Türker, K. S. (2004). Response of human jaw muscles to axial stimulation of a molar tooth. *Experimental Brain Research, 159*(2), 214–224.
Bruna, E., Fabianelli, A., Salin, M., & Scilla, M. (2013). *Occlusione in odontoiatria restaurativa Associazione*. Amici di Brugg Quaderno di aggiornamento odontoiatrico 31.
Chen, X. (1998). The instantaneous center of rotation during human jaw opening and its significance in interpreting the functional meaning of condylar translation. *American journal of Physical Anthropology: The Official Publication of the American Association of Physical Anthropologists, 106*(1), 35–46.
Choi, Y. H., Jang, W. H., Im, S. U., Song, K. B., Lee, H. K., Lee, H. D., Seo, Y. S., & Jang, S. H. (2017). The brain activation pattern of the medial temporal lobe during chewing gum: A functional MRI study. *Neural Regeneration Research, 12*, 812–814.
Christensen, F. T. (1959, September–October). The effect of Bonwill's triangle on complete dentures. *The Journal of Prosthetic Dentistry, 9*(5), 791–796.
Chung, G. Y., Choi, G. S., Shin, K. Y., & Park, J. S. (2016). Gait changes after using a temporomandibular joint exerciser in patients who underwent lower limb joint surgery. *Journal of Physical Therapy Science, 28*(5), 1584–1587.
Crompton, A. W. (1971). The origin of the tribosphenic molar. *Early Mammals, 50*, 65–87.
Crompton, A. W. (1989). The evolution of mammalian mastication. In D. B. Wake & G. Roth (Eds.), *Complex organismal functions: Integration and evolution in vertebrates* (pp. 23–40). Wiley.
Dean, C. (1990). *An introduction to human evolutionary anatomy*. Academic Press.
Deitch, J. S., & Rubel, E. W. (1989). Rapid changes in ultrastructure during deafferentation-induced atrophy. *Journal of Comparative Neurology, 281*(2), 234–258.
Dettori, S., Confaloni, A., & Pazzola, T. (2001). *RNO: evoluzione del rapporto dinamico funzionale tra occlusione ed ATM*. Mercurio editore.
Evans, A. R., & Sanson, G. D. (2003). The tooth of perfection: Functional and spatial constraints on mammalian tooth shape. *Biological Journal of the Linnean Society, 78*(2), 173–191.
Fish, D., & Mendel, F. (1982). Mandibular movement patterns relative to food types in common tree shrews (*Tupaia glis*). *American Journal of Physical Anthropology, 58*, 255–269.
Gans, C., de Vree, F., & Gorniak, G. C. (1978). Analysis of mammalian masticatory mechanisms: Progress and problems. *Anatomia, histologia, embryologia, 7*(3), 226–244.
Gerstner, G., Madhavan, S., & Crane, E. (2011). Mammalian oral rhythms and motor control. In V. Klika (Ed.), *Biomechanics in applications*. InTech.
Gomes, F., Guimarães, S., et al. (2010). Correlation of mastication and masticatory movements and effect of chewing side preference. *Brazilian Dental Journal, 21*(4), 351–355.
Grossnickle, D. M. (2017). The evolutionary origin of jaw yaw in mammals. *Scientific Reports, 7*, 45094.
Guichard, P., Mafart, B., & Orthlieb, J. D. (2001). Comparison of occlusion in medieval and present-day populations in southeast France. *American Journal of Orthodontics and Dentofacial Orthopedics, 120*(6), 585–587.
Gysi, A. (1930). *Handbuch der zahnheilkunde* (Vol. III). Verlag von J.F. Bergmann Munchen.
Hiiemae, K. M. (1978). Mammalian mastication: A review of the activity of the jaw muscles and the movements they produce in chewing. In P. M. Butler & K. A. Joysey (Eds.), *Development, function and evolution of teeth* (pp. 359–398). Academic Press.
Hylander, W. L. (1975). The human mandible: Lever or link? *American Journal of Physical Anthropology, 43*(2), 227–242.
Hylander, W. L. (1979). The functional significance of primate mandibular form. *Journal of Morphology, 160*(2), 223–239.

Hylander, W. L. (1980). Functional anatomy. In B. G. Sarnat & D. M. Laskin (Eds.), *The temporomandibular joint: A biological basis for clinical practice* (3rd ed., pp. 85–113). Thomas.

Hylander, W. L. (1992). Functional anatomy of the temporomandibular joint. In B. G. Sarnat & D. M. Laskin (Eds.), *The temporomandibular joint: A biological basis for clinical practice* (4th ed., pp. 60–92). Saunders.

Hylander, W. L. (2006). Functional anatomy and biomechanics of the masticatory apparatus. In *Temporomandibular disorders: An evidenced approach to diagnosis and treatment*. Quintessence.

Hylander, W. L., & Johnson, K. R. (1994). Jaw muscle function and wishboning of the mandible during mastication in macaques and baboons. *American Journal of Physical Anthropology, 94*, 523–547.

Ingawalé, S. M., & Goswami, T. (2012). Biomechanics of the temporomandibular joint. In T. Goswami (Ed.), *Human musculoskeletal biomechanics*. InTech.

Jacobson, A. (1986). Relationship of the muscles of mastication to the articular disc of the temporomandibular joint. *American Journal of Orthodontics and Dentofacial Orthopedics, 90*(6), 522.

Kaifu, Y., Kasai, K., Townsend, G. C., & Richards, L. C. (2003). Tooth wear and the "design" of the human dentition: A perspective from evolutionary medicine. *American Journal of Physical Anthropology: The Official Publication of the American Association of Physical Anthropologists, 122*(S37), 47–61.

Kolta, A., Westberg, K.-G., & Lund, J. P. (2000). Identification of brainstem interneurons projecting to the trigeminal motor nucleus and adjacent structures in the rabbit. *Journal of Chemical Neuroanatomy, 19*(3), 175–195.

Kondo, K., Niino, M., & Shido, K. (1995). A case-control study of Alzheimer's disease in Japan: Significance of life-styles. *Dementia, 5*(6), 314–326.

Koolstra, J. H., & van Eijden, T. M. (1996). Influence of the dynamical properties of the human masticatory muscles on jaw closing movements. *European Journal of Morphology, 34*, 11–18.

Koolstra, J. H., & van Eijden, T. M. (1999). Three-dimensional dynamical capabilities of the human masticatory muscles. *Journal of Biomechanics, 32*, 145–152.

Kopp, S. (1976). Topographical distribution of sulphated glycosaminoglycans in human temporomandibular joint disks. A histochemical study of an autopsy material. *Journal of Oral Pathology, 5*, 265–276.

Kopp, S. (1978). Topographical distribution of sulfated glycosaminoglycans in the surface layers of the human temporomandibular joint. A histochemical study of an autopsy material. *Journal of Oral Pathology, 7*, 283–294.

Korioth, T. W. P. (1997). Simulated physics of the human mandible. In C. McNeill (Ed.), *Science and practice of occlusion* (pp. 179–186). Quintessence.

Langenbach, G. E. J., & van Eijden, T. M. G. J. (2001). Mammalian feeding motor patterns. *American Zoologist, 41*(6), 1338–1351.

Laplanche, O., Orthlieb, J. D., Laurent, M., Vyslozil, O., & Dutour, O. (2010). Evolution of the incisal relationship in a Central European population (1870/1970). *International Journal of Stomatology & Occlusion Medicine, 3*(1), 2–9.

Lee, W.-Y., Okeson, J. P., & Lindroth, J. (1995). The relationship between forward head posture and temporomandibular disorders. *Journal of Orofacial Pain, 9*(2).

Levartovsky, S., Matalon, S., Sarig, R., Baruch, O., & Winocur, E. (2015). The association between dental wear and reduced vertical dimension of the face: A morphologic study on human skulls. *Archives of Oral Biology, 60*(1), 174–180.

Lewin, A., & Ramadori, G. (1985). *Electrognathographics: Atlas of diagnostic procedures and interpretation*. Quintessence.

Lucas, P. W. (1979). The dental-dietary adaptations of mammals. *Neues Jahrbuch für Geologie und Paläontologie, Monatshefte, 8*, 486–512.

Lucas, P. W. (1982). Basic principles of tooth design. In B. Kurtén (Ed.), *Teeth: Form, function and evolution*. Columbia University Press 15416.

Lucas, P. W. (2004). *Dental functional morphology: How teeth work*. Cambridge University Press.

Luke, D. A., & Lucas, P. W. (1983). The significance of cusps. *Journal of Oral Rehabilitation, 10*(3), 197–206.
Lund, J. P., & Kolta, A. (2006). Generation of the central masticatory pattern and its modification by sensory feedback. *Dysphagia, 21*(3), 167–174.
Luschei, E. S., & Goodwin, G. M. (1974). Patterns of mandibular movement and jaw muscle activity during mastication in the monkey. *Journal of Neurophysiology, 37*(5), 954–966.
Manns, A., Chan, C., & Miralles, R. (1987). Influence of group function and canine guidance on electromyographic activity of elevator muscles. *The Journal of Prosthetic Dentistry, 57*(4), 494–501.
Martinez Choy, S. E., Lenz, J., Schweizerhof, K., Schmitter, M., & Schindler, H. J. (2017). Realistic kinetic loading of the jaw system during single chewing cycles: A finite element study. *Journal of Oral Rehabilitation, 44*(5), 375–384.
McNeill, C. (1999). *L'Occlusione: basi scientifiche e pratica clinica*. Scienza e tecnica dentistica edizioni internazionali.
Miyawaki, S., Tanimoto, Y., Kawakami, T., Sugimura, M., & Takano-Yamamoto, T. (2001). Motion of the human mandibular condyle during mastication. *Journal of Dental Research, 80*, 437–442.
Molina, M., Magnano, G., Pisa, W., & Valsesia, C. (1980). *Concetti fondamentali di gnatologia moderna*. Ilic'.
Mongini, F. (1984). *L'Apparato stomatognatico: funzione disfunzione e riabilitazione*. Scienza e tecnica dentistica edizioni internazionali.
Mora, U. S. (2015). *The habitual chewing side syndrome* (Doctoral dissertation, Universidade de Santiago de Compostela).
Morquette, P., Lavoie, R., Fhima, M. D., Lamoureux, X., Verdier, D., & Kolta, A. (2012). Generation of the masticatory central pattern and its modulation by sensory feedback. *Progress in Neurobiology, 96*(3), 340–355.
Nikolopoulou, F., Xrysostomidis, A., Psari, X., Piagkou, M., & Troupis, T. (2019). Bonwill's triangle in Greek human mandibles. *Advances in Dentistry & Oral Health, 11*(4), 123–125.
Oberg, T., Carlsson, G. E., & Fajers, C. M. (1971). The temporomandibular joint. A morphologic study on a human autopsy material. *Acta Odontologica Scandinavica, 29*, 349–384.
Ogawa, T., Ogimoto, T., & Koyano, K. (2001). The relationship between non-working-side occlusal contacts and mandibular position. *Journal of Oral Rehabilitation, 28*(10), 976–981.
Ohm, E., & Silness, J. (1982). The size of the Balkwill angle and the height of the Bonwill triangle. *Journal of Oral Rehabilitation, 9*(4), 301–306.
Okeson, J. P. (2014). *Management of temporomandibular disorders and occlusion-E-book*. Elsevier Health Sciences.
Onozuka, M., Hirano, Y., Tachibana, A., Kim, W., Ono, Y., Sasaguri, K., ... & Watanabe, K. (2008). Interactions between chewing and brain activity in humans. In M. Onozuka & C.-T. Yen (Eds.), *Novel trends in brain science: Brain imaging, learning and memory, stress and fear, and pain* (pp. 99–113). Springer.
Osborn, J. W. (1982). Helicoidal plane of dental occlusion. *American Journal of Physical Anthropology, 57*(3), 273–281.
Oxilia, G., Menghi Sartorio, J. C., Bortolini, E., Zampirolo, G., Papini, A., Boggioni, M., Martini, S., Marciani, F., Arrighi, S., Figus, C., Marciani, G., Romandini, M., Silvestrini, S., Pedrosi, M. E., Mori, T., Riga, A., Kullmer, O., Sarig, R., Fiorenza, L., … Benazzi, S. (2021). Exploring directional and fluctuating asymmetry in the human palate during growth. *American Journal of Physical Anthropology, 175*(4), 847–864.
Papini, A., Botticelli, D., Omori, Y., Iida, T., Kawakami, S., Landini, L., & Cesaretti, G. (2018). Condylar retrusion on the horizontal plane associated with retrusive lateral excursion: A retrospective clinical axiographic study. *J Oral Science Rehabilitation, 4*(3), 16–23.
Papini, A., Cesaretti, G., & Defabianis, P. (2017). Kinesiographic analysis of lateral excursive movement on the horizontal plane: The retrusive component (pp. 60–67).

Pereira, F. J., Lundh, H., Eriksson, L., & Westesson, P. L. (1996). Microscopic changes in the retrodiscal tissues of painful temporomandibular joints. *Journal of Oral and Maxillofacial Surgery, 54*(4), 461–468.

Perrini, N. (1992). *Alfred Gysi: 70 anni di storia dell'odontoiatria: nel centenario della pubblicazione delle prime microfotografie di istologia dentaria: (1892-1992).* Fondazione Prof. Luigi Castagnola.

Piancino, M. G., & Kyrkanides, S. (2018). *La funzione masticatoria nei morsi incrociati monolaterali.* ed. edi ermes. ISBN 9788870516319.

Pileicikiene, G., & Surna, A. (2004). The human masticatory system from a biomechanical perspective: A review. *Stomatologija, 6*(3), 81–84.

Planas, P. (1994). *Rehabilitación neuro-oclusal.* Editorial Salvat.

Popowics, T. E., & Herring, S. W. (2006). Teeth, jaws and muscles in mammalian mastication. In V. L. Bels (Ed.), *Feeding in domestic vertebrates: From structure to behaviour* (pp. 61–83). CABI.

Posselt, U. (1962). *Physiology of occlusion and rehabilitation.* F.A. Davis. 41.

Price, W. A. (1945). *Nutrition and physical degeneration: A comparison of primitive and modern diets and their effects.*

Rangarajan, V., Gajapathi, B., Yogesh, P. B., Ibrahim, M. M., Kumar, R. G., & Karthik, P. (2015). Concepts of occlusion in prosthodontics: A literature review, part I. *The Journal of Indian Prosthodontic Society, 15*(3), 200–205.

Rathee, M., & Jain, P. (2020). *Anatomy, head and neck, lateral pterygoid muscle.* StatPearls [Internet].

Ravosa, M. J., Vinyard, C. J., Gagnon, M., & Islam, S. A. (2000). Evolution of anthropoid jaw loading and kinematic patterns. *American Journal of Physical Anthropology: The Official Publication of the American Association of Physical Anthropologists, 112*(4), 493–516.

Raya, Á., & Belmonte, J. C. I. (2006). Left–right asymmetry in the vertebrate embryo: From early information to higher-level integration. *Nature Reviews Genetics, 7*(4), 283–293.

Reilly, S. M., McBrayer, L. D., & White, T. D. (2001). Prey processing in amniotes: Biomechanical and behavioral patterns of food reduction. *Comparative Biochemistry and Physiology Part A: Molecular & Integrative Physiology, 128*(3), 397–415.

Saratti, C. M., Rocca, G. T., Vaucher, P., Awai, L., Papini, A., Zuber, S., Di Bella, E., Dietschi, D., & Krejci, I. (2021). Functional assessment of the stomatognathic system. Part 2: The role of dynamic elements of analysis. *Quintessence International (Berlin, Germany: 1985), 53*(1), 90–102. https://doi.org/10.3290/j.qi.b2091331. Epub ahead of print. PMID: 34595910.

Scarr, G., & Harrison, H. (2017). Examining the temporo-mandibular joint from a biotensegrity perspective: A change in thinking. *Journal of Applied Biomedicine, 15*(1), 55–62.

Schwenk, K. (Ed.). (2000). *Feeding: Form, function and evolution in tetrapod vertebrates.* Elsevier.

Sheppard, I. M., & Sheppard, S. M. (1968). Denture occlusion. *The Journal of Prosthetic Dentistry, 20*(4), 307–318.

Smith, R. J. (1978). Mandibular biomechanics and temporomandibular joint function in primates. *American Journal of Physical Anthropology, 49*(3), 341–349.

Stasolla, D., Mazza, A., Zaher, K., Maccioni, F., & Marini, M. (2004). Valutazione conRM delle alterazioni morfo-strutturali del tessuto retrodiscale nell'incoordinazione condilo-discale dell'ATM. *La Radiologia Medica, 107*, 261–268.

Tanaka, E., Detamore, M. S., Tanimoto, K., & Kawai, N. (2008). Lubrication of the temporomandibular joint. *Annals of Biomedical Engineering, 36*(1), 14–29.

Teaford, M. F., & Walker, A. (1983). Dental microwear in adult and still-born guinea pigs (*Cavia porcellus*). *Archives of Oral Biology, 28*(11), 1077–1081.

Thomason, J. J. (Ed.). (1997). *Functional morphology in vertebrate paleontology.* Cambridge University Press.

Todić, J., Lazić, D., & Radosavljević, R. (2011). Correlation analysis of craniomandibular index and gothic arch tracing in patients with craniomandibular disorders. *Vojnosanitetski pregled, 68*(7), 594–601.

Tsuboi, A., Kolta, A., Chen, C. C., & Lund, J. P. (2003). Neurons of the trigeminal main sensory nucleus participate in the generation of rhythmic motor patterns. *European Journal of Neuroscience, 17*(2), 229–238.

Türker, K. S., Yang, J., & Brodin, P. (1997). Conditions for excitatory or inhibitory masseteric reflexes elicited by tooth pressure in man. *Archives of Oral Biology, 42*(2), 121–128.

Tuthill, J. C., & Azim, E. (2018). Proprioception. *Current Biology, 28*(5), R194–R203.

Ungar, P. S. (2010). *Mammal teeth: Origin, evolution, and diversity*. JHU Press.

Vaahtoniemi, L. (2020). The reciprocal jaw-muscle reflexes elicited by anterior-and back-tooth-contacts—A perspective to explain the control of the masticatory muscles. *BDJ Open, 6*(1), 1–10.

Wall, C. E., & Smith, K. K. (2001). Ingestion in mammals. *e LS*. https://doi.org/10.1038/npg.els.0001837

Watanabe, K., Ozono, S., Nishiyama, K., Saito, S., Tonosaki, K., Fujita, M., & Onozuka, M. (2002). The molarless condition in aged SAMP8 mice attenuates hippocampal Fos induction linked to water maze performance. *Behavioral Brain Research, 128*(1), 19–25.

Weijs, W. A. (1994). Evolutionary approach of masticatory motor patterns in mammals. In V. L. Bels, M. Chardon, & P. Vandewalle (Eds.), *Biomechanics of feeding in vertebrates* (pp. 281–320). Springer.

Werner, J. A., Tillmann, B., & Schleicher, A. (1991). Functional anatomy of the temporomandibular joint. *Anatomy and Embryology, 183*(1), 89–95.

Xu, Y., Zhan, J. M., Zheng, Y. H., Han, Y., Zhang, Z. G., & Xi, C. (2013). Computational synovial dynamics of a normal temporomandibular joint during jaw opening. *Journal of the Formosan Medical Association, 112*(6), 346–351.

Yamada, H., & Evans, F. G. (1970). *Strength of biological materials*. Williams & Wilkins.

Yamamoto, T., & Hirayama, A. (2001). Effects of soft-diet feeding on synaptic density in the hippocampus and parietal cortex of senescence-accelerated mice. *Brain Research, 902*(2), 255–263.

Yang, J., & Turker, K. S. (1999). Jaw reflexes evoked by mechanical stimulation of teeth in humans. *Journal of Neurophysiology, 81*(5), 2156–2163.

Yang, L., Wang, H., Wang, M., Ohta, Y., & Suwa, F. (1992). Development of collagen fibers and vasculature of the fetal TMJ. *Okajimas Folia Anatomica Japonica, 69*(4), 145–155.

Conclusion

The physiology of our monolateral chewing is based on the alternation of the bolus treatment side, with masticatory cycles initially guided by the canine tooth. This guidance then transforms into a group function with progressive contralateral balancing contacts due to physiological wear. The explanation of why these essential functional parameters are fundamental to the physiology of our chewing has been the guiding thread of this work. It is written with the intention to view gnathology not merely as a field of different theories advocated by various schools but as a discipline with anatomical and functionally sound, scientifically shared roots.

When visiting patients, dentists should take into account the potential presence of monolateral chewing. Often, patients are unaware of this adaptation, but rehabilitation—whether orthodontic, prosthetic, or implant-prosthetic—should aim to maintain or restore chewing alternation and support the physiological dynamics of muscles and the temporomandibular joint.

Modern dentistry has developed highly refined procedures and techniques that allow for aesthetic results that were unimaginable a few decades ago. However, when addressing the functional aspect of rehabilitation by studying modern occlusal concepts, it becomes necessary to consider how these principles have evolved. Different approaches to gnathology developed over time by eminent individuals have created schools of thought with diversified foundations. The fundamental principles of occlusion are either accepted, rejected, or more often ignored by professionals in the field. This is often influenced by their past education (dental school), the personality and authority of the theory's presenter, the ease of the technique, scientific evidence, or the latest trends. However, all occlusal concepts are partly based on theory, and these theories draw from the past to some extent (Becker & Kaiser, 1993). Despite the different approaches to gnathological problems, each school of thought has its share of successes. This is likely because our masticatory system, thanks to its remarkable adaptability, can perform its functions within a broad functional range, well beyond the ideal homeostasis described by most dental schools. It is undeniable that our stomatognathic system demonstrates immense

A. Papini, *On the Origin of Human Masticatory Function*,
https://doi.org/10.1007/978-3-031-98664-2

adaptability daily, even in response to numerous dental interventions. However, this does not negate the equally undeniable fact that our species has evolved specific form and function, which must be considered therapeutically in light of paleoanthropological knowledge.

In this book, an attempt has been made to outline how the physiology of human mastication is the product of evolutionary development rooted in that of primates who were our ancestors, and even earlier, in that of mammals. Along this evolutionary path, masticatory function has shaped our stomatognathic apparatus with species-specific anatomical and functional characteristics, influenced not only by diet but also, and most importantly, by the unique posture of humans (erect posture and obligate bipedalism) combined with extreme global encephalization, a distinctive feature of *Homo sapiens.*

Just as in contemporary art, a proper understanding of forms requires knowledge of their evolutionary pathways, so to fully comprehend how we are constructed and why we function the way we do, it is essential to be aware of the successive anatomical and functional changes that have made us what we are today. However, when our intervention in the stomatognathic system becomes necessary, we must be aware that every treatment, no matter how minimal, is part of an incredibly sensitive and highly individual system. We must interact with it cautiously and delicately, seeking to interpret individualized treatment plans, often too rigid and standardized. The therapeutic trend that our field is at risk of adopting involves standardized protocols that have little to do with the task of resolving clinical problems optimally but are more connected to safeguarding against potential medical-legal disputes.

Just as every individual, while fitting within the characteristic patterns and parameters of the species, is unique and different from others, our approach to occlusal function must interpret each individual's functional uniqueness and respect it. We should always keep in mind the Hippocratic phrase: "*Life is short, and art (medicine) is long, opportunity (intuition) fleeting, experiment perilous, judgment difficult.*"

Editing: Carlo Adalberto Silenzi, Jacopo Falchi Picchinesi

English translation reviewed: Giuseppe Vincitorio

Reference

Becker, C. M., & Kaiser, D. A. (1993). Evolution of occlusion and occlusal instruments. *Journal of Prosthodontics, 2*(1), 33–43.

Afterword of the Author

Each newfound knowledge is not the end,
But a fresh beginning, my learned friend.
— Anonymous from Prato (twenty-first century)

With the hope that anyone who has taken the time to read this book may have experienced a bit of the wonder that I have often felt while writing it, I conclude by trying to be a bit lighter than the preceding pages: this desire arises from the fact that our fascinating work, and its necessary professionalism, does not exclude that touch of empathy and lightness (not meant as superficiality but as a joyful alternative to heaviness) that can alleviate our efforts and help our patients overcome their discomfort. This text was largely written during the sad period of the pandemic and the subsequent social confinement, as well as during the centenary of Gianni Rodari's birth: a Master of Lightness, always capable of combining depth with playfulness in his words, with the necessary levity for all ages, a delicacy that will bear fruit in all moments and in all continents, in beautiful moments but perhaps, most importantly, in the difficult ones.

A. Papini, *On the Origin of Human Masticatory Function*,
https://doi.org/10.1007/978-3-031-98664-2

Andrea Papini

The Communication

Homage to Gianni Rodari

The dentist kindly said unto the man:
"I'm sorry, but a hole I've found, forsooth,
A tiny breach within thy tooth, alas,
Which needs a gentle cure with utmost care."
With gentle touch and words so smooth he spoke,
The patient, calm, devoid of great distress,
Was cured with ease and perfect great success.
The dentist, rather brusque, did then exclaim:
"By Jove, thou art the cause of this, indeed!
Thy negligence hath brought this plight, I swear,
A crater gapes within thy molar, see!"
With stern resolve and words of power strong,
The patient quaked, in anxious fear he sat,
Was cured, though trembling, with a heavy heart.
There's no secret, this you ought to know:
When we are kind and calm in word and deed,
And show a gentle, soothing countenance,
The patients' fears will melt and fade away,
Their worries cease, their hearts no more in dread.
At end of our procedure, calm and sure,
With patience and with tact, a cure remains,
And gratitude doth follow in its wake

A. Papini, *On the Origin of Human Masticatory Function*,
https://doi.org/10.1007/978-3-031-98664-2

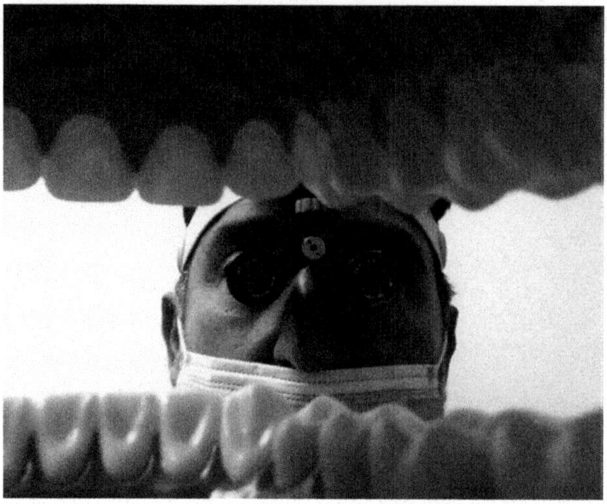

Drawing by Marco Boggioni

The title of this book may seem like an irreverent parody and a lack of respect toward the work of Charles Darwin; on the contrary it is merely a modest tribute to the Master's work, with the intention of drawing attention to how chewing, essential for the life of mammals, has developed over millions of years, constantly adapting in countless species, including our own. We "began to be humans" approximately 2 million years ago, and as the *Homo sapiens* species, our evolutionary journey started about 200,000 years ago. Throughout all this time, our masticatory system has not only adapted to a specific diet but has also been literally shaped by a posture and encephalization unique in the animal kingdom. Understanding these deep anatomical and functional interactions is essential for comprehending how our masticatory system should function and what the fundamental concepts are for regulating interventions on it in functional homeostasis.

Index

A
Abitbol, M., 243
Aboitiz, F., 250
Ahmed, I., 201, 203
Aiello, L., 118, 245, 248, 279, 282, 307, 309, 329, 435
Akobo, S., 120
Alghadir, A.H., 273
Al-Nimri, K.S., 385, 422
Alves, N., 292
Anastasi, M.R., 299
Anderson, P., 176
Angelo, D.F., 88
Anirudh, B.V.M., 337
Anochin, P.K., 383
Anson, B.J., 291
Anthwal, N., 46–48, 58, 79, 87, 93–95, 288
Arai, H., 299
Aristeguieta, L.M.R., 94
Aristotle, 7, 213
Arlegi, M., 231, 246, 249
Armfield, B.A., 4, 141
Arnold, P., 116, 237
Arshavsky, I., 357
Ashton, E.H., 279, 304
Atchley, W.R., 75, 80
Ayuse, T., 324

B
Balkwill, F.H., 413, 419
Baragar, F.A., 303
Baumel, J.J., 28
Bechtle, S., 177
Becker, C.M., 453
Beecher, R.M., 45, 153
Beek, M., 405
Been, E., 242, 246, 247
Begg, P.R., 145, 155
Bels, V., 18
Bemis, W.E., 14, 96
Benazzi, S., 143–146, 148, 161–164, 168, 209, 213, 394, 396
Benjamin, W., xix
Bennett, A.F., 91, 275, 328, 329, 407, 435, 437
Benoît, R., 212, 333, 345, 356–358
Benton, M.J., 47, 94
Berkovitz, B.K.B., 14, 16, 21, 22, 55, 98–100, 110–113
Berry, D.C., 145
Bertsen, W., 99
Bhullar, B.A.S., 4, 185
Biavati, P.S., 389
Biewener, A.A., 36, 42
Bimler, B., 195, 197, 199
Bodic, F., 76
Boggioni, M., 3, 21, 30, 72, 73, 82–85, 239, 241, 258, 259, 324, 458
Bono, D., 361
Bosshardt, D.D., 98–100
Boyde, A., 173, 179
Braly, A., 171
Bramble, D.M., 262, 268
Bresin, A., 45
Brinkworth, R.S., 389
Bromley, K.M., 177
Brown, E.C., 324
Brown, J.G., 341
Brown, W.A.B., 144
Bruna, E., 404

A. Papini, *On the Origin of Human Masticatory Function*,
https://doi.org/10.1007/978-3-031-98664-2

Bruner, E., 254, 256
Buffoli, B., 287, 291, 295
Burch, J., 303
Burrows, A.M., 27, 28
Butler, P.M., 18, 22
Byers, M.R., 113, 347

C
Cai, M., 323
Carlson, D.S., 45
Carlson, S.J., 140
Cash, R.M., 347
Casteyde, J.-P., 347, 352
Cawood, J.I., 46
Cerny, R., 56, 359
Chang, Y.-H., 160, 166
Chen, X., 271, 319, 393
Cheynet, F., 95, 302, 303
Choi, D.S., 45
Choi, Y.H., 447
Christensen, F.T., 192, 419
Chun, K.J., 112, 147
Chung, G.-Y., 410
Clark, G.T., 272
Cobourne, M.T., 358
Constantino, P.J., 16, 169
Cooper, J.S., 140
Corruccini, R.S., 45, 151–154
Cowin, S.C., 41
Craddock, F.W., 324
Crompton, A.W., 5, 15, 23, 26, 48, 65, 87, 107, 173, 189, 399, 404
Cuccia, A., 33, 319
Cui, F.-Z., 168, 177
Currey, J.D., 37
Cuy, J.L., 171

D
Daegling, D.J., 70
Dart, R.A., 367
Darwin, C., 18, 228, 342
Davis, B.M., 18
Dawson, P., 113
De Toffol, L., 91, 292
Dean, C., 5, 65, 87, 118, 245, 248, 279, 282, 307, 309, 317, 325, 327–329, 435
Deitch, J.S., 393
Del Palomar, A.P., 331
Deter, C.A., 209
Dettori, S., 54, 202, 316, 360, 362, 429
Di Vincenzo, F., 12, 230, 254, 262, 333, 336
Dickerson, W.G., 195, 275, 365, 366
Dickson, G.C., 154
Diekwisch, T.G.H., 75, 96, 98–100
Diogo, R., 27, 31
Dong, W.K., 347
Du Brul, E.L., 12, 324
Dumont, E.R., 108, 141, 149, 169, 171, 174, 179, 189
Dunbar, D., 269
Duncan, R.L., 37
Dupont, S., 37
Durband, A.C., 283, 304

E
Ebben, W.P., 273
Eckerdal, O., 300
Engelberth, S.A., 177
Eriksson, P.O., 271, 272, 325
Evans, A.R., 18, 145, 394

F
Fantner, G.E., 179
Fanucci, E., 328
Farella, M., 203
Federici, E., 366
Feldman, E., 203
Ferguson, M.W.J., 49
Fidler, B.C., 203
Fincham, A.G., 177, 179
Fiorenza, L., 86
Fish, D.R., 7, 22, 90, 181, 186, 398
Flower, S.S., 213
Foglio-Bonda, P.L., 328
Fricano, E.E.I., 288
Fritz, J., 7
Fueki, K., 203

G
Galis, F.D., 115, 238
Galli, M., 118, 125
Galton, P.M., 29
Galway-Witham, J., 256
Gangloff, P., 274
Gans, C., 181, 187, 383
Gao, X., 105
García-Castellano, J.M., 53, 361
Gaspard, M., 155
Ge, J.C.F.Z., 168, 177
German, R.Z., 357
Gerstner, G., 182, 358, 389

Gilbert, R.J., 30, 31
Goldberg, M., 169, 177
Gotthard, K., 44
Gould, S.J., 3, 46, 142, 225
Gow, B.S., 15
Granat, J., 212, 333, 337, 345, 357, 358
Grässel, S., 52, 361
Gregersen, G.G., 269
Grieznis, L., 274
Gross, M.D., 104, 155
Grossnickle, D.M., 311, 398
Guy, F., 110, 112
Gysi, A., 403, 413, 420, 435, 438

H

Häggman-Henrikson, B., 272
Hakkinen, T.J., 110, 141, 168, 170
Han, J., 3
Hanau, R.I., 195
Hannam, A.G., 268
Hansson, S., 76
Harcourt-Smith, W., 245, 342
Harris, C.K., 148
Hartstone-Rose, A., 25, 26, 90
Harvey, P.H., 250
Hashimoto, T., 293
He, L.H., 143, 173, 174, 177, 178, 186
He, T., 202
Heffner, H.E., 48
Heffner, M.A., 52, 361
Herring, S.W., 16, 24, 26, 39, 42, 43, 46, 65, 66, 78, 86, 87, 90, 91, 107, 143, 174, 184, 185, 201, 213, 392
Herzberg, F., 145
Hiiemae, K., 30, 181, 182, 184, 186, 357
Hildebrand, M., 50, 113
Hinton, R.J., 309
Hogervorst, T., 93, 142, 232, 243, 250
Holly Smith, B., 195
Hosoda, M., 274
Huang, X., 101
Humphrey, T., 292
Hunt, E.E. Jr., 151, 153, 158
Hylander, W.L., 5, 48, 60, 65, 87, 88, 90, 186, 187, 201, 331, 342, 363, 398, 402–405, 408, 415

I

Ichim, I., 70
Igarashi, N., 270
Ingawalé, S.M., 406
Inoue, T., 6
Iordansky, N.N., 23
Irish, J.D., 7, 8, 20, 22, 86, 107
Isberg, A., 48, 301
Ishiyama, M., 141
Isono, S., 323
Ito, H., 203, 309
Ivanovski, S., 98
Iwasaki, S.-i., 30
Izzo, R., 242

J

Jacob, F., 2
Janssen, M.M.A., 269
Jeng, Y.-R., 171
Jenkins, F.A., 93, 116, 120, 257
Jennings, J.S., 109
Jia, Y., 172, 175, 176
Jiang, Y., 175
Johanson, D., 348
Johnsen, S.E., 45, 361
Jones, D., 4
Jones, H.H., 37
Jones, S.J., 100
Jugessur, A., 51, 356

K

Kaidonis, J.A., 149, 164
Kaifu, Y., 5, 107, 152, 192–194, 387, 388
Karani, J., 201, 203
Kardong, K.V., 29, 264
Katona, T.R., 146
Katsavrias, E.G., 87, 192, 277, 305, 309, 320
Kaul, S.S., 152
Kay, R.F., 141
Kemp, A., 4, 141
Kemp, T.S., 5, 6, 48, 87, 141
Kermack, K.A., 48, 87
Kieser, J.A., 279, 307
Kikuchi, M., 104
Kiliaridis, S., 45, 154, 214
Kim, S.K., 166
Kim, Y., 155, 203
Kimbel, W.H., 279, 304
Kinzey, W.G., 347
Klein-MacPhee, G., 96
Klemetti, E., 46
Klineberg, I., 274
Kodaka, T., 181
Koebke, J., 121
Koenigswald, W. Von, 168, 172
Kohno, S., 271, 272, 325
Koldehoff, J., 173

Kolta, A., 4, 393
Kondo, K., 393
Kono, R.T., 169
Koolstra, J.H., 86, 268, 384, 408
Kopp, S., 405
Korioth, T.W.P., 405
Krahl, H., 37
Kuratani, S., 356

L
Langenbach, G.E.J., 384, 398
Lanyon, L.E., 37, 42, 46, 68
Larsen, C.S., 151, 152
Lautenschlager, S.P., 23–25
Lawn, B.R., 171
Le Douarin, N.M., 56, 352
Le Huec, J.C., 231
LeBlanc, A.R.H., 107
Lee, J.J.-W., 112, 169
Lee, T.C., 123
Lee, W.-Y., 410
Lesot, H., 174
Levartovsky, S., 396
Lewin, A., 413
Lieberman, D.E., 41, 262, 268
Liem, K.F., 4, 5, 23, 24, 29, 36, 50, 54, 55, 77, 78, 93, 115, 213, 232, 233, 236, 250, 270
Line, S.R.P., 141, 171
Loch, C., 141, 170
Lockwood, C.A., 304
Lombardi, A.V., 158
Lorenzon, G., 45
Loughner, B.A., 48, 300
Lovejoy, C.O., 248
Luan, X., 98, 100, 101
Lucas, P.W., 7, 17, 18, 109, 142, 143, 145, 149, 175, 394, 396
Luddleston, A.L., 37
Luke, D.A., 396
Lumsden, A.G.S., 5, 6
Lumsden, R.M., 269
Lund, J.P., 4, 393
Luo, Z.-X., 87
Luschei, E.S., 405
Lynch, C.D., 172, 175

M
Maas, M.C., 108, 149, 169, 171, 174, 179, 189
Macho, G.A., 175
Mack, P.J., 309
Maier, W., 46, 47, 50, 93
Manns, A., 389
Mantegazza, P., xx
Manzi, G., 16, 17, 151, 226–228, 230, 249, 251, 254, 256, 257, 259, 262, 265, 293, 339, 341, 364
Mao, J.J., 292
Mapelli, A., 319
Marchi, D., 39, 244, 250
Marshall, S.D., 197
Martinez Choy, S.E., 404
Matalová, E., 101, 102
Mayya, A., 99
McCollum, M.A., 240
McCormack, S.W., 99
McIntosh, J.E., 21, 100, 101
McNamara, J.A. Jr., 319
McNeill, C., 382
McNulty, K.P., 339
Menegaz, R.A., 54
Mérida-Velasco, J.R., 292, 298
Mew, J.R.C., 152
Mintenko, R., 203
Misra, A., 152
Miyahara, T., 273
Miyawaki, S., 184, 398
Mjor, I.A., 63, 98
Moggi-Cecchi, J., 227, 230, 367
Mohan, M., 203
Molina, M., 275, 385, 434
Molnar, S., 109, 144, 145, 152
Mongini, F., 312, 316, 435
Monson, G.S., 195, 207
Montemurro, N., 39, 63, 123
Moon, H.S., 63
Moore, W.J., 154
Moradian-Oldak, J., 176, 177
Morquette, P., 357, 393
Moss, M., 152
Moss, M.L., 45, 79
Mossey, P.A., 357
Motoyoshi, M., 274
Mrör, I.A., 110, 210
Mudry, A., 287
Murray, G.M., 274
Myers, L.J., 48, 301

N
Nakamura, Y., 357
Naples, V.L., 87, 181
Nietzsche, F., xix
Nikolopoulou, F., 419

Nishikawa, S., 174
Norman, D.B., 5

O
Oberg, T., 405
Ogawa, T., 404
Ohm, E., 419
Okeson, J., 443
Onozuka, M., 393
Osborn, J.W., 5, 6, 19, 92, 145, 189, 195, 196, 199, 203, 208, 209, 303, 317–320, 366, 367, 394
Ota, M.S., 22
Owen, C.P., 309
Oxilia, G., 35, 154, 199, 398

P
Padoa, E., 289, 357
Panek, H., 351
Pang, D., 120, 260
Papini, A., 82, 426, 429, 430, 432, 456
Parfitt, G.J., 155
Parilov, V.V., 192
Pearson, O.M., 41
Pekala, J.R., 293
Penfield, W., 390, 391
Pereira, C.B., 155
Pereira, F.J., 152, 302, 419
Pérez-Pérez, A., 3
Perrini, N., 20, 135, 420
Piancino, M.G., 185, 358, 384
Picq, P., 348
Pileicikiene, G., 86, 104, 105, 154, 155, 384, 394, 405, 407
Planas, P., 53, 352, 354, 360, 362, 382, 397, 399, 413, 425
Plomp, K.A., 244
Pompa, G., 366
Popowics, T.E., 16, 24, 39, 65, 66, 78, 87, 91, 107, 143, 174, 184, 185, 201, 213, 392
Posselt, U., 185, 328, 329, 335, 413, 414, 419–425, 428–431, 440, 441, 445, 446
Powell, V., 27, 250
Price, W.A., 144, 401
Proença, H.H.F., 153
Proops, D., 303
Puech, P.-F., 343

Q
Qiao, Y., 52
Quam, R., 297

R
Raia, P., 249, 345, 364, 367
Rak, Y., 78, 226, 240
Rangarajan, V., 399
Rasch, L.J., 14
Rashed, F., 46
Ravosa, M.J., 44, 402
Raya, A., 398, 409
Rees, L.A., 298, 318
Reif, W.E., 14
Reilly, S.M., 383
Richards, L.C., 20
Richman, J.M., 50, 53
Riga, M., 48, 300
Rodriguez Vázquez, J.F., 291, 298
Ross, C.F., 13
Rowe, T., 5
Rowicki, T., 299
Rubidge, B.S., 23, 24
Rubin, C.T., 37, 42, 46
Ruff, C.B., 37, 337

S
Sahni, A., 107, 142
Saito, P., 145, 155
Sakaguchi, K., 203
Samman, T., 270
Sánchez-Villagra, M.R., 79
Sanders, W.J., 172
Santana-Mora, U., 443
Sarnäs, K.V., 145
Sarrafpour, B., 145
Scarr, G., 383, 408
Scheidegger, R., 316
Schindler, H.J., 272
Schroeder, H.E., 100
Schwartz, G.T., 112, 169
Schwenk, K., 29, 141, 213, 357, 409
Scott, J.E., 343
Scott, N.A., 7, 8, 20, 22, 86, 107
Sedwick, C., 177
Sencimen, M., 299, 302
Senter, P., 245, 249, 251
Seo, H., 22
Sergi, S., 304, 309
Sforza, C., 274

Shellock, F.G., 120
Sheppard, I.M., 404
Shimazaki, T., 274
Sicher, H., 145, 277, 324
Silvertown, J., 142
Silvestri, A., 199, 201
Simpson, G.G., 17
Sisask, G., 53, 361
Slavicek, R., 144
Smith, P., 27, 208, 223
Smith, R.J., 181, 324
Sockol, M.D., 228
Spencer, M.A., 22
Sperber, G.H., 292
Spoor, F., 277
Stasolla, D., 302, 418
Stedman, H.H., 254
Stelzer, S., 343, 364
Stem, D.N., 107
Stern, D., 179
Stewart, D., 113
Ström, D., 25
Su, D.F., 344

T
Takada, Y., 273
Tanaka, E., 418
Tardieu, C., 244
Teaford, M.F., 145, 171, 345, 394
Terhune, C.E., 77, 92, 304, 305, 310
Thielemann, K., 366
Thomason, J.J., 49, 50, 402
Throckmorton, G.S., 5
Tiwari, B., 143, 224
Tjäderhane, L., 110
Todić, J., 429
Tomer, G., 79
Tomlinson, R.E., 53
Torii, S., 173
Torisu, T., 271
Towle, I., 111
Trevizan, M., 353, 356
Trinkaus, E., 37
Trulsson, M., 45, 361
Tsuboi, A., 393
Tucker, A.S., 87, 277, 287, 288
Tuisku, F., 113
Türker, K.S., 389
Turner, C.H., 37
Tuthill, J.C., 393

U
Uhthoff, H.K., 37, 42
Ungar, P.S., 6, 15, 17–19, 23, 24, 50, 55, 110, 140, 143, 148, 149, 173, 213, 345, 398
Urban, D.J., 93, 94

V
Vaahtoniemi, L., 389, 399
Valkenburgh, B. Van, 337
Varrela, J., 152
Veit, O., ix
Veneziano, A., 257, 363
Villmoare, B.A., 51, 59
Vishwakarma, A., 102, 103
von Spee, F.G., 155, 195, 196, 199, 201, 203, 204, 207, 353
Vrba, E.S., 3, 46

W
Wall, C.E., 29, 50, 65, 186, 187, 196, 201, 279, 282, 316, 317, 319, 396
Waltimo, A., 146
Ward, C.V., 344
Washburn, S.L., 340
Watanabe, K., 393
Watson, P.J., 76
Weidenreich, F., 304
Weijs, W.A., 24, 26, 91, 181, 398
Werner, J.A., 405
Werth, A.J., 141
White, S.N., 174
Wilken, A.T., 44
Will, M.J., 152
Williams, S.A., 232, 251
Williamson, E.H., 48, 203, 301
Williamson, L., 109
Woda, A., 164
Wolpoff, M.H., 155, 264
Woo, S.-L., 41
Woronowicz, K.C., 94

X
Xia, J., 173, 176–179
Xu, H.H.K., 174
Xu, Y., 419

Y
Yamada, H., 415
Yamamoto, T., 100, 393
Yang, J., 389
Yang, L., 302, 418
Yilmaz, E.D., 151, 171
Yoshizawa, H., 361
Young, W.G., 178, 179

Z
Zafar, H., 272, 273
Zaytsev, D., 179
Zeredo, J.L., 270
Zhang, M., 152
Zheng, S.Y., 176–178
Zhou, Z.-R., 142, 145, 151, 172

The manufacturer's authorised representative in the EU is Springer Nature Customer Service Centre GmbH, Europaplatz 3, 69115 Heidelberg, Germany. If you have any concerns regarding our products, please contact ProductSafety@springernature.com

Printed and bound by CPI Group (UK) Ltd, Croydon, CR0 4YY
01/10/2026
02225872-0015